Advances in Electronics and Electron Physics

EDITED BY
L. MARTON and C. MARTON
Washington, D.C.

Assistant Editor
CLAIRE MARTON

EDITORIAL BOARD

T. E. Allibone
H. B. G. Casimir
W. G. Dow
A. O. C. Nier
F. K. Willenbrock
E. R. Piore
M. Ponte
A. Rose
L. P. Smith

VOLUME 52

1979

ACADEMIC PRESS
New York · London · Toronto · Sydney · San Francisco
A Subsidiary of Harcourt Brace Jovanovich, Publishers

ADVANCES IN ELECTRONICS AND ELECTRON PHYSICS

VOLUME 52

Photo-Electronic Image Devices

PROCEEDINGS OF THE SEVENTH SYMPOSIUM HELD
AT THE BLACKETT LABORATORY, IMPERIAL COLLEGE,
LONDON, SEPTEMBER 4–8, 1978

EDITED BY

B. L. MORGAN

The Blackett Laboratory, Imperial College, University of London

AND

D. McMULLAN,

Royal Greenwich Observatory, Herstmonceux, Sussex

1979

ACADEMIC PRESS
London · New York · Toronto · Sydney · San Francisco
A Subsidiary of Harcourt Brace Jovanovich, Publishers

Copyright © 1979 by ACADEMIC PRESS INC. (LONDON) LTD.

ALL RIGHTS RESERVED

NO PART OF THIS BOOK MAY BE REPRODUCED IN ANY FORM
BY PHOTOSTAT, MICROFILM OR ANY OTHER MEANS
WITHOUT WRITTEN PERMISSION FROM THE PUBLISHERS

ACADEMIC PRESS INC. (LONDON) LTD.
24–28 Oval Road
London NW1

U.S. Edition
Published by
ACADEMIC PRESS INC.
111 Fifth Avenue
New York, New York 10003

British Library Cataloguing in Publication Data

Advances in electronics and electron physics.
Vol. 52
1. Electronics
I. Morgan, B L   II. McMullan, Dennis
537.5   TK7815   LCCCN 79-40961
ISBN 0-12-014652-5
ISSN 0065-2539

Set in Northern Ireland by
Universities Press, Belfast

Printed in Great Britain by
Galliard Printers, Ltd., Gt. Yarmouth

LIST OF CONTRIBUTORS

M. C. ADAMS, *Blackett Laboratory, Imperial College, Prince Consort Road, London SW7 2BZ, England* (p. 265)

J. R. P. ANGEL, *Steward Observatory, University of Arizona, Tuscon, Arizona 85721, U.S.A.* (pp. 183 & 347)

R. J. APSIMON, *Rutherford Laboratory, Chilton, Didcot, Oxon OX11 0QX, England* (p. 189)

U. W. ARNDT, *MRC Laboratory of Molecular Biology, Hills Road, Cambridge, England* (p. 209)

J. E. BATEMAN, *Rutherford Laboratory, Chilton, Didcot, Oxon OX11 0QX, England* (p. 189)

Y. BEAUVAIS, *Thomson-CSF, Electron Tube Division, 92100 Boulogne-Billancourt, France* (p. 217)

E. A. BEAVER, *Physics Department, University of California, San Diego, La Jolla, California 92037, U.S.A.* (p. 463)

A. BOKSENBERG, *Department of Physics and Astronony, University College, University of London, London WC1E 6BT, England* (p. 355)

J. BOULESTEIX, *Observatoire de Marseille, 2 Place Le Verrier, 13004 Marseille, France* (p. 379)

D. J. BRADLEY, *Blackett Laboratory, Imperial College, Prince Consort Road, London SW7 2BZ, England* (p. 265)

J. M. BREARE, *Department of Physics, University of Durham, Durham DH1 3LE, England* (p. 431)

A. W. CAMPBELL, *Department of Physics, University of Durham, Durham DH1 3LE, England* (p. 431)

B. R. CAPONE, *Rome Air Development Center, Deputy for Electronic Technology, Hanscom AFB, Massachussets 01731, U.S.A.* (p. 495)

G. R. CARRUTHERS, *E. O. Hulburt Center for Space Research, Naval Research Laboratory, Washington D.C. 20375, U.S.A.* (p. 283)

F. H. CHAFFEE, JR., *Smithsonian Institution, Mount Hopkins Observatory, Amado, Arizona, U.S.A.* (p. 415)

L. W. CHOU, *North Industries Corporation, P.O. Box 2137, Peking, China* (p. 119)

C. I. COLEMAN, *Department of Physics and Astronomy, University College, University of London, London WC1E 6BT England* (pp. 89 & 355)

A. CONDAL, *Department of Geophysics and Astronomy, University of British Columbia, Vancouver, BC, Canada* (p. 453)

E. R. CRAINE, *Steward Observatory, University of Arizona, Tucson, Arizona 85721, U.S.A.* (p. 339)

R. H. CROMWELL, *Steward Observatory, University of Arizona, Tucson, Arizona 85721, U.S.A.* (pp. 183, 339 & 397)

J. C. DAINTY, *The Institute of Optics, The University of Rochester, Rochester, New York 14627, U.S.A.* (p. 481)

C. L. DAVIES, *The Blackett Laboratory, Imperial College, Prince Consort Road, London SW7 2BZ, England* (p. 481)

LIST OF CONTRIBUTORS

W. A. DELAMERE, *Ball Aerospace Systems Division, Boulder, Colorado 80306, U.S.A.* (p. 89)
N. J. DIONNE, *Raytheon Company, Microwave and Power Tube Division, Waltham, Massachusetts 02154, U.S.A.* (p. 89)
B. DRIARD, *Thomson-CSF, Electron Tube Division, 38120 St Egrève, France* (p. 227)
M. J. ELLIS, *Mount Stromlo and Siding Spring Observatories, Australian National University, Canberra, Australia* (p. 389)
R. EVRARD, *Riber S.A., B.P. 231, 92505 Rueil, France* (p. 133)
G. G. FAHLMAN, *Department of Geophysics and Astronomy, University of British Columbia, Vancouver, Canada* (p. 453)
R. J. GELUK, *Oldelft Research Laboaratories, Delft, The Netherlands* (p. 237)
D. J. GILMORE, *MRC Laboratory of Molecular Biology, Hills Road, Cambridge, England* (p. 209)
P. J. GRIBOVAL, *Department of Astronomy, The University of Texas at Austin, Texas 78712, U.S.A.* (p. 305)
M. A. R. HARDWICK, *Blackett Laboratory, Imperial College, Prince Consort Road, London SW7 2BZ, England* (p. 329)
A. B. HARRISON, *Blackett Laboratory, Imperial College, Prince Consort Road, London SW7 2BZ, England* (p. 329)
Y. HATANAKA, *Research Institute of Electronics, Shizuoka University, Hamamatsu 432, Japan* (p. 31)
A. R. HEDGE, *Department of Physics, University of Durham, Durham DH1 3LE, England* (p. 431)
E. K. HEGE, *Steward Observatory, University of Arizona, Tucson, Arizona 85721, U.S.A.* (p. 397)
K. HELBROUGH, *John Hadland (P.I.) Ltd., Bovingdon, Hemel Hampstead, Herts HP3 0EL, England* (p. 253)
R. G. HIER, *Physics Department, University of California, San Diego, La Jolla, California 92037, U.S.A.* (p. 463)
G. R. HOPKINSON, *Department of Physics, University of Durham, Durham DH1 3LE, England* (p. 431)
J. J. HOUTKAMP, *Oldelft Research Laboratories, Delft, Netherlands* (p. 159)
A. HUMRICH, *Department of Physics, University of Durham, Durham DH1 3LE, England* (p. 431)
A. E. HUSTON, *John Hadland (P.I.) Ltd., Bovingdon, Hemel Hampstead, Herts HP3 0EL, England* (p. 253)
A. R. JORDEN, *Royal Greenwich Observatory, Herstmonceux Castle, Hailsham, East Sussex BN27 1RP, England* (p. 109)
W. KAMMINGA, *Laboratorium voor Technische Natuurkunde, Rijksuniversiteit te Groningen, Netherlands* (p. 89)
T. KAWAMURA, *NHK Technical Research Laboratories, Setagaya, Tokyo, Japan* (p. 51)
Y. KIUCHI, *Department of Electronic Engineering, Tokyo University of Agriculture and Technology, Tokyo, Japan* (p. 75)
E. S. KOHN, *RCA Laboratories, David Sarnoff Research Center, Princeton, New Jersey 08540, U.S.A.* (p. 495)
W. F. KOSONOCKY, *RCA Laboratories, David Sarnoff Research Center, Princeton, New Jersey 08540, U.S.A.* (p. 495)
B. LEFÈVRE, *Observatoire de Paris, 92190 Meudon, France* (p. 295)

LIST OF CONTRIBUTORS

G. LELIÈVRE, *Observatoire de Paris, 92190 Meudon, France* (p. 295)
D. C. LONG, *Department of Astrophysical Sciences, Princeton University Observatory Princeton, New Jersey 08540, U.S.A.* (p. 89)
J. L. LOWRANCE, *Department of Astrophysical Sciences, Princeton University Observatory Princeton, New Jersey 08540, U.S.A.* (pp. 89, 421 & 441)
H. G. LUBSZYNSKI, *"Cala-na-Sith", Stronafian, Glendaruel, Argyll, Scotland* (p. 11)
W. V. MCCOLLOUGH, *Systems Engineering Laboratory, Raytheon Company, Portsmouth, Rhode Island 02871, U.S.A.* (p. 63)
D. MCMULLAN, *Royal Greenwich Observatory, Herstmonceux Castle, Hailsham, East Sussex BN27 1RP, England* (pp. 109 & 315)
J. MAGNER, *Steward Observatory, University of Arizona, Tucson, Arizona 85721, U.S.A.* (p. 347)
A. MATHIOT, *C.E.N.G., Grenoble, France* (p. 217)
S. W. MOCHNACKI, *Department of Geophysics and Astronomy, University of British Columbia, Vancouver, BC, Canada* (p. 453)
B. L. MORGAN, *Blackett Laboratory, Imperial College, Prince Consort Road London SW7 2BZ, England* (pp. 329 & 481)
H. MULDER, *Oldelft Research Laboratories, Delft, Netherlands* (p. 159)
T. N'GUYEN-TRONG, *Laboratoire d'Astronomie Spatiale, Traverse du Siphon, 13012 Marseille, France* (p. 369)
R. NISHIDA, *Research Institute of Electronics, Shizuoka University, Hamamatsu 432, Japan* (p. 31)
S. OKAMOTO, *Research Institute of Electronics, Shizuoka University, Hamamatsu 432, Japan* (p. 31)
S. OKAZAKI, *NHK Technical Research Laboratories, Setagaya, Tokyo, Japan* (p. 51)
S. OKUDE, *NHK Technical Research Laboratories, Setagaya, Tokyo, Japan* (p. 51)
D. J. PEDDER, *Plessey Research (Caswell) Ltd., Towcester, Northants, NN12 8EQ, England* (p. 23)
A. J. PENNY, *Royal Greenwich Observatory Herstmonceux Castle, Hailsham, East Sussex BN27 1RP, England* (p. 109)
R. POLAERT, *Laboratoires d'Electronique & de Physique Appliquée, 3 avenue Descartes, 94450 Limiel Brévannes, France* (p. 369)
P. J. POOL, *English Electric Valve Company Ltd., Chelmsford, Essex, CM1 2QU England* (p. 23)
J. R. POWELL, *Royal Greenwich Observatory, Herstmonceux Castle, Hailsham, East Sussex BN27 1RP, England* (p. 315)
C. PRITCHET, *Department of Geophysics and Astronomy, University of British Columbia, Vancouver, BC, Canada* (p. 453)
L. RENARD, *Observatoire de Paris, 92190 Meudon, France* (p. 295)
G. RENDA, *Department of Astrophysical Sciences, Princeton University Observatory, Princeton, New Jersey 08540, U.S.A.* (p. 441)
J. RING, *Blackett Laboratory, Imperial College, Prince Consort Road, London SW7 2BZ, England* (p. 275)
A. W. RODGERS, *Mount Stromlo and Siding Spring Observatories, Australian National University, Canberra, Australia* (p. 389)
S. A. ROOSILD, *Rome Air Development Center, Deputy for Electronic Technology, Hanscom AFB, Massachussets 01731, U.S.A.* (p. 495)

LIST OF CONTRIBUTORS

J. C. ROSIER, *Laboratoires d'Electronique & de Physique Appliquée, 3 avenue Descartes, 94450 Limiel Brévannes, France* (p. 369)

H. ROUGEOT, *Thomson-CSF, Electron Tube Division, 38120 St Egrève, France* (p. 227)

G. ROZIÈRE, *Thomson-CSF, Electron Tube Division, 38120 St Egrève, France* (p. 227)

T. SAKUSABE, *Department of Electronic Engineering, Tokyo University of Agriculture and Technology, Tokyo, Japan* (p. 75)

R. J. SCADDAN, *The Blackett Laboratory, Imperial College, Prince Consort Road, London SW7 2BZ, England* (p. 481)

G. D. SCHMIDT, *Steward Observatory, University of Arizona, Tucson, Arizona 85721, U.S.A.* (p. 463)

G. W. SCHMIDT, *Physics Department, University of California, San Diego, La Jolla, California 92037, U.S.A.* (p. 463)

B. SERVAN, *Observatoire de Paris, 92190 Meudon, France* (p. 295)

F. D. SHEPHERD, *Rome Air Development Center, Deputy for Electronic Technology, Hanscom AFB, Massachussets 01731, U.S.A.* (p. 495)

W. SIBBETT, *Blackett Laboratory, Imperial College, Prince Consort Road, London SW7 2BZ, England* (p. 265)

B. SIDORUK, *Laboratoire d'Astronomie Spatiale, Traverse du Siphon, 13012 Marseille, France* (p. 369)

L. H. SKOLNIK, *Rome Air Development Center, Deputy for Electronic Technology, Hanscom AFB, Massachussets 01731, U.S.A.* (p. 495)

T. SONODA, *Electron Tube and Device Division, Toshiba Corporation, 72 Horikawacho, Sawai-ku, Kawasaki City, Kanagawa 210, Japan* (p. 201)

T. E. STAPINSKI, *Mount Stromlo and Siding Spring Observatories, Australian National University, Canberra, Australia* (p. 389)

R. W. TAYLOR, *Rome Air Development Center, Deputy for Electronic Technology, Hanscom AFB, Massachussets 01731, U.S.A.* (p. 495)

G. O. TOWLER, *The Research & Development Establishment, British Relay Ltd., Leatherhead, Surrey, England* (p. 143, 167)

G. A. H. WALKER, *Department of Geophysics and Astronomy, University of British Columbia, Vancouver, BC, Canada* (p. 453)

H. WASHIDA, *Electron Tube and Device Division, Toshiba Corporation, 72 Horikawacho, Sawai-ku, Kawasaki City, Kanagawa 210, Japan* (pp. 51 & 201)

W. L. WILCOCK, *Department of Physics, University College of North Wales, Bangor, Gwynedd, Wales* (p. 1)

G. WLÉRICK, *Observatoire de Paris, 92190 Meudon, France* (p. 295)

N. J. WOOLF, *Steward Observatory, University of Arizona, Tucson, Arizona 85721, U.S.A.* (p. 397)

K. YANO, *Electron Tube and Device Division, Toshiba Corporation, 72 Horikawacho, Sawai-ku, Kawasaki City, Kanagawa 210, Japan* (p. 51)

O. YOSHIDA, *Electron Devices Lab., Toshiba R & D Center, Toshiba Corporation, 72 Horikawacho, Sawai-ku, Kawasaki City, Kanagawa 210, Japan* (p. 39)

P. ZUCCHINO, *Department of Astrophysical Sciences, Princeton University Observatory, Princeton, New Jersey 08540, U.S.A.* (p. 441)

P. VAN ZUYLEN, *Technische Physische Dienst, Delft 2208, Netherlands* (p. 89)

FOREWORD

We are delighted that once again the Proceedings of the Symposium on Photo-Electronic Image Devices appears as a volume of Advances in Electronics and Electron Physics. The size of the volume and the nature of the contributions is a clear indication of the vigour of the field. Many of the advances reported in the earlier Proceedings have already manifested themselves as commonplace applications. No doubt, in a few years, this same statement may be aptly applied to this volume.

We extend our compliments to Drs Morgan and McMullan for the excellent volume they have put together and our thanks to them for the work they have done.

L. Marton
C. Marton

PREFACE

The Seventh Symposium on Photo-Electronic Image Devices was held in the Blackett Laboratory, Imperial College, University of London, from September 4th to 8th 1978. As in previous years the Proceedings are now published in the series "Advances in Electronics and Electron Physics" and we would like to express our thanks to Drs C. and L. Marton and Academic Press.

It is now twenty years since Professor J. D. McGee organized the First Symposium in 1958 and it is therefore a particular pleasure to be able to thank him for giving the opening address at this, the Seventh, Symposium. Professor McGee's remarks started sadly with a tribute to Professor André Lallemand who had died earlier in 1978. Professor Lallemand began development of his electronographic camera in 1936 and his pioneering work established the technique of electronography as an astronomical tool. He participated in every previous Symposium and will be greatly missed by all his colleagues.

The Editors have been told on many occasions that the earlier volumes of Proceedings constitute a convenient and comprehensive source of information in this field. However, the comment is frequently made that the delay which inevitably occurs between a Symposium and publication of its Proceedings is undesirable, and particularly frustrating to those who were unable to attend the meeting. On this occasion the Editors decided to overcome the latter objection by publishing, at the Symposium, a soft-backed volume containing short preprints of the papers. After the Symposium a committee selected about half the contributions, and the authors of these were asked to write full versions for inclusion in this volume of Advances in Electronics and Electron Physics. Four review papers are also included, reflecting on recent developments and mentioning important aspects of those papers which, because of lack of space, could not be included here. Professor W. L. Wilcock has written an introductory paper giving an overall view of the changes which have taken place in the field of Photoelectronics during the twenty years since he assisted Professor McGee in organizing the First Symposium.

Finally we would like to thank Professor D. J. Bradley, Head of the Blackett Laboratory and of the Optics Group, and Professor J. Ring, Head of the Astronomy Group, who made it possible to hold the Symposium at Imperial College and acted as members of the organizing

Committee and hosts to the participants. Thanks are also due to the members of the Astronomy and Optics Groups who gave enthusiastic help in running the Symposium.

Encouraged by the interest which has been shown, the organizers have made preliminary plans for an Eighth Symposium in September 1982.

August 1979

B. L. MORGAN
D. MCMULLAN

ABBREVIATIONS

For the most part the Editors have tried to keep to the units and terminology currently accepted and to adopt consistent abbreviations following Système Internationale usage wherever possible.

References cite journals abbreviated as recommended in Science Abstracts. Citation of earlier Symposia is frequent and the Editors have sought to simplify by the use of "Adv. E.E.P." for "Advances in Electronics and Electron Physics" followed by the appropriate volume and page numbers. References to the Symposium preprint (see Preface) are in the form "Seventh Symp. P.E.I.D. Preprints" followed by the appropriate page numbers.

CONTENTS

LIST OF CONTRIBUTORS v
FOREWORD ix
PREFACE xi
ABBREVIATIONS xii

Photo-Electronic Imaging 1958–1978. By W. L. WILCOCK 1

Camera Tubes and Electron Optics

Review of TV Camera Tubes and Electron Optics. By H. G. LUBSZYNSKI   11
A Pyroelectric Vidicon with Reticulated Target. By P. J. POOL AND D. J. PEDDER 23
Hard Vacuum Tube with a Pyroelectric Target of PVF_2 Film. By Y. HATANAKA, S. OKAMOTO AND R. NISHIDA 31
Recent Chalnicon Developments. By O. YOSHIDA 39
Proximity Focused SEC Vidicon with Porous MgF_2–Ag Target. By T. KAWAMURA, S. OKUDE, S. OKAZAKI, H. WASHIDA AND K. YANO   51
A Model for Signal Generation in Silicon Target Camera Tubes. By W. V. McCOLLOUGH 63
Current Density Distribution in a Magnetically Focused Low-Velocity Electron Beam. By Y. KIUCHI AND T. SAKUSABE 75
Recent Developments in Permanent Magnet Focusing Assemblies for Image Intensifiers and Camera Tubes. By C. I. COLEMAN, W. A. DELAMERE, N. J. DIONNE, W. KAMMINGA, D. LONG, J. L. LOWRANCE AND P. VAN ZUYLEN 89
Tolerance Analysis of Magnetically Deflected Photon Counting Detector Systems. By W. A. DELAMERE AND E. A. BEAVER 101
Step Deflection of Electron Images. By D. McMULLAN, A. R. JORDEN AND A. J. PENNY 109
Electron Optics of Concentric Spherical Electromagnetic Focusing Systems. By L. W. CHOU 119
Catadioptric Electron Optics Using a Retarding Electrostatic Field and its Application to the Development of Short Image Tubes of Very High Performance. By R. EVRARD 133

Image Intensifiers and Converters

Review of Image Intensification and Conversion. By G. O. TOWLER . . . 143
Stray Light in Proximity Focused Image Intensifiers. By J. J. HOUTKAMP AND H. MULDER 159

A Miniature Magnetic Intensifier. By G. O. TOWLER 167
Elimination of Corona and Related Problems with Astronomical Image Tubes. By R. H. CROMWELL AND J. R. P. ANGEL 183
A New Photocathode for X-ray Image Intensifiers Operating in the 1–50 keV Region. By J. E. BATEMAN AND R. J. APSIMON 189
High Resolution Phosphor Screen for X-ray Image Intensifier. By H. WASHIDA AND T. SONODA 201
Performance of an X-ray Television Detector for Crystallography. By U. W. ARNDT AND D. J. GILMORE 209
X-ray Topography with Scintillators Coupled to Image Intensifiers or Camera Tubes. By Y. BEAUVAIS AND A. MATHIOT 217
Image Device for Gamma Cameras Incorporating a Solid-State Localizer. By H. ROUGEOT, G. ROZIÈRE AND B. DRIARD 227
The Image Intensifier as a Convolution Processing Device. By R. J. GELUK 237
Photography with Gated Microchannel Plate Intensifiers. By A. E. HUSTON AND K. HELBROUGH 253
Electron Optical Picosecond Streak Camera Operating at 140 MHz and 165 MHz Repetition Rates. By M. C. ADAMS, W. SIBBETT AND D. J. BRADLEY 265

Applications in Astronomy

A Review of Astronomical Applications. By J. RING 275
High Resolution Large Format Electronographic Cameras for Space Astronomy. By G. R. CARRUTHERS 283
Etudes Extragalactiques Avec la Caméra Electronique "Grand Champ". By G. WLÉRICK, G. LELIÈVRE, B. SERVAN, L. RENARD ET B. LEFÈVRE . 295
The UT Electronographic Camera: Present Status, Astronomical Performance and Future Developments. By P. J. GRIBOVAL 305
Operational Experience with the RGO Electronographic Cameras. By D. MCMULLAN AND J. R. POWELL 315
Electronographic Photometry of NGC 3379. By M. A. R. HARDWICK, A. B. HARRISON AND B. L. MORGAN 329
Laboratory Tests and Astronomical Application of an Image Intensifier with a 146 mm Diameter Photocathode. By E. R. CRAINE AND R. H. CROMWELL 339
An Intensified Storage Vidicon Camera for Finding and Guiding at the Telescope. By J. R. P. ANGEL, R. H. CROMWELL AND J. MAGNER . . 347
Image Photon Counting Detectors for Spaceborne Applications. By A. BOKSENBERG AND C. I. COLEMAN 355
An Image Tube with a Curved Microchannel Plate and its Use in a Photon Counting Imaging System. By J. C. ROSIER, R. POLAERT, T. N'GUYEN-TRONG AND B. SIDORUK 369
First Observations of Faint Extended Emission Sources with an Image Photon Counting System. By J. BOULESTEIX 379
Photon Counting with Intensified Solid State Arrays. By T. E. STAPINSKI, A. W. RODGERS AND M. J. ELLIS 389

Quantum Noise Limited Readout of Spectrographic Data Using Image Intensifiers and Reticon Photodiode Arrays. By E. K. HEGE, R. H. CROMWELL AND N. J. WOOLF 397
Echelle Spectroscopy with Electronographic and Solid State Detectors. By F. H. CHAFFEE, JR 415

Solid State Image Detectors

A Review of Solid State Image Sensors. By J. L. LOWRANCE 421
A Novel Photodiode Array System for Direct Astronomical Spectroscopy. By A. W. CAMPBELL, A. R. HEDGE, G. R. HOPKINSON, A. HUMRICH AND J. M. BREARE 431
ICCD Development at Princeton. By J. L. LOWRANCE, P. ZUCCHINO, G. RENDA AND D. C. LONG 441
Astronomical Imagery with Solid-State Arrays. By G. G. FAHLMAN, S. W. MOCHNACKI, C. PRITCHET, A. CONDAL AND G. A. H. WALKER . . . 453
Photon Detection Experiments with Thinned CCDs. By R. G. HIER, E. A. BEAVER, G. W. SCHMIDT AND G. D. SCHMIDT 463
Astronomical Applications of a CCD. By C. L. DAVIES, B. L. MORGAN, R. J. SCADDAN, R. W. AIREY AND J. C. DAINTY 481
Schottky IRCCD Thermal Imaging. By F. D. SHEPHERD, R. W. TAYLOR, L. H. SKOLNIK, B. R. CAPONE, S. A. ROOSILD, W. F. KOSONOCKY AND E. S. KOHN 495

AUTHOR INDEX 513

SUBJECT INDEX 521

Introductory Review

Photo-Electronic Imaging 1958–1978

W. L. WILCOCK

*Department of Physics, University College of North Wales,
Bangor, Gwynedd, Wales*

It is now twenty years since the First symposium on Photo-Electronic Image Devices was held at Imperial College; but the real origin of that meeting can be traced further back to the establishment, or more accurately the revival, of the Chair of Instrument Technology at the College in 1954. I happen to have had a peripheral contact with this event during its gestation period. At that time I was in the Physics Department at the University of Manchester, playing about among other things with designs of photoelectric spectrophotometers intended for astronomical use and exploring various possible ways of increasing their optical efficiency, but slowly recognizing that however carefully we selected our photomultipliers and suppressed the unwanted noise, the time-sequential observations dictated by their lack of image-preserving properties were a crippling limitation. One day I was summoned by P. M. S. Blackett, then Head of the Department of Physics at Manchester but already elected to the corresponding position at Imperial College. He said B. V. Bowden had drawn his attention to a vacuum physicist working at the EMI Research Laboratories who had ideas for making devices of the kind we needed, and told me that I had better arrange to talk to him. So I went to Hayes to see J. D. McGee, and returned to Manchester filled with enthusiasm that I had met someone with not only the scientific insight to grasp the problem and propose solutions, but also the technical experience to give some assurance that the proposed solutions were practically feasible. Not so long afterwards McGee was appointed to be Professor of Instrument Technology at Imperial College in October 1954, and two years later I joined the small group he had by then gathered round him there.

Apart from McGee and his chief technician, S. Dowden, none of us has had previous experience of the design and fabrication of photoelectronic devices, and all must soon have realized that we had entered a scientific environment quite different from the one to which our various university research backgrounds had accustomed us. Photoelectronics was a relatively new subject which had been developed intensively in a small number of industrial laboratories, firstly during the thirties because of commercial interest in electronic television, and then during and after the Second World War in response to military interest in light-sensitive devices for guiding and surveillance. As a result the subject was surrounded by a veil of secrecy: commercial rivalry combined with the Official Secrets Act to produce a situation in which all information that was not classified seemed to be proprietary. There were of course patents, but they are addressed to those "skilled in the art", not to novices. There were also some books, for instance Klemperer on electron optics, Zworykin et al. on photoelectricity and television, Bruining on secondary emission and Leverenz on phosphors, but these were not up-to-date nor did they provide the technical and engineering know-how essential for the construction of successful devices. For instance, the recipe in Zworykin and Ramberg for the preparation of the "most sensitive" cathode reads "after the tube has been baked ... antimony is evaporated to a thickness which has by previous tests, been found to give the most favorable result". The operative words are "previous tests": these were part of the tradition and experience of laboratories where phototubes were manufactured for sale, and newcomers had no option but to carry out these and similar experiments for themselves.

It is a measure of the keenness and energy with which McGee's inexperienced colleagues applied themselves to this learning process that after less than four years he felt the group was sufficiently strong and independent to throw open its laboratories and act as host to an international conference. The date proposed for the meeting was deliberately chosen to follow the 10th General Assembly of the International Astronomical Union in Moscow, because representatives of groups in the USA that were also engaged in developing photoelectronic image devices for use in astronomy would then be passing through Europe, and it was hoped they would be willing to attend. In the event the meeting was fully representative of current activity inspired by astronomical needs. What was not initially foreseen was that it would attract the interest of research workers in other fields, notably nuclear physics and X-ray fluoroscopy, who were also concerned to exploit the possibilities of photoelectronic imaging for scientific observation, and that it would also be supported by major industrial laboratories, which sent representatives and contributions. Thus, somewhat unexpectedly, the First Symposium on Photo-Electronic Image Devices, held at the Imperial College from September 3 to 5, 1958, dealt with a wide variety of devices and applications, and so established a pattern which has persisted in later meetings. Besides this catholicity of content the first meeting also firmly established the tradition, inevitable since it was sponsored by a University group, of open discussion. It was made clear to intending contributors that after presentation of a paper there would be time for discussion, and for questions to be asked that would need to be answered: what could not be talked about freely was not to be talked about at all. With only a few embarrassing exceptions this convention has been observed over the years, and the published discussion has often been as valuable as the papers themselves.

In the late fifties the only photoelectronic image devices generally available were picture-signal generating tubes designed for entertainment television, and electrostatically focused image intensifiers intended for visual use. Some efforts were being made to apply these devices to scientific purposes, but essentially on the grounds that they would provide experience in handling this new kind of image detector. There was no real prospect that these experiments would yield useful scientific observations, and so it is not surprising that the main thrust of the First Symposium was towards new devices adapted to the more exacting demands of physical measurements.

The conviction motivating work in the astronomical field was stated by W. A. Baum at the beginning of the opening paper of the meeting: "Sixty years ago the advent of photography completely revolutionized astronomy.... We now regard image tubes as another revolution in the making". In his classic paper Rose had shown that photographic emulsions were inefficient detectors, and over typical astronomical exposures it was estimated that the information gathered corresponded to an average quantum efficiency of around 10^{-3}. By comparison photoemissive cathodes could be made with peak quantum efficiency around 0·1, so if all the information available at such a photocathode could be utilized there would be a gain of 100 times. This was a spectacular number, which even engendered loose talk that the new detectors would convert 10 in. telescopes into 100 in. telescopes—a vision of Mount Wilson in every backyard. The prerequisite that all the information available at the photocathode would have to be utilized to achieve this gain was all too easily glossed over. In extenuation it is only fair to recall that, although Clark Jones had some years earlier emphasized the difference between responsive and detective quantum efficiency, it was still common not to make the distinction between them. Maybe this distinction would have been better appreciated had attention been paid to results already published showing that the responsive quantum efficiency of the primary photographic effect was of the same order as that of photocathodes. Yet perhaps it is well this was disregarded: new enterprises are not commonly undertaken by people with too strong presentiments of likely difficulties. In the event it took a great deal of work to translate the responsive quantum efficiencies of photocathodes into comparable detective quantum efficiencies of practical devices.

The doyen of contributors to the first meeting was A. Lallemand. In his pioneering "Caméra Electronique", which had already been under development for more than twenty years, the photoelectrons were accelerated from the photocathode and imaged directly on an electron-sensitive emulsion where each was recorded as a short track of developable grains. This system gave a close approach to ideal transfer of information from the photocathode, as evidenced by astronomical observations already made with it, which showed speed gains upwards of 50 times compared with direct photography, with better resolution and higher storage capacity. Its application was restricted by the relatively small field of the electrostatic lens, about 20 mm diameter at the photocathode; but the chief practical drawback was limited life. The photocathode was poisoned by gases given off by the emulsion. In order to make the camera work at all the photocathode had to be kept in an evacuated glass ampoule which was not broken until the last moment, and by refrigeration and gettering it was then possible to retain useful photosensitivity for one night, long enough to expose the eight electron-sensitive plates carried in a magazine inside the camera. After that the camera, which was a complicated and relatively fragile glass structure, had to be dismantled and put through a lengthy cleaning and restorative process before it could be used again.

Because of the level of technical skill required for this preparation there was scepticism that the Lallemand camera would find general acceptance amongst astronomers. Yet electronography had very attractive features: recording of single photoelectrons and hence lack of reciprocity failure, high resolution, large storage capacity and dynamic range; and these prompted schemes to prolong the life of the photocathode by isolating it from the emulsion with a thin membrane impermeable to gas but easily penetrated by the photoelectrons. The membranes envisaged were not mechanically strong enough to withstand more than a very small pressure differential without bursting, so it was necessary to keep both sides of the membrane under vacuum at all times, which meant that plates had to be loaded and unloaded through a vacuum lock. The arrangements thus still had the character of demountable laboratory vacuum systems, and in consequence they were not attractive to those who believed that the future lay with completely sealed-off vacuum tubes.

There seemed to be only two approaches consistent with a sealed-off device: either an image intensifier, in which the photoelectrons were reconverted to an amplified optical image at a luminescent screen, which could then be photographed; or a picture-signal generating tube, in which the photoelectron image was stored inside the tube as a charge pattern, which could then be scanned and read out by an electron beam. The chief problem with image intensifiers was the inefficiency with which the optical image on the luminescent screen could be transferred to a photographic emulsion. The screen was essentially a Lambertian source and the fastest lenses available were not able to collect more than about 10% of the light emitted. The number of photons generated from a single photoelectron was determined by the conversion efficiency of available phosphors and by the accelerating voltage applied to the tube, which in practice was limited by the onset of field emission effects. As a result a photoelectron could not be made to give rise to more than about 1000 photons at the screen, which meant about 100 at the photographic emulsion, too few to make up for the low quantum efficiency of the emulsion and allow each photoelectron to register an individual contribution in the developed image.

A way to increase the optical efficiency was to make the luminescent screen on a thin window of mica which could be strong enough to support atmospheric pressure while still thin enough to allow a near-contact print to be taken of the image. In this way all the emitted light reached the emulsion, but the necessary thickness of mica made for a heavy penalty in loss of resolution compared with what could be obtained when focusing, and this sacrifice of information seemed bound to be unacceptable in the long run. The alternative was to seek a further increase in the number of photons rising from each photoelectron. There were two ways of going about this. One was to use two image intensifiers with the output image of the first transferred to the photocathode of the second. In principle external optical coupling might have been adequate in spite of its inefficiency, but everything favoured internal coupling inside a single vacuum envelope, with the first luminescent screen and the second photocathode on opposite sides of a very thin transparent membrane. Two-stage cascaded intensifiers of this kind had already been made and operated with light amplification 800 for blue light, corresponding to an output of about 8000 photons per primary photoelectron. This gain was about as much as was needed, and while the resolution was no better than had been obtained with thin output window tubes there were obvious technical possibilities for improvement.

The internal photocathode-phosphor sandwich of a cascade tube acts as an electron multiplier. The other way envisaged to increase the number of photons arising from each photoelectron was to multiply the photoelectrons directly by secondary electron emission in an image-preserving manner. One arrangement of this kind being tried consisted of a two-dimensional array of discrete dynode structures, similar to those used in ordinary photomultipliers in which photoelectrons from separate elements of the image were multiplied by secondary emission. The design problem with such a system is to shape the electric fields between multiplying stages in such a way that secondaries from one are extracted efficiently and accelerated to become efficient primaries at the next. Unfortunately scaling down such channelled structures leads to increased voltage gradients, so it appeared inevitable that attempts to preserve useful resolution would encounter severe technical problems. These difficulties, which arose essentially from the fact that the secondary electrons left the emitter on the same side as the incident primary, could be avoided by following a then recent proposal to use transmission secondary emission. The secondary emitter was deposited on a thin conducting support film, so that primary electrons incident on the support film with sufficient energy to penetrate it could give rise to secondaries emerging on the opposite side. With this configuration an electron image could be transferred by an electron lens from one multiplying stage to the next without any need for lateral confinement. Some promising experiments with thin film dynodes had been made. There were, however, difficulties in supporting them which made it still uncertain whether they could be successfully incorporated in useful image intensifiers.

As already mentioned, another avenue to the use of photoelectronic detection in a sealed-off system was offered by picture-signal generating tubes. Those that existed had been developed to work at the scanning frequencies and relatively high levels appropriate to broadcast television. At the very much lower light levels characteristic of astronomical images the noise added in the process of extracting a usable signal from the tube was much greater than the noise intrinsic to the electron image on the storage target. There were two ways in which this shortcoming might be removed. One was to raise the signal level on the target by essentially noise-free amplification so that the noise from subsequent readout would be negligible. This approach was exemplified by the development of the intensifier orthicon, and image orthicon preceded by internal cascaded intensifier stages so that the primary photoelectron image was multiplied before it reached the target. The noise in the output was in this way made to be predominantly photoelectron shot noise, and in this sense the system approached ideal performance over a storage time of the order of a second, which was all the target permitted. There was, however, serious loss of resolution, so that in spite of large speed gains over photography the information rate gain was much less. There was also still left the problem of finding external means of storing the picture signal if longer exposure times were to be used. The other possible way to reduce the significance of readout noise was to increase the time interval during which the charge image could be accumulated on the target before it was scanned. The limitation to this in existing tubes was set by the electrical conductivity of the glass targets which caused the electron image to leak away in a few seconds. Experiments indicated that integration for much longer periods could be achieved if the target was refrigerated. Apparently promising work was also in progress aimed at developing special purpose tubes with highly insulating targets which could integrate the electron image for many hours in linear fashion before readout.

So much for the situation, and the enterprises being undertaken, in the astronomical field. There is no need to make apology for describing them at length because the demands of astronomers were most exacting, and consequently the problems of satisfying them were severe. Astronomy is, after all, an observational rather than an experimental science, and detection and accurate measurement of radiation is the only practical activity. The other groups of users, or would-be users, of photoelectronic image devices present at the First Symposium, radiologists and nuclear physicists, had by comparison much less stringent requirements and there was a correspondingly greater prospect that they might be met quickly. The radiologists were seeking to reduce the X-ray dosage required for fluoroscopic examination to the ultimate minimum set by photon shot noise; but their photons were X-ray photons, more than three orders of magnitude more energetic than optical photons, and the images to be observed visually were only shadows without the wealth and significance of photometric detail in, for example, astronomical spectra. The nuclear physicists were excited by the potentialities of the scintillation or luminescent chamber, which was a device for photographing the tracks of charged particles passing through scintillation material. The amount of light emitted was much too small to allow the tracks to be photographed directly, so the idea was to amplify it by means of an image intensifier. The particular virtues of such a chamber in high energy particle physics were very short sensitive time and recycling time, and the possibility of counter control by gating the image tube, which meant that it could be operated in high intensity beams, and desired events picked out from a high background of uninteresting ones. The number of photons available in these transient images was so small that a useful device would need to register every photoelectron as an identifiable spot on the final photograph, and this called for an image intensifier with light gain 10^5 or more. Such tubes did not exist, at least not in the Western World, but they had been made in the USSR and used for this very purpose. The knowledge that they could be made gave the firmest possible assurance that sooner or later they would be made in the West, and the dynamism of nuclear physicists, stimulated by group competitiveness and fed by long purses, made it likely this would be sooner rather than later. Parenthetically it is interesting to recall that although Soviet scientists were known to have made important pioneering contributions to the development and use of photoelectronic devices in astronomy as well as in nuclear physics, their activities were largely shrouded in mystery and in spite of persistent efforts through the proper channels there was no response to invitations for them to attend the early meetings. That was a matter for regret since it impaired the international character of the Symposia.

This, then, is a vignette of the activity in applying photoelectronic image devices to scientific observations at the time of the First Symposium. The dramatic expansion of activity in the field since then is chronicled in the nine volumes, totalling nearly five thousand pages, of published Proceedings of the five further Symposia that have been held in the intervening twenty years. It would be tedious to offer even the briefest summary of this mass of material, and invidious to select highlights; but re-reading the volumes in sequence brings home to me what tremendous technological changes have occurred during the period they span, changes which take place gradually enough to be scarcely noticeable at the time but whose cumulative effect is striking, just as a familiar face seen frequently seems not to grow older until an early photograph proves otherwise. New semiconducting materials for photocathodes, photoconducting targets and electron multipliers; new circuit techniques based on solid-state circuit elements, and now miniaturization through integrated circuit technology; new high vacuum techniques with bakeable clean metal systems, ion and sorption pumps, and cold vacuum seals; new optical materials transparent to a wider range of wavelengths, and fibre optic elements for windows; all these have led to a level of sophistication in the devices now current and of the peripheral equipment with which they are associated that shows up much of the hardware of the First Symposium as quaint period pieces. Many of these devices have been made up from the wide range of commercially manufactured image intensifiers, converters and picture-signal generating tubes now on the market, and there is no doubt that the ready availability of high quality items of this kind is responsible for the greatly increased variety of observational contexts in which photoelectronic image devices are now being used. In some cases the purpose is to increase the sensitivity or the accuracy or the speed or the convenience of observations that can already be made by other methods; but in others, for example applications to photo-biology, the field of study has been created by the development of the new instrumentation. Numerous instances of such uses occur in the

Proceedings. They surface for a brief period and then are not mentioned again; either because the need ceased to exist, as happened with the luminescent chamber, or more usually, as with fluoroscopes, because they have become part of common practice in the field in question. Amidst the descriptions of the technical development of devices and components these examples provide a welcome reminder that large numbers of people now actually do make scientific observations with photoelectronic image devices. That is certainly a change from the situation twenty years ago when promises for the future abounded but there was little to show in the way of scientific accomplishment.

Yet whatever may have happened in other fields the continuing accounts of the development of devices for astronomical use bear witness that the needs of astronomy, the origin of the Symposia, have not yet been fully satisfied. This is perhaps not a matter for surprise. The goal declared by A. Lallemand in his first words to the first meeting: "Nous avons cherché à réaliser un récepteur idéal des photons" is so ambitious and far-reaching that its attainment is all too likely to recede indefinitely into the future. What we have seen over the years has been not so much a revolution as an evolution; and what is surprising, to me at least, is that so many of the different approaches described twenty years ago have recognizably survived, even if in technically different clothes. The electronographic cameras, either of the Lallemand open type, or with photocathode protection and plate loading through a vacuum lock, are obvious descendants of earlier versions, though improved out of recognition in the matter of life, field size, convenience of operation and robustness; and they provide scope still for those who value the personal satisfaction of making, rather than just assembling, devices. But I think it is not fanciful to believe also that fibre optic window intensifiers are the modern descendants of thin mica window phosphor output tubes; that the channel plate is the current version of two-dimensional dynode structures; that the image photon counting systems are in direct line from the intensifier orthicon; and that the attempt to use silicon vidicons as large-area photometers is the latest expression of experiments with storage targets. These relationships between past and present illustrate how far invention runs ahead of technology.

What of the future? For faint sources and small fields the image photon counting system is essentially as perfect a detector as the photocathode allows. The electronographic camera, too, is probably as near perfect as it will ever be, with its photometric properties limited largely by deficiencies of the available emulsions. Whether storage target tubes will establish themselves as better photometers remains to be seen. But now solid state array devices have appeared, which alone or in hybrid combinations seem certain to have an important future role. What the impact of this will be on photoelectronic instrumentation for scientific purposes cannot yet be foreseen. I remember G. A. Morton once remarking to me that since optical photons have so little energy it cannot be desirable to remove the photoelectrons from the solid into vacuum. That was many years ago, but perhaps we really are now on the threshold of the change he envisaged; in any case, we can be sure there will be plenty of material forthcoming to keep the Symposia flourishing for another twenty years.

Review of TV Camera Tubes and Electron Optics

H. G. LUBSZYNSKI

"Cala-na-Sith", Stronafian, Glendaruel, Argyll, Scotland

TV CAMERA TUBES

Television camera tubes can now look back on almost 50 years of development and their performance has improved beyond the wildest dreams of their originators. Yet the requirements confronting the researchers and designers are still largely the same, though on a greatly elevated level. They are mainly:

Sensitivity
Resoultion, generally expressed as MTF
Spectral response
Light transfer function or "γ".
Speed of response
Signal to noise ratio, S/N
Tolerance to light overload
Size and weight of the tube, its associated circuits and cameras
Ease of operation.

The priorities of some of these characteristics depend on the applications which can be roughly divided between entertainment and closed circuit television. The latter comprises industrial applications, surveillance, military applications and scientific applications such as astronomy, space research, biology, etc.

To take entertainment television first, the last few years have seen an almost complete change-over from camera tubes based on photoemission to tubes based on photoconduction. The much higher quantum efficiencies obtainable, at present, from photoconductors as compared with photoemitters made possible the use of small area targets and thus considerably smaller and lighter tubes, lenses and cameras. However, because of their lag, the early vidicon tubes were useful only in industrial applications and, in the entertainment field, for film scanning, where the high target illumination obtainable made the lag negligibly small.

The advent of the Plumbicon, using as the target a layer of suitably doped lead oxide forming a reverse biased p–i–n junction changed matters completely. This tube has completely displaced the image orthicon for both studio use and outside broadcasts because it is as sensitive and it has a linear light transfer function. This is important for the colorimetry in colour television, but it makes the output signal more vulnerable to light overloads. A solution to the problem was found[1] by scanning the target during line return with a strong, defocused beam while pulsing the gun cathode positively so as to lap off all charges on the target which had caused voltage excursions above the height of the pulse, while leaving the useful signal charges below this voltage untouched.

The resolution in lead oxide tubes is, at present, more limited by the target itself than by the electron optics of the scanning beam, especially at the red end of the spectrum. Recent developments have resulted in improved target layers which have made it possible to reduce the tube diameters from 30 mm to 25 mm with virtually no loss in resolution. Understandably, the manufacturers have not been very forthcoming in describing how it is done. Some of it is doubtless due to making the layers thinner. Treatment of the layers with H_2S has improved the response at the red end of the spectrum. Capacitance or discharge lag can become troublesome at very low light levels because of the reduced discharging power of the beam for small voltage excursions on the target.[2] Going to smaller tube sizes reduces the target area and thus target capacitance and lag. The application of small, controlled amounts of bias light, causing a small voltage pedestal on the target also reduces this lag considerably.[3] This feature is now standard in lead oxide and other cathode potential stabilized tubes.

In recent years, targets made of a variety of amorphous layers of chalcogenide compounds have been developed. Some of them have Schottky barriers at the signal plate side, others incorporate reverse biased $p-p^+$ junctions or heterojunctions in order to keep dark currents low. They all have linear light transfer characteristics. Most of them are about equal to or more sensitive than lead oxide, one of them considerably so. Three types of tube, developed in Japan and known under their trade names Saticon, Newvicon and Chalnicon have come into use.[4] The Saticon uses amorphous Se doped with As and Te, followed by a layer of Sb_2S_3 on the scanned side. It is claimed to be a reverse biased $n-p-p^+$ junction. The Newvicon uses a double layer consisting of ZnSe, followed by a further layer of ZnCdTe on top. The Chalnicon target consists of a layer of CdSe, followed by a layer of As_2S_3. Of these tubes, the Newvicon has a quantum yield greater than unity and is nearly twice as sensitive as a silicon diode array target, although rather laggy. The other two have quantum yield near unity and are under intensive development to reduce lag to tolerable limits. They have, in fact, already been used in colour cameras for news gathering and outside broadcasts.

Up to now, they all had one feature in common: the layers had to be kept very thin, between 1 and 3 μm. Attempts to increase the thickness substantially in order to reduce the capacitance have led to either longer photoconductive lag or reduced sensitivity or both. The capacitance of such thin, solid layers is particularly high because of the high dielectric constants of these materials. O. Yoshida† reports considerable reductions in the capacitance in Chalnicons by replacing the solid As_2S_3 layers with As_2Se_3 and modifying the physical structure of the target so as to incorporate a solid–porous–solid structure. Moreover, by doping it with Te, he has been able to extend the response of the layer further into the red. The use of bias lighting has further improved the lag. In general, however, it can be said that any long, low level lag, even as low as 3 to 4% is found to be intolerable for entertainment television. As the light in these targets is almost entirely absorbed near the surface of entry, and as the layers are rather thin, there is practically no spread of light and no halation. Also, the carriers hardly diffuse laterally in such thin layers. Therefore resolution is not limited by the target, but by the electron optics of the beam. Hence the diameters of some of these tubes can be reduced to 18 mm while still giving satisfactory resolution. The smaller size target then gives another bonus in lower target capacitance and lower discharge lag. Another approach to the capacitance lag problem is to improve the gun design. This will be dealt with in the section on electron optics below.

The use of these new materials is very tempting, because most of them are largely unaffected by exposure to the atmosphere and their deposition and processing are greatly simplified as compared with the manufacture of lead oxide targets.

To reach ultimate sensitivity in very low light level applications such as are encountered in astronomy, surveillance, etc., it is essential to intensify the incoming signal before storage to such an extent that the noise in the signal itself is large compared with the noise introduced by the reading process. There are two ways of achieving this at present. One is to use an image intensifier in front of the camera tube. The other is to carry out the intensification in the target itself. The two main representatives of the latter approach are the SEC tubes[5] and the SIT tubes,[6] based on secondary electron conduction and intensification in silicon diode arrays respectively. SEC tubes have been around for a considerable number of

†See p. 39.

years. They have even been taken to the moon. However, a number of factors have militated against their being used except in very specialized applications, where their great operating sensitivity is of paramount importance. The drawback in their use is that the targets consisting of porous layers of KCl or MgO have a low first crossover of the secondary emission characteristic. They become easily unstabilized under excess illumination, leading to voltage breakdown inside the layer. When such a breakdown occurs, the target is permanently damaged. Hence a stabilizing mesh must be introduced in front of the target, held at a potential lower than the first crossover. The tubes are then stable, but the low field gradient in front of the target, caused by the mesh, greatly reduces the resolution. Bringing the stabilizing mesh closer to the target increases the field gradient, but introduces a high signal plate to ground capacitance, reducing the S/N ratio. T. Kawamura *et al.*† describe an interesting development which seems to have succeeded, where many have failed before. They have developed a target consisting of a double layer of porous MgF_2 and a second layer of porous Ag; by this means the first crossover is raised to over 100 V. This enables them to discard the stabilizing mesh and allows them to increase the field gradient in front of the target. It also reduces the signal plate to ground capacitance. The result is greatly improved resolution and S/N ratio. Thus the tube stands a good chance of becoming usable in entertainment television having practically no lag and an operating sensitivity of 3750 to 6250 $\mu A\,lm^{-1}$ which is about ten times that of the lead oxide tube.

Returning to the SIT tube, the target gains obtainable can be quite considerable and may reach several thousand. V. J. Santilli[7] has described such a tube having an 80 mm diameter S·20 photocathode on a concave fibre optic plate. The photoelectrons are focused on to a circular array of silicon diodes 30 mm in diameter. Their linear density is 80 mm^{-1}. The bombarding photoelectrons are accelerated to between 4 and 12 kV giving a range of target gains varying by 20:1. Target gains of 5000 have been obtained when stressing the tube up to 15 kV in the image section. Typically, the tube responsivity is 1160 $\mu A\,lx^{-1}$ (in absolute terms 25 mA W^{-1} for radiation from a black body at 5900 K). The light transfer function is linear. The photoelectron image can be gated and zoomed from the full 80 mm cathode being imaged onto the 32 mm target down to only 32 mm of the cathode being so imaged. The reading section contains a conventional electron gun using a small deflection angle. The square wave MTF is shown in Fig. 1, plotted against number of TV lines per picture height.

†See p. 51.

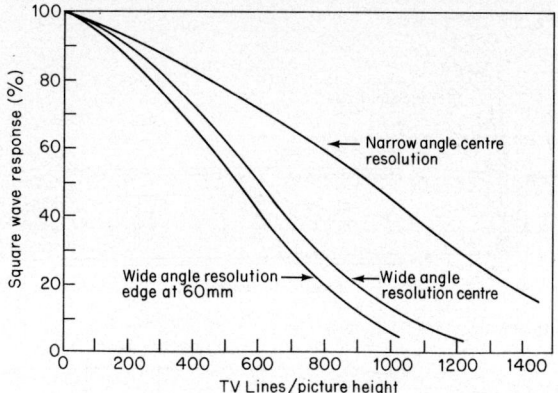

Fig. 1. Square wave MTF of Ebsicon camera tube given by Santilli.[7] Response measured at $I_B = 1000$ nA and $I_{sig} = 800$ nA; 1×1 raster on 32 mm diameter target.

Measurements of blooming for a 60 μm point illumination on the cathode show that for a light overload of 100:1 above peak white the spot size in the picture is only increased by four times. Line oscillograms through two selected adjacent stars having an amplitude ratio of 5:1 show only one star in the wide angle mode, but they are clearly resolved as two stars in the narrow angle mode.

W. V. McCullough† has studied a simplified, one-dimensional array of reverse-biased silicon diodes. Charge removal rate, point spread function and phase transfer function are calculated and the equations fed into a computer. Measured beam acceptance curves were also fed into the computer. The computed MTF and PTF curves agree closely with curves measured on a commercial SIT tube.

Suzuki et al.[8] have developed a camera tube sensitive in the vacuum ultraviolet, but which is solar blind, for the study, in daylight, of the aurora over the North Pole. The camera is to be satellite-borne at an altitude of 4000 km. A diagram of the tube is shown in Fig. 2. Photoelectrons from an evaporated, solid photocathode of KBr are intensified in two microchannel plates in tandem, giving a gain of about 10^5. On leaving the second channel plate, the electrons are accelerated on to a storage target consisting of an insulating layer deposited on that side of a metal mesh facing the second channel plate. Secondary electrons released from the insulating target surface are collected on the mesh itself

†See p. 63.

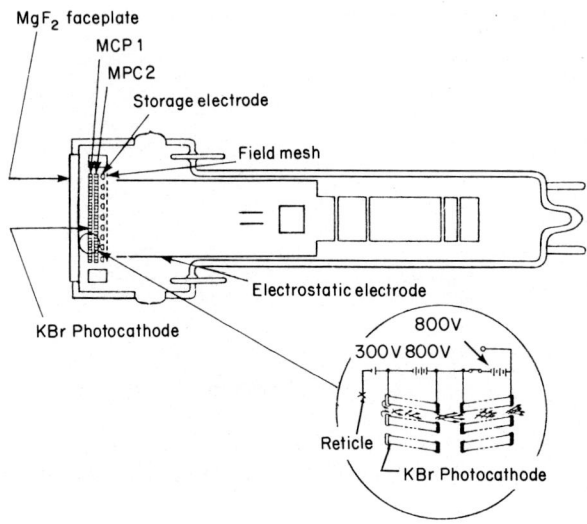

Fig. 2. Construction of the vacuum ultraviolet pickup tube described by Suzuki et al.[8]

and on the adjacent field mesh of the scanning section. The scanning beam, on penetrating through the storage mesh, is modulated by the charges on the insulating layer and collected on the output electrode of the second channel plate which also serves as signal plate. The photocathode is deposited on the entrance surface of the first channel plate. On the inside face of the tube input window of MgF_2 a thin, transparent metal layer is deposited. It is held at a negative potential of −300 V with respect to the photocathode, thus forcing the emitted photoelectrons to turn round and enter the first channel plate. A voltage of 800 V is maintained across each channel plate. Writing, reading, erasing and priming of the target surface are sequential and are obtained by applying appropriate potentials to the relevant electrodes. Erasing and priming take about 2 sec each, while reading is slow and takes about 100 sec. For writing, an electronic shutter switches off the voltage on the second channel plate when sufficient charge has passed; this prevents overloading of the storage target. In fact, the voltage on the second channel is only applied during the writing cycle. Resolution is stated to be about 200 TV lines and about 7 grey steps have been observed. A picture of the aurora, taken by the tube mounted in the Japanese satellite Kyokko, has been published.

There has been much recent interest in pyroelectric vidicons for thermal imaging. The peculiar character of the pyroelectric effect governs the way in which such tubes are operated. The theory and methods of operation have been described in a number of papers.[9-11] The majority of these tubes use thin, large area crystals of triglycine sulphate (TGS). A more recent refinement has been to treat the TGS targets with deuterium. These deuterated targets are referred to as DTGS targets. Their main advantage is that the dielectric constant is only 24 compared with 76 for plain TGS at the operating temperature in the camera. This means a greatly increased voltage change at the target for a given pedestal current.[10] A number of other materials have been investigated and one of the more promising ones seems to be polyvinylidene (PVF_2). All these crystals are very good insulators. The charges read off the scanned surface by the scanning beam must therefore be restored or after a few scans the scanning beam will no longer reach the target. The method mostly used at present is to leave or generate small amounts of gas in the tube. The scanning beam electrons ionize some of the gas molecules and these positive ions rain on to the target, raising its surface potential continuously to more positive values so that the scanning beam can continue to reach it. Tubes with a hard vacuum require either slightly leaky targets, which are difficult to make, or a series of sequential operations such as separate reading, erasing and priming scans.

As the heat from a picture point spreads through the thickness of the target, it also diffuses laterally, reducing resolution and thereby contrast. The latter, in turn, reduces the S/N ratio. P. J. Pool and D. J. Pedder† describe a pyroelectric target which is reticulated by cutting deep grooves through the DTGS crystal in the line and frame directions to prevent the spread of heat between elements. Such reticulated targets had previously been made by chemical etching which results in undercutting. Mechanical methods resulted in trapezium-shaped elements presenting reduced surface areas to the scanning beam. Pool and Pedder describe a process of cutting by means of ion beams which results in much improved grooves with straight, vertical sides. Both resolution and resolvable temperature differences are much improved, but the mechanical strength would still appear to be a problem.

Y. Hatanaka *et al.*‡ describe a pyroelectric vidicon with a hard vacuum in which the current, necessary for replacing the scanned-off charges, flows by lateral leakage over the surface of the target and not through the thickness of the crystal. For this purpose, a system of narrow, parallel aluminium strips is deposited on the scanned surface. These are

†See p. 23.
‡See p. 31.

joined at one end and connected through a common high resistance to an electrode held at a suitable positive potential. By adjusting the potential, the required pedestal current can be set up. The target material in this tube is PVF_2. Though it has a much lower pyroelectric coefficient than TGS, this is compensated for by a much lower thermal diffusivity making the tube about as sensitive as tubes with TGS targets

Electron Optics

In the electron optics field, progress has been reported on various fronts. They include improvements in electron guns for use in cathode potential stabilized camera tubes, permanent magnet focusing assemblies (PMFAs), computer-assisted design of electron optical systems for image intensifiers and for line arrays of sensors, improvements in continuously pumped, demountable electron optical systems and in the step deflection of electron images from extended areas.

As mentioned in the review on TV camera tubes above, there are now a number of such tubes available with evaporated, amorphous target layers consisting of chalcogenide compounds. The outstanding problem in these tubes is, at present, the long discharge lag caused by their high target capacitance. In addition to the measures taken to reduce this capacitance, there is another approach to the lag problem, i.e., to design better guns. The speed of the discharge process being dependent on the slope of the beam acceptance curve, it is important to raise it as much as possible towards the theoretical limit given by the Maxwellian distribution of emission energies of the electrons. Early hopes to reduce these energies in camera tube guns by using cathodes working with negative electron affinities and running at room temperature have not been realized, so far, in practical tubes because of the delicate processing required and the very high vacua which must be maintained in such tubes during the whole of their life.

The other approach is to avoid imparting any additional radial energy components to the beam electrons in the focusing process. Electrons with such components will be unable to land on target elements at low positive potential excursions. They will thus not participate in the discharging process of small signals and will give rise to low level lag. These considerations led to the design of guns with laminar flow in which no crossover occurs. The modulator electrode in front of the cathode is held at a positive potential to accelerate the electrons and the equipotentials in front of the cathode are flat. With this type of gun shorter capacitance lags have been achieved, almost making the chalcogenide target tubes usable in colour studios.

A number of papers, both experimental and theoretical, have dealt with focusing of extended area electron beams in image tubes. G. K. Bhide et al.[12] describe a continuously pumped, demountable electron optical system. The photocathode consists of a thin, transparent gold layer of 25 nm thickness deposited over an opaque test pattern of the Westinghouse test chart ET–1332A which is, in turn, produced on the inside surface of a curved silica window. Under illumination with an ultraviolet lamp, photocurrent densities of the order of 10^{-7} to 10^{-8} A cm^{-2} were obtained, but the emission energies of the photoelectrons range from 0 to 1·5 eV which is much more than encountered from normal photocathodes irradiated with visible light. Hence better figures can be expected for the MTF in normal, sealed-off tubes than those measured in the demountable gear. In addition, the density curves of the master masks used to produce the test pattern were only modulated to a depth of 50% at 60 lp mm^{-1}. Conventional diode image tube systems of magnification 1 and 0·75 were investigated. Variations of the resolution with radial distance from the centre of the image were measured as well as the tolerance of resolution as a function of the axial displacement of the anode cone.

A. Choudry[13] has studied the possibilities of improving the resolution in proximity focused systems. He claims to have found by mathematical investigations that by introducing spatial variations of potential by means of a multi-electrode system with a non-uniform dielectric constant, the point spread function could be improved by an order of magnitude, but "at the sacrifice of some of the intensity and of the simple uniform electrode structures". No examples of such structures or arrangements are given.

In a theoretical paper, Chou Li-Wei† has investigated the focusing properties of electric and magnetic fields between concentric spheres. He has derived formulae for the cases of electric fields, magnetic fields and for a combination of the two. By letting the radii of the two spheres tend to infinity the formulae for plane parallel proximity focusing and for focusing in a homogeneous, rectilinear magnetic field, such as a long solenoid, appear as special cases in his equations.

Great advances have been made in designs for permanent magnet focusing assemblies (PMFA). The development of very high flux density permanent magnets of rare earth/metal alloys such as SmCo has considerably reduced the weight of the magnets. This is of special importance in the field of space applications and light weight image intensifiers. The

†See p. 119.

elimination of the heavy solenoid coil with its power requirements as well as the associated cooling equipment has resulted in great savings in weight, power and size.

Three of the new assembly designs are described by C. I. Coleman *et al*.† They are the Profiled Reluctance, Reluctor Ring and Flux Compensated assemblies. Another paper presented at the Symposium described a reluctor ring assembly, specially designed for a 35mm SEC tube to be used in space for a Coronagraph/Polarimeter.[14]

R. Evrard‡ describes an electron optical arrangement for image intensifiers where photoelectrons from a concave fibre optic cathode are brought to a crossover in a fine hole in the anode, then decelerated by an electron mirror formed between a transparent, conducting layer applied to the inner surface of the flat end window of the tube (which is at a low potential), and the final anode. A thick fluorescent screen is deposited on this side of the anode and is observed through the end window. The luminous output and the halation are reported to be much superior to those from a conventional screen.

In most photoelectronic devices, the electron optical problems are static ones, i.e. the picture elements on the photocathode have a fixed spatial relationship with corresponding elements on either a screen or a storage target. The problems become more complex in devices where the whole electron image is scanned across a line array of sensors or over a two-dimensional sensor such as a fluorescent screen or a photographic emulsion. In these cases, not only is geometric fidelity required, but also the times of flight of all the electrons from a single picture element on the photocathode to the corresponding target element must be as nearly as possible the same. A number of papers deal with this aspect of the design and evaluation of electron optical systems, by taking into account the range of emission energies and emission angles of the electrons.

W. A. Delamere and E. A. Beaver§ deal with geometric and focus aberrations in tubes where an electron image from a flat cathode is scanned across line arrays for photon counting, as is done in Digicon tubes and some intensified CCD arrays. Some of the factors affecting resolution and distortion are discussed. The tolerances permissible to keep within desired limits of performance are calculated by numerical integration of the electron trajectories. The investigation includes the effects of details of tube design such as the thickness of the indium end-window seals and of the intermediate potential rings, the uniformity

---

†See p. 89.
‡See p. 133.
§See p. 101.

of the wall coating and the shape of the mounting flanges of the line array. As mentioned above, in a dynamic system the transit times are important. They are tabulated for electrons of various emission energies.

Further problems encountered in dynamic systems are discussed by D. McMullan et al.† Step deflection is required in some high-speed cameras and also for step and repeat operations in the production of integrated circuits etc. To ensure the desired resolution of 50 lp mm^{-1}, the accuracy and stability of the magnetic focusing and deflecting fields must be very high out to the edges of the picture over extended periods. An elegant way to reduce the high requirements for long time stability is to connect the focusing and deflecting coils in series to a constant current power supply. This maintains the deflection angle to a first approximation. The step deflection is carried out by tapping the scanning coils and effecting the steps by switching the taps, and hence the deflecting field, by means of a computer and associated driving circuits.

K. W. Jones et al.[15] have reported a suite of computer programs developed for the design and analysis of electrostatically focused electron optical systems. Two-dimensional and three-dimensional systems of cylindrical symmetry can be dealt with.

The first program reads the electrode structures into a fine grid system and calculates the electrostatic fields by a reiterative relaxation routine. After twenty-five iterations, the final accuracy is about 0.1%. Three further programs calculate and display equipotentials, electron trajectories and differences in the time of flight along the axis for electrons of different emission energies. A final program allows the shape, position and voltages of the electrodes to be altered and the suite of programs to be rerun without the need to prepare new input data. An analysis of the electron trajectories in the Photochron II streak camera tube shows reasonable agreement with experimental values for magnification, image spread from point sources and time dispersion. Similar analyses done on commercial tubes agree well with published data.

Y. Kiuchi and T. Sakusabe‡ have measured the current density in the focused spot of a conventional vidicon gun and focusing system and found the half-width of the final spot diameter to be about 25 to 30 μm. They calculated the trajectories of electrons with various emission energies. Based on these calculations, they propose a new focusing method which they call the Concentrated Magnetic Flux (CMF) method. The magnetic focusing and deflection fields increase linearly along the axis, while the electrons are accelerated by a voltage which increases with the square of

†See p. 109.
‡See p. 75.

the distance along the axis. Their calculations show that in this way demagnification of the scanned spot can be achieved. They find that a reduction of 50% in spot size can be expected.

REFERENCES

1. Van Roosmalen, J. H. T., *In* "Adv. E.E.P." Vol. 28A, p. 281 (1969).
2. Webley, R. S., Lubszynski, H. G. and Lodge, J. A., *Proc. Inst. Electr. Eng.* **102,** 403 (1955).
3. Lubszynski, H. G., Taylor, S. and Wardley, J., *J. Brit. I.R.E.* **20,** 331 (1960).
4. *Japan Electric Engineering* p. 28, Feb (1974).
5. Goetze, G. W., *In* "Adv. E.E.P." Vol. 22A, p. 219 (1966).
6. Rodgers, R. L., *In* Proc. I.E.E.E. Intercont. Conf. New York (1973).
7. Santilli, V. J., *In* "Seventh Symp. P.E.I.D. Preprints" p. 31 (1978).
8. Suzuki, Y., Kinoshita, K., Makamaya, M., Kaneda, E. and Niwa, N., *In* "Seventh Symp. P.E.I.D. Preprints" p. 19 (1978).
9. Holeman, B. R. and Wreathall, W. M., *J. Phys. D.* **4,** 1898 (1971).
10. Watton, R., *Ferroelectrics* **10,** 91 (1976).
11. Watton, R., Burgess, D. and Harper, B., *J. Appl. Sci. & Eng. Sect. A* **2,** 47 (1977).
12. Bhide, G. K., Rangarajan, L. M. and Singh, S., *In* "Seventh Symp. P.E.I.D. Preprints" p. 169 (1978).
13. Choudry, A., *In* "Seventh Symp. P.E.I.D. Preprints" p. 181 (1978).
14. Bailey, S., Delamere, W. A. and Dionne, N. J., *In* "Seventh Symp. P.E.I.D. Preprints" p 183 (1978).
15. Jones, K. W., Sibbett, W. and Bradley, D. J., *In* "Seventh Symp. P.E.I.D. Preprints" p. 159 (1978).

A Pyroelectric Vidicon with Reticulated Target

P. J. POOL

English Electric Valve Company Ltd., Chelmsford, England

and

D. J. PEDDER

Plessey Research (Caswell) Ltd., Towcester, Northants, England

Introduction

In the EEV P8092 pyroelectric vidicon an infrared image is focused onto a thin pyroelectric target of deuterated triglycine sulphate (DTGS). This image produces a heat pattern which in turn generates a charge pattern which is read out by the electron beam.[1] The resolution is degraded by thermal diffusion of the heat image.

An improvement in the thermal modulation transfer function would be expected if the lateral thermal conductivity of the target were reduced. A reduction in thermal conductivity can be achieved by dividing the target into an array of discrete islands as shown in Fig. 1. This process is known as reticulation.[2,3] A support film is required, of course, to maintain the integrity of the structure. This paper describes the process developed for the reticulation of DTGS. Measurements will be presented of the performance of these targets in sealed-off tubes.

Ion Beam Etching

The reticulation process, by removing pyroelectric material, reduces the target response. At high spatial frequencies this is more than offset by the increased MTF resulting from the reduction in thermal diffusivity of the target. The loss becomes apparent at low spatial frequencies where the reduced thermal diffusivity of the reticulated target has less effect on the MTF.

Target reticulation by chemical etching would normally be subject to undercutting and excessive loss of material; for a reticulation pitch which is of the same order as the target thickness this undercutting could make reticulation impossible. To minimize the loss of material, targets have therefore been reticulated by ion beam etching. This has allowed the etching of vertical sided grooves of controlled width, with no undercutting. Etching has been carried out in a Veeco "Microetch" unit at the Plessey Company. DTGS was found to etch very rapidly and useful etch rates have been obtained using low energy ions.

The Support Film

For a reticulated target the limiting resolution is set by the pitch of the islands. The MTF is determined by the pitch and the degree of thermal isolation between islands. For a particular groove width the heat flow between islands is determined by the thermal conductivity of the support layer multiplied by its thickness.

In his analysis of the thermal properties of reticulated layers, Watton[4] has derived the MTF at high spatial frequency as a function of the conductivity–thickness product of the support film. These results are given in Fig. 2 where the dashed line represents the MTF for an unreticulated target. The MTF is seen to fall off steadily for a conductivity–thickness product greater than $0.2\ \mu W\ K^{-1}$. However, very little is to be gained by reducing this product below $0.2\ \mu W\ K^{-1}$.

Although plastics do not, in general, have a particularly low thermal conductivity, very thin plastic films (1–5 μm) can be prepared which have adequate mechanical strength and a thermal conductivity–thickness product in the required range.

For optimum performance, it is necessary to reticulate the target to its full thickness. Target thickness is not always uniform, and in obtaining full reticulation of the thicker areas, the signal plate on thin areas can suffer prolonged exposure to the ion beam. To prevent perforation of the thin nichrome signal plate the layer of oxide shown in Fig. 1 is formed between the nichrome and the DTGS.

Masks for Etching

Ion beam masks for etching have been made in evaporated aluminium. The etch rate of aluminium appears to be reduced in the presence of DTGS etch products. Consequently, an aluminium film of 100 nm is sufficient to mask the etching of 25 μm of DTGS. The masking patterns have been formed by the evaporation of aluminium from a small source through a fine metal mesh held in contact with the target surface.

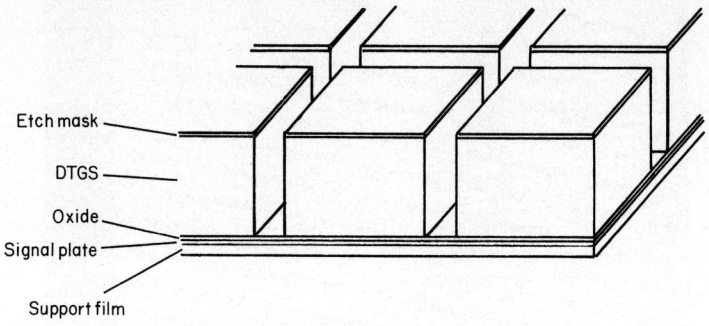

Fig. 1. Reticulated target cross section.

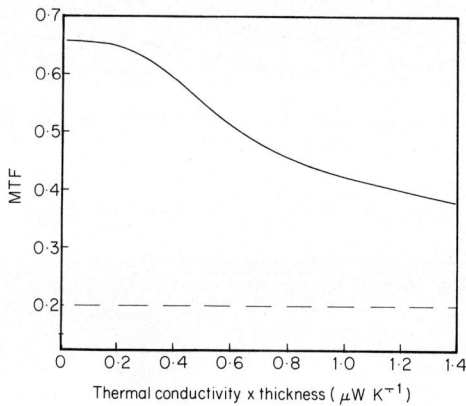

Fig. 2. The variation of MTF (high spatial frequency) with the thermal conductivity–thickness product of the support film.

Unfortunately, the mesh can easily become distorted when heated by radiation from the evaporation filament. This leads to uncontrolled variation of groove width, which can produce background shading in the tube.

As an alternative, photoresist techniques may be used to produce the etch masks, which involves more handling of the target, but does allow good dimensional control.

The equipment for silicon microcircuit photolithography has been adapted to the processing of DTGS targets. These targets are significantly less robust than silicon slices and have, therefore, required the development of special handling techniques. Apart from these handling difficulties, few problems have been encountered with photolithography on

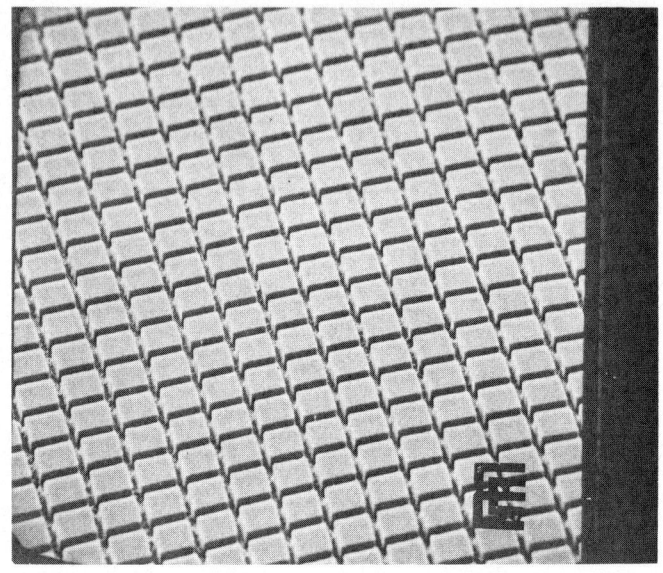

FIG. 3. Scanning electron micrograph of a reticulated T.G.S. target. Element pitch: 34 µm. Magnification: 180×.

FIG. 4. Scanning electron micrograph of a reticulated T.G.S. target. Element pitch: 34 µm. Magnification: 1800×.

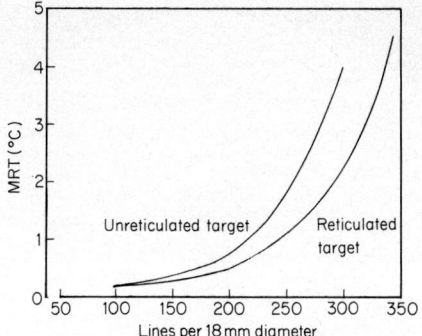

Fig. 5. Comparison of minimum resolvable temperature for tubes with reticulated and unreticulated targets.

DTGS. Adhesion is adequate and pattern delineation good. Photoresist has excellent masking properties against low-energy ions and has been used for most of our reticulation work. Figures 3 and 4 show scanning electron micrographs of a reticulated target. The etch mask is photoresist and the reticulation pitch is 34 μm. Most cosmetic blemishes have been traced to handling damage or imperfect photolithography.

Tube Performance

Figure 5 compares the minimum resolvable temperature (MRT) for reticulated and continuous targets. These measurements were taken with the camera operating in the panned mode with a panning rate of 3 mm sec^{-1}. The MRT for the reticulated target at 300 lines is half that for the continuous target. The difference in MRT is less marked than may be expected. It is thought that this could be the result of noise contributed by the reticulation structure and by variations in that structure.

Figure 6 demonstrates the spatial resolution which can be achieved. The tube is operating in the chopped mode, chopping frequency 25 Hz. The maximum resolution on the fan is 400 lines per diameter.

The measurements of tube response (Fig. 7) were taken with the tube operating at a chopping frequency of 25 Hz. The tubes are gas filled and are running at a mean pedestal current of 40 nA. These curves each represent the average performance of six tubes. As expected the low-frequency response of the reticulated target is reduced because of material loss; about 35% of the DTGS is removed during reticulation and the

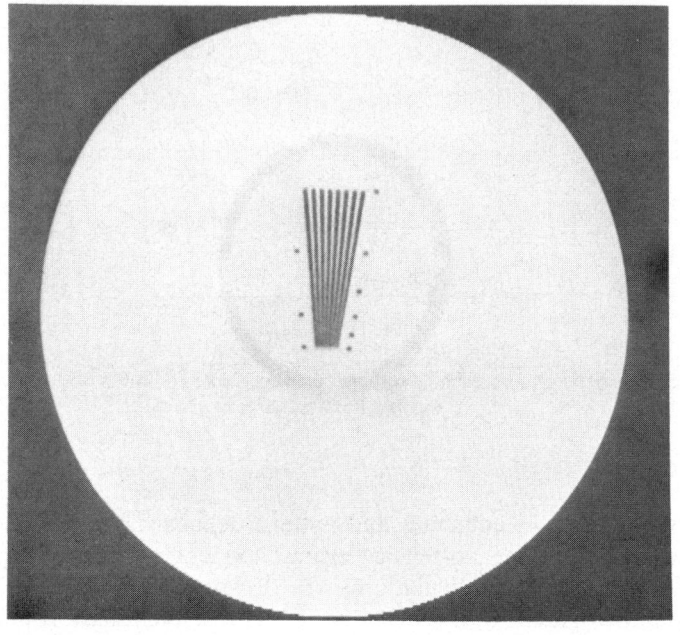

Fig. 6. Resolution bar chart imaged with a reticulated target pyroelectric vidicon. Maximum fan resolution: 400 lines per diameter.

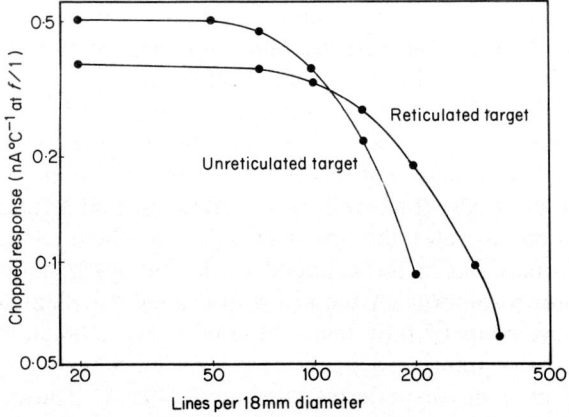

Fig. 7. Comparison of system response using reticulated and unreticulated targets.

FIG. 8. Typical thermal image obtained using a reticulated target pyroelectric vidicon.

response is down by about 25%. The curves cross over at about 100 lines per diameter and at 200 lines reticulation gives a two-fold improvement in response. The reponse is measured up to 350 lines (10 lp mm^{-1}).

It should be noted that with reticulated targets the performance of the system is limited by the MTF of the lens.

A photograph taken from a television monitor (Fig. 8) indicates the quality of thermal image available from the reticulated target pyroelectric vidicon. The image is flicker free, operating at standard television frame rate.

Summary

Reticulated targets of deuterated tryglycine sulphate have been made on a plastic support film using photoresist masked ion beam etching. Tubes made with these targets show significantly improved resolution.

Acknowledgments

The authors wish to acknowledge contributions made by P. D. Nelson and J. J. Harris at English Electric Valve Company Ltd, and D. J. Warner and P. Douglas at Plessey Company

Ltd, and would also like to thank the above and other colleagues at R.S.R.E., Plessey and English Electric Valve Company for stimulating discussions.

This work has been carried out with the support of the Procurement Executive, Ministry of Defence.

References

1. Holeman, B. R. and Wreathall, W. M., *J. Phys. D* **4,** 1898 (1971).
2. Watton, R., *Ferroelectrics* **10,** 91 (1976).
3. Singer, B. and Lalak, J., *Ferroelectrics* **10,** 103 (1976).
4. Watton, R., *Infrared Phys.* **18,** 73 (1978).

Hard Vacuum Tube with a Pyroelectric Target of PVF$_2$ Film

Y. HATANAKA, S. OKAMOTO and R. NISHIDA

Research Institute of Electronics, Shizuoka University, Hamamatsu 432, Japan

INTRODUCTION

Infrared sensitive vidicon camera tubes using the pyroelectric effect in triglycine sulfate (TGS) have been described by Holeman[1] and Charles.[2] TGS is a difficult material to handle, and in particular to prepare in the form of a thin plate of a sufficiently large area for a vidicon target. We have made an experimental study of the application of polyvynilydene fluoride (PVF$_2$) which is readily available in the form of thin film (9 μm) from Kureha,† Japan, to the pyroelectric vidicon target.[3-4] Using the PVF$_2$ film, we have devised a vidicon target structure that is suitable for operation in the cathode potential stabilization (CPS) mode in hard vacuum. The experimental hard tube was capable of forming a thermal image of a human hand against the room temperature background.

This paper describes the structure of the experimental hard tube and its target and experimental results obtained with the tube.

PVF$_2$ FILM

Figure 1 illustrates the structure of the experimental vidicon camera tube that employs the PVF$_2$ pyroelectric film target. Availability of thin film (9 μm) and the ease in handling are the major reasons for our choice of this material. In addition, the low heat diffusion makes up for the apparent loss in sensitivity due to the small pyroelectric coefficient, as can be seen from Table I.

One side of the PVF$_2$ film is covered with a black layer of bismuth to aid the infrared absorption. Bismuth on PVF$_2$ has acceptable infrared absorption[6] while having a lower evaporation temperature than gold. The time required for the heat to diffuse through the bismuth layer (2 μm)

---

†Kureha Chemical Industry Co., 8 Nikombashi, Horrido-me-cho 1-Chome, Chuo-ku, Tokyo, Japan.

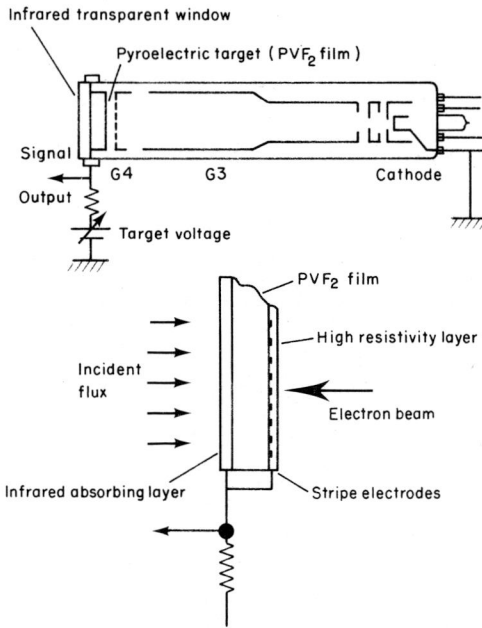

FIG. 1. The structure of an experimental hard vacuum vidicon tube having a PVF$_2$ pyroelectric target. The target is designed to operate in the CPS mode in hard vacuum.

TABLE I

Material constants relevant to the pyroelectric effect[5]

	Pyroelectric coefficient $\lambda \times 10^8$ (C cm^{-2}·°C^{-1})	Density ρ (g cm^{-1})	Specific heat C_p (J g^{-1}°C^{-1})	Thermal diffusivity $k \times 10^3$ (cm^2 sec^{-1})
PVF$_2$	0·3	1·76	1·3	0·5
TGS	3·5	1·7	1·5	3·0

was measured as 0·5 msec in air and 0·9 msec in vacuum. This is sufficiently small compared with one TV frame time, 16 msec.

The pyroelectric current of the PVF$_2$ target was measured at various temperatures using four samples which had different temperature histories. Results are presented in Fig. 2, where sample No. 1 was not pretreated by heating while samples Nos. 2–4 had been previously heated at 80°C, 100°C and 120°C respectively, for an hour. In this measurement,

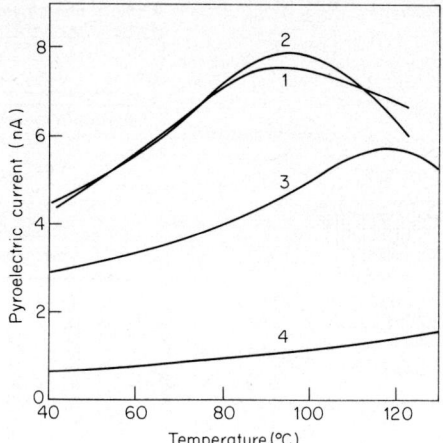

FIG. 2. Temperature effect on the pyroelectric current generated by 4 samples of PVF_2 film having different heat pretreatments under He–Ne laser irradiation of 2 mW. Rate of temperature rise to $2°C\,min^{-1}$. Curve 1, no pretreatment; curves 2–4, pretreated at 80°C, 100°C and 120°C respectively.

the temperature of the sample was increased at a rate of $2°C\,min^{-1}$ under irradiation (about 2 mW) by a He–Ne laser and the photoexcited pyroelectric current measured. Based on the results of this measurement, we restricted the temperature of the target to below about 90°C during the heating for degassing in the tube evacuation process.

Vidicon Target

The structure of the target in our experimental vidicon is illustrated in Fig. 3, where the thicknesses of the black bismuth layer, PVF_2 film, aluminium stripe electrodes, and a high-resistivity layer are 2 μm, 9 μm, 100 nm and 0·3 μm, respectively. The width and separation of the aluminium stripe electrodes have been arbitrarily chosen to be 40 μm and 80 μm respectively. The metal electrodes are covered with the high resistivity layer. The combination of the electrodes and this layer provides leakage paths for the charges accumulated on the PVF_2 target which is a highly insulating material; the currents in these leakage paths produce a pedestal current. Without the leakage paths, the negative charges generated by the pyroelectric effect during the periods of decreasing temperature (with the chopped optical signal off) would accumulate on the surface of the insulating PVF_2 and prevent the electron beam from landing on the target, making it impossible to read out the signal. With this structure, it is possible to operate the target in the CPS mode.

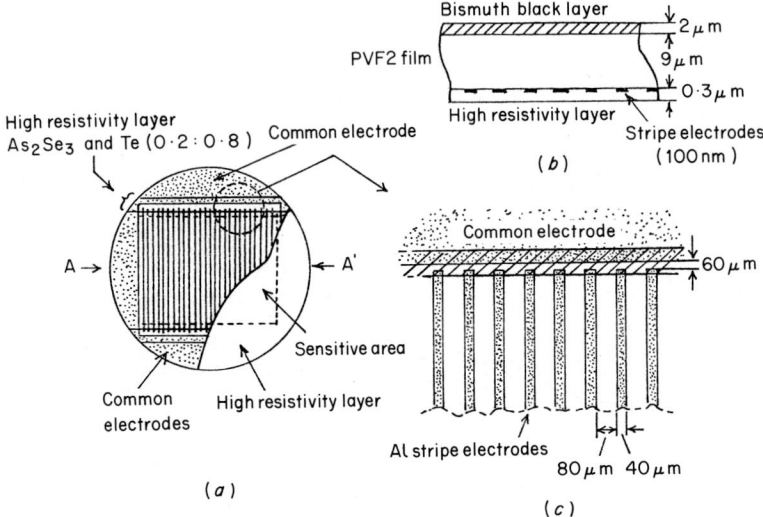

FIG. 3. The structure of the PVF$_2$ pyroelectric vidicon target designed to operate in the CPS mode in hard vacuum. (*a*) Target structure; (*b*) an enlarged cross section of the target along A-A'; (*c*) Al stripe electrodes, series resistors, and the common electrode.

As shown in Fig. 3(*c*), all the stripe electrodes are connected to a common electrode through a high resistance ($10^{13}\Omega$), which suppresses the pseudo-signal current generated when the electron beam lands on the metal electrodes. Our experiments show that a series resistance of 10^{13} Ω, formed by an evaporated layer (the hatched area in Fig. 3(*c*)) of a mixture of As$_2$Se$_3$ and tellurium (0·2 : 0·8 by weight) and covered with the high resistivity layer ($\sim 10^{15}$ $\Omega\,\square^{-1}$) gives acceptable results.

We would like to mention that we have experimented with two other types of PVF$_2$ target: one without the series resistance and the other with an insulating layer on top of the high resistivity layer. The unsatisfactory results obtained with these structures led to the target described above.[7]

PERFORMANCE CHARACTERISTICS

Performance characteristics of our experimental vidicon with the PVF$_2$ target were measured and the results are presented in Figs. 4 and 5. The electronic window technique was used to scan the sensitive area of the target ($14 \times 9\cdot5$ mm^2). Figure 4 shows the measured signal and pedestal currents versus the target voltage. The thermal signal was obtained through a germanium single lens (*f*/0·89) by viewing a hot plate at a temperature 10°C above the ambient temperature. At target voltages

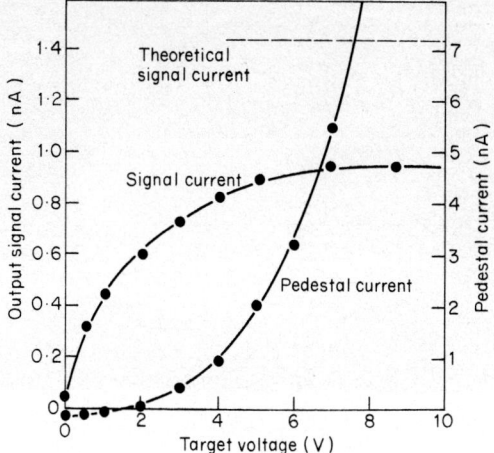

FIG. 4. Measured signal and pedestal currents versus the target voltage of the experimental vidicon.

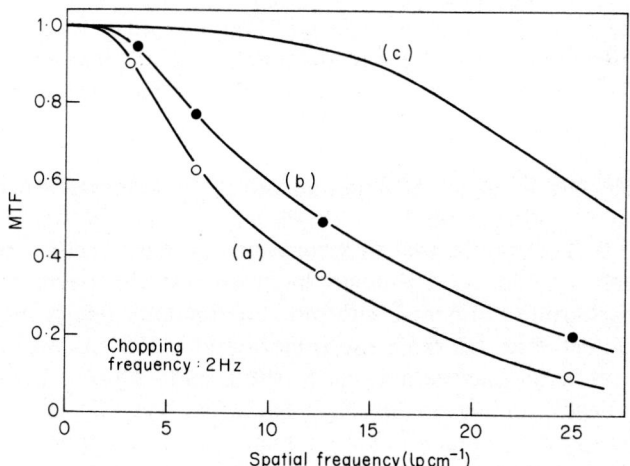

FIG. 5. The MTF characteristics of pyroelectric vidicons with PVF_2 film target of different types. (a) The PVF_2 film target with the metal stripe electrodes covered with high resistivity layer for CPS operation in hard vacuum. (b) The PVF_2 film target without metal electrodes for CPS operation in soft vacuum. (c) Theoretical MTF curve for the PVF_2 film.

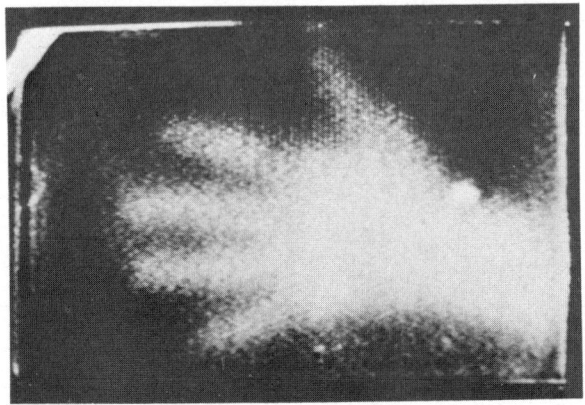

Fig. 6. A thermal image of a human hand against the room temperature (10°C) background. The object is at 2 m from the camera and is in motion (20 cm sec^{-1}).

above 5 V, the signal current is saturated while the pedestal current increases quickly, as shown in Fig. 4, implying that the optimum target voltage is about 5 V for this particular tube. The saturation value of the signal current is about 64% of the theoretical value obtained for PVF_2 film with no metal electrodes. The difference is accounted for by the dead area under the stripe electrodes.

The measured MTF curve is given in Fig. 5 together with, for comparison purposes, the curve of a soft tube having a PVF_2 target with no metal electrodes, and the theoretical curve for PVF_2 which has been calculated assuming that all the pyroelectric charges are read out by electron beam scanning.

Possible effects of residual gas in our hard tube were examined by adjusting the voltages on G3 and G4 (see Fig. 1). The ion current was found to be as small as that in a conventional hard vacuum tube.

A thermal image of a human hand against the room temperature (10°C) background obtained with the experimental tube is shown in Fig. 6. The object is at 2 m from the camera and is in motion (20 cm sec^{-1}). The stripes in the picture are due to the pseudo-signal produced by the electron beam landing on the electrodes. We consider that this can be improved to an acceptable level for many practical applications of thermal imaging with the device by making the stripes thinner.

Conclusion

We have described an experimental pyroelectric vidicon camera tube for thermal imaging that employs the PVF_2 film target. The target is designed to operate in the CPS mode in hard vacuum. The hard tube avoids the problems of gas pressure control and cathode degradation, resulting in stable operation and long life.

A shortcoming of our experimental tube is the pseudo-signal current generated when the electron beam lands on the stripe metal electrodes which are necessary to provide the paths for the pedestal current. The series resistance incorporated into the target of our present tube minimizes the pseudo-signal. This scheme works fairly well, and a further improvement can be expected by making the stripes thinner. Another possibility is to deposit fine insulating stripes on the high-resistivity layer above the metal stripes. Negative charges accumulated on the insulator stripes would prevent the electron beam from landing on the metal stripes, resulting in elimination of the pseudo-signal.

Acknowledgment

The authors would like to acknowledge the guidance and encouragement of Dr E. Fukada of The Institute of Physical and Chemical Research, the help of Professor S. Mizushina in preparation of the manuscript, and Dr T. Ando for useful discussions. Thanks are also due to the Kureha Chemical Industry Co. Ltd. for providing PVF_2 film and Messrs T. Urata and M. Aoyama for their help in the experiments.

References

1. Holeman, B. R. and Wreathall, W. M., *J. Phys. D* **4**, 1898 (1971).
2. Charles, D. R. and Carvennec, F. L., *In* "Adv. E.E.P." Vol. 33A, p. 279 (1972).
3. Yamaka, E. and Teranishi, A., Technical group on image technology applications of Inst. TV Engineers of Japan, Tokyo (1976).
4. Okamoto, S., Hatanaka, T., Kato, T. and Nishida, R., Paper No. 3-3, Meeting Inst. TV Engineers of Japan, Tokyo (1976).
5. Garn, L., *IEEE Trans. Parts, Hybrids & Packag.* **PHP-10**, 208(1974).
6. Hatanaka, Y., Okamoto, S. and Nishida, R., *In* Proc. 1st meeting on Ferroelectric Materials and their Applications, Kyoto (1977).
7. Okamoto, S., Hatanaka, Y., Ando, T. and Nishida, R., Paper No. 2-5, Meeting Inst. TV Engineers of Japan, Hiroshima (1977).

Recent Chalnicon Developments

O. YOSHIDA

Electron Devices Lab., Toshiba R & D Center, Toshiba Corporation, Kawasaki, Japan

Introduction

Characteristics of a new camera tube with a CdSe photoconductive target were reported at the 5th Symposium in 1971.[1] The tube was successfully marketed under the name of "Chalnicon" in the autumn of 1972 in two sizes, 25 mm (E5001) and 18 mm (E5022), both magnetically focused and deflected.[2] At present, there are over forty five types of Chalnicon; Figure 1 shows a selection of them.

This paper describes new Chalnicons with improved spectral response and lag characteristics. Sensitivity in the ultraviolet region is sufficiently high for spectroscopic analysis or for ultraviolet microscopy. It is also possible to extend the spectral response into the infrared region, resulting in a tube suitable for an industrial surveillance TV camera at low light levels.

The lag of recent Chalnicons has been reduced by a factor of three, thus greatly extending their fields of application.

Standard Chalnicon Target

A standard Chalnicon has a double layer target of cadmium selenide (CdSe) and arsenic trisulphide (As_2S_3) as shown in Fig. 2. There is also a cadmium selenite layer ($CdSeO_3$) between the CdSe layer and the As_2S_3 layer but this can be neglected as it is thin and does not affect the spectral response and image lag.

The CdSe photoconductive layer determines the spectral response characteristics. The energy gap of 1·7 eV of CdSe corresponds to 730 nm wavelength and as a result the sensitivity peak for the standard Chalnicon occurs at 700 nm, as shown in Fig. 3, the peak sensitivity being about 0·5 $\mu A \mu W^{-1}$. The sensitivity is high over the whole visible range from blue to red with quantum yields close to unity.

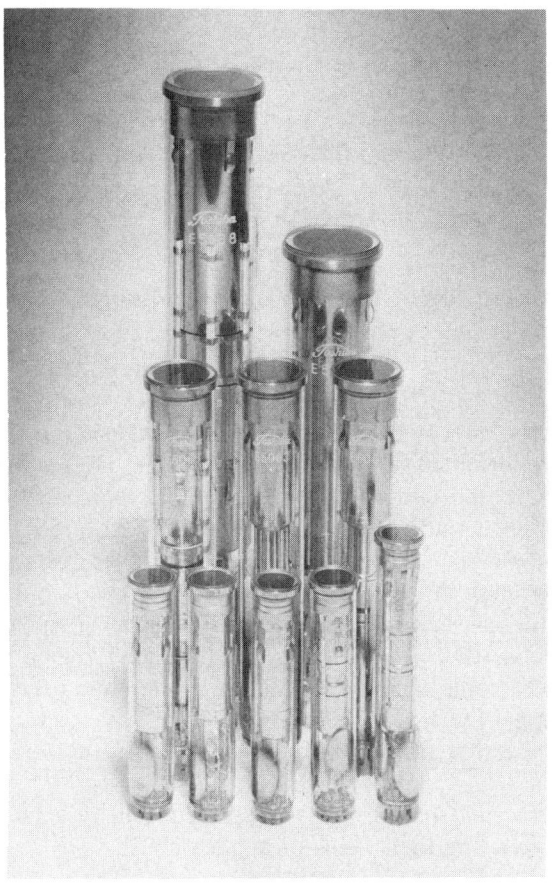

Fig. 1. Chalnicons of various sizes: 18 mm (front row), 25 mm (middle row) and 38 mm (back row).

The As_2S_3 amorphous layer determines the dark current and resolution characteristics and also, as we found later, the image lag.

Improvements in Spectral Response

One recent development is the extension of the spectral response either into the ultraviolet or into the infrared regions; these tubes are referred to as either the UV-Chalnicon or the IR-Chalnicon.

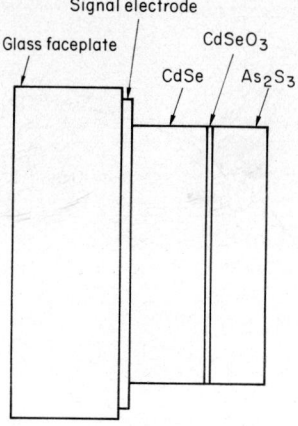

FIG. 2. Standard Chalnicon target (CdSe-As_2S_3 double layer target).

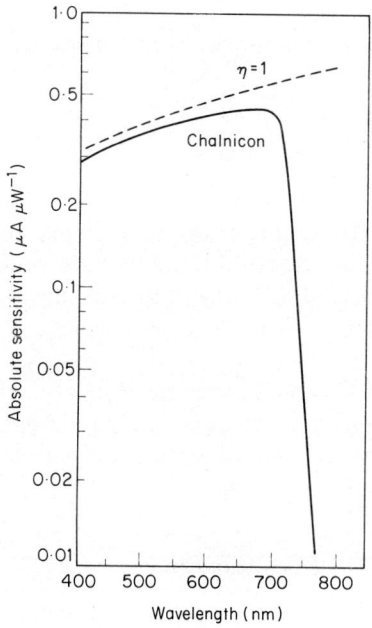

FIG. 3. Standard Chalnicon spectral response. η is the responsive quantum efficiency.

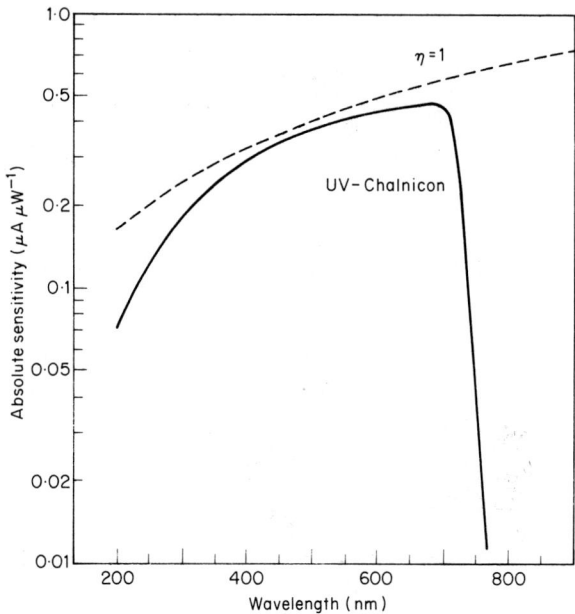

Fig. 4. UV-Chalnicon spectral response. η is the responsive quantum efficiency.

UV-Chalnicon

As was mentioned, the sensitivity of the standard CdSe-As$_2$S$_3$ Chalnicon target is high at 400 nm; this suggests that the sensitivity in the ultraviolet region can also be expected to be high. In order to confirm it, the glass faceplate was replaced with one of quartz having high ultraviolet transmission and for the same reason the SnO$_2$ signal electrode layer was made as thin as possible.

Figure 4 shows the UV-Chalnicon spectral response. The sensitivity at 250 nm wavelength is about 0·12 μAμW^{-1}, high enough for ultraviolet microscopy, while the spectral response is unaffected in the visible region which is convenient for optical focusing.

IR-Chalnicon

A high infrared sensitivity can be useful for industrial surveillance low light applications where the infrared intensity may be higher than that of the visible light.

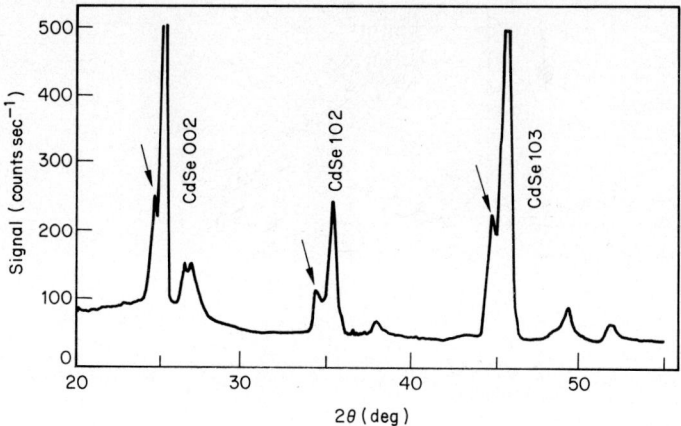

FIG. 5. X-ray diffraction pattern for CdSe(Te) film (the small sub-peaks indicated by arrows are CdTe).

In order to extend the infrared sensitivity without affecting other Chalnicon features the CdSe photoconductive layer is modified by heat treating the CdSe layer in tellurium vapour and an inert gas, such as nitrogen. The resulting CdSe(Te) layer is shown by X-ray diffraction to be composed of a mixture of CdSe and CdTe. Figure 5 shows a typical X-ray pattern, its high peaks being CdSe and the small sub-peaks that appear after tellurium heat treatment being CdTe. From lattice spacing calculations it was found that about 30% of CdSe was converted into CdTe by tellurium heat treatment.

The IR-Chalnicon with the CdSe(Te) photoconductive layer has the spectral response shown in Fig. 6; the sensitivity peak is located at around 750 nm wavelength and the sensitivity extends to nearly 900 nm.

The extension of spectral response into the infrared region doubles the total sensitivity of the tube for 2856 K tungsten illumination as shown in Table I. Other IR-Chalnicon characteristics are similar to those for the standard Chalnicon, except that, as shown in Table I, the dark current is larger than that of the standard Chalnicon although it is smaller than the values for ordinary Sb_2S_3 or silicon vidicons.

The IR-Chalnicon can be used not only in surveillance camera work, but also in special cameras such as those for the study of radiation patterns of semiconductor lasers emitting infrared light or for imaging hologram pictures synthesized by laser beams. In these applications although the IR-Chalnicon has a lower infrared sensitivity than the silicon vidicon it has an advantage in that infrared light is absorbed in the

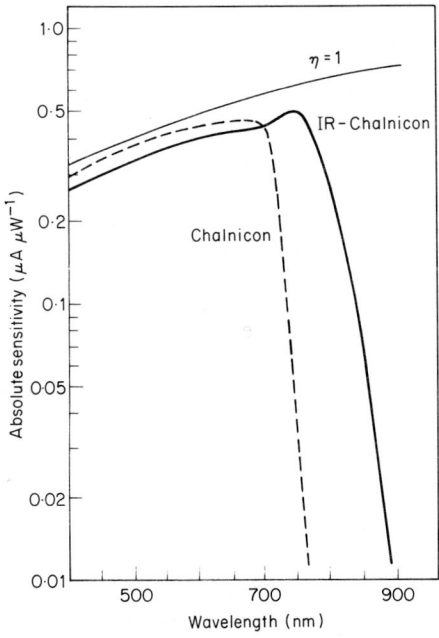

Fig. 6. IR-Chalnicon spectral response. η is the responsive quantum efficiency.

CdSe(Te) layer without the multiple optical reflections that occur in a silicon target and result in image degradation.

IMPROVEMENTS IN LAG CHARACTERISTICS

The image lag of a camera tube and the resulting picture smearing determine the lowest useful camera sensitivity in low light level or studio colour TV cameras. Improvement of the image lag is therefore very important. In the Chalnicon, the image lag has been improved by modification of the scanned surface of the target without affecting the other photoelectronic properties.

Standard Chalnicon Lag Characteristics

A standard 18 mm Chalnicon with $CdSe\text{-}As_2S_3$ double layer target has the lag characteristics shown in Fig. 7. When a 0·4 μm thick As_2S_3 film is used the image lag at the third field following a 50 nA signal current is just over 20%. The lag can be reduced by bias illumination as shown in

TABLE I
Tube characteristics (25 mm tube)

| Tube | Target | Signal currents† | | Dark current (nA) |
		White (no filter) (nA)	IR (IR–DIB) (nA)	
Chalnicon	CdSe	160	0	0·5
IR-Chalnicon	CdSe(Te)	310	3·6	2·0

† Faceplate illumination: 0·5 L; light source temperature: 2856 K.

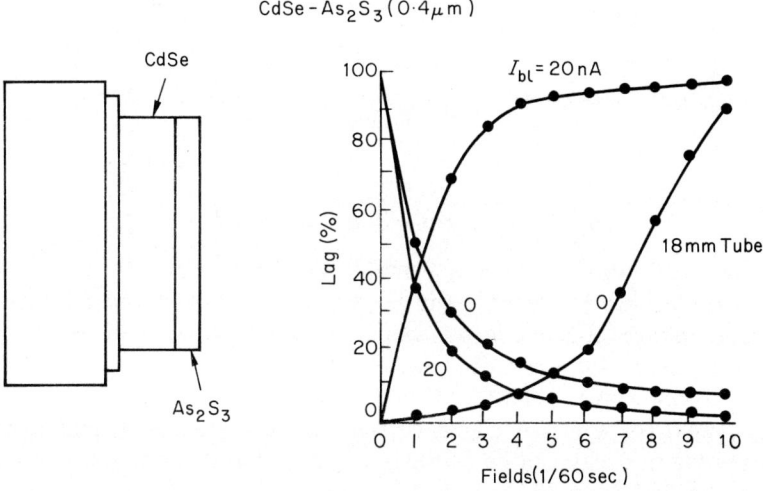

FIG 7. Standard CdSe-As$_2$S$_3$ target structure and its image lags with bias illumination $I_{bl} = 0$ and 20 nA. Signal current $I_{sig} = 50$ nA.

Fig. 7. The lag is capacitive in nature and calculation shows that for the 0·4 μm target the capacitance is 3 nF.

Reducing the target capacitance by increasing the thickness is a simple way to improve lag characteristics. A target with a 0·8 μm As$_2$S$_3$ layer gives smaller lag values than that with one 0·4 μm thick, but the required target operating voltage rises to at least 70 V, exceeding the normally useful target voltage of 50 V or less.

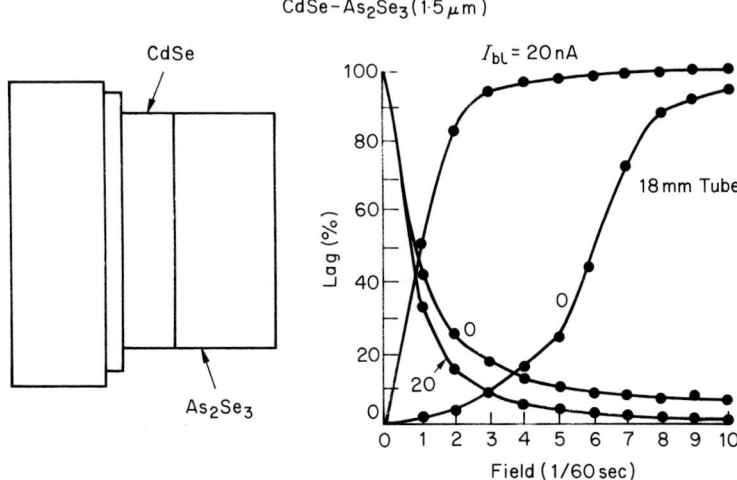

FIG. 8. CdSe-As$_2$Se$_3$ target structure and its image lags with bias illumination $I_{bl} = 0$ and 20 nA. Signal current $I_{sig} = 50$ nA.

Lag Improvement with a CdSe-As$_2$Se$_3$ Target

The substitution of As$_2$Se$_3$, a similar V–VI compound, for As$_2$S$_3$ makes it possible both to increase the target layer thickness and maintain the target voltage at a moderate value. Figure 8 shows typical image lag characteristics for the CdSe-As$_2$Se$_3$ double layer target with a 1·5 μm thick As$_2$Se$_3$ layer. The lag in decay is improved to about 15% at the third field for a 50 nA signal current. By using a 20 nA bias light the image lag in decay can be decreased to 8%. Optimum target voltages lie between 25 V and 30 V.

Increasing the As$_2$Se$_3$ layer thickness, e.g., to 3·0 μm, does not appreciably improve image lags. On the contrary, it takes rather longer for transient currents to saturate to 100% level or to fall to zero. The target voltage cannot be increased in proportion with the thickness and it is concluded that the decrease in electric field across the target prevents an image lag improvement in spite of the decrease in capacitance.

A New Target Structure for Low Lag

An attempt to further improve the image lag characteristics was made by employing a new target structure, with a porous layer (made by evaporation in a low vacuum) as shown in Fig. 9.

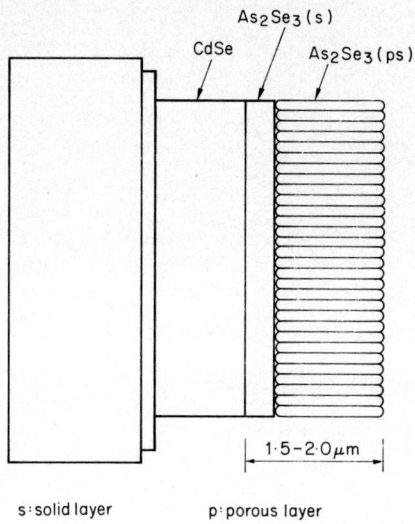

FIG. 9. CdSe-As_2Se_3 (SPS) target structure.

The scanned surface layer is composed of three different As_2Se_3 layers. The first layer is solid As_2Se_3, and the second is a porous As_2Se_3 layer, evaporated in argon at 0·1 Torr pressure; a third solid As_2Se_3 layer is evaporated onto the porous layer. Thus, the scanned layer has a solid/porous/solid (SPS) structure similar to that of an ordinary vidicon target.

However, a scanning electron microscopic observation reveals that the surface structure is not a simple combination of the porous layer and the solid layer. SEM photographs of the SPS layers, shown in Fig. 10, present the following information. Evaporation of the third layer does not form a solid layer on the porous layer, but instead a composite structure appears consisting of rice-grain-like particles which stand substantially vertically to the plane, the second porous layer serving as nuclei for these particles. When the porous layer is thin, the scanned surface layer is essentially composed of two layers, the first solid layer and the composite layer, as shown in Fig. 9. The total thickness of the scanned surface layer thus formed is from 1·5 μm to 2·0 μm, depending on the evaporation conditions of the porous layer.

The addition of the porous layer results in an increase in usable target voltage, from several volts to ten volts, and a considerable inprovement in image lag characteristics. Typical image lags are shown in Fig. 11. Values of the image lag in decay at the third field for 50 nA signal current and

FIG. 10. SEM photographs of As_2Se_3 SPS layer at 10 000 times magnification. (top: bird's eye view; middle, viewed normally; bottom, cross section).

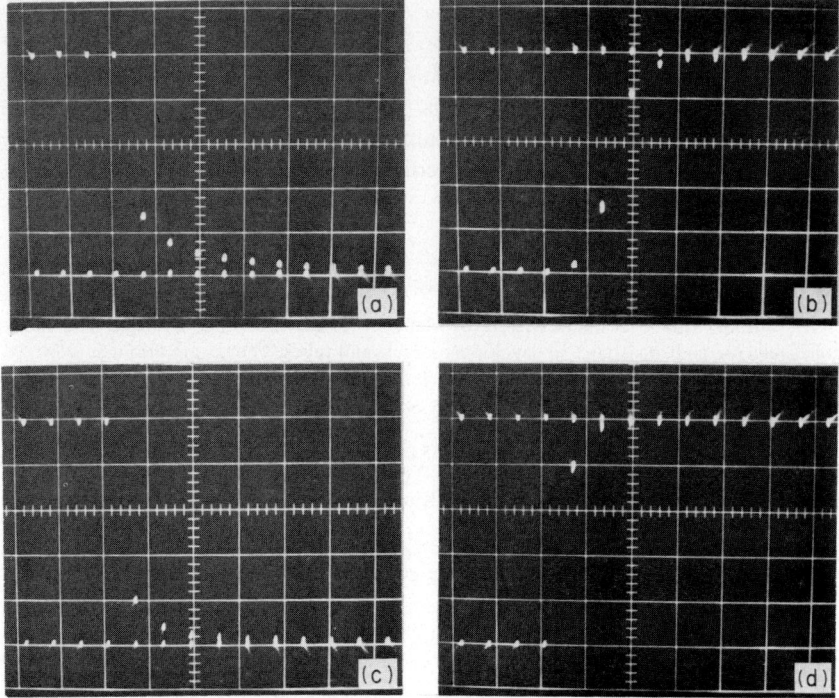

FIG. 11. Typical image lags for CdSe-As_2Se_3 (SPS) target. The upper curves in each frame correspond to 50 nA signal current and the lower are for zero signal. Signal is switched on or off at the fourth point from the left. (a) Signal switched off, no bias; (b) signal switched on, no bias; (c) signal switched off, bias light 20 nA; (d) signal switched on, bias light 20 nA.

for an 18 mm tube are about 5% and 10% with and without 20 nA bias illumination, respectively. The image build-up rate is also improved. The target capacitance of this SPS structure is calculated to be nearly 1 nF.

Conclusion

A Chalnicon with a quartz faceplate can be used not only in the ultraviolet, but also as a radiation resistant camera tube for surveillance use in atomic power stations, because the quartz faceplate is not impaired by exposure to radiation. The IR-Chalnicon is a suitable tube for low light level surveillance cameras. Both UV- and IR-Chalnicons require set-up procedure similar to those of the standard Chalnicon.

Image lag characteristics have been improved by the adoption of As_2Se_3 in place of the As_2S_3 layer or by forming an As_2Se_3 SPS layer.

The CdSe-As$_2$Se$_3$ target modification has been put onto the market under the name of Chalnicon-FR. The As$_2$Se$_3$ SPS layer is still being studied and promises even lower image lags.

With these improvements the Chalnicon can be applied in wider fields, including industrial surveillance, medical, and sophisticated broadcasting uses.

Acknowledgments

The author wishes to thank Messrs S. Manabe, Y. Hayashimoto, I. Nagae and many colleagues for their assistance in developing targets and tubes. Thanks are also due to Mr M. Kajimura for encouragement and to Dr K. Shimizu for helpful discussions.

References

1. Shimizu, K., Yoshida, O. Aihara, S. and Kiuchi, Y., *In* "Adv. E.E.P." Vol. 33A, p. 293 (1972).
2. Yoshida, O. and Shimizu, K., *J. SMPTE* **34,** 11 (1975).
3. Yoshida, O. and Kiuchi, Y., *Jpn. J. Appl. Phys.* **10,** 1203 (1971).

Proximity Focused SEC Vidicon with Porous MgF$_2$-Ag Target

T. KAWAMURA, S. OKUDE and S. OKAZAKI

NHK Technical Research Laboratories, Tokyo, Japan

and

H. WASHIDA and K. YANO

Electron Tube and Device Division, Toshiba Corporation, Kawasaki, Japan

INTRODUCTION

The SEC vidicon[1] has sensitivity ten times higher than the Plumbicon and has therefore been used for special applications. If it is to replace the Plumbicon as a broadcast TV camera improvements are needed in ease of operation, S/N ratio, stability and so on.

In this paper we shall discuss a proximity focused SEC vidicon with a newly developed, porous MgF$_2$-Ag target, which results in a compact and light tube having good stability and S/N ratio. A photocathode transfer technique is employed.

CONSTRUCTION AND DESIGN OF THE TUBE

Figure 1 shows a photograph of a proximity focused SEC vidicon and a $1\frac{1}{4}$ in. Plumbicon. In Fig. 2 the internal construction and dimensions of the SEC vidicon are shown.

In the image section the photocathode and the target are located 1·5 mm apart from each other, which enables a compact tube to be constructed. The size of the tube is approximately the same as that of a $1\frac{1}{4}$ in. Plumbicon and, if minor modifications are made to the Plumbicon colour camera, this tube and the Plumbicon are interchangeable. Moreover, development of the new, stable target has eliminated the need for the suppressor required in the conventional SEC tube.

THE MANUFACTURING PROCESS

In order to fabricate the photocathode in the proximity configuration an improved technique using photocathode transfer has been developed.[2-4]

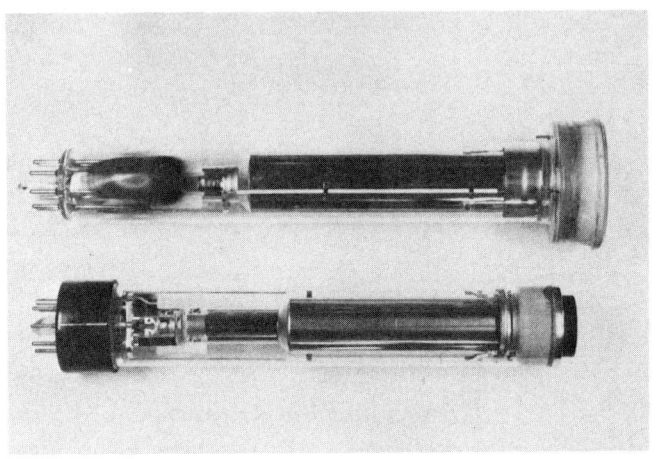

FIG. 1. Proximity focused SEC vidicon (above) and $1\frac{1}{4}$ in. Plumbicon (below).

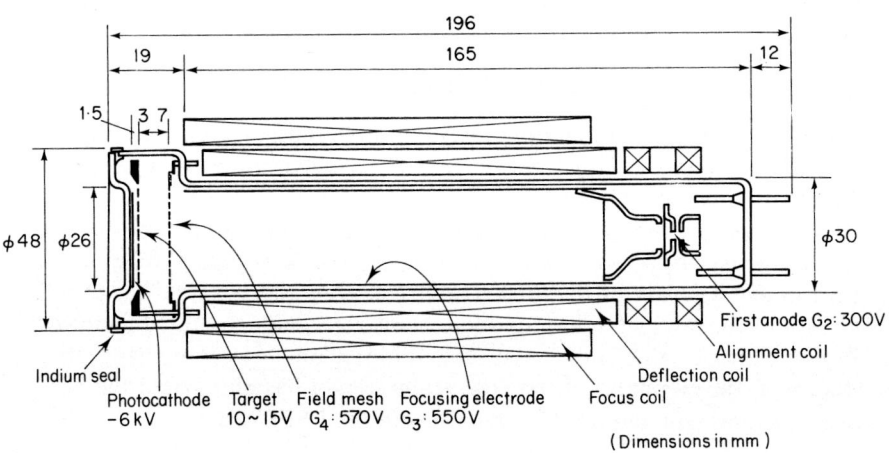

FIG. 2. Internal construction of the proximity focused SEC vidicon.

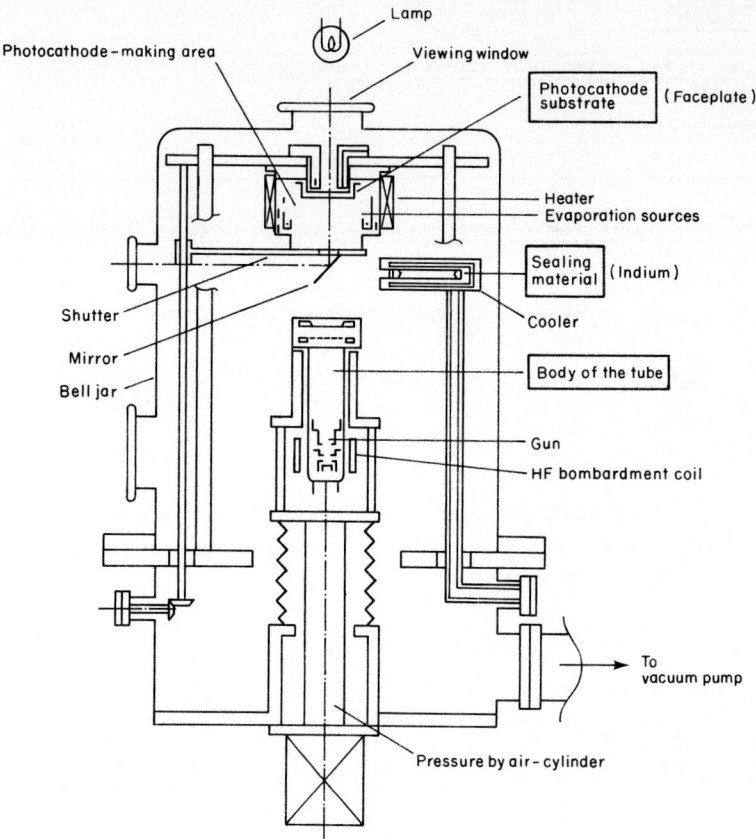

FIG. 3. The photocathode transfer unit.

Figure 3 shows a cross section of the transfer unit. A flow diagram of manufacturing process is shown in Fig. 4. The process is as follows: firstly parts such as a faceplate, an indium ring and a tube body are set in the belljar together with the evaporation sources for making the photocathode. The system is evacuated by an oil diffusion pump. The belljar is baked and degassed by the outer electric oven at a temperature of 350°C, while the photocathode-making space is brought to between 350°C and 400°C by the additional inner heater. After bake-out, the electron gun is activated and the photocathode is processed. A special processing method has been developed in order to make the photocathode within the large volume with its high heat capacity. Figure 5 shows the processing sequence for

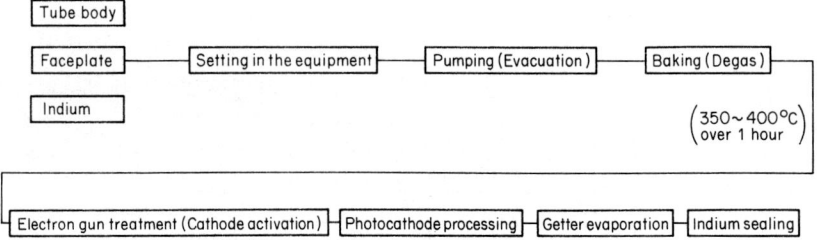

Fig. 4. Flow diagram of manufacturing process.

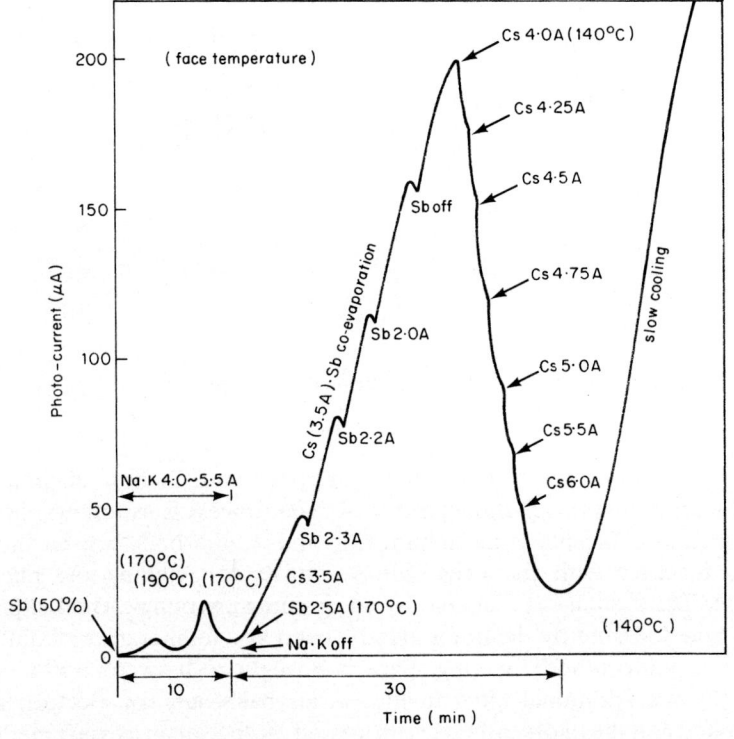

Fig. 5. The processing sequence for an S·20 photocathode.

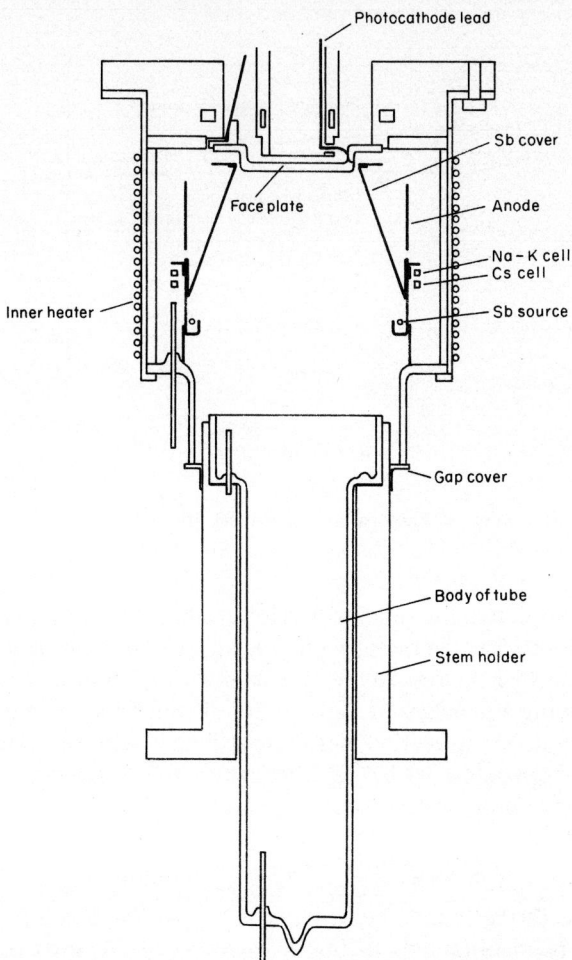

FIG. 6. Photocathode-forming area and tube body during processing.

an S·20 photocathode. In this process the photocathode is formed while the temperature of the faceplate is decreasing gradually. A special activation schedule was introduced to suit the rather slow temperature response. At the mid-point of the photocathode processing, a rotary mechanism removes the shutter from the bottom of the photocathode-forming area and the tube body is elevated into the lower entrance of the area as shown in Fig. 6. The arrangement is then close to that of non-transferred image type tubes, and alkali metal vapour is allowed into the tube body. When photocathode

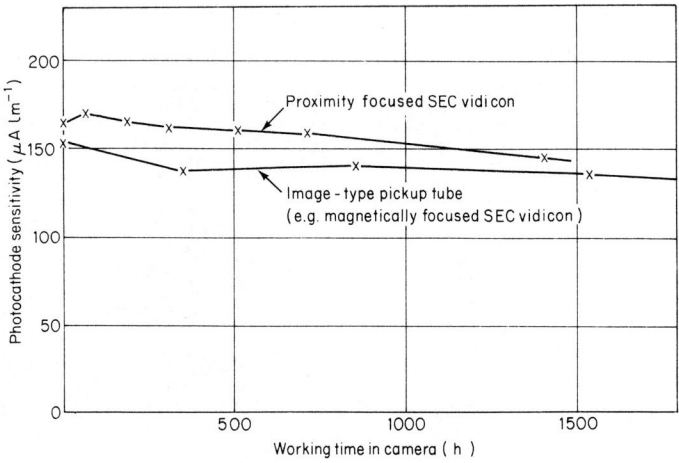

FIG. 7. Photocathode sensitivity life tests.

formation is completed, the tube body is lowered and the indium ring, which has been previously kept in the cool region, is moved into position. The tube body is again pushed up toward the faceplate with the indium ring between them. By applying a pressure of 100 kg cm^{-2} through the air cylinder, as in conventional vidicon fabrication, hermetic sealing is achieved. The pressure in the belljar is kept below 3×10^{-5} Pa throughout these processes. The whole procedure takes about seven hours.

PROPERTIES OF THE PHOTOCATHODE

S·20 photocathodes with a sensitivity of 150 to 250 μA lm^{-1} and high stability have been obtained. In Fig. 7 operational life tests of the tube are compared with those for a conventional SEC vidicon. Hitherto, photocathodes made by transfer techniques have been found to have rather poor operational life characteristics, but these results show that there is no difference between the life characteristics of SEC tubes made by our photocathode transfer technique and those made by ordinary manufacturing methods.

THE MgF$_2$-Ag DOUBLE LAYER TARGET

In this device, the KCl target previously used in the SEC tube has been replaced by a newly-developed MgF$_2$–Ag target. This contributes to the improved overload characteristic and life expectancy.

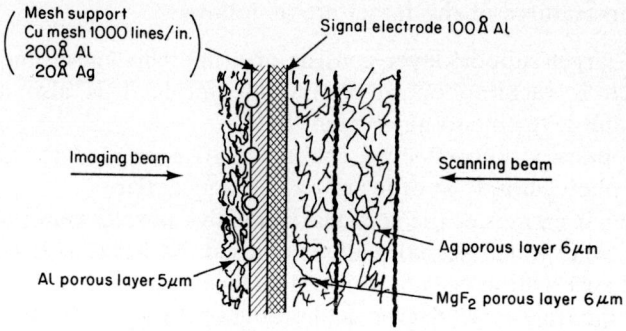

FIG. 8. Structure of the SEC target.

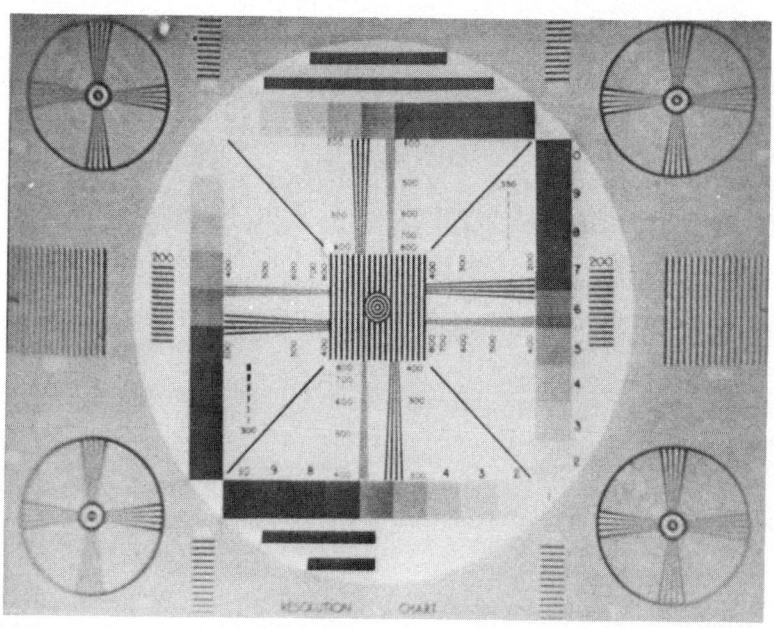

FIG. 9. A monitor picture obtained using the tube.

The main features of the target are as follows:

(1) The target-support layer consists of a fine wire-mesh and a metal layer which is vacuum-deposited onto the mesh. This also acts as a burn-resistant layer in strong incident light.

(2) An optically non-reflecting coating of Al is applied to the surface facing the photocathode in order to reduce optical flare.[5]

(3) MgF_2 is chosen as the secondary emissive porous material, instead of KCl or MgO which has normally been used. KCl porous layers absorb water and aggregation occurs, whilst MgO layers suffer shrinkage which is caused by the transverse electric field during operation. Thus both materials are regarded as inadequate for SEC targets because of the resulting decline in picture quality. The MgF_2 porous layer does not change its properties substantially in normal operating conditions.

(4) The new SEC target consists of double porous layers of MgF_2 and Ag. Each porous layer is evaporated in an inert atmosphere of about 100 to 1000 Pa and has a thickness of 6 μm. A schematic model of the structure of the target is shown in Fig. 8. The porous layer of Ag has a low secondary electron emission ratio so that the first crossover voltage of the target surface is higher than 100 V.[6] The suppressor mesh electrode, which is located near the target in conventional SEC vidicons to prevent excessive excursions of the surface potential, can therefore be omitted. The advantage of doing this is that it enables the target capacity to be lowered from 28 pF to 11 pF, which gives a large improvement in the S/N ratio of the tube. The target gain was 25 at a target voltage of about 15 to 20 V.

TABLE I
Tube performance

Photocathode sensitivity (S·20)	150–200 μA lm^{-1}
Peak signal current	300 nA
Limiting resolution	800 TV lines per picture height
Amplitude response (MTF) 200TV lines	80%
400TV lines	>35%
Lag (residual signal in the 3rd field)	<2%
Signal-to-noise ratio	42 dB
Dark current	<1 nA
Geometric distortion	<1%
Shading (dark)	negligible
(highlights)	<15%
Optical flare	<10%
Overall sensitivity (target gain: 25)	3750–6250 μA lm^{-1}

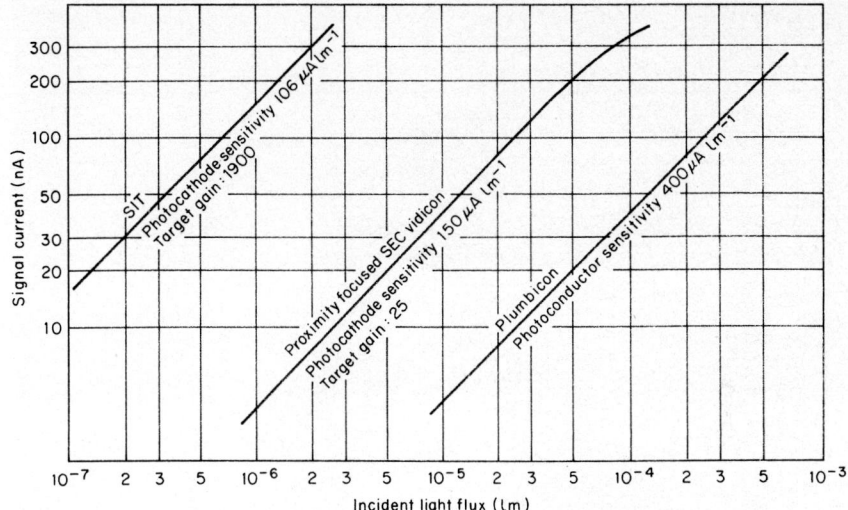

FIG. 10. Light transfer characteristics of three pickup tubes.

TUBE CHARACTERISTICS

When the tube is installed in the coil assembly, the target capacity increases by only 1–2 pF to 13 pF. This increase is small because the image section in not completely inserted into the coil.

Figure 9 shows a monitor picture of a pattern obtained using the proximity focused SEC vidicon tube. We also used the tube in a modified Plumbicon colour camera (Toshiba PK-31) and compared to a Plumbicon it had high sensitivity, reduced lag, low dark current, low comet tailing, good stability to signal overload, and superior colour rendition. Table I summarizes various characteristics measured using this camera. The improvements in stability, uniformity and S/N ratio when compared to SEC vidicons with KCl or MgO targets and a suppressor mesh electrode can be readily seen.

Figure 10 shows the light transfer characteristics of the SIT, the Plumbicon and the proximity focused SEC tubes. In our tube, the slope of the logarithmic light transfer characteristic curve, γ, is unity over a wide range of illumination and shows a saturation tendency for larger inputs, which is considered to be effective in preventing the occurrence of comet tails. The overall sensitivity of the tube in the colour camera is such that the lens aperture setting is $f/11$ at an illumination of 2000 lx under typical conditions. This value corresponds to a sensitivity about eight

FIG. 11. A monitor picture obtained using the proximity focused SEC vidicon.

times higher than that of the Plumbicon camera, which requires an aperture setting of $f/4$ under the same conditions. The proximity tube has little distortion so that colour registration is easily adjusted. Fig. 11 shows a monitor picture obtained by the colour camera.

CONCLUSIONS

Through the development of a new tube-making technique, the application of a proximity focusing system to the image section of a SEC vidicon has been achieved. This advance, coupled with the improvement of the tube characteristics by the use of a newly developed MgF_2-Ag target, has resulted in an experimental proximity focused SEC vidicon with improved characteristics. Colour pickup experiments conducted with these tubes showed high sensitivity and low lag. To use the tube in practice it will be necessary to improve some characteristics such as the highlight shading, optical flare and target granularity.

ACKNOWLEDGMENTS

The authors wish to express their gratitude to Dr T. Ninomiya, Dr I. Oishi and Dr M. Takahashi of NHK Technical Research Laboratories for their support and helpful discussions. We also give thanks to Mr M. Ishizaka, Mr M. Kajimura and Mr A. Onoe of the Electron Tube and Device Division, Toshiba Corporation for their support and useful suggestions.

REFERENCES

1. Frank, K., *Fernseh & Kino-Technik* **9,** 323 (1970).
2. Dolizy, P. and Legoux, R., *In* "Adv. E.E.P." Vol. 28A, p. 367 (1969).
3. Holeman, B. R., Conder, P. C. and Skingsley, J. D., *In* "Adv. E.E.P." Vol. 40A, p. 1 (1976).
4. Van Huyssteen, C. F., *In* "Adv. E.E.P." Vol. 40B, p. 419 (1976).
5. Toyonaga, R. and Sato, K., *NHK Tech. Monogr.* **14,** 513 (1967).
6. McMullan, D. and Towler, G. O. *In* "Adv. E.E.P." Vol. 28A, p. 173 (1969).

A Model for Signal Generation in Silicon Target Camera Tubes

W. V. McCOLLOUGH

Systems Engineering Laboratory, Raytheon Company, Portsmouth, Rhode Island, U.S.A.

INTRODUCTION

The operation of a camera tube cannot be described simply as the convolution of the input charge image and the readout electron beam because of the charge replacement function of the beam. Previous investigators[1-3] have modelled various aspects of this process. The work described here is both an extension and refinement of this earlier work. Its principal features are that the interaction of the beam and target are modelled more accurately than before, and that all parameters used can be readily measured in television systems.

It was not intended, at least initially, that every nuance of tube operation be accounted for in the model although it has been designed to make the incorporation of additional features relatively easy. The principal use which was envisioned for the model was to explain some of the interesting, and unpredicted, measurements which had been obtained in the camera tube analysis laboratory at the University of Rhode Island.[4] The model proved to be robust however, and its usefulness is being extended beyond explaining experimental measurements to areas such as the simulation and analysis of detection problems.

TUBE OPERATION

A diagram of the target which was modelled is shown in Fig. 1. This target is the type found in the RCA 4826 SIT Tube and RCA 4532 silicon vidicon, both of which were used to obtain the data presented here. There is nothing inherent in the model which would prevent its being directly applied to targets of other manufactures.

The details of the operation of the tubes are readily available in the literature[5] and will not be discussed in detail here. The two most

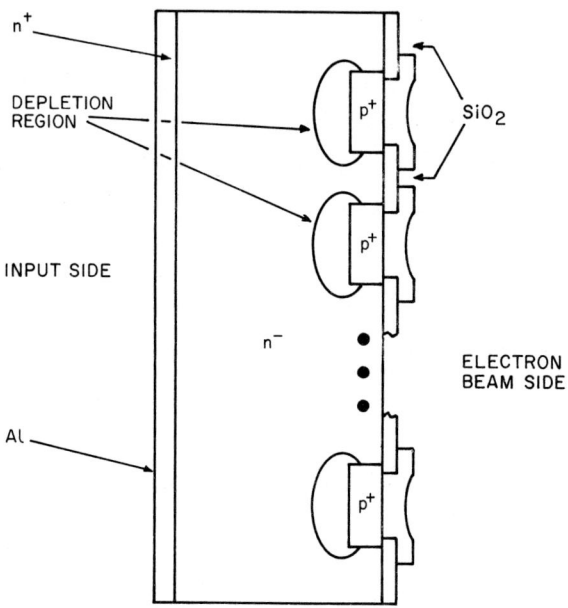

FIG. 1. Silicon target structure.

important target parameters in the model are the capacitance of the junctions and the beam–target interaction curve (signal current as a function of target voltage). Because of differences in doping levels, a one-sided junction, or diode, is formed under each island with the depletion region extending into the substrate. The capacitance of this one-sided junction is important because it relates the charge stored to the target voltage. It is given by:

$$C(V_T) = C_0\left(1 + \frac{(V_B - V_T - \phi_W)}{V_{bi}}\right)^{-1/2}, \qquad (1)$$

where V_B is the target bias voltage, V_T is the target voltage, ϕ_W is the difference in work functions between the silicon and thermionic cathode, and V_{bi} is the built-in junction voltage for silicon. C_0 is the zero-bias capacitance of a one-sided junction given by the well known expression

$$C_0 = \left(\frac{e\epsilon_s N_B}{2V_{bi}}\right)^{1/2}. \qquad (2)$$

An analytical discussion of the physics of the beam–target interaction can be found in the literature.[6–8] For the work described here, the

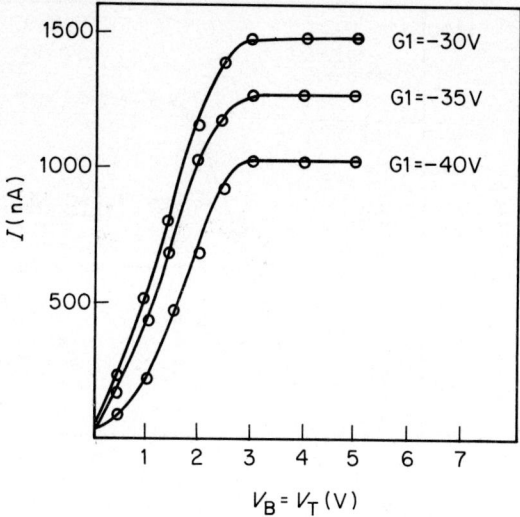

FIG. 2. Measured V–I characteristics for a SIT tube.

beam–target characteristic (also called the V–I characteristic or $f(V_T)$) was measured experimentally by saturating the target with a high input light level so that the diodes were essentially discharged even during the beam's dwell time.

Since the target voltage V_T is given by:

$$V_T = V_B - V_C - \phi_W, \qquad (3)$$

discharging the diodes with a high light level enables V_T to be determined as a function of measurable or constant parameters, namely

$$V_T = V_B - \phi_W. \qquad (4)$$

Thus, V–I characteristics for targets were measured by varying V_B and recording the output signal current I_S. Examples of measured curves for the SIT tube and silicon vidicon are shown in Figs. 2 and 3. The apparent shift of axes between the two figures is due to different values of ϕ_W in the two tubes. Also note how the G1 setting determines the value of beam saturation current, but does not significantly change the shape of the curves.

THE MODEL

A one-dimensional scan was considered because it was efficient to implement and sufficed to enable point spread functions, modulation

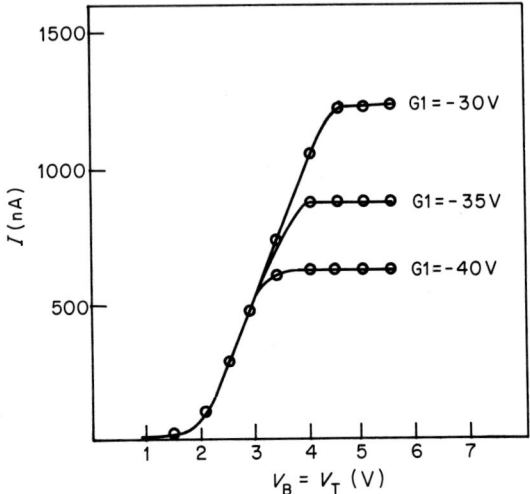

Fig. 3. Measured V–I characteristics for a silicon vidicon.

transfer functions (MTF) and phase transfer functions (PTF) to be evaluated.

If the linear charge density is denoted by q and the target current density by J_T, then the rate of charge removal at location x when the beam is located at α, is given by

$$\frac{dq(x,\alpha)}{dt} = -J_T(x,\alpha), \quad (5)$$

where

$$J_T(x,\alpha) = f(V_T)J_B(x,\alpha), \quad (6)$$

$f(V_T)$ is the V–I characteristic described above, and J_B is the beam current density. If the beam is assumed to be Gaussian with width σ then J_B is given by:

$$J_B(x,\alpha) = \frac{I_0}{\sqrt{2\pi}\sigma} \exp\left(-\frac{(x-\alpha)^2}{2\sigma^2}\right). \quad (7)$$

Substituting Eqs. (6) and (7) in (5) and noting that $q = cV_T$ we obtain

$$\frac{dq(x,\alpha)}{dt} = -f\left(\frac{q}{c}\right)\frac{I_0}{\sqrt{2\pi}\sigma}\exp\left(-\frac{(x-\alpha)^2}{2\sigma^2}\right). \quad (8)$$

Since the target is discrete it is natural to rewrite Eq. (8) as a difference equation, with the differential terms referenced to a diode. Thus we let Q represent the charge on a diode, ΔQ the charge removed from the diode in Δt, and I_B the current contribution from the beam discharging the jth

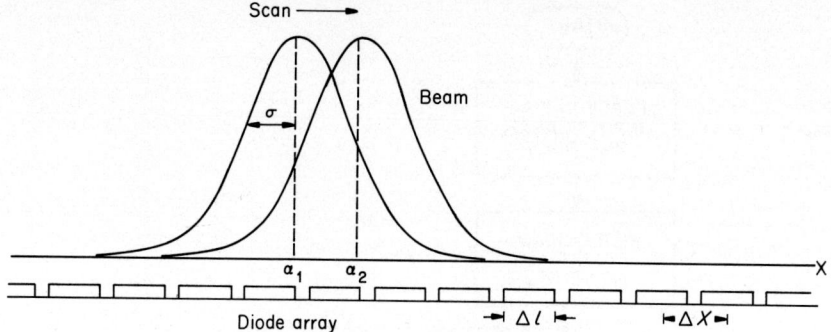

FIG. 4. Beam scan geometry and scanning parameters.

diode. If the diode length is Δl and the diode spacing is Δx then the following substitutions can be made in Eq. (8):

$$x = j\,\Delta x, \qquad (9a)$$

$$dt = \Delta t, \qquad (9b)$$

$$C = c\,\Delta l, \qquad (9c)$$

$$\Delta Q(j\,\Delta x, \alpha) = dq(x, \alpha)\,\Delta l, \qquad (9d)$$

$$I_B(j\,\Delta x, \alpha) = J_B(x, \alpha)\,\Delta l, \qquad (9e)$$

to finally obtain

$$\frac{\Delta Q(j\,\Delta x, \alpha)}{\Delta t} = -\mathrm{f}\!\left(\frac{Q(j\,\Delta x, \alpha)}{C}\right)\frac{I_0\,\Delta l}{\sqrt{2\pi}\sigma}\exp\!\left(-\frac{(j\Delta x - \alpha)^2}{2\sigma^2}\right). \qquad (10)$$

Figures 4 and 5 show how Eq. (10) is interpreted and evaluated. The variable α is incremented to simulate the beam scan, and $j\,\Delta x$ is the variable which identifies the location of the jth diode. Figure 4 shows the beam centre located at α_1. The ΔQ removed from each diode is a function of the beam magnitude at that diode and the charge already present, as reflected in the terms of Eq. (10). As the scan continues and the beam moves to α_2 the magnitude of the beam at each diode will change as shown. The amount of charge removed from each diode when the beam is at α_2 will again depend on the beam magnitude and the charge present.

Figure 5 is a flow chart which shows how Eq. (10) is evaluated. An initial charge distribution is placed on the diodes and the $\Delta Q/\Delta t$ values from each diode are computed using Eq. (10). The output current for the beam at position α is obtained by summing the $\Delta Q/\Delta t$ from each diode:

$$I(\alpha) = \sum_{j=1}^{N} \frac{\Delta Q(j\,\Delta x, \alpha)}{\Delta t}. \qquad (11)$$

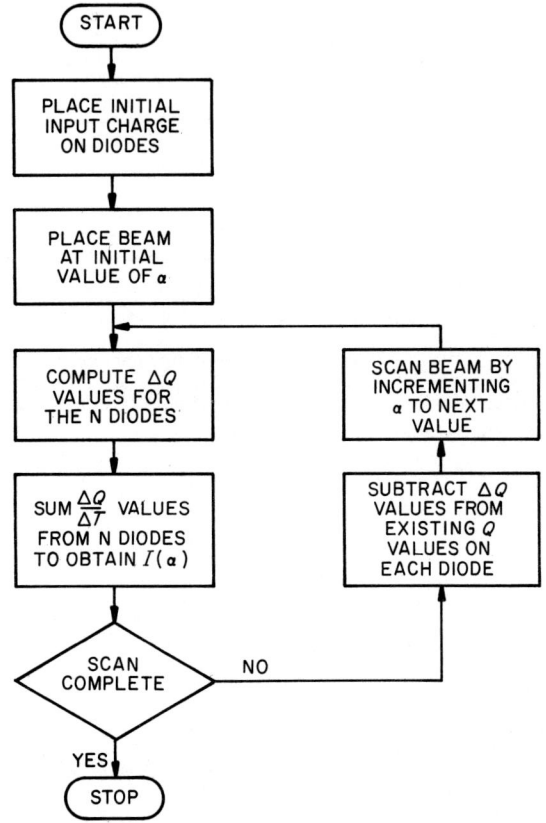

Fig. 5. Flow chart of model implementation.

The charge on each diode is updated by subtracting the ΔQ values from the present Q values. The beam is scanned by incrementing α and the above process is repeated using the new Q values on the diodes as the basis for computing the next set of ΔQ. Note that the scan velocity is given by the ratio of the incremental value of α and Δt, and thus can be easily varied.

The model also inherently includes the relative active area of the target in the ratio of Δl to Δx. It is assumed that the charge and potential are uniform over a diode surface; considering the conductive nature of the diode this is reasonable.

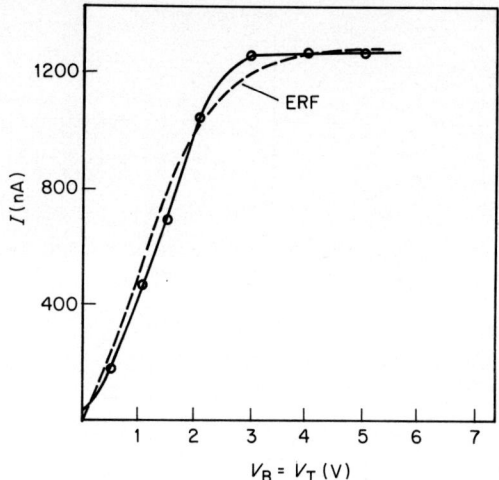

FIG. 6. The ERF function approximation to the V–I characteristic of a SIT tube.

Model Implementation

The model was implemented using a FORTRAN program on the same computer that services the camera tube laboratory. For reasons of efficiency, the function $f(V_T)$ was approximated using the ERF function. Figure 6 shows a comparison of ERF function and a measured $f(V_T)$ curve for the SIT tube. The principal differences occur at small values of V_T where the ERF function becomes zero and the experimental curve does not. This makes the model somewhat inaccurate for small signals and it will be changed in the future. This was not a problem for the data presented here.

An array consisting of 22 diodes was modelled. Each diode was 10 μm long, and the diodes were spaced 15 μm apart. The sampling interval, Δt, was 10 μsec. To simulate a conventional TV scan rate, the increment in α for time Δt was 2.75 μm for the SIT tube and 2.4 μm for the silicon vidicon. (The scanned areas of the SIT and silicon vidicon targets are taken as 10.8×14.4 mm^2 and 9.5×12.7 mm^2, respectively.)

C_0, the unbiased diode capacitance, was assumed to be 5×10^{-15} F. Point charge inputs of any desired magnitude can be placed on any of three centrally located diodes. All other parameters of the model can be entered as inputs to the program. The scan direction is left to right.

Modelled and Experimental Results

The model was used to calculate the point spread functions which would be generated by a camera tube in response to point source inputs

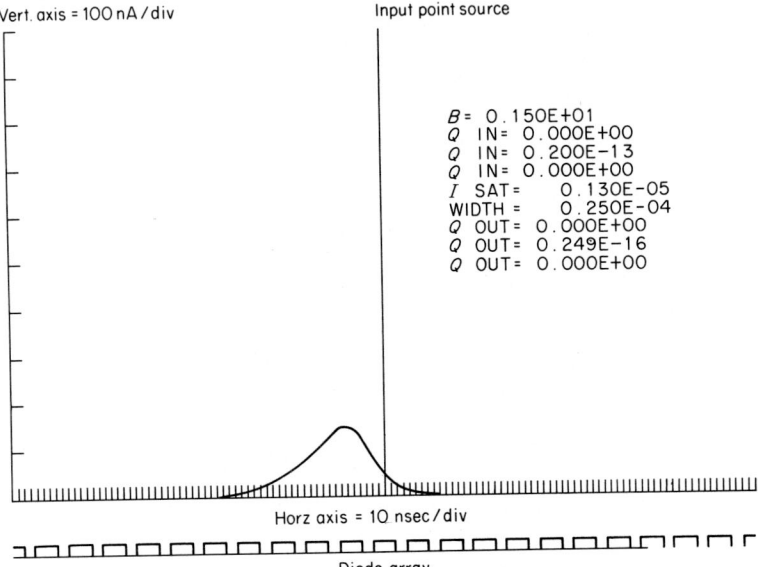

Fig. 7. Point spread function of SIT tube; $-Q = 2 \times 10^{-4}$ C.

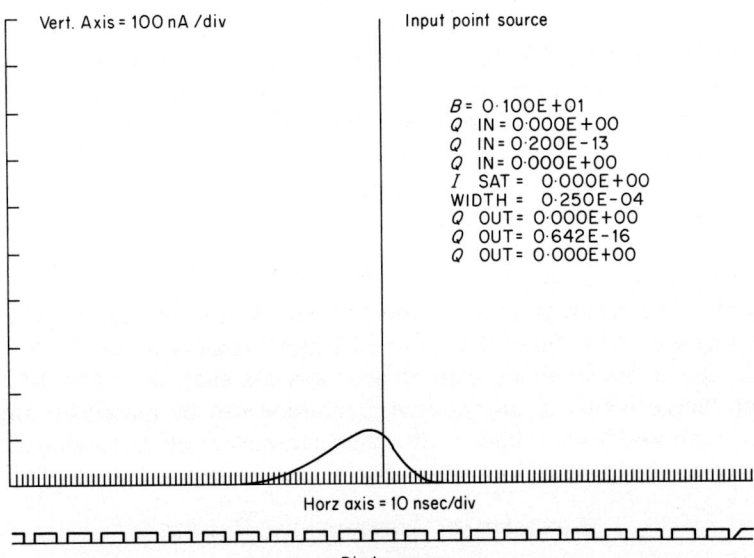

Fig. 8. Point spread function of silicon vidicon; $-Q = 2 \times 10^{-4}$ C.

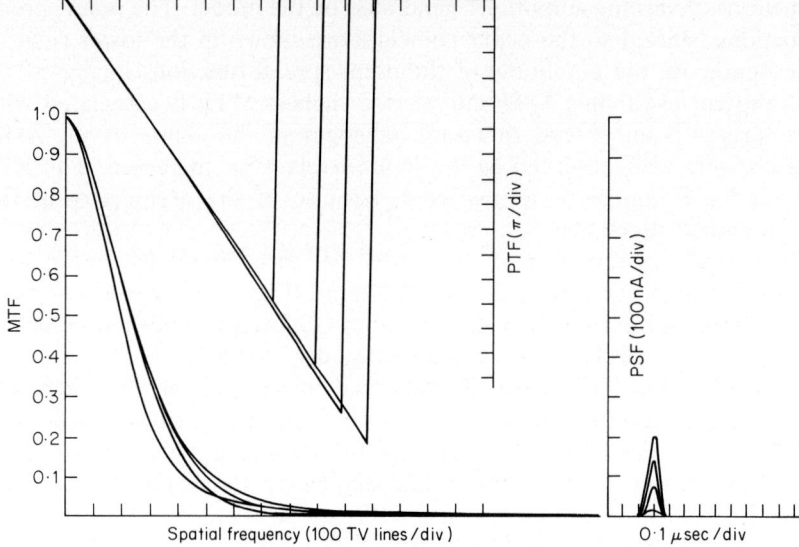

FIG. 9. Computed modulation transfer, phase transfer and point spread function for point source inputs of various amplitudes (SIT tube).

of varying amplitudes placed on one of the diodes. From the point spread function data, MTF and PTF curves were calculated and compared to MTF and PTF curves experimentally measured under equivalent conditions.

Examples of the point spread functions which were obtained for point source inputs are shown in Figs. 7 and 8 for the SIT tube and silicon vidicon respectively. The vertical line in the figures marks the location of the input point object. The beam width σ is shown as WIDTH in metres on the figures. Also shown are the initial charge on the central diodes (in coulombs) and the charge remaining after the scan is completed. The parameters B and I SAT are not of importance here. In all cases the peak value of the output signal occurred to the left of the actual location of the input, and the amount of the shift was found to depend inversely on the input signal level. Also note that the point spread functions are not symmetric.

There are several implications to the data. When viewed by a camera tube, the apparent location of a point source will depend on its magnitude—a nontrivial consideration in applications such as astronomy. Also, if the MTF and PTF of a camera tube are calculated from the point spread function, they will depend on the amplitude of the point spread function used.

Figure 9 shows MTF and PTF curves computed from point spread functions of varying amplitudes generated by the model. The point spread functions (shifted so the peaks coincide) are shown in the lower right of the figure. As the amplitude of the point spread functions increased, so did the corresponding MTF curves (the highest MTF is associated with the largest point spread function). Changes in the shape of the MTF curves were also observed as the input levels were increased. The PTF curves were nonlinear, as expected, because of the asymmetry of the point spread functions.

The most rigorous test of the model was to compare its output with measured data for similar input conditions. This comparison was most easily done in the frequency domain where differences between modelled and experimental data were more readily discernible.

The MTF and PTF of the SIT tube and silicon vidicon were therefore measured experimentally as a function of input point source level. The MTF curves were found to increase as the signal level increased although the PTF curves were nonlinear and similar to those predicted in the model.

Conclusions

Despite its simplicity, the model described here has been found to be one which is accurate in predicting tube outputs for a wide range of input and scan conditions. Future work will involve applying the model in two dimensions. When this is completed, a number of areas can be addressed, including comparing the effects of non-interlaced and interlaced scans on the two-dimensional point spread function, and questions relating to RMS noise values and power density spectra. The other aspect of the work involves using the model to investigate systems-related problems such as the detection of point objects in the presence of background and other point objects. Some initial work in the latter area has recently been presented.[9]

Acknowledgment

This work was performed at the University of Rhode Island, Electrical Engineering Dept, under grant ENG74-21930 from the National Science Foundation and contract DAAK 02-74-6-0234 from the U.S. Army Night Vision Laboratory.

References

1. Miller, A. and Izatt, J. R., *Appl. Opt.* **5,** 12 (1966).
2. Selke, L. A., *IEEE Trans. Electron Devices,* **EDS-16,** 7 (1969).
3. Muller, K. F. and Sauerman, G. O., "Intensified Silicon Diode Array Target Camera Tubes", MITRE Corp. Bedford, Mass. (1976).

4. McCollough, W. V., "Computer-Aided Analysis of Camera Tubes" PhD Thesis, University of Rhode Island, Kingston, R. I. (1977).
5. Biberman, L. and Nudelman, S., "Photoelectronic Imaging Devices" Plenum Press, New York (1973).
6. Meltzer, B., *J. of Electronics and Control* **3**, 355 (1957).
7. Spangenberg, K. R., *In* "Vacuum Tubes", McGraw-Hill, New York (1948).
8. Pierce, J. R., "Theory and Design of Electron Beams" Van Nostrand, New York (1954).
9. McCollough, W. V., "A Comparison of Three Methods of Detecting Point Objects Using a LLLTV System". Presented at "Electro-Optics 78", Boston, Mass. (1978).

Current Density Distribution in a Magnetically Focused Low-Velocity Electron Beam

Y. KIUCHI and T. SAKUSABE

Department of Electronic Engineering,
Tokyo University of Agriculture and Technology, Tokyo, Japan

Introduction

With advances in the application of television engineering, image pick-up devices are required to have a much higher resolution than in the past. For example, in order to record an A4 size document printed in Japanese, a resolution of 2000 to 3000 TV lines is required over the whole area of the paper.

At the present time, solid state image sensors cannot produce such a high resolution picture, because of the insufficient number of picture elements. Although it is possible to obtain a very high resolution picture from the Return Beam Vidicon[1] developed in the laboratories of the RCA company, this is a rather special tube. We have tried to determine the resolution limit of conventional vidicon camera tubes.

It is well known that the resolution of vidicons changes with the beam focusing voltage or focusing magnetic field; this suggests that the resolution depends on the scanning beam spot size. On the other hand, it is also known that the resolution is lowered to some extent by charge leakage in the photoconductive target as well as by light scattering and reflection in the target, faceplate, and field mesh region.

Figure 1 shows a cross sectional schematic diagram of a magnetically focused and deflected vidicon. The triode section of the gun determines the size, current density distribution and divergence angle of the electron beam at the crossover[2] and in front of the target; beam bending,[3] and the effect of progressive erasure of the stored charges as the scanning electron beam progresses (the so-called self-sharpening effect[4,5]) are other important factors which determine the size and shape of the scanning beam.

The present discussion is, however, restricted only to the beam focusing section indicated in Fig. 1. Measurements have been made of the actual

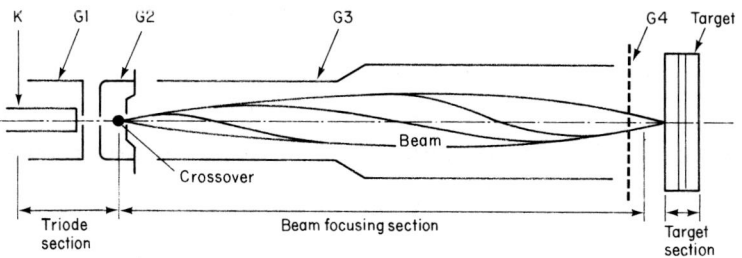

FIG. 1. Schematic of vidicon.

spot size and current density distribution in magnetically focused low velocity electron beams in vidicon camera tubes, and theoretical and experimental investigations on improved methods of beam focusing have been carried out.

MEASUREMENT OF CURRENT DENSITY DISTRIBUTION OF ELECTRON BEAM

The construction of the experimental tube is shown schematically in Fig. 2. The electron gun is that of a conventional 1 in. vidicon but instead

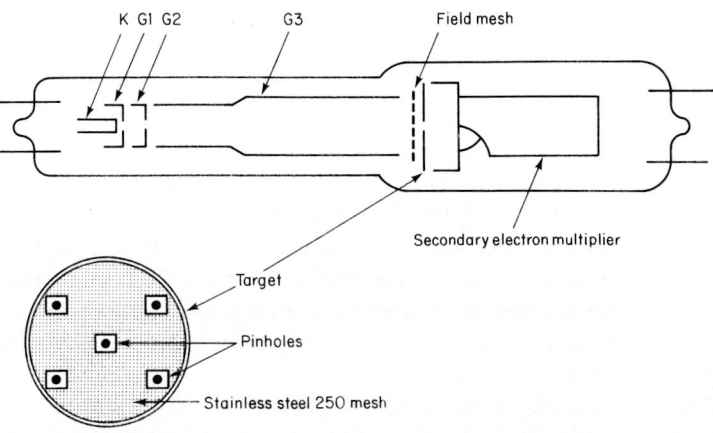

FIG. 2. Schematic of the experimental tube and target.

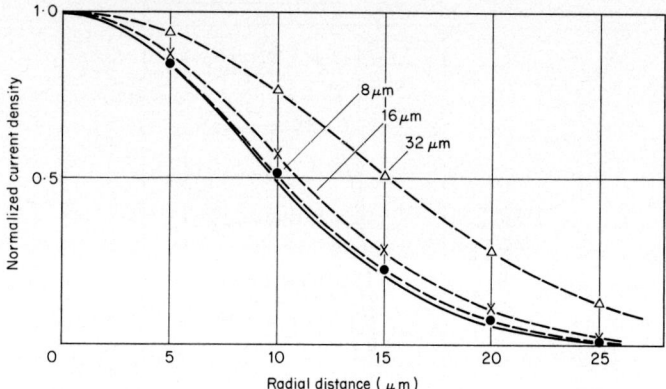

Fig. 3. Comparison between actual distribution in the beam (20 μm diameter) (solid line) and distribution computed from the signal which passes through pinholes of 8, 16, and 32 μm diameters (dashed lines). The density distribution of the beam is assumed to be Gaussian and the half width of the spot is 20 μm.

of a photoconductive target, a metallic target having several pinholes about 5 μm diameter located at the centre and some corner positions is used. An electron multiplier is attached just behind the target, so as to detect the small proportion of the beam electrons that pass through a pinhole.

By scanning the target around one of the pinholes with a small amplitude scan, the output signal obtained from the multiplier can be displayed on an oscilloscope as a three-dimensional current density distribution in the beam.

The accuracy of this measurement is expected to be mainly determined by the diameter of the pinhole, linearity of the scan and read out error of the oscilloscope wave form. Figure 3 shows the comparison between an actual beam distribution having a half width of 20 μm and the calculated distribution in output of the signal with pinholes of 8, 16 and 32 μm diameter. In our experiment the diameter of the pinhole used was 5 to 8 μm and the beam spot measured as 20 to 30 μm. Therefore, the error due to the diameter of the pinhole may be considered to be negligible.

The calibration of the scanning amplitude and linearity was carried out by comparing the pitch of a fine mesh attached to the target. It was confirmed that the distortion of the scan was less than 3% and the read-out error less than 4%. The total error was, therefore, estimated to be within ±8%.

Figure 4 shows the beam landing curves and the current passing

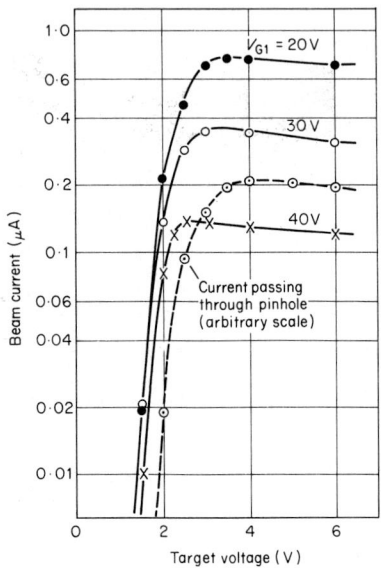

Fig. 4. Beam landing curves and the current passing through a pinhole, as a function of target voltage.

through a pinhole as a function of target voltage including the contact potential difference from the cathode. Little difference can be seen in the velocity distributions shown by these curves.

The current density distribution in the best focused beam at the centre of the target in a magnetically focused and deflected, separate mesh type 1 in. vidicon is approximately Gaussian as shown in Fig. 5. The beam spot is almost round if the beam is focused on the target.

The spot in an electrostatically focused beam at the corner of the target is always deformed such that the elongation is in the direction of deflection; the reason for this is astigmatic aberration. In contrast, the deformation of a magnetically focused beam cannot be explained by the use of the ordinary theory of aberration. It can be shown by calculation that the direction of the elongation, or major axis, caused by astigmatism is rotated by a focusing magnetic field. The rotation angle of the major axis depends on the length of the deflection region and distribution of the focusing magnetic field.

An observed current density distribution of the beam at the target can be considered to be an image of the density distribution at the object projected by the focusing lens. Therefore the current density distribution

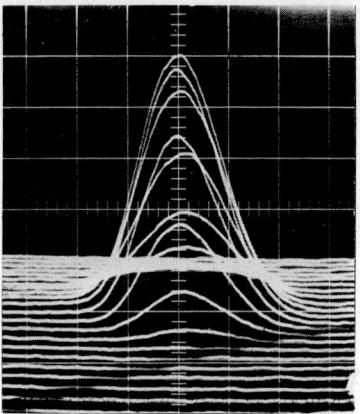

Fig. 5. Three-dimensional current density distribution of the best focused beam at the centre of the target. Horizontal scale is 15 μm/div. and the beam current is 0·3 μA.

of the beam in the neighbourhood of the crossover can be viewed by changing the focusing condition as shown in Fig. 6. Among the images in Fig. 6, image No. 6, which has the sharpest beam edge, can be identified as the image of the anode G2 aperture.

The radial length is determined by comparison with the 30 μm diameter G2 aperture and the axial distances of each cross section of the beam are computed from the focusing magnetic field strength and the focusing voltage. The solid curves in Fig. 6 are not electron trajectories in the usual sense but are the normalized equal current density curves. The dashed curve is the line that connects the half width points of each cross section. There is a weak electrostatic lens action at the G2 aperture

The diameter of the crossover is estimated at about 20 μm. This seems to be the size of the crossover seen through the anode G2 aperture from the target side. With another electron gun from which the G2 aperture had been removed, the size of the crossover was as large as ~40 μm, as shown in Fig. 7.

CALCULATION OF BEAM SPREADING

Calculations of beam spreading are made by tracing trajectories of electrons having various initial velocities. Then the outline of the spot composed of a group of electrons which passed through a point of the crossover with the specified initial velocity is obtained.

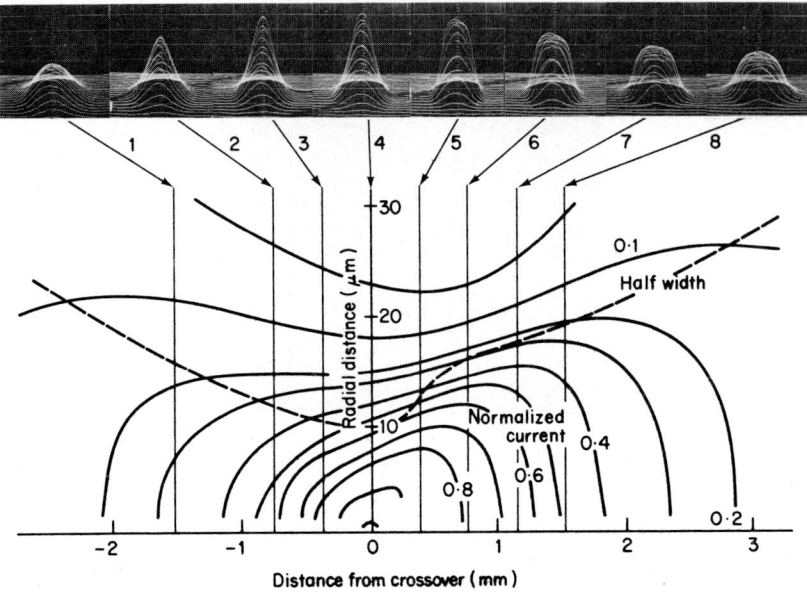

FIG. 6. Current density distribution in the neighbourhood of the crossover point. Diameter of the anode G2 aperture is 30 μm and the beam current is 0·3 μA.

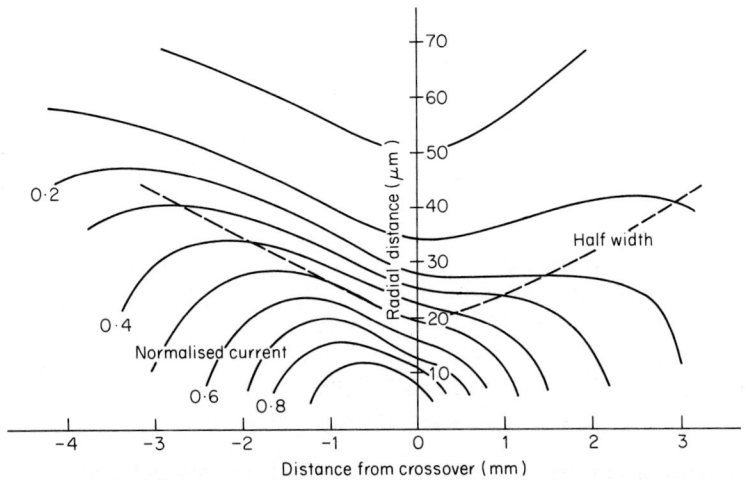

FIG. 7. Current density distribution of the beam in the neighbourhood of the crossover. In this experiment, the G2 aperture of the electron gun was removed and the grid drive voltage was kept equal to that of other normal experimental tubes. Accelerating voltage is 300 V.

In these calculations we assumed that: (1) there is no space charge effect in the beam; (2) there is no collimating lens action at the field mesh (mesh not separate); and (3) distribution functions of axial and radial initial velocities of electrons are expressed in the form of $\exp(-aV_z)$ and $\exp(-bV_r)$ respectively where eV_z and eV_r are equivalent axial and radial electron energies, and a is a constant which can be determined from the beam landing characteristics. The radial initial velocity has never been measured, so that b is assumed to be equal to a in this calculation.

We denote the distances from the crossover to the field mesh and to the target l_{cm} and l_{ct}, the distance from the field mesh to the target l_{mt}, the anode G3 voltage V and the target surface voltage V_t. If the focusing magnetic field is uniform and there is no deflecting magnetic field, the spread radius r can be expressed approximately:

$$V_t = V, \quad r \cong \frac{l_{ct}\sqrt{V_r(V_z)}}{2V^{3/2}}, \tag{1}$$

$$V_t \geq 0, \quad r \cong \frac{2l_{mt}\sqrt{V_r}(\sqrt{V_t+V_z}-\sqrt{V_t})}{V} + \frac{l_{cm}\sqrt{V_rV_z}}{2V^{3/2}}, \tag{2}$$

$$V_t < 0, \quad r \cong \frac{2l_{mt}\sqrt{V_r}\sqrt{V_t+V_r}}{V} + \frac{l_{cm}\sqrt{V_r(V_t+V_z)}}{2V^{3/2}} \quad \text{(provided } V_z \geq -v_t\text{)}. \tag{3}$$

The current ΔI which passes through a point of the crossover having the components of initial velocity between V_z and $V_z + \Delta V_z$ and between V_r and $V_r + \Delta V_r$ is

$$\Delta I = \frac{I_0 ab}{2\pi} \exp(-aV_z) \exp(-bV_r) \Delta V_z \Delta V_r \Delta\theta, \tag{4}$$

where I_0 is the total beam current and θ is an angle of the cylindrical coordinate; an axially symmetric distribution of the initial velocity is assumed.

The spread radius r is a function of V_z and V_r as shown in Eqs. (1), (2) and (3). Considering V_z and V_r to be independent of each other, we can rewrite these relations

$$V_r = G(r, V_z) \tag{5}$$

and

$$\Delta V_r = G'(r, V_z) \Delta r, \tag{6}$$

so that

$$\Delta I = \frac{I_0 ab}{2\pi} \exp(-aV_z) \exp[-bG(r, V_z)] G'(r, V_z) \Delta V_z \Delta r \Delta\theta. \tag{7}$$

The current density J is written

$$\Delta J = \frac{\Delta I}{r \cdot \Delta r \, \Delta \theta} = \frac{I_0 ab}{r} \exp(-aV_z) \exp[-bG(r, V_z)] G'(r, V_z) \Delta V_z. \quad (8)$$

However, ΔJ diverges at $r = 0$; this is unavoidable because of the assumption that finite numbers of electrons come from a point in the crossover where the current density is infinity. Therefore we introduce a new current density ΔJ^*:

$$\Delta J^* = \frac{\Delta I_r(r+W)}{2\pi[(r+W)^2 - r^2]} = \int_r^{r+W} \frac{I_0 ab}{2\pi W(2r+W)} \exp(-aV_z)$$
$$\times \exp[-bG(r, V_z)] G'(r, V_z) \Delta V_z \, \Delta r. \quad (9)$$

ΔJ^* is the current flowing into the ring of radius r and with W at the target, in other words W can be considered as the aperture and we set it at $0{\cdot}001$ μm to be much smaller than the desired accuracy of the calculation.

The current density J^* as a function of the spread radius r can be expressed approximately:

$$V_t > 0, \quad J_+^* \cong \frac{2I_0}{\pi W(W+2r)} \int_0^\infty \exp(-x)$$
$$\times \left[\exp\left(-\frac{V^2 abr^2}{4l_{mt}^2(\sqrt{aV_t + x} - \sqrt{aV_t})^2}\right) \right.$$
$$\left. - \exp\left(-\frac{V^2 ab(r+W)^2}{4l_{mt}^2(\sqrt{aV_t + x} - \sqrt{aV_t})^2}\right) \right] dx, \quad (10)$$

$$V_t = 0, \quad J_0^* \cong \frac{2I_0}{\pi W(W+2r)} \int \exp(-x) \left[\exp\left(-\frac{V^2 abr^2}{4l_{mt}^2 x}\right) \right.$$
$$\left. - \exp\left(-\frac{V^2 ab(r+W)^2}{4l_{mt}^2 x}\right) \right] dx, \quad (11)$$

$$V_t < 0, \quad J_-^* \cong \exp(aV_t) J_0^*. \quad (12)$$

An example of the beam spread curve derived from above equations is shown in Fig. 8. The beam spreading is reduced considerably by high velocity beam operation, while at low velocity, spreading is mostly caused by the chromatic aberration originating from the axial initial velocity dispersion of the beam.

Concentrated Magnetic Flux Scanning

To obtain a smaller spot size, it is necessary to minimize (1) crossover size, (2) beam spreading, and (3) lens magnification.

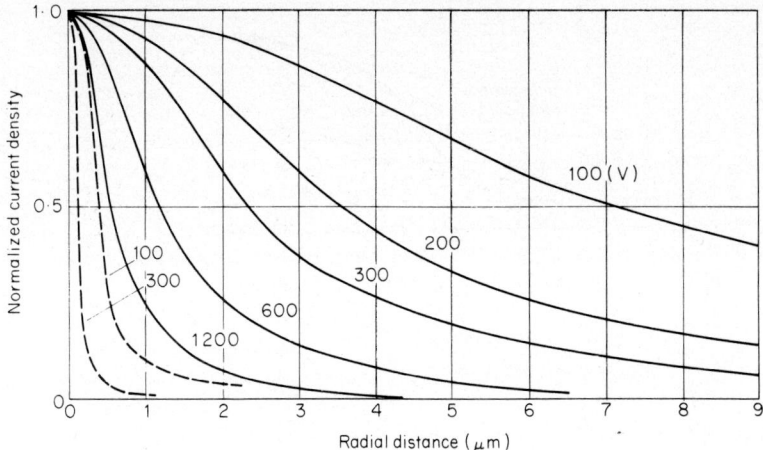

FIG. 8. Computed beam spread curves of a low velocity beam ($V_t = 0$ V) at the centre of the target (solid lines) for several accelerating voltages. Dashed lines show that of a high velocity beam with $V_t = 100$ and 300 V.

The crossover size has already been discussed. Beam spreading may determine the ultimate limit of the spot size of a low velocity beam, but it is calculated to be less than a few microns. However the overall spot size of an actual low velocity beam is about 30 μm and is mainly governed by the lens magnification of the beam focusing system.

The uniform magnetic field and constant accelerating voltage present the best set of focusing conditions. However, the lens magnification cannot be made less than unity. The increased beam spreading is caused by a sharp change in the radial component of non-uniform magnetic field in combination with the initial velocity of the electrons.

We propose a new focusing method which we call "Concentrated Magnetic Flux" scanning (CMF). The concept of CMF scanning is shown in Fig. 9. The concentrated magnetic flux formed by the combination of focusing and deflecting magnetic fields scans the target. The beam electrons spiral around the magnetic flux lines and are focused at the target.

For simplicity, the focusing and deflecting magnetic fields are made proportional to the axial distance and the accelerating voltage is at first kept constant, but properly speaking, it should be made proportional to the square of the axial distance to simulate the focusing condition of a uniform magnetic field.

The basic equations which determine the electron motion are

$$\left. \begin{aligned} m\ddot{x} &= e(H_y \dot{z} - H_z \dot{y} - E_x), \\ m\ddot{y} &= e(H_z \dot{x} - H_x \dot{z} - E_y), \\ m\ddot{z} &= e(H_x \dot{y} - H_y \dot{x} - E_z), \end{aligned} \right\} \quad (14)$$

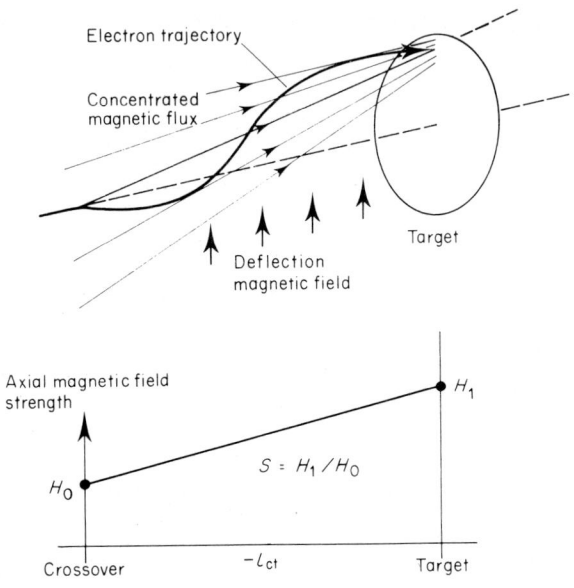

FIG. 9. Concept of the CMF scanning method.

where e and m are the electron charge and mass, H_x, H_y, H_z and E_x, E_y, E_z are the components of magnetic field strength and electric field strength in the x, y, and z directions.

The axial and radial components of the magnetic field strength are expressed in the form of Taylor expansions:

$$H_z = \sum_0^\infty \frac{(-1)^n H_a^{2n}}{n!} \left(\frac{r}{2}\right)^{2n}, \tag{15}$$

and

$$H_r = \sum_0^\infty \frac{(-1)^{n+1} H_a^{2n+1}}{(n+1)!n!} \left(\frac{r}{2}\right)^{2n+1}, \tag{16}$$

where $H_r^2 = H_x^2 + H_y^2$ and H_a is the axial magnetic field strength along the z axis.

In the case of the CMF scanning method the magnetic field strength of the z and r components are simply

$$H_z = H_0 + \frac{H_1 - H_0}{l_{ct}} z \tag{17}$$

and

$$H_r = -\frac{H_1 - H_0}{2l_{ct}} z, \tag{18}$$

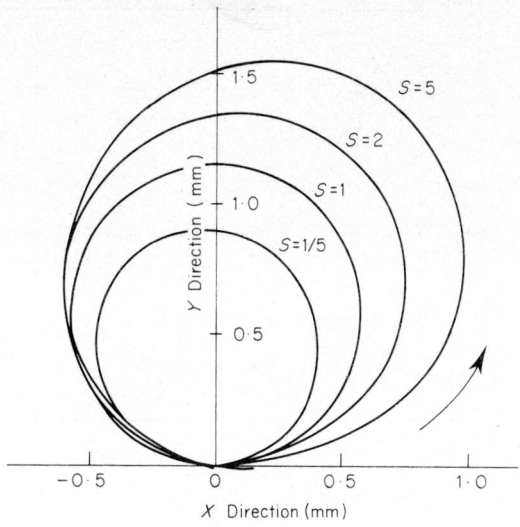

FIG. 10. Computed electron trajectories in the CMF field. $V_2 = 300$ V, $V_r = 0.5$ V. No deflection.

TABLE I

Computed characteristics of the CMF scanning beam at the centre of the target. Accelerating voltage is 300 V. $V_r = 0.5$ V.

S	H_0(G)	H_1(G)	M	Landing error (V)		
				x	y	z
5	14	70	0.55	0.29	−0.046	−1.0 ×10⁻³
4	17	68	0.58	0.23	−0.027	−0.7 ×10⁻³
3	21	63	0.65	0.155	−0.012	−0.4 ×10⁻³
2	28	56	0.74	0.062	−0.002	−0.1 ×10⁻³
1	41.5	41.5	1.0	0	0	0
1/2	52.4	26.2	1.32	−0.03	−0.0004	0.038×10⁻³
1/3	57	19	1.49	−0.055	−0.0015	0.063×10⁻³
1/4	60	15	1.58	−0.069	−0.0025	0.075×10⁻³
1/5	61.5	12.3	1.65	−0.08	−0.003	0.084×10⁻³

where H_0 and H_1 are the axial magnetic field strengths on the z-axis at the crossover and the target respectively. Figure 10 shows the computed electron trajectories from the centre of the crossover in the CMF field. The trajectory is slightly deformed from a circular orbit compared to that in a uniform magnetic field ($S = 1$ in Fig. 10), but electrons return to the point of origin at nearly the same angle as they had initially. Table I

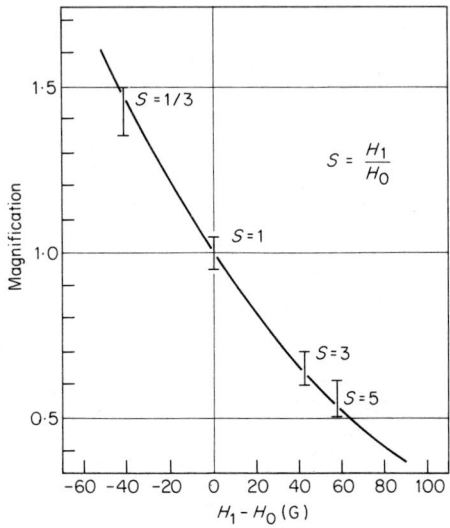

FIG. 11. Magnification of the CMF scanning beam. Solid line shows calculated curve and vertical bars show fluctuations of experimental errors.

TABLE II

Computed characteristics of the CMF scanning beam at the corner of the target. Accelerating voltage is proportional to the square of the axial distance. Deflection length is 8 mm.

Accelerating voltage (V)	Magnification M	Landing error (ratio)	Beam spot size on the target (μm^2)	Disc of least confusion Diameter (μm)	Disc of least confusion Distance from target (mm)
300 const.	0·64	−0·0443	320×200	110	2·15
300～600	0·63	−0·0931	200×120	60	1·8
100～900	0·62	0·136	160×120	40	1·5
50～450	0·62	0·314	230×150	70	1·5

shows the computed characteristics of the CMF scanning beam at the centre of the target. Figure 11 shows the lens magnification of the CMF scanning beam as a function of the difference between the focusing field strengths at the crossover and at the target.

When the beam is deflected to the corner of the target the beam spot size and beam landing error becomes larger. By making the accelerating voltage proportional to the square of the axial distance, the calculated

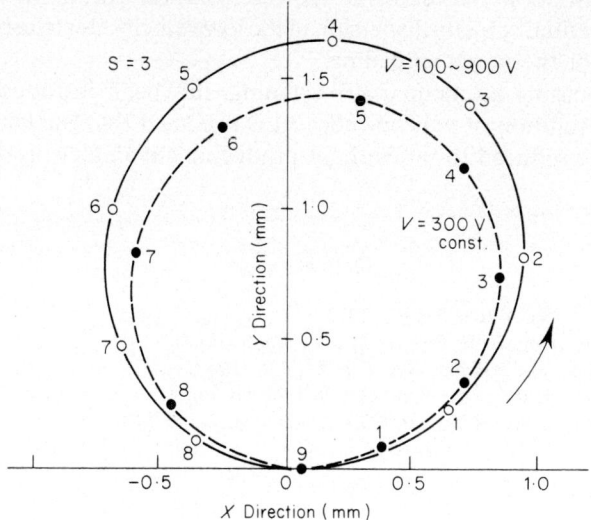

FIG. 12. Computed electron trajectory (dashed line) in the CMF field. Accelerating voltage is proportional to the square of the axial distance (solid line). Numbers on the curve show the axial distance (cm) from the point of origin. No deflection. $V_r = 0.5$ V.

spot size of the beam is improved considerably as shown in Table II; Fig. 12 shows the electron trajectories. The position of the disc of least confusion of the beam is still off the target plane. Therefore, dynamic focusing is necessary to focus the beam exactly on the target plane at the corners.

At the present time, we have not tried the ideal CMF condition. However we have used both a constant accelerating voltage and also a voltage approximately proportional to the square of the axial distance, and have shown that the lens magnification can be reduced to 0·6 without any serious changes in the beam characteristics.

Conclusion

The beam spot size, current density distribution, and beam landing error in vidicon camera tubes have been measured experimentally. The spot size of the beam in the 1 in. vidicon using a conventional coil assembly is 25 to 30 μm in diameter and the current density distribution is nearly Gaussian. The crossover is estimated at about 40 μm in diameter but the actual size seen from the target side is reduced to 20 μm due to interception by the anode G2 aperture.

The beam spread was calculated taking into consideration the initial velocity distribution and the result indicates that the chromatic aberration arising from initial velocity dispersion in the low velocity electron beam is the main cause of the beam spreading.

A new focusing method, CMF scanning, has been introduced to improve the resolution of vidicon tubes. It is expected that the lens magnification can be reduced to 0·6 without producing any defects in the vidicon operation.

REFERENCES

1. Schade, O. H., *RCA Rev.* **31,** 60 (1970).
2. Moss, H., *In* "Adv. E. E. P." Supplement No. 3 (1968).
3. Van de Polder, L. J., *Philips Res. Rep.* **22,** 178 (1967).
4. Miller, A., and Izatt, J. R., *Appl. Opt.* **5,** 1940 (1966).
5. Kurashige, M., *Conv. Rec. of Inst. TV Engin. Jap.* **2,** (1978) (in Japanese).

Recent Developments in Permanent Magnet Focusing Assemblies for Image Intensifiers and Camera Tubes

C. I. COLEMAN

Department of Physics and Astronomy, University College, University of London, England.

W. A. DELAMERE

Ball Aerospace Systems Division, Boulder, Colorado, U.S.A.

N. J. DIONNE

Raytheon Company, Microwave and Power Tube Division, Waltham, Massachusetts U.S.A.

W. KAMMINGA

Laboratorium voor Technische Natuurkunde, Rijksuniversiteit te Groningen, Netherlands

D. LONG and J. L. LOWRANCE

Princeton University Observatory, Princeton, New Jersey, U.S.A.

and

P. VAN ZUYLEN

Technische Physische Dienst, Delft, Netherlands

INTRODUCTION

There are two principal reasons for using magnetically focused intensifiers or camera tubes rather than electrostatically focused systems: first, the resolution performance of the former is typically better by a factor of two (linearly) and, second, for use in the far ultraviolet where fibre optic windows are unavailable, it is necessary to utilize a device with a plane photocathode.

In certain applications, provision of the required magnetic field by means of a solenoid poses several difficulties, such as the mass of the coil and its associated power supply, the power consumption and the attendant heat dissipation; the last can result in serious thermal, mechanical and optical stability problems. Whilst there is considerable scope for power versus mass design trade-offs, the power dissipation of a solenoid is typically around 90 W for single loop focusing of a four stage intensifier, and around 12 W for large SEC camera tubes. There has, therefore, been considerable interest in the use of permanent magnet focusing assemblies (PMFAs), which also offer improved system reliability. Both EMI (UK) and RCA (USA) offered PMFAs for intensifiers, and several designs using arrays of Alnico bars were reported in the literature.[1-3] All of these magnets, however, suffered from problems such as excessive mass and bulk, difficult optical access, inadequate uniformity and trimmability of the field, poor shielding, etc. It is only more recently that practical PMFA designs have emerged. This review paper describes three separate PMFA development studies funded by NASA or ESA for Space Telescope and other applications; the studies were carried out by Princeton University (for the 35 mm and 70 mm Westinghouse magnetic image-section SEC camera tubes), by Ball Brothers Research Corporation[5] (also for SEC tubes), and by Technische Physische Dienst[6] (for the EMI 48 mm image intensifiers used in the Space Telescope Faint Object Camera).

Before describing the three PMFAs, brief consideration is given to the principal requirements for a sucessful design for spaceborne application:

(i) Stable and adequately uniform magnetic field of the required magnitude, namely ~ 8 mT for the SEC tubes using a single loop focused image section and a four loop focused readout section, $\sim 14 \cdot 2$ mT for the EMI four stage intensifier at single loop focus, and $\sim 15 \cdot 6$ and $11 \cdot 5$ mT for the three and two stage intensifiers respectively; short-term stability should be $\sim 0 \cdot 1\%$. Stability requirements are reviewed by Long et al.[7] and Coleman.[8]

(ii) Minimum mass and size. This generally dictates the choice of single loop focus (e.g. for the "Flux Compensated" design discussed below, an extra 5 kg of magnetic material would be required for double loop focus).

(iii) Shielding against perturbing external fields; shielding factors of up to 1000 against transverse fields of $\leq 0 \cdot 3$ mT may be necessary. The requirements have been analysed by various authors.[7,9,10]

(iv) Minimum external magnetic moment, so that the torque exerted by the Earth's magnetic field is small compared with the gravitational torque (this is also important in ground-based use, where motion of a PMFA relative to the telescope structure can change the reluctance of the external field path and, therefore, affect the internal PMFA field).

(v) Mechanical-thermal stability of photocathode position with respect to the focal plane of the input optics, and unvignetted optical access.

(vi) Thermal design allowing cooling of detector tubes to temperatures below ambient.

(vii) Compatibility with space environment (non-outgassing materials, resistance to degradation in Earth's trapped radiation belts, ability to withstand thermal extremes).
(viii) Survival of launch acceleration and vibration levels.

THE PROFILED RELUCTANCE (PR) PMFA

This design, the principle of which is described elsewhere,[11] is illustraded in Fig. 1. It was adapted and developed at Princeton University for the magnetically focused and deflected SEC tube.

The symmetrical assembly consists of two axially magnetized Alnico V annular magnets located at opposite ends of a tapered low-carbon iron "profile tube", together with two iron end rings and an outer cylindrical shunt (this is also profiled to minimize mass). The latter provides a low reluctance flux return which prevents variation of the internal field caused by external magnetic coupling variations; partial shielding against external fields is also provided by this, and is supplemented by an outer Conetic cylinder. The profile tube is shaped according to computer calculations, so that the saturation flux density in the iron is constant, thus producing a uniform field in the interior of the PMFA. Figure 2 is a scale diagram of the PMFA; an unvignetted optical image of $f/2·7$ can be accepted. A prototype for the 35 mm SEC tube has been fabricated and tested, giving approximately the designed field of 8 mT on axis; the field non-uniformity in one half (see Fig. 7) is attributed to inhomogeneities in

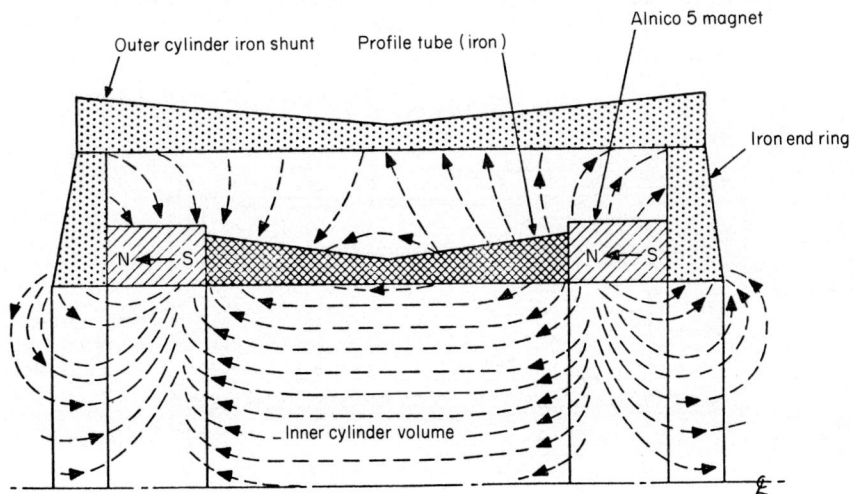

FIG. 1. Half-section diagram illustrating the Profiled Reluctance PMFA design concept.

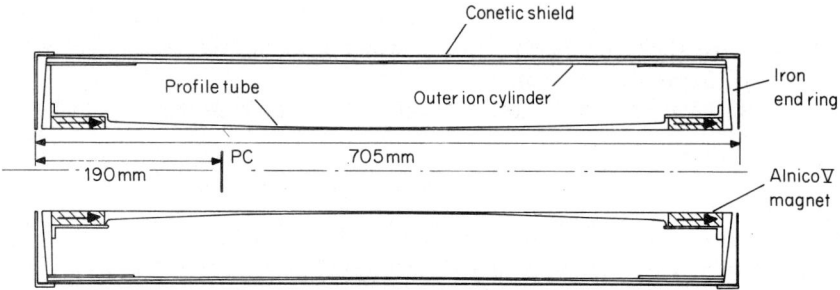

Fig. 2. Scale diagram of the Profiled Reluctance PMFA, showing the approximate SEC photocathode location (PC).

the profile tube material. Further data on this and the other two PMFA designs are compiled in Table I.

The Reluctor Ring (RR) PMFA

The Reluctor Ring (also known as Floating Pole) type of PMFA originated at the Raytheon Company, in a design for focusing a 160 mm ITT image intensifier.[12] With Ball Corporation funding, the Raytheon Company developed a version suitable for the 70 mm all magnetic SEC tube. The design concept is illustrated in Fig. 3. The reluctor circuit is composed of rings of soft magnetic material (mild steel) spaced apart by non-magnetic material (air gaps). The axial length and the spacing of the rings were designed, according to the results of electrolytic trough simulations and digital calculations, such that the magnetic scalar potential along

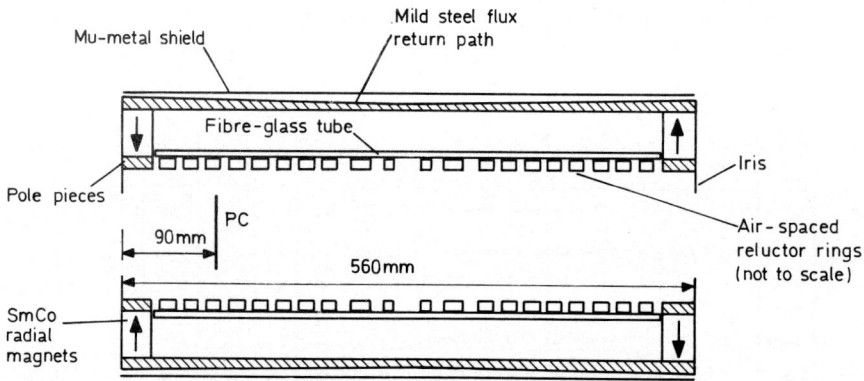

Fig. 3. The Reluctor Ring PMFA (schematic).

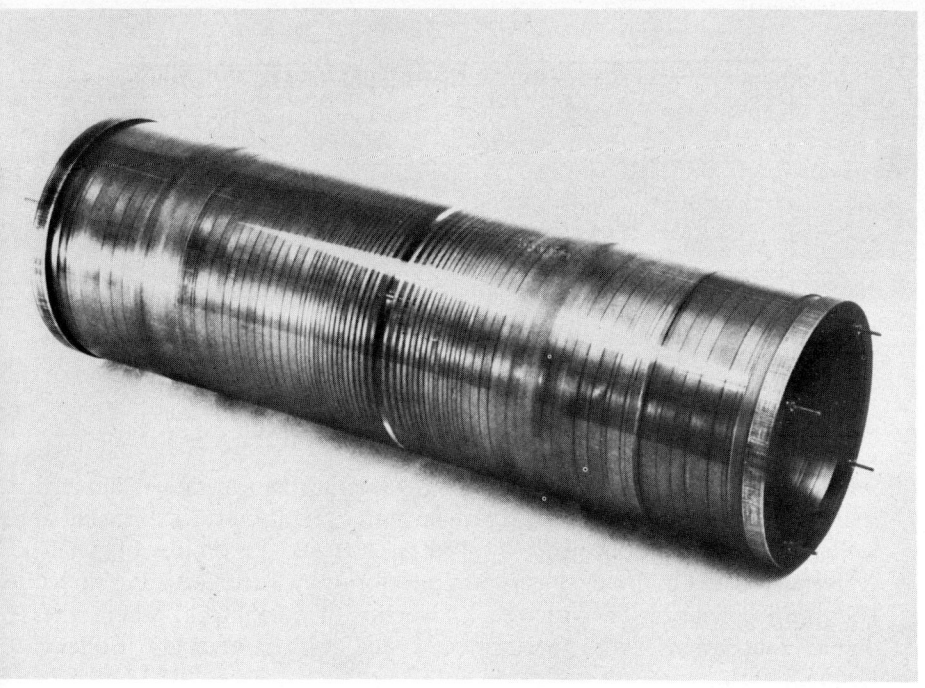

FIG. 4. Photograph of the Reluctor Ring structure.

the reluctor is controlled to achieve a sufficiently uniform magnetic field within the prescribed active volume (the ring spacings can be adjusted during manufacture, in order to trim the field uniformity—indeed, it is conceivable with this design that the ring spacings should be adjusted in order to tailor the field to compensate for the large disturbances caused by Kovar parts which are often present in image tubes). The reluctor ring structure is shown in the photograph in Fig. 4. The radially polarized magnets supplying the flux at the ends of the reluctor ring assembly are each constructed of 25 cubes of side 25 mm of sintered samarium–cobalt alloy; this rare earth material has high energy-product, high coercivity and a low temperature coefficient. A functional flux return path is provided by an outer mild steel cylinder, and a mu-metal canister gives additional shielding. Details of construction and performance are given in Table I. Figure 7 shows the high uniformity achieved. 70 mm SEC tubes have been successfully operated in this PMFA; resolution measurements are in excess of 50% modulation at 20 lp mm^{-1} over the central 25×25 mm^2. Tests aimed at optimizing the system have shown that performance was limited more by the SEC tube used than by the PMFA.

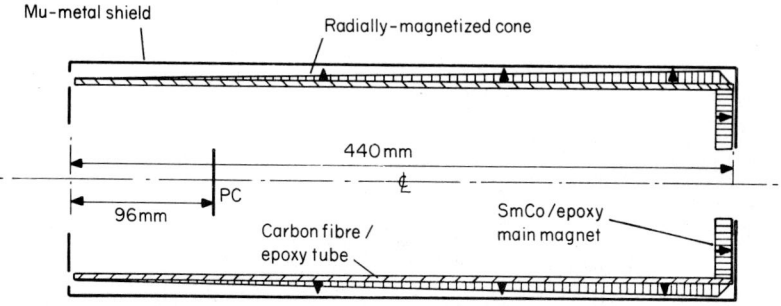

FIG. 5. The Flux Compensated PMFA (schematic).

THE FLUX COMPENSATED (FC) PMFA

This design, essentially different from the other two in that it directly shapes the main magnetic flux into near homogeneity instead of using a leakage field, was conceived by Professor J. B. Le Poole of Delft University. The Le Poole concept was developed by Technische Physische Dienst in collaboration with the University of Groningen, where the digital calculations were performed,[13] and with University College London.

Two prototypes have been fabricated. The first, symmetric, design[6] was built for a four stage EMI intensifier; the second, asymmetric, design (Fig. 5) will be described in more detail. The main flux is provided by an axially magnetized disc, and is shaped to a homogeneous field by a radially magnetized cone; the angle of the cone is determined by the ratio of the coercivity of the magnetic material to the desired field intensity. The return path is provided by a mu-metal shield. The PMFA construction is slim and light because no space is needed between the field shaping element (the radial magnet) and the shield. The magnets are made from blocks of HERA, a SmCo alloy in an epoxy matrix; this material is easy to work, and the magnets are trimmed by machining. The field of the PMFA was trimmed until its shape at the front approached closely to that of the standard solenoid; at the rear it was kept more homogeneous. Modulation of 75% at $10 \, \text{lp mm}^{-1}$ and 36% at $20 \, \text{lp mm}^{-1}$ has been measured with a two stage EMI intensifier. This is identical to that obtained in a solenoid. The magnification was 0·89, and distortion was <2%. The finished PMFA is shown in Fig. 6. A fine-tuning coil, providing ±5% field variation with maximum power dissipation <1 W, is incorporated; with the two stage intensifier, this permits gain control from 75% to 125% of nominal.

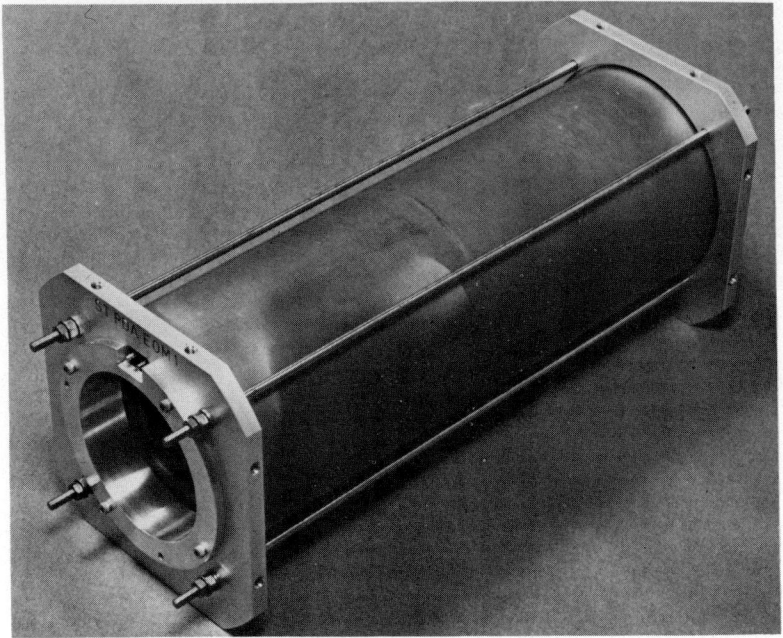

Fig. 6. Photograph of the prototype Flux Compensated PMFA.

Conclusion

Basic data on the three PMFAs are collated in Table I, and axial field uniformity data are presented in Fig. 7. Any intercomparisons must be made with great care because the magnets are still developmental and because they are built to different specifications for different applications. Each of the designs can meet optical access, structural and thermal requirements; regarding the latter, temperature coefficients are all about -0.03% $°C^{-1}$ (note that even better thermal performance can now be obtained with temperature compensated SmCo), and the magnets can work over the likely operating temperature range of any image tube system. Long-term stability for RR and FC designs is better than 0·1% per annum; for PR it is nearer to 1% per annum.

The PR design is very susceptible to inhomogeneities in the profile tube material (especially as it is working at saturation) and this accounts for the problems in attaining field uniformity. The RR structure is easiest to trim (by adjusting the positions of the rings) and has the advantage that the reluctor circuit is unsaturated. The presently achieved uniformity with the

TABLE I.
Properties of the three PMFA designs

PMFA	Profiled Reluctance	Reluctor Ring	Flux Compensated
Designed for	35 mm SEC	70 mm SEC	2–stage intensifier
Axial induction (mT)	8	8	11·5
Uniformity (% pk–pk)			
axial	1[a]	1·5	5
radial	—	2	<2
rotational	—	<0·5	<1
Shielding factor[b]			
axial	—	3	10000
transverse	50	40	2000
Magnetic dipole moment (A m^2)	≤20	320	≤1
Active volume			
length (mm)	420	430	250
diameter (mm)	36	70	>40
Overall length[c] (mm)	705	560	440
Overall diameter[c] (mm)	254	219	150
Mass[c] (kg)	18	30	11

Notes: [a] Uniformity for PR applies to good half only.
[b] Shielding factors are for DC fields, except for FC which was measured at 50 Hz.
[c] Including shields.

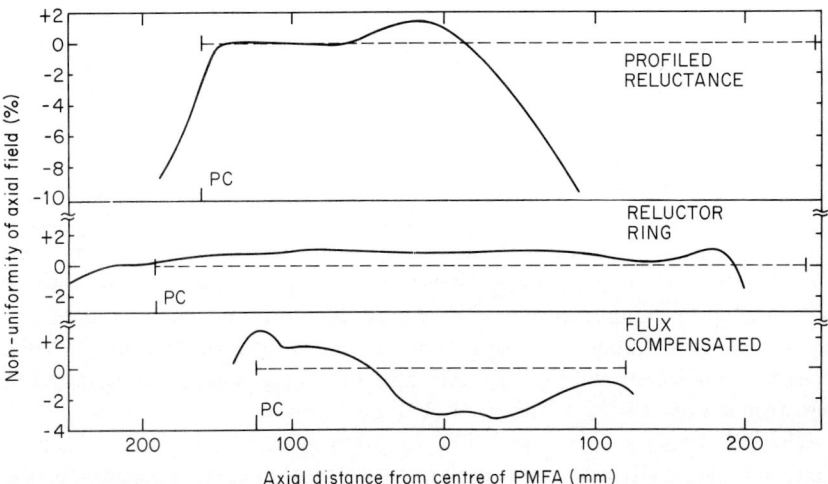

FIG. 7. Axial field uniformity for the three PMFA designs. In each case, PC denotes the photocathode position for the appropriate tube, and the broken line shows the axial extent of the required active volume.

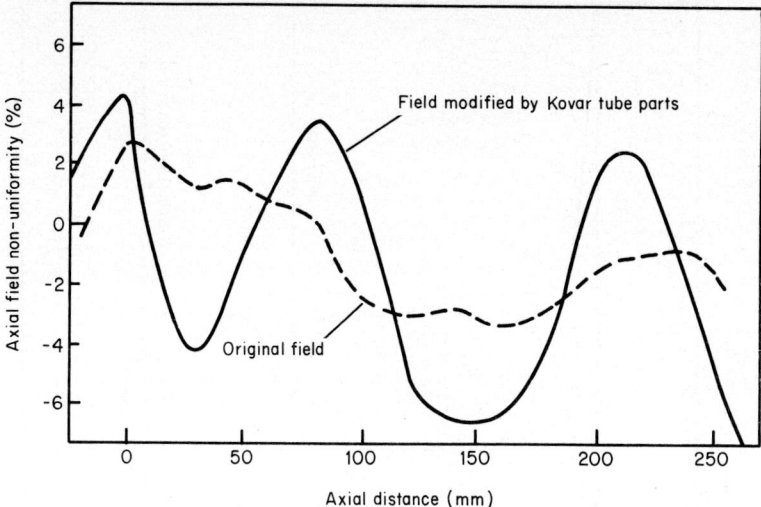

FIG. 8. Measured effect of Kovar parts of two stage intensifier on axial field uniformity of Flux Compensated PMFA

FC design (without further trimming) is not as good as RR but, judging by the performance figures with the intensifier, is certainly sufficient; in any case, it should be noted that, as shown in Figs. 8 and 9, the introduction of a tube body with Kovar parts results in far worse non-uniformity, amounting to $\sim \pm 6\%$ on-axis, and even more off-axis. There is thus no point in attempting further improvements to the uniformity of FC or RR PMFAs, until image tubes constructed from non-magnetic materials are available. As regards shielding performance and external magnetic moment, the FC configuration is undoubtedly superior (at least in part because $\sim 50\%$ of the flux is utilized in the active volume, whilst the flux efficiency of the other designs is only ~ 5–10%); the next RR model will, however, incorporate much improved shielding, although probably with a significant length and mass penalty.

As explained above, the FC design can have a much smaller diameter than the others, whilst the RR design can be shortest for a given task. Mass comparisons are difficult; the mass and volume of PR and RR designs are dominated by the flux return shunt and shielding, which must be sufficiently distant from the reluctor to avoid "shorting out" the stray flux. The present RR design has not been "lightweighted", and its mass bears an inverse relationship to its diameter. For any given application, a detailed trade-off study must be undertaken. As an example, an optimized RR development for a 35 mm SEC tube has been described.[14]

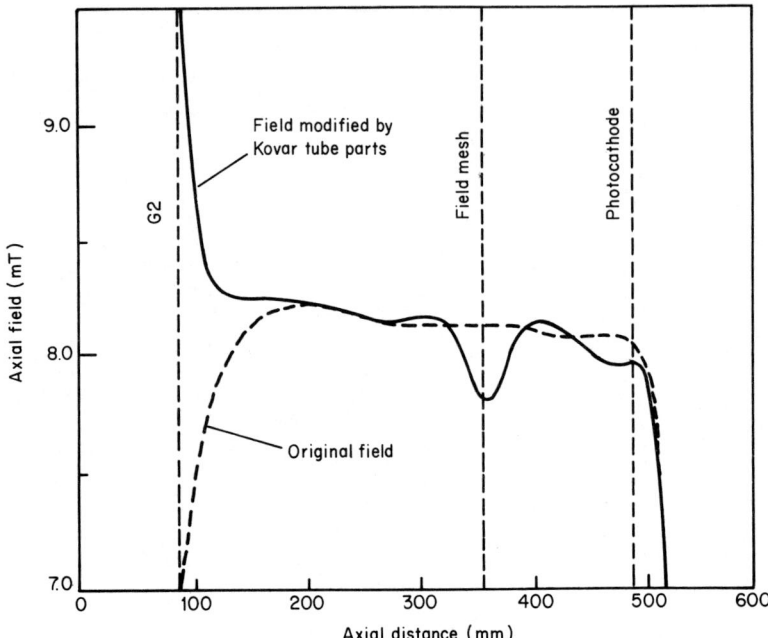

FIG. 9. Measured effect of Kovar parts of SEC tube on axial field uniformity of Reluctor Ring PMFA.

This paper demonstrates that PMFAs are now attractive alternatives to electromagnetic focusing coils, and indeed that they permit the use of magnetically focused tubes in situations where this was previously impossible. A contributory factor to this progress is the availability of the new high efficiency rare earth magnets. Development of both Flux Compensated and Reluctor Ring PMFAs is being actively pursued, and both are candidates for the Space Telescope instrument complement.

REFERENCES

1. Baum, W. A., *In* "Adv. E.E.P." Vol. 22A, p. 617 (1966).
2. Dennison, E. W., Schmidt, M. and Bowen, I. S., *In* "Adv.E.E.P." Vol. 28B, p. 767 (1969).
3. Carruthers, G. R., *In* "Adv.E.E.P." Vol. 33B, p. 881 (1972).
4. Long, D. C. and Lowrance, J. L., "Study of Permanent Magnet Focusing for Astronomical Camera Tubes", Final Rept., NASA Contract: NAS–5–20507 (1975).
5. Wooley, R. P. and Delamere, W. A., "An Evaluation Study of the Reluctor Ring Type of Permanent Magnet Focus Assembly", Final Report on NASA Contract No: NAS–5–24048 (1977).

6. van Zuylen, P., "Design of a PMA for Focusing a 4-stage Image Intensifier", Final Report on ESTEC Contract No: 2489/75 PP, (1977).
7. Long, D. C., Zucchino, P. and Lowrance, J., "Study of Magnetic Perturbations on SEC Vidicon Tubes", Final Report on NASA Contract No: NAS-5-23254 (1973).
8. Coleman, C. I., "Review of Specification for Permanent Magnet Assembly", Technical Note on ESTEC Contract No: 2489/75 PP, (1975).
9. Boksenberg, A., Bowen, P. J., Coleman, C. I. and Petford, A. D., "Study and Evaluation of Image Photon Counting Systems for a Spaceborne Astronomical Instrument", Final Report on ESTEC Contract No: 2322/74 PP, (1975).
10. Coleman, C. I., "Image Stability and Photometric Accuracy in the IUE Spectrograph Cameras", IUE Technical Note No. 36 (1976).
11. Cioffi, P. P. and Hagelbarger, D. W., *IEEE Trans. Magn.* **2,** 122 (1966).
12. Lapp, H. S. and Ceckowski, H. D., Proc. Conf. on Imaging in Astronomy, A.A.S., June 1975, Cambridge, Mass.
13. Kamminga, W., *J. Phys. D* **8,** 841 (1975).
14. Bailey, S., Delamere, W. A. and Dionne, N. J., *In* "Seventh Symp. P.E.I.D. Preprints" p. 183 (1978).

Tolerance Analysis of Magnetically Deflected Photon Counting Detector Systems

W. A. DELAMERE

Ball Aerospace Systems Division, Boulder, Colorado, U.S.A.

and

E. A. BEAVER

University of California at San Diego, La Jolla, California, U.S.A.

INTRODUCTION

Magnetically focused and deflected photon counting detectors have the potential for extremely high geometric precision when operated in uniform fields. The advent of the reluctor type of permanent magnet focus assemblies makes possible highly uniform focus fields with no power dissipation.[1,2] Of particular interest are the magnetically deflected Digicon[3-5] and the Intensified Charge Coupled Device (ICCD). Basically, these detectors consist of a photocathode on a flat window and a silicon array, 10 to 20 cm away. Each photoelectron is accelerated across 20 kV to the silicon array where it generates about 5000 electron–hole pairs. In order to use the full area of the photocathode the electrons are deflected with a transverse magnetic field. Deflection is also used to enhance the resolution of the silicon array by stepping in increments of a fraction ($\frac{1}{4}$ to $\frac{1}{8}$) of a pixel.

For two of the Space Telescope instruments a 512 Digicon is being developed from the existing 212 Digicon design.[5] To keep the size and weight of the focusing magnet small, it is desirable for the diameter of the tube to be as small as possible. Therefore, an electron optical analysis has been performed to determine the level of performance that can be achieved with only minor design changes.

TUBE ELECTROSTATIC DESIGN PARAMETERS

The Digicon tube geometry, shown in Fig. 1, represents a possible modification of the 212 Digicon. The inner diameter of the electrode

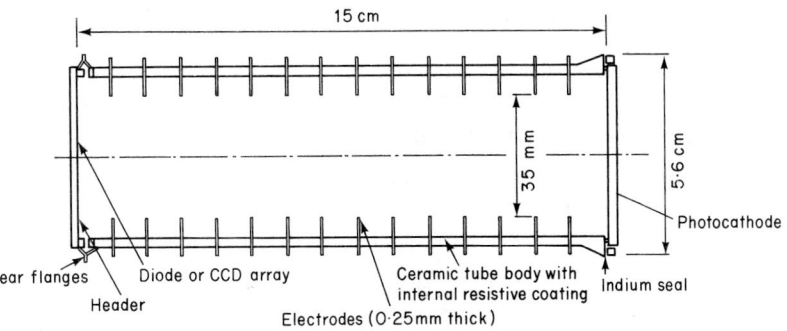

Fig. 1. Tube geometry.

rings has been increased from 25 mm to 35 mm and the method of mounting the faceplate has been changed to produce a more uniform electrostatic field at the photocathode.

Geometric and focus aberrations are caused by the photocathode indium seal, the thickness of the electrode rings, the uniformity of the resistive wall coating, the silicon array mounting flanges, tube mechanical tolerances, electrode potential tolerances and voltage gradients within the photocathode. An analysis of the effects produced by the first four causes has been completed and the principal results are presented in this paper. The tube mechanical and electrode potential tolerances will be dealt with during the final development stage. The photocathodes will have a transparent metal undercoating with a resistivity of less than $1000 \, \Omega\square^{-1}$. As the tube is used as a photon counting detector the photocathode current density will be less than $10^{-10} \, \text{A cm}^{-2}$ and so there is no significant current flow.

Magnetic Fields

Focus Field Parameters

The principal cause of focus and geometric aberrations is variation of the focusing field in the plane perpendicular to the tube axis, i.e., the transverse direction. Reluctor ring permanent magnet focus assemblies have been built with transverse non-uniformities of less than $0.05 \, \text{G cm}^{-1}$ in an 80 G field.[2] This uniformity can only be maintained if the tube is constructed of non-magnetic materials.

Deflection Field

The electron trajectory is completely immersed in the magnetic deflection field. This mode of deflection permits one to approach uniform deflection fields with practical coils. The electron deflection is approximately, $y = dB_y/B_z$, where B_y is the deflection field, B_z is the focus field, and d is the tube length. This simple equation is sufficient for roughly determining the field needed but, because the electron is accelerating along the trajectory, a small deflection in the x direction occurs. This x deflection is corrected by rotating the direction of the deflection field through an appropriate angle, $\tan^{-1}(1/\pi) = 17\cdot6°$.[6]

Equations of Motion for the Electron

Simple Model

In uniform electrostatic and magnetic fields the following equations can be used.

The transit time T from photocathode to diode array is given by

$$T = (2d^2 m/Ve)^{1/2}, \tag{1}$$

where e and m are the electron charge and mass. V is the potential between photocathode and silicon array, d is the distance between photocathode and silicon array.

The cyclotron radius for an electron with a transverse velocity component v, is

$$R = mv/Be, \tag{2}$$

where B is the magnetic focus field.

The time t for one loop is

$$t = 2\pi R/v = 2\pi m/Be. \tag{3}$$

This shows that the loop time is independent of the transverse velocity a necessary condition for focus. The other condition for focus is that the transit time equal the loop time. Equating Eqs. (1) and (3) gives

$$B = (\pi/d)(2Vm/e)^{1/2}. \tag{4}$$

Computer Model

The calculation of the electron trajectory uses numerical integration techniques. Our comprehensive models operate in three dimensions with relativistic effects included. The starting point is the acceleration on the electron which can be expressed as

$$\frac{d\boldsymbol{v}}{dt} = -\frac{e}{m}(\boldsymbol{E} + \boldsymbol{v} \times \boldsymbol{B}), \tag{5}$$

where v is the velocity, E the electric field and B the magnetic field vectors.

Equation (5) may be rewritten in Cartesian coordinates:

$$\ddot{x} = -\frac{e}{m}(E_x + v_y B_y - v_z B_z), \qquad (6)$$

$$\ddot{y} = -\frac{e}{m}(E_y + v_z B_x - v_x B_z), \qquad (7)$$

$$\ddot{z} = -\frac{e}{m}(E_z + v_x B_y - v_y B_x). \qquad (8)$$

When the electric and magnetic fields are known as a function of (x, y, z) the electron trajectory can be calculated as follows: each time increment δt, the velocity change is $\delta v_x = \ddot{x}\delta t$, and the positional change is

$$\delta x = v_x \delta t + \tfrac{1}{2}\ddot{x}\delta t^2.$$

The equations for y and z directions are similar.

Known Field Analysis Program

Using the above equations a computer program has been developed to determine the landing position of the electron under various conditions.

Four trajectories were calculated: for zero and full deflection, and zero and 1 eV initial transverse velocity. The approximate transit time T is calculated using Eq. (1) and is divided by the number of steps desired. Four thousand steps give better than 1 μm accuracy.

The motion of an undeflected electron with 1 eV transverse velocity is shown in Table I. From these data it can be seen that the electron spends over 25% of the transit time covering the first centimeter of travel from the photocathode. The last centimeter is covered in 3·5% of the time. The cyclotron diameter is 0·64 mm with the point of maximum displacement 4 cm from the photocathode.

Variation of Tube Voltage and Magnetic Focus Field

For uniform electrostatic and magnetic fields, variations of ±160 V and ±8 G resulted in tube defocusing of 0·046 μm V^{-1} and 20.7 μm G^{-1} for a 1 eV transverse electron with no deflection and with 7·5 mm deflection. These numbers represent the radius of the blur circle for the 1 eV transverse electron. The deflection sensitivity changes at $4·2 \times 10^{-4}$% per volt and 1·2% per gauss. Rotation of the image occurs at a rate of 13·7 μradians per volt and $-0·005$ radians per gauss.

TABLE I
1 eV transverse velocity trajectory

z(mm)	x(mm)	y(mm)	θ(deg)	T(nsec)
10	0·32	0·338	93	0·83
20	0·24	0·53	131	1·24
30	0·10	0·627	160	1·52
40	−0·034	0·64	185	1·76
50	−0·15	0·60	208	1·76
60	−0·239	0·539	228	2·16
70	−0·29	0·45	246	2·33
80	−0·32	0·36	263	2·49
90	−0·319	0·27	279	2·64
100	−0·29	0·19	293	2·78
110	−0·25	0·12	308	2·92
120	−0·199	0·067	322	3·05
130	−0·13	0·028	335	3·17
140	−0·068	0·006	347	3·29
150	−0·00	−0·002	360	3·41

Deflection Field Non-uniformities

A version of the program was developed that enabled an approximation of the deflection field non-uniformities to be incorporated. Measurements on a candidate deflection coil for use with the Digicon tube suggested that a fourth power law might be a good starting point to establish field uniformity requirements.

The magnitude of the deflection field is assumed to follow a symmetrical fourth power law for all three directions. The equations take the general form,

$$B = B_0\left[1 - F\left(\frac{P-d}{d}\right)^4\right],$$

where B_0 is the maximum value of the deflection field, P is the electron position, d is the maximum deflection, F is the reduction in the field at maximum deflection.

This equation has been used for three axes of non-uniformity in the field. In the axial direction, a 40% reduction at each end of the trajectory was used. The deflection sensitivity decreased from $1·503$ mm G^{-1} to $1·447$ mm G^{-1} and the coil angle changed from $17·6°$ to $15·7°$.

In the transverse direction a 1% reduction for the x and y directions produced a geometric error of 59 μm at the corner of the format.

Magnetic Focus Field Non-uniformities

The effects of non-uniformities in the magnetic focus field have not been modelled. However, they will produce geometric errors similar to those produced by the deflection field. Measurements of the field uniformity of a typical permanent magnet focus assembly yielded transverse variations in the range 0·02%–0·01% per cm between the centre and corner of the format. On this basis geometric errors of less than 7 μm can be expected.

SLAC ELECTRON TRAJECTORY PROGRAM

This program is widely used throughout the USA. It was originally created by W. B. Herrmannsfeldt[7] at the Stanford Linear Accelerator Center (SLAC) and is operational on IBM 370 and CDC 6600 computers. We have modelled the geometry of the Digicon and performed trajectory analysis of undeflected electrons in a uniform magnetic focus field.

The following cases have been modelled.

(a) Ideal Digicon—no electrostatic non-uniformities near the photocathode.
 —4 μm thick rings.
 —0·5 cm guard ring near diode array.
(b) Indium —0·5 mm of indium seal protruding into tube.
(c) Indium 2 —0·25 mm of indium seal.
(d) Rings —0·25 mm thick rings.
(e) Real —0·5 mm indium.
 —0·25 mm thick rings.
 —other parameters same as ideal.
(f) Real 3 —as "Real" except with 0·25 mm indium.

The effect of the indium seal and the thick rings on the electrostatic fields can be seen by observing the landing location errors given in Table II. Plots of the geometric error as a function of radius is given in Fig. 2 for the indium cases. It is interesting to note that it is theoretically possible to partially cancel the effect of the indium with the effect of the rings. The 0·25 mm indium seal and the 0·25 mm rings appear to be close to optimum, resulting in worst case blur circles of less than 4 μm and geometric distortion less than 16 μm.

DISCUSSION AND CONCLUSIONS

We have established a baseline analytical design for a tube capable of high resolution and low geometric distortion when used in near perfect

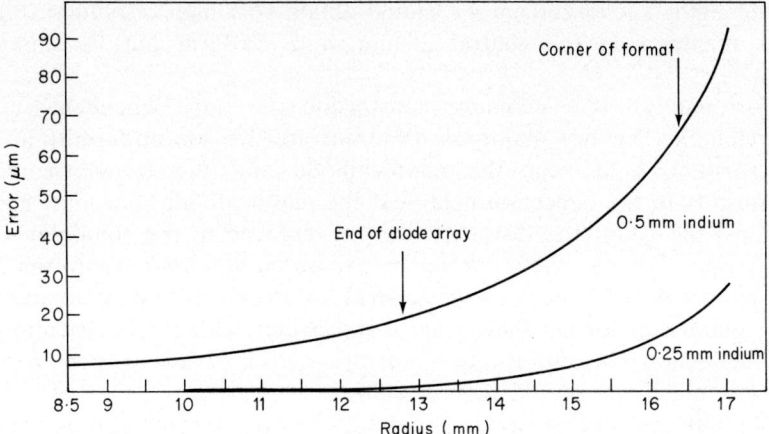

FIG. 2. Geometric position error as function of radius for 0·25 and 0·5 mm thick indium + 0·25 mm thick rings.

TABLE II

Landing location errors (μm) for rays starting at 8·5, 14 and 16·2 mm radius

Ray†	Starting radius (mm)	Rear flange	0·25 mm Rings	0·5 mm Indium	0·25 mm Indium	Rings + 0·5 mm Indium	Rings + 0·25 mm Indium
1	8·5	0	−5	12	5	7	0
2	8·5	0	−5	13	6	7	0
3	8·5	0	−5	11	5	6	0
4	14	−1	−16	42	18	27	3
5	14	−1	−17	46	20	30	4
6	14	−1	−15	41	18	26	4
7	16·2	−1	−23	82	36	60	14
8	16·2	−1	−25	91	40	67	16
9	16·2	−1	−23	82	36	60	14

† Rays 1, 4 and 7 leave normal to photocathode with 1 eV velocity.
Rays 2, 5 and 8 leave at 45° radially.
Rays 3, 6 and 9 leave at 45° tangentially.

magnetic fields. The actual resolution obtainable depends upon the transverse velocities of the photoelectrons, which are in turn dependent on the photocathode for the tube and the wavelength. In the vacuum ultraviolet we have not found a good set of data for the electron transverse velocity distribution for the CsI and Cs_2Te photocathodes. The "1 eV at 45°" photoelectron represents a reasonable number at 250 nm wavelength. At wavelengths above 250 nm we should obtain very high resolution: 8 μm blur diameter in the central 28 mm, and <18 μm out to 32·4 mm diameter.

Fortunately, the geometric distortion is not dependent upon wavelength. The two major contributions are the non-uniformity in the electrostatic field near the photocathode and the transverse non-uniformity in the deflection field near the photocathode. For most applications, the geometric distortion can be removed in the computer data processing. For an echelle format spectrograph, the order separation and the height of the diodes in a linear array are dependent upon the amount of geometric distortion. This analysis shows that, with careful attention to the deflection coil uniformity, geometric errors of less than 60 μm at 7·5 mm deflection should be obtainable.

The Digicon tubes generally use diodes on 50 μm centres so one might question why resolution of the order of 10 μm is important. The resolution of the diode array is limited by aliasing occurring in the sampling process. To reduce this characteristic the image is stepped across the array in small increments. A significant improvement in resolution is achieved if many exposures at the same signal to noise ratio are obtained. This improvement is limited by the sharpness of the transition from diode to diode and by the image resolution of the tube.

In an intensified CCD the diodes can be as small as 13 μm square, so high image resolution is mandatory. The aliasing characteristic can also be improved by substepping. The CCD has extremely good geometric precision so we are interested in preserving this performance in the ICCD.

References

1. Lapp, H. S. and Ceckowski, D. A. "A 162 mm Image Amplifier Tube for Photographic Applications", Proc. Conf. on Imaging in Astronomy, A.A.S. June 1975, Cambridge, Mass.
2. Delamere, W. A. and Woolley, R. P., An Evaluation Study of the Reluctor Ring Type of Permanent Magnet Focus Assembly, Ball Final Report F76–23 (1977).
3. Beaver, E. and McIlwain, C., *Rev. Sci. Instrum.* **42,** 1321 (1971).
4. Beaver, E., McIlwain, C., Choisser, J. and Wysoczanski, W., *In* "Adv. E.E.P." Vol. 33B, p. 863 (1972).
5. Beaver, E. A., Harms, R. J. and Schmidt, G. W., *In* "Adv. E.E.P." Vol. 40B, p.745 (1976).
6. Coleman, C. I., Doctoral Thesis, University of London (1974).
7. Herrmannsfeldt, W. B., "Electron Trajectory Program" SLAC Report No. 166 (1973).

Step Deflection of Electron Images

D. McMULLAN, A. R. JORDEN and A. J. PENNY

Royal Greenwich Observatory, Herstmonceux, Sussex, England

INTRODUCTION

The step deflection of an electron image is required in a number of electron optical devices including step–and–repeat electron lithographic cameras and high speed framing cameras. One application of the latter in astronomy is the detection and light curve measurement of optical pulsars using an electronographic camera:[1] the electron image of the small area of the sky where the pulsar is expected to be found is deflected cyclically to say twenty-five different positions on the nuclear emulsion with a cycle period exactly equal to that of the radio period of the pulsar. Dwell times of the order of a millisecond may be necessary and both the geometrical as well as the temporal stability must be high because exposures may need to be of the order of hours.

IMAGE DEFLECTION

The electron image in an image tube focused by parallel homogeneous electric and magnetic fields can be deflected in two dimensions by cross magnetic fields produced by currents through two sets of orthogonal deflection coils. For small deflection angles, e.g., the 7 deg required to deflect a 10 mm diameter image from the centre to the edge of the field of a 85 mm electronographic camera,[2]† the deflection is nearly proportional to the current in the coil. It is evident that high stability in the deflecting current is essential if the resolution is to be maintained at even these rather small deflections: at least 5 μm positional stability is required for 50 lp mm^{-1} resolution and in the above example the deflecting current stability (38 mm maximum deflection) would have to be better than $5 \times 10^{-3}/38$, i.e. about ±0·01%. The focusing solenoid current would also need to be equally stable.

† See p. 315.

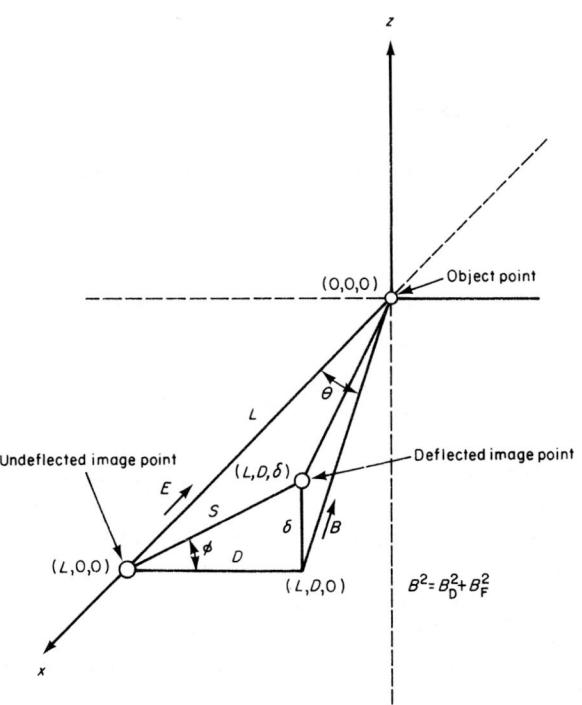

Fig. 1. Object point and image point locations in deflection system (adapted from Johnson and Hallam[3]).

Suitable high accuracy, high slew rate programmable power supplies are obtainable but are rather expensive. A simpler solution if only step deflections are required will now be described.

Tapped Deflection Coils

It has been shown by Johnson and Hallam[3] that, for an image tube in which the magnetic field is inclined at an angle θ to the electric field (assumed normal to the photocathode), the displacement of the focused image from the ($\theta = 0$) position is given by the resultant of two components, a radial deflection D and a tangential shift δ due to a drift motion perpendicular to the electric and magnetic fields (see Fig. 1):

$$D = L \tan \theta, \qquad (1)$$

and

$$\delta = (2m/e)^{1/2}[V_A/(B_D^2 + B_F^2)]^{1/2} \tan \theta, \qquad (2)$$

where L is the distance between the photocathode and the image plane, V_A is the accelerating potential, B_F is the focusing field, B_D is the transverse deflecting field, e and m are the electronic charge and mass, and $\theta = \tan^{-1}(B_D/B_F)$.

The condition for the electron image to be in focus is that

$$(B_D^2 + B_F^2)^{1/2} = (2m/e)^{1/2} V_A^{1/2} L^{-1} \pi N \cos \theta, \qquad (3)$$

where N is the number of focus loops. Substitution of Eq. (3) into Eq. (2) gives

$$\delta = (L \tan \theta)/(\pi N \cos \theta), \qquad (4)$$

and the resultant deflection

$$\begin{aligned} S &= (D^2 + \delta^2)^{1/2} \\ &= L \tan \theta [1 + (\pi N \cos \theta)^{-2}]^{1/2}. \end{aligned} \qquad (5)$$

The deflection S is thus dependent on the ratio of the fields and if the focusing solenoid and the deflection coils are connected in series and driven from a common constant current source the deflection will depend on the number of turns in the deflection coil. By the use of tapped coils and suitable switching, step deflections can be obtained.

Stability of Power Supplies

The expression Eq. (5) for the deflection S is for the focused condition and assumes that $V_A/(B_D^2 + B_F^2)$ is constant. In assessing the effect of variations in V_A and the constant current supply on the deflection, the two components must be considered separately. From Eq. (1) D will be unaffected, but δ depends on both $V_A^{1/2}$ and $(B_D^2 + B_F^2)^{1/2}$ as is shown by Eq. (2).

In the above example of a 7 deg deflection in an 85 mm image tube $\delta \approx 0.3S$ for single loop focusing and the required stabilities are $\pm 0.08\%$ for V_A and $\pm 0.04\%$ for the constant current supply. These are about ten times higher than the stabilities needed for an undeflected image tube,[4] but if a separate programmable supply were used not only would its stability need to be about $\pm 0.01\%$ but the focusing supplies would have to be equally stable.

The stability requirements can be relaxed still further by using multi-loop focusing. For example with $N = 3$, $\delta \approx 0.1S$, and the required stabilities would be $\pm 0.24\%$ and $\pm 0.12\%$. However the power dissipated in the focusing and deflection coils would be nine times higher.

Focusing

The magnetic field required for focusing a deflected image will be slightly less than for an undeflected one because the time of flight of the electrons will be increased. If B_0 is the focusing field for $\theta = 0$, then from Eq. (3)

$$B_F = B_0 \cos^2 \theta.$$

For $\theta = 7$ deg, $B_F/B_0 = 0.985$, so that step-programming of the current supply will be necessary if optimum resolution in the deflected image is to be maintained.

Distortion

The deflection vector S makes an angle ϕ with the direction of the deflecting field (see Fig. 1):

$$\phi = \tan^{-1}(\delta/D)$$

and for the focused condition, from Eqs. (1) and (4),

$$\phi = \tan^{-1}(1/\pi N \cos \theta).$$

For small deflections ($\theta \to 0$) and $N = 1$, $\phi = 17.66$ deg; as θ increases ϕ will also increase producing S-distortion in the matrix of images (but not in the individual images). For $\theta = 7$ deg, $\phi = 17.78$ deg so that the amount of S-distortion at maximum deflection (38 mm) in the 85 mm image tube will be 38 tan $(17.78 - 17.66)$ mm = 80 μm. Again with multi-loop focusing the distortion will be less: 30μm for $N = 3$.

Driving Circuits

The first application of the deflection system has been for pulsar detection using an electronographic camera with a 40 mm diameter window.[2] Twenty-five images in a 5×5 matrix are required. The basic elements of the circuit for driving the deflection coils are twelve transistor switches contained in one CAMAC-compatible module (Fig. 2). These switches may be controlled by a computer via CAMAC to pass the fixed focus coil current through the centre-tapped X and Y deflection coils; using either the whole winding or only half with the current either in the forward or reverse direction gives five deflection positions (one being zero deflection) for X and for Y making twenty-five possible positions in all. For this particular application the maximum resolution is not needed and provision for modulating the focus current has not been made.

If a computer (and CAMAC) are not available the system may be used in a "stand-alone" mode with a preprogrammed read-only memory controlling the transistor switches in a set sequence. The period of the repetition is controlled by an external clock signal fed directly into the

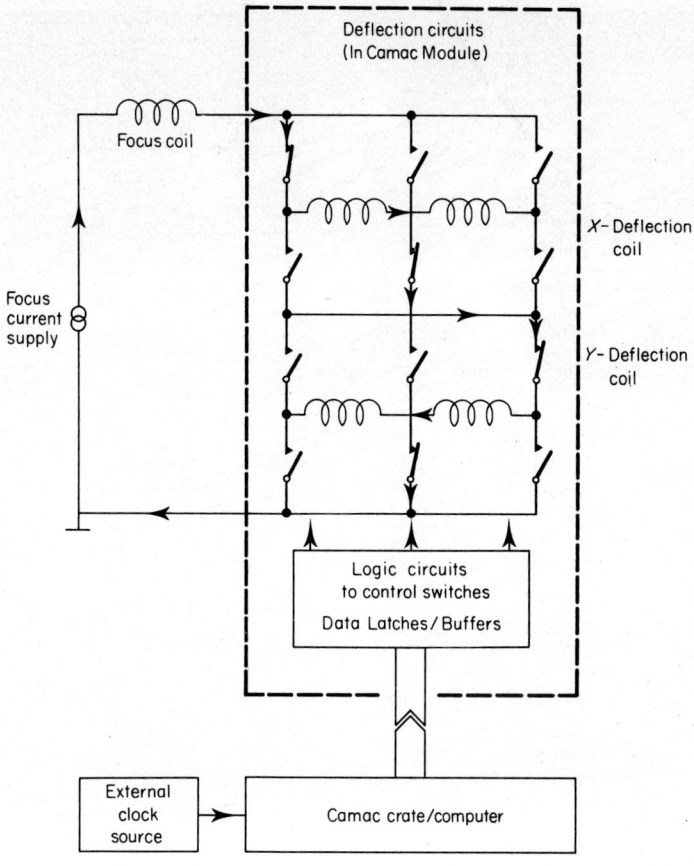

FIG. 2. Schematic diagram of deflection circuits.

circuit module instead of into the computer. The switching rate is limited by the characteristics of the deflection coils rather than by the switching circuit.

LABORATORY TESTS

Static deflections of an image in an 85 mm electronographic camera fitted with saddle deflection coils extending over the full length of the focus coil have confirmed that good resolution and negligible distortion are obtained. However tests with the 40 mm camera and dynamic deflection system described above have shown that some image smearing and a loss of resolution occur as a result of rapid deflections if the dwell time is

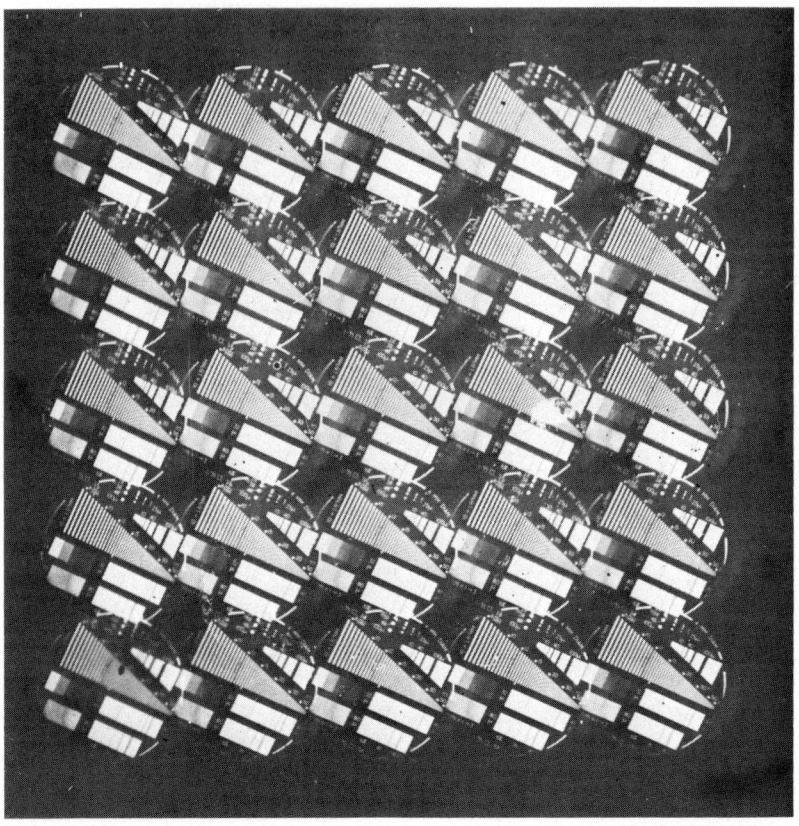

Fig. 3. Baum pattern deflected to 25 positions in a 40 mm electronographic camera. Step length 3·7 mm, step frequency 500 sec^{-1}.

3 msec or less in each position. This is believed to be partly the result of "ripple" on the focus current due to the constant current supply being unable to cope with the rapid change in load during switching although the inductance of the focus coil assists in maintaining the current at a constant value. More important causes are that the deflection field does not follow exactly the "ampere turns" in the deflection coils and the actual current in the coils takes a finite time to change when the switches operate. Induced currents in the magnetic shields and coil assembly (a non-conducting former is essential), and the inductance and capacitance of the coils themselves are the obvious factors that limit the speed but their relative importance has not yet been determined. Measurements at a 1000 sec^{-1} switching rate show that a resolution of greater than

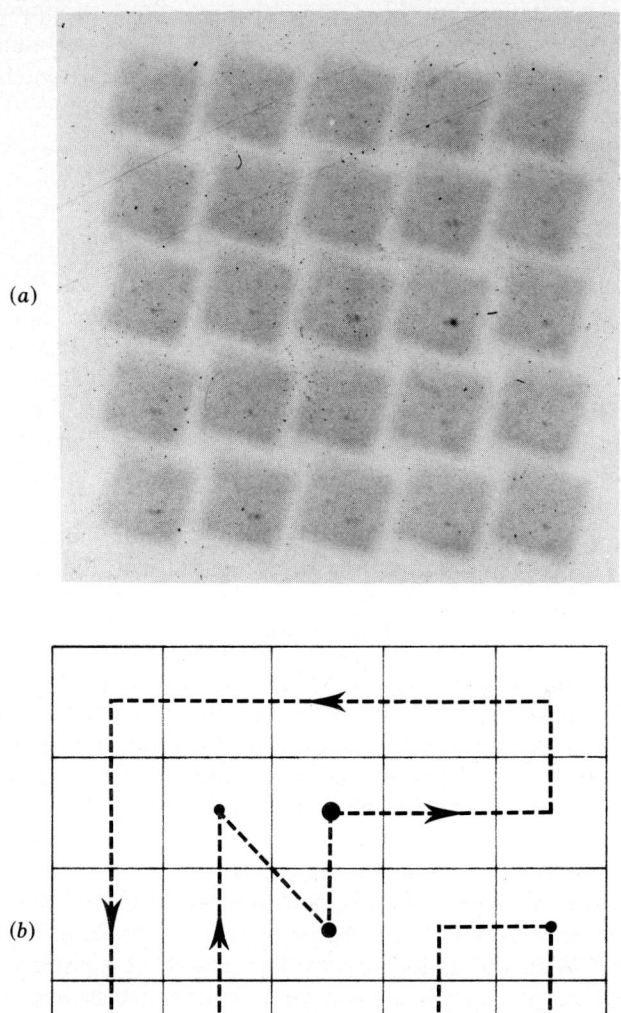

Fig. 4. (a) Electonograph of Crab pulsar observations on RGO 36 in. telescope. 10 min on G5 through light cloud. Step frequency ~825 sec^{-1}. (b) Scanning sequence for pulsar observation. The two groups of dots indicate position of the main (centre group) and inter-pulses (lower right).

20 lp mm^{-1} is attained in the centre nine positions dropping to 10–15 lp mm^{-1} at some outer positions. Figure 3 shows a Baum pattern deflected in a 5×5 matrix, 3·7 mm steps, and 500 sec^{-1} switching rate; the resolution at maximum deflection is about 25 lp mm^{-1}. The focus current was not modulated.

Astronomical Test

The 40 mm electronographic camera and deflection system have been tested at the f/15 Cassegrain focus of the 36 in. telescope at Herstmonceux and observations of the Crab pulsar have been made.[5] The period of this pulsar is ~33 msec so that the switching rate with twenty-five frames was ~825 sec^{-1}. As mentioned above there is some image smearing at this switching rate but it is less than the image spread caused by the rather poor seeing (image scale 1 arcsec: 66 μm). A typical electronograph is shown in Fig. 4(a); the observing conditions were poor with light cloud, and the exposure time was 10 min on Ilford G5 emulsion. Figure 4(b) shows the scan sequence, and the positions of the main pulse and interpulse are indicated and should be compared with Fig. 4(a).

Conclusions

The deflection system described enables electron images to be deflected statically in steps up to 7 deg with no loss in resolution and little distortion. At smaller angles (~1·5 deg) the image can be stepped at rates of up to a few hundred per second with little loss of resolution, and a maximum of 1000 sec^{-1} has been attained but with some smearing. Faster rates are probably possible with careful design of the coil assembly to reduce the capacitance of the windings and induced currents.

It is hoped to use the 85 mm camera stepped in a 7×7 matrix as the detector for an imaging Fabry–Perot spectrograph being built at the Observatory. With the deflection coil taps placed at $\frac{1}{3}$ instead of at the centre it will be possible to use the same driving circuits and obtain the 7×7 matrix instead of 5×5 by directing the current either through $\frac{1}{3}$ or $\frac{2}{3}$ of the windings. This principle can be extended to give binary addressing with the number of digits limited by the number of taps and transistor switches that can be provided.

Acknowledgments

The authors wish to thank Dr K. F. Hartley and Mr D. J. King who were responsible for programming the computer for the pulsar experiment, and Mr W. E. Matthews who constructed the electronic circuits and carried out some of the laboratory tests. The paper is published by kind permission of the Director of the Royal Greenwich Observatory.

REFERENCES

1. Walker, M. F., *In* "Adv. E.E.P." Vol. 40B, p. 829 (1976).
2. McMullan, D., Powell, J. R. and Curtis, N. A., *In* "Adv. E.E.P." Vol. 40B, p. 627 (1976).
3. Johnson, C. B. and Hallam, K. L., *In* "Adv. E.E.P." Vol. 40A, p. 69 (1976).
4. Hartley, K. F., *J. Phys. D* **7,** 1612 (1974).
5. Peterson, B. A., Murdin, P., Wallace, P., Manchester, R. N., Penny, A. J., Jorden, A., Hartley, K. F. and King, D., *Nature* **276,** 475 (1978).

Electron Optics of Concentric Spherical Electromagnetic Focusing Systems

L. W. CHOU†

Optical Department, North Industries Corp., Peking, China

Introduction

A system composed of two concentric spherical electrodes and their corresponding potentials is called a bi-electrode concentric spherical electrostatic focusing system. If we add to this system a concentric radial magnetic field parallel to the electric field, we get a concentric spherical electromagnetic focusing system, in other words, the system has spherically symmetric radial electric and magnetic fields.

From the viewpoint of electron optics, it is interesting to study the concentric spherical electromagnetic focusing system as an imaging system, as well as the more familiar concentric spherical electrostatic focusing system. This is because the system can meet both magnifying and demagnifying requirements; the magnetic and electric field distributions and the "paraxial" electron trajectories can be written as analytic forms and the focusing characteristics and the aberrations can be investigated quantitatively. With a concentric spherical electromagnetic focusing system, whether realized accurately or approximately, good image quality can be obtained. Only spherochromatic aberrations are present.

The concentric spherical electromagnetic focusing system as an imaging system was described at the Second Symposium on Photoelectronic Image Devices,[1,2] but the electron trajectories were solved by digital computer. An analytic solution for the electron trajectories in this system has not yet been given, and the focusing characteristics and the aberrations have not been discussed fully.

In this paper, the concentric spherical electromagnetic focusing system is studied together with three other systems, namely, the electromagnetic focusing system using parallel homogeneous magnetic and electric fields, the concentric spherical electrostatic focusing system and the proximity

† Present address: Optical Engineering Department, Peking Institute of Technology, Peking, China.

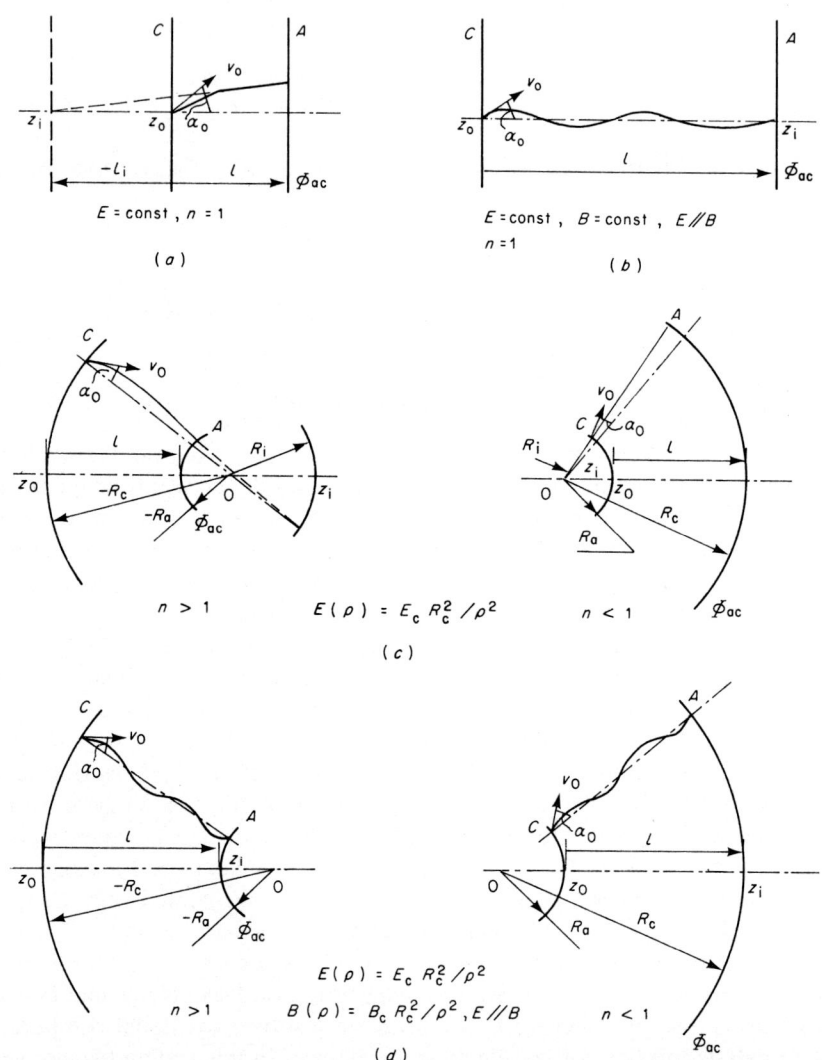

FIG. 1. Four focusing systems: (a) the proximity focusing system; (b) the electromagnetic focusing system using parallel, homogeneous electric and magnetic fields; (c) the concentric, spherical electrostatic focusing system; (d) the concentric, spherical electromagnetic focusing system.

focusing system (Fig. 1). The method for solving the electron trajectories in these systems is explored; the analytic solution for "paraxial" trajectories, the focusing conditions and the spherochromatic aberrations are given; the demagnifying tube suggested by Zacharov[1] and the corresponding magnifying tube are verified; and the case of the tri-electrode concentric spherical electrostatic focusing system and the tri-electrode electromagnetic focusing system using parallel, homogeneous electric and magnetic fields are discussed.

THE CONCENTRIC SPHERICAL ELECTROMAGNETIC FOCUSING SYSTEM

The Solution for "Paraxial" Trajectories and the Focusing Condition

For the concentric spherical electromagnetic focusing system using spherically symmetric magnetic and electric fields (Fig. 2), principal trajectories are straight lines perpendicular to the photocathode surface. Now we determine the position of neighbouring trajectories relative to the principal trajectory $P_0 P_i$. A plane normal to the principal trajectory at a point N intersects a neighbouring trajectory at a point N^*. The position of N^* can be expressed by the vector $\mathbf{p}$ from N to N^*. Let s be the linear distance along the principal trajectory from the initial point P_0

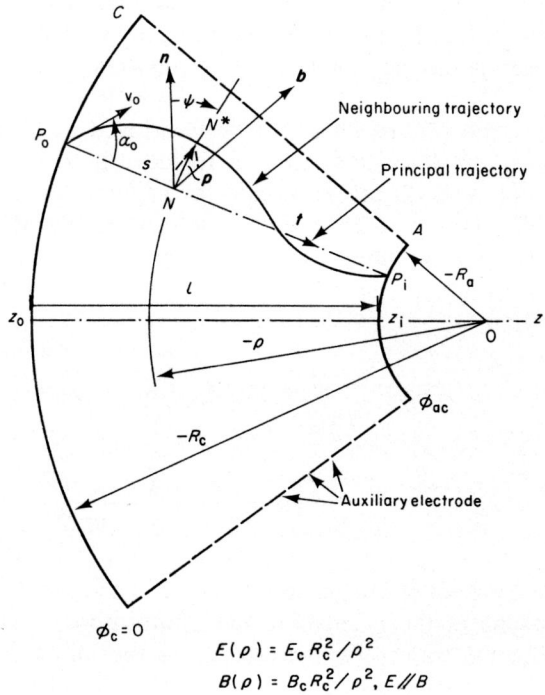

FIG. 2. The concentric, spherical electromagnetic focusing system ($n > 1$).

on the photocathode and a prime denote differentiation with respect to s. We introduce an orthogonal coordinate system with coordinate axes assigned to the tangential, normal and bi-normal directions t, n, b at point N. Then the trajectory of an electron from the photocathode with angle of emission α_0 and velocity of emission v_0 can be solved from the curvilinear "paraxial" equations, in which the reference axis is the principal trajectory whose curvature and torsion are equal to zero:

$$(p_2'\sqrt{\phi_*})' = -\frac{1}{4}\frac{\phi''}{\sqrt{\phi_*}}p_2 - \sqrt{\frac{e}{2m_0}}\left(\frac{B'}{2}p_3 + Bp_3'\right), \tag{1}$$

$$(p_3'\sqrt{\phi_*})' = -\frac{1}{4}\frac{\phi''}{\sqrt{\phi_*}}p_3 + \sqrt{\frac{e}{2m_0}}\left(\frac{B'}{2}p_2 + Bp_2'\right), \tag{2}$$

where p_2, p_3, ϕ and B are functions of the independent variable s; B and ϕ are the magnetic induction and the potential along the principal trajectory respectively; e and m_0 are the charge and mass of the electron; $\phi_* = \phi + \varepsilon_0 \cos^2 \alpha_0$, ε_0 is the so called "initial potential" of the emitted electron, $\varepsilon_0 = m_0 v_0^2/2e$; p_2 and p_3 are the projections of the neighbouring trajectory vector p on the normal n and binormal b of the principal trajectory.

Equations (1) and (2) are similar to those derived by Greenberg[3] and Sturrock,[4] but in our derivation $\phi + \varepsilon_0$ is replaced by $\phi + \varepsilon_0 \cos^2 \alpha_0$, and therefore Eqs. (1) and (2) are also valid for cathode lenses with axially symmetric fields whose reference axis is the central axis of the system.

If we express the modulus of vector p by p, the angle of rotation between p and n by ψ, and let $p_2 = p \cos \psi$, $p_3 = p \sin \psi$, then Eqs. (1) and (2) can be written as

$$\psi' = \sqrt{\frac{e}{2m_0}}\frac{1}{\sqrt{\phi_*}}\left(\frac{B}{2} + \frac{C}{p^2}\right), \tag{3}$$

$$\sqrt{\phi_*}(\sqrt{\phi_*}p')' = -p\left(\frac{1}{4}\phi'' + \frac{e}{8m_0}B^2\right) + \frac{e}{2m_0}\frac{C^2}{p^3}, \tag{4}$$

where C is the constant of integration. Because of the spherical symmetry of the system, any of the normals to the photocathode surface may be taken as a principal trajectory; we take z_0 as the initial point emitting electrons and let z be the distance from the initial point z_0. In this case, the replacement of $p(s)$, $\phi(s)$ and $B(s)$ by $r(z)$, $\phi(z)$ and $B(z)$ respectively will not affect the generality of the results.

Substituting the field equations of the said system:

$$\phi(z) = \frac{z}{nl - (n-1)z} \phi_{ac}, \tag{5}$$

$$B(z) = \frac{n^2 l^2}{[nl - (n-1)z]^2} B_c, \tag{6}$$

into Eqs. (3) and (4), we can obtain the solution for the "paraxial" electron trajectory with initial conditions v_0, α_0, $z_0 = 0$, $r_0 = 0$, $\psi_0 = 0$:

$$\psi(z) = \sqrt{\frac{e}{2m_0} \frac{B_c}{-E_c}} \left(\sqrt{\frac{z}{nl-(n-1)z} \phi_{ac} + \varepsilon_z} - \sqrt{\varepsilon_z} \right), \tag{7}$$

$$r(z) = 2\sqrt{\frac{2m_0}{e} \frac{\sqrt{\varepsilon_r}}{B_c}} \left(1 - \frac{n-1}{n}\frac{z}{l}\right) \sin\psi(z), \tag{8}$$

where $n = R_c/R_a$, R_c and R_a are the radii of the spherical photocathode C and the spherical anode (target) A, R_c and R_a are taken as positive from left to right, starting from the centre of curvature 0; B_c and E_c ($= -\phi_{ac}/nl$) are the magnetic induction and the electric field intensity at the photocathode respectively; l is the distance from the photocathode to target; $\varepsilon_z = \varepsilon_0 \cos^2 \alpha_0$ and $\varepsilon_r = \varepsilon_0 \sin^2 \alpha_0$ are the so called initial axial and radial potentials.

From $\sin\psi = 0$, namely, $\psi = m\pi$ ($m = 1, 2 \ldots\ldots i$, $i =$ number of loops), for the various stages of the imaging positions z_m, from Eq. (7) we have

$$z_m = \frac{2\sqrt{\frac{2m_0}{e}} \frac{m\pi}{B_c} \sqrt{\varepsilon_z} - \frac{2m_0}{e} m^2 \pi^2 \frac{E_c}{B_c^2}}{1 + \frac{2(n-1)}{\phi_{ac}} \left(\frac{m_0}{e} m^2 \pi^2 \frac{E_c^2}{B_c^2} - \sqrt{\frac{2m_0}{e}} m\pi \frac{E_c}{B_c}\sqrt{\varepsilon_z}\right)}. \tag{9}$$

From $\psi = (2m-1)\pi/2$, $r(z)$ in Eq. (8) tends to a minimum r_i at z_i, for which we may substitute $(\mu - 1)/2$ for m in Eq. (9). It follows that the electron trajectory is a conical helix, and the radius of rotation of the conical helix, inversely proportional to b_c or E_c, gradually decreases when $n > 1$, and increases when $n < 1$.

Let $m = i$, for the imaging distance z_i or the distance l from the photocathode to the target corresponding to the initial axial potential ε_{z1} of electrons emitted from the photocathode, we obtain

$$z_i = l = \frac{i\pi}{nB_c}\sqrt{\frac{2m_0}{e}}(\sqrt{\phi_{ac} + \varepsilon_{z1}} + \sqrt{\varepsilon_{z1}}). \tag{10}$$

In actual practice, owing to the fact that a very strong accelerating electric field is usually applied, the velocity of electrons arriving at the target is much greater than the initial electron velocity, i.e., $\phi_{ac} \gg \varepsilon_{z1}$. Then l can be expressed approximately by

$$l \cong i\pi \sqrt{\frac{2m_0}{e}} \frac{\sqrt{\phi_{ac}}}{nB_c}. \tag{11}$$

This is the focusing condition for the system. From Eq. (11), the value of $\sqrt{\phi_{ac}}/B_c$ can be fully determined when l, n, i are given.

Spherochromatic Aberrations

In what follows, the aberrations of the imaging system caused by different emission angle and different emission velocities are called "spherochromatic aberrations".

We define the focusing position $z_i = l$ expressed by Eq. (10) as an ideal image position corresponding to the initial axial potential ε_{z1}. Another "paraxial" electron with angle of emission α_0 and velocity of emission v_0 intersects the axis at point $z_{i\alpha 0}$. The axial deviation from the ideal image position is the so called longitudinal "paraxial" spherochromatic aberration Δz given by

$$\Delta z = z_{i\alpha 0} - l. \tag{12}$$

From Eq. (9) and considering Eq. (10), $z_{i\alpha 0}$ can be expressed by

$$z_{i\alpha 0} = \sqrt{\frac{2m_0}{e}} \frac{i\pi}{nB_c} \sqrt{\phi_{ac}}$$
$$\times \frac{\sqrt{1+(\varepsilon_{z1}/\phi_{ac})} - \sqrt{(\varepsilon_{z1}/\phi_{ac})} + 2\sqrt{(\varepsilon_z/\phi_{ac})}}{1-[2(n-1)/n](\sqrt{1+(\varepsilon_{z1}/\phi_{ac})})-(\sqrt{(\varepsilon_{z1}/\phi_{ac})})(\sqrt{(\varepsilon_{z1}/\phi_{ac})})}. \tag{13}$$

Substituting Eqs. (10) and (13) in Eq. (12) and neglecting the higher order terms of $(\varepsilon_0/\phi_{ac})^{3/2}$, we have

$$\Delta z = \Delta z_1 + \Delta z_2, \tag{14}$$

where

$$\Delta z_1 = \frac{2M^2}{E_c} \sqrt{\phi_{ac}} (\sqrt{\varepsilon_{z1}} - \sqrt{\varepsilon_z}), \tag{14a}$$

and

$$\Delta z_2 = \frac{2M^2}{E_c} \left\{ \frac{2(n-1)}{n} (\sqrt{\varepsilon_{z1}} - \sqrt{\varepsilon_z})^2 - \sqrt{\varepsilon_{z1}} (\sqrt{\varepsilon_{z1}} - \sqrt{\varepsilon_z}) \right\}, \tag{14b}$$

in which $M = 1/n$ is the linear magnification.

It can be seen from Fig. 3 that if a "paraxial" electron with initial conditions (v_0, α_0) in the ideal image plane at $z_i = l$ is rotated through an angle $\psi(l)$, the transverse deviation of the electron trajectory from the ideal image position in the image plane, i.e. the so called lateral "paraxial" spherochromatic aberration Δr can be written as

$$\Delta r = 2\sqrt{\frac{2m_0}{e}} \frac{\sqrt{\varepsilon_r}}{nB_c} \sin \psi(l). \tag{15}$$

Neglecting the higher order terms of $(\varepsilon_0/\phi_{ac})^2$, from Eq. (15) we have

$$\Delta r = \Delta r_1 + \Delta r_2, \tag{16}$$

where

$$\Delta r_1 = -\frac{2M}{E_c} \sqrt{\varepsilon_r}(\sqrt{\varepsilon_{z1}} - \sqrt{\varepsilon_z}), \tag{16a}$$

and

$$\Delta r_2 = \frac{M}{E_c\sqrt{\phi_{ac}}} \sqrt{\varepsilon_r}(\varepsilon_{z1} - \varepsilon_z). \tag{16b}$$

In Eqs. (14) and (16), Δz_1, Δz_2 and Δr_1, Δr_2 are the first order and the second order "paraxial" longitudinal and lateral spheromatic aberrations, which are divided according to the order of $(\varepsilon_0/\phi_{ac})^{1/2}$. Numerical analysis shows that Δz_1 and Δr_1 account for most of the total spherochromatic aberrations, while the second order "paraxial" spherochromatic aberrations Δz_2 and Δr_2 are higher order terms compared with Δz_1 and Δr_1.

It should be noted that when we expand Eq. (15), we must ensure

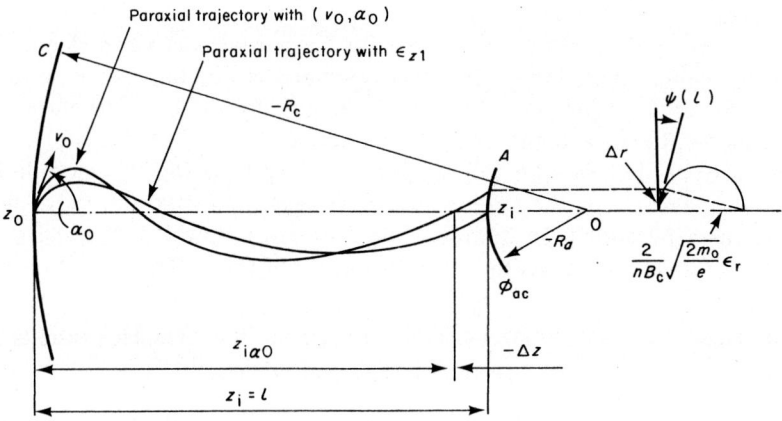

FIG. 3. The longitudinal and lateral paraxial spherochromatic aberrations.

that the angles of over- or under-rotation $\psi(l)$ are less than $\pi/4$. From Eq. (15), we obtain the conditional expression

$$\frac{B_c}{-E_c} < \frac{1}{2}\sqrt{\frac{m_0}{2e\sqrt{\varepsilon_0}}}\,\pi. \quad (17)$$

It can be proved from Eqs. (14a) and (16a), that the electrons emitted from photocathode with monochromatic ε_0 form a dense defocusing surface in the region between the image planes with $\varepsilon_{z1}=0$ and $\varepsilon_{z1}=\varepsilon_0$ separated by

$$\Delta z_{max} = \frac{-2M^2\sqrt{\phi_{ac}\varepsilon_0}}{E_c}. \quad (18)$$

In the plane of best focus of electrons, which corresponds to $\varepsilon_{z1}=0.3\varepsilon_0$, the minimum radius Δr_{min} of the circle of confusion in the electromagnetic focusing system can be expressed by

$$\Delta r_{min} = 0.6\frac{M}{-E_c}\varepsilon_0 = 0.6\frac{l}{\phi_{ac}}\varepsilon_0. \quad (19)$$

It may be seen that whether for the magnifying type or the demagnifying type of electromagnetic focusing system, the size of the circle of confusion only depends on l, ε_0 and ϕ_{ac}. This can be explained as follows. In the demagnifying system, as B_c is small, the initial radius of rotation is rather big, while the radius of rotation of the helix gradually decreases. In the magnifying system, as B_c is large, the initial radius of rotation is rather small, while the radius of rotation of the helix gradually increases. Finally, the size of the circle of confusion at the target surface in the two cases is the same.

The demagnifying tube suggested by Zacharov[1] was verified and the corresponding magnifying tube was calculated from the said equations. The magnifying tube resembles the demagnifying tube in structure, but the photocathode and the target exchange positions (Fig. 1(d)).

It is clear that we can follow Hartley[5] in calculating the modulation transfer function of the system, but it is suggested that one more term (16b) may be added. In addition, we also can follow Picat[6] and Beurle and Wreathall[2] in calculating the resolution of the system.

THE ELECTROMAGNETIC FOCUSING SYSTEM USING PARALLEL HOMOGENEOUS MAGNETIC AND ELECTRIC FIELDS

If both R_a and R_c tend to infinity, i.e., $n=1$, $E=$ const. and $B=$ const., all the formulae mentioned above can be applied to the case of the electromagnetic focusing system using parallel homogeneous magnetic and electric fields (Fig. 1(b)). It is well known that in this system, the electrons move with a constant acceleration in the axial direction, the transverse orbit is a circle returning to the axis in a period $T = 2\pi m_0/eB$, and the radius of rotation is $(m_0/e)(\dot{r}_0/B)$. Therefore, the electron trajectory describes a cylindrical helix whose pitch gradually increases. It is obvious that the solution for the real trajectories in this system is in accordance with the solution for the "paraxial" trajectories and the "geometric" spherochromatic aberrations δz and δr are equal to zero.

THE CONCENTRIC SPHERICAL ELECTROSTATIC FOCUSING SYSTEM

In the absence of a magnetic field, i.e., $B(z) = 0$, the concentric spherical electromagnetic focusing system becomes a purely electrostatic focusing system (Fig. 1(c)). As $\psi(z) = 0$, the electron trajectories are plane curves. Because

$$\lim_{\psi(z)=0} \frac{\sin \psi(z)}{\psi(z)} = 1,$$

from Eqs. (7) and (8) we obtain the solution for the "paraxial" trajectories under the initial condition (v_0, α_0) as

$$r(z) = 2[nl - (n-1)z] \frac{\sqrt{\varepsilon_r}}{\phi_{ac}} \left(\sqrt{\frac{z}{nl-(n-1)z} \phi_{ac} + \varepsilon_z} - \sqrt{\varepsilon_z} \right). \qquad (20)$$

Suppose the anode is an ideal "electron-transparent" mesh, and the space behind the anode is free from fields, when electrons pass through the mesh, their slope is unchanged. Then from Eq. (20) the distance from the centre of curvature 0 to the image position for electrons with ε_{z1} can be obtained:

$$R_i = -R_c \frac{1}{n-2} \left(1 + \frac{2(n-1)}{n-2} \sqrt{\frac{\varepsilon_{z1}}{\phi_{ac}}} + \frac{2n(n-1)}{(n-2)^2} \frac{\varepsilon_{z1}}{\phi_{ac}} \right), \qquad (21)$$

where R_i is taken as positive from left to right, starting from centre 0. It may be seen from Eq. (21) that a real image can be obtained only when $n > 2$. If $\varepsilon_{z1} = 0$, we can get the approximate expression for the image position

$$R_i \cong \frac{-R_c}{n-2} \qquad (22)$$

derived by Schagen et al.[7]

For the longitudinal "paraxial" spherochromatic aberration Δz, we obtain

$$\Delta z = \Delta z_1 + \Delta z_2, \qquad (23)$$

where

$$\Delta z_1 = \frac{2M^2}{E_c}\sqrt{\phi_{ac}}(\sqrt{\varepsilon_{z1}} - \sqrt{\varepsilon_z}), \qquad (23a)$$

$$\Delta z_2 = \frac{2M^2}{E_c}\left(\frac{n}{n-2}(\varepsilon_{z1} - \varepsilon_z) - \frac{4(n-1)}{n-2}\sqrt{\varepsilon_{z1}}(\sqrt{\varepsilon_{z1}} - \sqrt{\varepsilon_z})\right), \qquad (23b)$$

and for the lateral "paraxial" spherochromatic aberration Δr,

$$\Delta r = \Delta r_1 + \Delta r_2, \qquad (24)$$

where

$$\Delta r_1 = -\frac{2M}{E_c}\sqrt{\varepsilon_r}(\sqrt{\varepsilon_{z1}} - \sqrt{\varepsilon_z}), \qquad (24a)$$

$$\Delta r_2 = \frac{2M}{E_c\sqrt{\phi_{ac}}}\sqrt{\varepsilon_r}(\varepsilon_{z1} - \varepsilon_z), \qquad (24b)$$

in which M is the linear magnification, $M = R_i/R_c$.

If we use the solution for the real trajectories $r(z)$ under the initial conditions (v_0, α_0) we have

$$r(z) = \frac{2(n-1)\sqrt{\varepsilon_r}}{\sqrt{\phi_{ac}}\left(1 - 4(n-1)^2\frac{\varepsilon_r\varepsilon_z}{\phi_{ac}^2}\right)}\left[(z+R_c)\sqrt{\frac{\varepsilon_z}{\phi_{ac}}} - \frac{2z(n-1)\sqrt{\varepsilon_z\varepsilon_r}}{\phi_{ac}^{3/2}}\right.$$

$$\left. -(z+R_c)\left(\frac{-z}{(n-1)(z+R_c)} + \frac{(z+R_c)^2\varepsilon_z + z^2\varepsilon_r}{(z+R_c)^2\phi_{ac}}\right)^{1/2}\right]. \qquad (25)$$

The following "geometric" spherochromatic aberrations δz and δr, which are respectively, the axial deviation and the transverse deviation in the image plane of the real trajectory from the "paraxial" trajectory with the same initial conditions should be added to Eqs. (23) and (24) respectively.[8]

$$\delta z = \frac{2M^2}{E_c}(n-1)\varepsilon_r, \qquad (26)$$

$$\delta r = -\frac{2M}{E_c\sqrt{\phi_{ac}}}(n-1)\varepsilon_r^{3/2}. \qquad (27)$$

It is clear that δz, Δz_2 and δr, Δr_2 are of the same order of magnitude respectively.

In Table I the calculated results for the "paraxial" trajectory and the

TABLE I

The solutions for the "paraxial" and real trajectories in two concentric electrostatic focusing systems

Tube parameters	Distance from the photocathode z(mm)	Potential along axis $\phi(z)$ (v)	"Paraxial" trajectory		Real trajectory	
			$r(z)$ (mm)	$r'(z)$	$r(z)$ (mm)	$r'(z)$
$R_c = -25$ mm	4	476·1705	0·3666056	0·0370970	0·3666200	0·0371337
$R_a = -5$ mm	8	1176·4705	0·4664752	0·0154348	0·4665200	0·0154706
$\phi_{ac} = 10\,000$ V	12	2307·6923	0·4995984	0·0016013	0·4996921	0·0016205
$\varepsilon_0 = 1$ V	16	4444·4444	0·48	−0·0116666	0·4801206	−0·0116827
$\alpha_0 = 90°$	20	10 000·0000	0·4	−0·03	0·4003198	−0·0300886
$R_c = 5$ mm	4	5555·5555	0·1073312	0·0193792	0·1073332	0·0193771
$R_a = 25$ mm	8	7692·3076	0·1824272	0·0184181	0·1824326	0·0184107
$\phi_{ac} = 10\,000$ V	12	8823·5294	0·2554992	0·0181604	0·2555067	0·0181529
$\varepsilon_0 = 1$ V	16	9523·8095	0·3279008	0·0180540	0·3279124	0·0180437
$\alpha_0 = 90°$	20	10 000·0000	0·4	0·0180000	0·4000128	0·0179810

real trajectory in two electrostatic focusing systems ($R_c = -25$ mm, $R_a = -5$ mm, $\phi_{ac} = 10\,000$ V and $R_c = 5$ mm, $R_a = 25$ mm, $\phi_{ac} = 10\,000$ V) with the initial conditions ($\alpha_0 = 90°$, $\varepsilon_0 = 1$ V) are given. Numerical calculation shows that the solution for the "paraxial" trajectories (Eq. (20)) and solution for the real trajectories (Eq. (25)) are very close to each other.

THE PROXIMITY FOCUSING SYSTEM

For the concentric spherical electrostatic focusing system, we assume that both R_a and R_c tend to infinity, i.e. $n = 1$. Equation (20) then reduces to the solution for the proximity focusing system (Fig. 1(a)). It is obvious that the electron trajectory is a parabola and the electrons emitted from the photocathode with ε_{z1} form a virtual image behind the cathode. Let l_i be the distance from the photocathode to the virtual image point, we have

$$l_i = |(R_i - R_c)|_{R_c \to \infty, \, R_a \to \infty}, \quad (28)$$

where l_i is taken as negative from right to left, starting from the photocathode.

Substituting Eq. (21) into Eq. (28), and using $R_a - R_c = l$, $n = R_c/R_a = (1 - l/R_a)_{R_a \to \infty}$, we obtain

$$l_i = -l\left(\sqrt{1 + \frac{\varepsilon_{z1}}{\phi_{ac}}} - \sqrt{\frac{\varepsilon_{z1}}{\phi_{ac}}}\right)^2. \quad (29)$$

The spherochromatic aberration formulae, Eqs. (23) and (24) of the concentric spherical electrostatic focusing system are entirely suitable to the proximity focusing system with $n = 1$. Under this condition, δz in Eq.

(26) and δr in Eq. (27) will both equal zero. This means that the solution for real trajectories in this system is in accordance with the solution for "paraxial" trajectories.

THE TRI-ELECTRODE FOCUSING SYSTEMS WITH GRID

The Tri-electrode Concentric Spherical Electrostatic Focusing System with Spherical Grid

If we insert a spherical grid (mesh) with radius of curvature R_s in front of the photocathode of the bi-electrode concentric spherical electrostatic focusing system, it becomes a tri-electrode concentric spherical electrostatic focusing system. Based on the above discussions, the distance R_i from the centre of curvature to the image positions for electrons with ε_{z1} can be written as

$$R_i = R_c M_0 \left(1 - 2r(y-1)M_0 \sqrt{\frac{\varepsilon_{z1}}{\phi_{sc}}} + [M_0 C_0 + 4M_0^2 r^2(y-1)^2]\frac{\varepsilon_{z1}}{\phi_{sc}}\right), \quad (30)$$

where $y = R_c/R_s$, $r^2 = \phi_{ac}/\phi_{sc}$, ϕ_{sc} is the potential of the grid relative to the photocathode, M_0 is the linear magnification of zero order, derived by Linden and Snell[9] as

$$M_0 = -\frac{r+1}{r(n-2)+2r^2(y-1)-n}, \quad (31)$$

and C_0 is a parameter dependent on the structure and applied potentials given by

$$C_0 = r(y-1) - \frac{n-y}{r+1} + \frac{(n-1)+r(y-1)}{r(r+1)}. \quad (32)$$

From Eq. (30) we can study the focusing characteristics and the aberrations. For example, from Eq. (31) we can derive the following condition, in which a real image will be formed:

$$n > y > 1 + \frac{n - r(n-2)}{2r^2}. \quad (33)$$

In addition, we can prove that the first order "paraxial" spherochromatic aberrations Δz_1 (Eq. (14a)) and Δr_1 (Eq. (16a)) are also valid in the tri-electrode electrostatic focusing system.

Finally, it should be pointed out that we can follow Schagen et al.[7] to discuss the case of electrostatic focusing systems in which the anode has an aperture.

The Tri-electrode Electromagnetic Focusing System using Parallel, Homogeneous Magnetic and Electric Fields with a Plane Grid

For the tri-electrode electromagnetic focusing system using parallel, homogeneous magnetic and electric fields with a plane grid (mesh), as mentioned above, good focus may be obtained at the target if the angle of rotation ψ is an integral multiple of π, i.e., $\psi(l_{Cs}) + \psi(l_{sa}) = i\pi$. Thus we have the focusing condition for this system. The magnetic induction B can be written as:

$$B = i\pi \sqrt{\frac{2m_0}{e}} \left[\frac{l_{cs}}{\sqrt{\phi_{sc}}} \left(\sqrt{1 + \frac{\varepsilon_{z1}}{\phi_{sc}}} - \sqrt{\frac{\varepsilon_{z1}}{\phi_{sc}}} \right) \right.$$
$$\left. + \frac{l_{sa}}{\sqrt{\phi_{as}}} \left(\sqrt{1 + \frac{\phi_{sc} + \varepsilon_{z1}}{\phi_{as}}} - \sqrt{\frac{\phi_{sc} + \varepsilon_{z1}}{\phi_{as}}} \right) \right]^{-1}, \quad (34)$$

where ϕ_{sc} is the potential of the mesh relative to the photocathode, ϕ_{as} is the potential of the target relative to the mesh, l_{cs} is the distance from the photocathode to the mesh and l_{sa} is the distance from the mesh to the target. Let $\varepsilon_{z1} = 0$, transform the unit of B from Wb m^{-2} to Gauss, and leave the other values in SI units, then we get

$$B = \frac{0 \cdot 106 i}{\frac{l_{cs}}{\sqrt{\phi_{sc}}} + \frac{l_{sa}}{\sqrt{\phi_{as}}} \left(\sqrt{1 + \frac{\phi_{sc}}{\phi_{as}}} - \sqrt{\frac{\phi_{sc}}{\phi_{as}}} \right)} G \quad (35)$$

as was derived by Linden and Snell.[9]

It can be proved that the first order "paraxial" spherochromatic aberrations, Eqs. (14a) and (16a), are also valid for this system.

Conclusions

The analytic solution of the paraxial trajectories and the expressions of the spherochromatic aberrations for four typical electrostatic and electromagnetic focusing systems are derived. The forms of the first order "paraxial" spherochromatic aberrations for electrostatic and electromagnetic focusing imaging systems are proved to be the same.

The results of this paper can be used to design electromagnetic focusing image tubes and to calculate the initial electron trajectories in image tubes.

Acknowledgments

The author wishes to thank the Director of the North Industries Corporation for permission to publish this paper. Thanks are also due to the Optical Engineering Department of the Peking Institute of Technology for support and cooperation.

REFERENCES

1. Zacharov, B., *In* "Adv. E.E.P." Vol. 16, p. 99 (1962).
2. Beurle, R. L. and Wreathall, W. M., *In* "Adv. E.E.P." Vol. 16, p. 333 (1962).
3. Greenberg, G. A., *Reports of the Academy of Science USSR.* **37,** 295 (1942); **38,** 89 (1943).
4. Sturrock, P. A., *Philos. Trans. R. Soc. London A* **245,** 155 (1952).
5. Hartley, K. F., *J. Phys. D* **7,** 1612 (1974).
6. Picat, J. P., *Astron. & Astrophys.* **11,** 257 (1971).
7. Schagen, P., Bruining, H. and Francken, J. C., *Philips Res. Rep.*, **7,** 119 (1952).
8. Chou, L. W., *In* Abstracts of the reports on the 21st Science-technical Conference, Leningrad, p. 137 (1966).
9. Linden, S. R. and Snell. P. A., *Proc. IRE.* **45,** 513 (1957).

Catadioptric Electron Optics Using a Retarding Electrostatic Field and its Application to the Development of Short Image Tubes of Very High Performance

R. EVRARD[†]

National Physical Research Laboratory, C.S.I.R., P.O. Box 395, Pretoria, South Africa

General Description of the Tube

The system (Fig. 1), which has been named the "Fountain tube", was first described in 1975.[1] Basically it consists of a spherical cathode with a radius of curvature R_c, an L-shaped auxiliary electrode, a first flat anode fitted with a circular axial aperture having a radius R_a, a second flat anode having a small axial hole (diameter 0·3 mm) and an output window coated with a transparent semiconductive layer, located at a distance D from the second anode. The centre of the aperture in the first anode coincides with the centres of curvature of the cathode. The axial hole of the second anode is located at the "crossover" of the electron optical system. A "naked" (uncoated) fluorescent screen is deposited on the second anode, on the side facing the output window.

The cathode and the output window are earthed when the second anode (screen) is at the maximum positive potential, V_c. The potential V_a of the first anode is adjusted in such a way that the corresponding equipotential surface across the aperture is a part of a spherical surface so as to form a virtual spherical anode concentric with the cathode. This ensures that the resultant equipotential surfaces between the cathode and the *virtual* anode are almost perfectly spherical, at least in the space where the electrons travel.

The electrons emitted by the cathode pass through the screen hole at the "crossover" and are forced back onto the screen by the uniform retarding field E, applied between the output window and the screen. For a given value of V_c, the value of E is determined by the focusing conditions. D is then given by $D = V_c/E$.

[†] Now with Riber S.A., B.P. 231, 92505 Rueil, France.

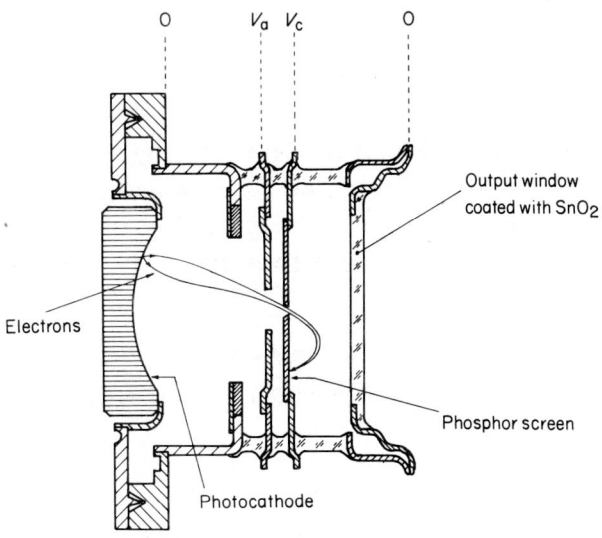

Fig. 1. Catadioptric image intensifier tube ("Fountain" tube).

Electron Optics

Detailed trajectory computations have already been published.[1,2] The method and the results will be summarized here.

Between the cathode and the virtual anode, the equipotential surfaces are concentric spheres. The electrons travel in a central field of force and the trajectories are determined by Kepler's law of equal areas, the energy equation and the initial conditions. This part of the system acts as a thick convergent lens.

Between the virtual anode and the screen, the potential distribution along the axis can be approximated by a parabolic function and the trajectories in the Gaussian approximation can be easily computed. This part of the system is again a convergent lens. In addition, the strong axial component of the electric field reduces the angle of the trajectories with respect to the axis. The hole in the screen acts as a thin divergent lens and increases the depth of focus. Its effect can also be easily computed. Between the screen and the output window, the retarding field E is uniform and the trajectories are parabolic.

Simplified results of the computations are as follows. The crossover (i.e., the screen) is located at a distance Z_c from the cathode.

$$Z_c = R_c + R_a(l-1) \tag{1}$$

where l is a dimensionless constant ($l \simeq 1\cdot 20$).

The ratio V_c/V_a between the screen and the anode potentials is given by

$$\frac{V_c}{V_a} = 1 + \frac{nl}{n-1}(l+1), \tag{2}$$

where $n = R_c/R_a$.

The retarding field E, the distance D between the screen and the output window, and the magnification factor M are given by

$$E = 4P\frac{(V_c V_a)^{1/2}}{R_c}, \tag{3}$$

$$D = \frac{R_c}{4P}\left(\frac{V_c}{V_a}\right)^{1/2}, \tag{4}$$

and

$$M = \frac{1}{P}\left(1 + \frac{n}{2(n-1)}\right)^{1/2}, \tag{5}$$

where

$$P = \left(\frac{V_a}{V_c}\right)^{1/2}\frac{n^2(2l+1)}{2(n-1)} - (2n-1). \tag{6}$$

Discussion

With the exception of the L-shaped electrode, the geometry of the tube is fully computable starting from chosen values for the magnification factor M and the photocathode radius R_c, as follows: P and the ratio $n = R_c/R_a$ are derived from Eqs. (5) and (6). The crossover position, i.e., the screen position, is then determined by Eq. (1) and the voltage ratio V_s/V_a by Eq. (2). Finally, D is given by Eq. (4).

The expression for Z_c is computed for the electron leaving the cathode at any point (R_c, ψ_0). The initial coordinate ψ_0 does not appear in the equation and the crossover position is well defined. The same remark applies to M. More detailed computations show that the image surface is nearly spherical and since the value of M does not depend on the initial position of the electron, this image is similar to the input image on the spherical cathode. Actually, the screen is flat and the image on the phosphor is a projection of the true spherical image. Due to the accelerating field between the anode and the screen the angles that the trajectories make with the axis are small at the screen surface. Thus the screen image is a simple transfer, parallel to the axis, of a sphere onto a plane. The distortion introduced by this transfer cancels the inverse distortion corresponding to the transfer of the plane optical image at the input onto the spherical cathode through the fibre optics.

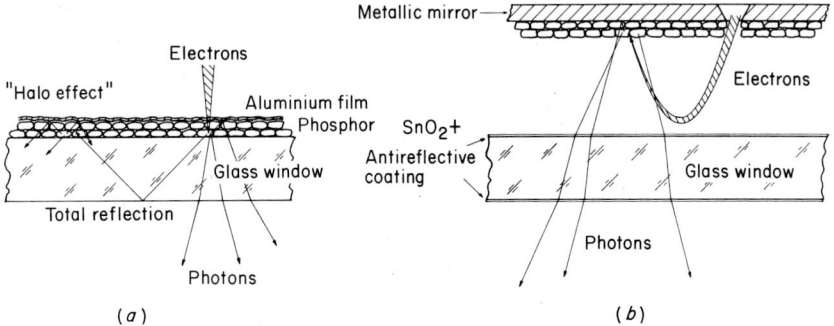

Fig. 2. Fluorescent screen structures. (a) Conventional screen deposited on the output window. (b) Screen on polished metallic substrate in Fountain tube.

It must be stressed that the above formulae are only a first approximation. The results were optimized using numerical methods.[3]

Electron Transmission Through the Screen Aperture

The total relative number of electrons transmitted through a hole, having a radius ρ_s and located at the crossover, can be computed easily, assuming a Gaussian distribution of the initial energies and a Lambertian emission. Taking into account the spectral distribution of the light assuming starlight conditions and the cathode spectral sensitivity we find approximately

$$I \simeq \frac{W_t - 0.1}{0.8} + \frac{1}{0.8(\pi - 2)} \int_{W_t}^{0.9} \left[2\cos\left(\frac{\pi}{2}\frac{W_t}{W_m}\right) + \pi \frac{W_t}{W_m} - 2 \right] dW_m,$$

where

$$W_t = \frac{V_a}{R_c^2}\left(1 + \frac{R_c}{2d}\right)\rho_s^2.$$

For $\rho_s = 0.15$ mm $I = 98\%$; for $\rho_s = 0.1$ mm $I = 80\%$.

There are unavoidable tolerances in centring the electrodes, and it seems that the minimum value acceptable for ρ_s is $\simeq 0.12$ mm.

Fluorescent Screen

Theoretical considerations as well as measurements show that most of the light produced in a screen phosphor emerges on the beam side.[4] In a conventional screen (Fig. 2(a)), the phosphor is deposited on the output window of the tube and is backed with an aluminium film, preventing

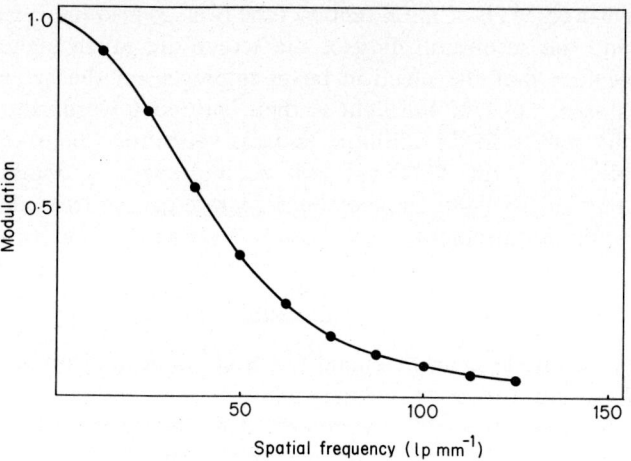

FIG. 3. MTF of the Fountain tube.

light feedback to the cathode. The electrons lose $\simeq 10\%$ of their kinetic energy in penetrating this aluminium layer. Most of the light has to be reflected towards the observer and as the aluminium film is rather a poor reflector, nearly 20% is absorbed. Furthermore, 10% more losses are due to reflections at the output window. It should be noted that the contrast of the image in this type of screen is limited by, among other factors, the "halo" effect, which is partly due to total reflections at the output interface (Fig. 2(a)).

In the Fountain tube, the screen is deposited on a thick metallic substrate (Fig. 2(b)). The phosphor is uncoated and separate from the output window. The consequences are obvious: the electrons are not scattered and they do not lose any kinetic energy before entering the phosphor. As the metallic substrate can be highly polished, the light is almost perfectly reflected towards the observer. Total reflection cannot occur on the output window and the "halo" effect is reduced. Partial reflection can be easily prevented using an anti-reflective coating on the window. (Such a coating would be inefficient or even harmful in the conventional structure since the phosphor particles are not in continous contact with the glass substrate.)

Due to all these factors, the light output of the Fountain tube screen should be improved by at least 40%. The prototypes have a brightness gain four times larger than similar conventional first generation image intensifiers having the same cathode sensitivity. The matter is still being investigated.

The measured MTF of the Fountain tube (Fig. 3) also indicates that the contrast and the resolution limit of the screen are much higher, mainly due to the fact that the electron beam impinges on the screen on the observer's side; most of the light is then emitted towards the observer without any reflection. In addition, there is very little "halo" effect. For some applications, the MTF can still be improved by eliminating the reflected part of the light: the metallic substrate can be treated according to any specific requirement.

Technology

The photocathode is an extended red S·20, deposited on a fibre optic window by using a transfer technology.[5]

The screen phosphor (P·20) is deposited by the "wet settling" method on a carefully polished metallic electrode. Special care is taken to eliminate any phosphor particles in the screen hole.

The output window's inner face is coated with a transparent semiconductive layer of SnO_2.[6] A transmission of $\simeq 97\%$ for $\lambda = 550$ nm is obtained with anti-reflective coatings. In order to eliminate the light transmitted through the cathode, all the electrodes are blackened by oxidation.

Applications

The simplified theory of the Fountain tube indicates that the magnification factor M can be, in principle, chosen from 0 to ∞. Actually, the minimum value for M is fixed by practical considerations: the "blind" area corresponding to the central aperture in the screen must be very small, compared with the useful area of the image. For most applications, it seems that for $M = 0.5$, the central black spot is still relatively negligible.

The characteristics and typical performance of the prototypes having this magnification factor are listed in Table I.

Due to its very short length, high gain, and high resolution, a Fountain tube with a magnification factor equal to 0·5 is well adapted to the development of night vision goggles. It would be also interesting to apply the Fountain tube concept to the development of image intensifiers with a very large magnification factor ("display" tubes). With a cathode having a diameter equal to 20 mm and a magnification factor equal to 5, the image would have a diameter of 100 mm and the total length of the tube would be only 50 mm, several times smaller than for conventional tubes having the same magnification factor. As the diameter of the hole in the screen

TABLE I
Parameters of the prototype Fountain tube

Magnification factor M	0·5
Radius of curvature of the cathode	16 mm
Useful diameter of the cathode	20 mm
Optical length of the tube (from fibre optics input to screen)	$\simeq$20 mm
Total length	$\simeq$25 mm
Cathode sensitivity	$\simeq$300 μA lm^{-1}
Brightness gain at 15 kV	$\simeq$550 cd m^{-2} lx^{-1}
Resolution limit on the screen (2 mm off centre)	$\simeq$140 lp mm^{-1}
Maximum distortion	$\simeq$2%

does not depend on the magnification factor and could still be of the order of 0·3 mm, the central black spot would hardly be noticeable.

It must be emphasized that the Fountain tube can be easily zoomed by applying a negative voltage on the output window, thus reducing at will the effective distance D. The anode potential V_a has to be adjusted for each value of M.

In considering the other possible applications, one should keep in mind that the Fountain tube can only be used in a single stage system or as the last tube of a cascaded system.

Conclusion

The use of a catadioptric electron optical system drastically shortens an image intensifier tube: the cathode–screen distance is equal to the crossover distance (18 mm if the radius of curvature of the cathode is 16 mm).

The combination of a spherical and a parabolic distribution of the field allows for a distortion-free image with a high resolution (Figs. 4, 5 and 6). The magnification factor can be chosen from a very wide range and the tube can be zoomed.

As the design allows the use of an uncoated fluorescent screen deposited on a perfect mirror, the brightness gain is much larger than for a conventional first generation tube. The MTF is also very much improved with a resolution limit of 150 lp mm^{-1} on the screen. It should be noted that the uncoated screen is simpler to make, stronger, and better protected against accidental flashes than the conventional aluminium backed screen. The loss of information due to the central black spot on the image, corresponding to the screen hole ($\sim$0·3 mm diameter) is negligible, at least if the magnification factor is equal to or larger than 0·5.

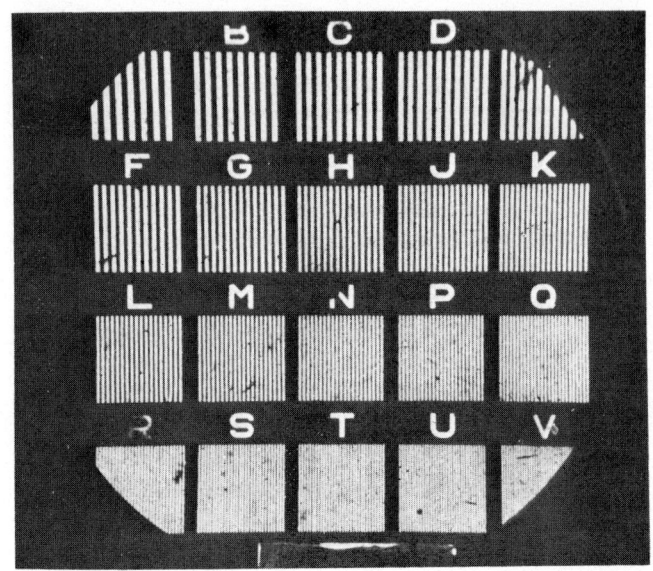

Fig. 4. Resolution test.

Fig. 5. Distortion test.

FIG. 6. Low light level image taken with the Fountain tube.

ACKNOWLEDGMENTS

The author wishes to thank Drs G. J. Ritter and M. McDowell, Messrs C. F. van Huyssteen, W. van der Berg, W. de Beer, D. Chinnery, and A. G. du Toit for their assistance in the development of the prototypes.

REFERENCES

1. Evrard, R. Thèse de doctorat d'Université n° 186 Université Paris-Sud. (1975).
2. Schagen, P., Bruining, H., and Francken, T. C., *Philips Res. Rep.*, **7,** 119 (1952).
3. Du Toit, A. G., *In* "Adv. E. E. P." Vol. 40A, p. 485 (1976).
4. Bril, A., and Klasens, H. A., *Philips Res. Rep.* **7,** 401 (1952).
5. Van Huyssteen, C. F., *In* "Adv. E. E. P." Vol. 40A, p. 419 (1976).
6. Gomer, R., *Rev. Sci. Instrum.* **24,** 993 (1953).

Review of Image Intensification and Conversion

G. O. TOWLER

The Research & Development Establishment, British Relay Ltd, Leatherhead, Surrey, England

IMAGE INTENSIFIERS

I will begin this section by referring to the paper by Cromwell and Angel† which will no doubt be consulted by many as a source of practical techniques to eliminate corona discharge and high voltage tracking. While many of these techniques are familiar to those associated with the tube industry the article fills a void in an area where such information is seldom available on manufacturer's data sheets. However, some of the points such as whether it is better to insulate high voltage surfaces or to electrically stress them in a well defined manner with electrostatic shielding, may be controversial. It is interesting to see how the techniques were so effective in holding off the overall tube voltage on a double sided, tin oxide coated faceplate, and in the partial holding off of the stage potentials across the interstage coupling.

The paper by Houtkamp and Mulder‡ on scattered light in proximity focused image amplifiers was presented at the Symposium in a detailed way that attracted considerable interest. It deals with three well known sources of light induced background which produce considerable contrast loss in proximity diodes. viz.:

(1) Light transmitted through the photocathode and reflected off the aluminium backing of the screen.

(2) Light from the phosphor transmitted through the aluminium backing and picked up by the photocathode.

(3) Photoelectrons backscattered from the screen backing and returned to the screen by the retarding field with a considerable displacement.

It points out that effects (1) and (2) can be reduced by a light absorbent layer on the aluminium backing. To this end, layers of aluminium smoke

† See p. 183.
‡ See p. 159.

and silicon were tried, and in addition to their optical absorption the electron reflection from these layers was measured by an indirect method.

An interesting example of an image intensifier being used for image processing is described by Geluk.† Here a special image intensifier was used to sharpen blurred images by deflecting them across a mask corresponding to the point spread function of the degrading system in the manner of an image dissector. The mean output current from a photomultiplier placed behind the mask generated the video signal for the sharpened image which was displayed on a picture monitor. The illustration given of typescript is impressive.

The limitations of such a system are well explained by the author who points out that because the device is in essence a Farnsworth image dissector (with a special aperture) providing no storage, it is necessary to operate at a high current level to obtain a satisfactory signal to noise ratio. However, it is not made clear that this level could be prohibitively high. The output phosphor of the intensifier is yttrium silicate to provide a fast decay; obviously any appreciable afterglow will introduce an unwanted asymmetrical aperture response into the system.

The ultimate limitation of the system will be set by the modulation transfer function of the image intensifier and the corresponding point spread function. To a lesser extent the lens will introduce a similar effect. Accurate MTF measurements could be made for both the intensifier and the lens, and these could be used to correct the point spread function used in the mask.

The system depends on accurate knowledge of the degrading process and the ability to physically reproduce this in the form of a mask. There are problems where the response takes on both positive and negative values, and in this situation the image is split and separate masks and photomultipliers are used. There are obvious limitations to this approach.

The use of this convolution technique in transverse analogue tomography is mentioned. This offers an alternative approach to computer tomography which is referred to in the next section.

X-Ray Applications of Image Intensification and Conversion

Judging by the numbers of papers in this field presented at the Symposium, there is a general interest at the present time in X-ray applications. The subjects fall under three headings:

(1) X-ray crystallography.

(2) Conventional X-ray radiology.

(3) X-ray computer tomography.

† See p. 239.

X-Ray Crystallography

The papers presented in this field are typical of the interest over recent years in the application of image intensification to this particular problem. Unfortunately some workers in the field were expecting the same resolution and image quality from an intensification system used in real-time as they were used to getting from X-ray film such as Ilford Industrial G, where exposures may have taken several hours. A simple calculation based on the X-ray fluxes from typical generators shows that the quantum noise limited resolution is relatively poor for the integration times involved in real-time operation. However, I still believe that there is a useful role for image intensification systems in real-time applications where resolution can be sacrificed for speed. A good example is as an alignment or monitoring aid used to orientate the specimen. If this has to be done photographically it can require a number of rather lengthy exposures just to get the specimen correctly positioned. High resolution is not required for monitoring because quantitative results are not asked of it, and only the barest outline of an image may be sufficient.

Conventional X-ray phosphor screens are unsuitable for this application where the X-ray quantum energy is rather low (≈ 8 keV for CuKα characteristic radiation) and a rather thin, high resolution screen is necessary. Arndt and Gilmore† do not claim to have optimized their screen, but have used a specially prepared polycrystalline ZnS(Ag) phosphor. Beauvais and Mathiot‡ have investigated a number of thin screens based on P·20, P·43, etc., and the best quantum yield they quote is for a 16 μm P·43 screen where the figure was 86% when coupled to an S·20 photocathode. Unfortunately no resolution figures are given for the screen, but reference is made to a 12 μm thick CsI(Tl) layer where a resolution of 30 μm was recorded. It was inferred that the resolution was MTF limited and that this could be improved to about 10 μm with a magnifying coupling of times 2 or 3.

The paper by Arndt and Gilmore gives a clear and concise exposition of the problems that can be encountered in developing a system. It is clear that they have considered most aspects of the problem very carefully. The system presented at the Symposium in 1974 used an isocon, overcoming the inherent integration problems by employing a quantum counting system where the integration was done in digital memory after analogue to digital conversion. A 1 : 1 fibre optically coupled intensifier was added to the isocon to provide additional gain so that the camera tube was operating in the quantum counting mode. It is interesting to see

† See p. 209
‡ See p. 217

that in the current paper the intensifier has been changed from unit magnification to a 2:1 demagnification type with an 80 mm input. The increase in format size is desirable so that the operational distance from the specimen can be increased. The choice of an intensifier with an even larger format is proposed (100 mm input). This device will contain a microchannel plate so that less sensitive camera tubes can be used.

While Beauvais and Mathiot in their X-ray topography application have chosen to magnify from a small format on to the input intensifier or camera tube, Arndt and Gilmore use as large a screen format as possible for ease of working at a reasonable distance from the X-ray specimen. Beauvais and Mathiot claim that it is desirable to use magnifying optics whenever the X-ray flux is high enough to permit it, because they maintain that the resolution of their X-ray screens is such that the limitation is set by the MTF of either the image intensifier or the camera tube. They report that a resolution of about 30 μm has been achieved using a 12 mm thick CsI(Tl) layer coupled to a TH 9659 Nocticon tube and claim that the flux of the synchrotron radiation was sufficiently high that they would have been able to magnify by 2 to 3 times and achieve real time imaging of 10 μm details. Since the input field of the TH 9659 is only 17 mm this would have reduced the input field at the screen to about 5·5 mm which might pose operational problems.

Messrs Beauvais and Mathiot are to be congratulated on the high resolution and efficiency of their phosphor screens, but unless they are operating with a very intense X-ray source such as synchrotron radiation, the quantum noise limited resolving power will be well below these resolution figures. Although the equivalent quantum efficiency of a photoelectronic system can be higher than that of an X-ray emulsion by say a factor of five,[1] as mentioned earlier the detail that can be obtained on an X-ray emulsion after a long exposure cannot be expected in a real time application if the exposure time is decreased by more than the above factor.

The paper by Bateman and Apsimon† describes a new photocathode for X-ray image intensifiers operating in the 1–50 keV region. The device is based on the use of low density alkali halide layers for the direct conversion of the X-ray energy into electron energy, and because of the low stopping power of the layers it is most attractive for soft X-rays such as CuKα radiation. To achieve good quantum efficiencies at these relatively low X-ray energies the authors have found it necessary to use a low density layer ~0·5 mm thick and of 5% bulk density; similar layers used for 7 keV electrons are typically 20 μm thick and 2% of the bulk density.

† See p. 189.

A relatively thick support membrane (5 μm of aluminium foil) can be used compared with the 50 nm of aluminium oxide typically employed on electron sensitive layers. It is interesting to see that the Bateman and Apsimon favoured CsI in preference to KCl because the layers could be demounted to moist atmosphere with little poisoning. I would like to point out that a similar use of CsI was reported by Edgecumbe and Garwin[2] who produced demountable low density layers for relativistic electron detectors.

The predicted resolution of 10 lp mm^{-1} for a 0·5 mm thick layer looks impressive, and it will be interesting to see the outcome of the experimental investigation of how spatial resolution varies with layer thickness. While it seems to be an efficient high resolution cathode for X-ray crystallography there may be some problems in using the large area layer (120 mm diameter) if the membrane has to be stressed in a field of 3 kV mm^{-1}.

Conventional Medical X-ray Radiology

Driard et al.[3] described a system which, superficially, appears to be another X-ray image converter system with a vidicon-type TV readout. However, the fact that the large area X-ray image intensifier and the EBIC silicon diode array vidicon are constructed in one vacuum envelope makes the system technologically interesting. This approach is designed to overcome the shortcomings of relay lens coupling and the problems associated with fibre optic defects. The electron optical image is decelerated between the anode and the target so that a high value of accelerating voltage can be used to produce a sharp image and yet the silicon target can be operated at optimum charge gain. Although only a modest spatial frequency resolving power is required at the input, this translates in the video readout to a high number of TV lines per picture height, which raises the question of whether the silicon diode matrix target is best suited to provide the high line numbers demanded of these very large input fields. The MTF of the different sections of the device are reproduced in Fig. 1 (from the paper by Driard et al.[3]). It was reported that the design was only in the development stage, and it will be interesting to see if the device becomes commercially viable.

Although the paper by Rougeot et al.† on an image device for gamma cameras is not exactly in the field of X-ray radiology, it is so closely related that it seems appropriate to mention it here. The authors have been

† See p. 227.

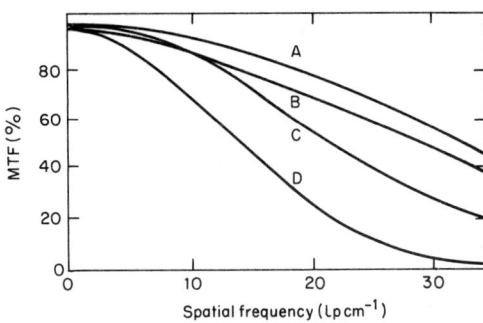

Fig. 1. MTFs (referred to input) of the sections of the X-ray image convertor described by Driard et al.[3] A, silicon target; B, image section; C, electron beam; D, overall MTF.

associated with these developments over a number of years and papers have been presented at earlier Symposia.[4,5]

The first stage of the gamma camera is very similar to that reported[5] in 1976, namely an electrostatic image convertor tube with a large diameter input photocathode faceplate approximately 35 cm diameter against which a single crystal scintillator is brought into contact. The demagnifying electron optics produce an image on a phosphor screen which has a fast time response. The output faceplate is a fibre optic for coupling to the second stage which is an electrostatic inverter tube with a large area silicon target at the output. It is the novel way in which this silicon diode is utilized that is the main feature of the device. The target not only provides a charge gain of approximately 4000 for the bombarding photoelectrons but, by the utilization of pairs of perpendicular conducting strips on the two surfaces of the diode, it has been possible to determine the coordinates and amplitude of a given scintillation. This method has replaced the multiple photomultiplier location technique reported in earlier papers. The advantages achieved by this method are higher spatial resolution and energy resolution. However, to achieve good energy resolution, the gain uniformity must be good. This has been accomplished by correcting the non-uniformity between stages by a compensation filter. Some degradation in performance has been introduced because the compensating filter is not perfect, and in the future there is the prospect of improving this by doing the compensation electronically.

The silicon diode localizer also offers perfect energy linearity and better multiplication statistics for the charge gain than the channel plate multipliers used in earlier devices. The time constant of the localizer is of the order of that of the tube (approximately 3 μsec) and this should allow 10^5 counts sec^{-1} with less than 20% of the counts lost.

From a practical standpoint it has been considered necessary to perform the inter-stage coupling through a thick fibre optic block. Although this must introduce some image impairment the fact that it can hold off the EHT of the combined tube supplies means that both the scintillator input and the localizer output can be at earth potential.

In conclusion, the introduction of the localizer into the system appears to have greatly simplified the operation of the camera.

Lastly, in this section I would like to refer to the paper by Washida and Sonoda† on a high resolution X-ray phosphor screen. This screen is based on sodium activated CsI that was first reported in thin film form by Bates,[6] although earlier work has been carried out on bulk materials for scintillators.

The main improvement made in the screens by Washida and Sonoda is the MTF performance. This has been improved by cutting down the transverse spread of light by forming a block structure in the CsI layer so that the light is piped through these blocks by total internal reflection. The blocks are less than 100 μm in size so high resolution properties are achievable.

There is an elegant technique for forming the CsI blocks reminiscent of the technique used to produce S·1 photomosaics in early camera tubes. The secret is to create a crazed substrate layer on which the evaporated CsI is deposited in block form. There are two methods for forming the "crazy paving" substrate.

The first method consists of producing an anodized aluminium faceplate, which when baked develops surface cracks in the form of a 50–100 μm mosaic with groove widths of 5 μm. The second method is to nickel plate this structure and then remove the anodizing so that a cobweb-like layer of nickel remains. Although in both cases the CsI layer is cracked into similarly sized blocks, the latter substrate is preferred for thick layers because the cracks grow larger with thickness instead of vanishing as happens with the first method.

The results show that the phosphor MTF has improved over that of the conventional screen: for example, the spatial frequencies at which the modulation depth falls to 50% are 11 lp cm^{-1} for the conventional screen, 16 lp cm^{-1} for the "blocked" screen based on the first substrate, and 27 lp cm^{-1} for the screen based on the second substrate. In addition to the improvement in resolution, it is shown that the brightness of a screen formed on the first substrate can be 50% up on a conventional screen if it has been optimally baked to about 300°C.

It is interesting to note that Washida and Sonoda deposit an intermediate preservative layer on to the CsI before the photocathode is activated. They claim that such a layer improves the photocathode

† See p. 201.

sensitivity. In contrast, Bates reported[6] that an S·9 photocathode could be formed directly on the CsI(Na) surface, and there was no harmful effect to the phosphor.

X-ray Computer Tomography

The paper given at the Symposium by Herstel[7] outlined the principles of X-ray axial tomography and referred to the pioneering work of Hounsfield[8] at E.M.I. Central Laboratories. He went on to conclude that in this relatively new field there are no accepted standards for image quality assessment.

It is evident that although X-ray tubes and photomultipliers play a part in providing the signal inputs to the system, by far the most significant part of the system is the data processing facilities. The ability of the computer to handle vast numbers of simultaneous equations that can be solved rapidly to give a matrix of absorption values and processed in such a way as to display a wide dynamic range of signal levels, illustrates that we are not dealing with a simple photoelectric image device, and as such we should not expect to be able to describe the process by the normally defined parameters of photoelectronics.

ELECTRON EMITTERS AND PHOSPHORS

Photocathodes and Secondary Emitters

The interest in low light level imaging under night sky radiation conditions in the early seventies stimulated activity in the near infrared region. This was largely because there are 300 times as many photons available per unit wavelength at 1 μm than there are at 300 nm.

Although great advances had been made in extending the response of the multialkali photocathodes into the near infrared there has been the promise of high quantum efficiency surfaces operating out to about 1 μm, based on negative electron affinity emitters. Since the early work in 1965 by Scheer and van Laar,[9] who activated a GaAs photocathode with caesium, and the subsequent work by Turnbull and Evans[10] who produced an improvement by activating with caesium and oxygen, many workers have been actively engaged in this field.

At the last Symposium considerable interest was shown in the GaAs photocathode[11] and its application to image tubes.[12] Commercial products were emerging and in addition to a number of reflective mode photomultipliers, at least one proximity focused image tube was released based on a transmission mode GaAs photocathode.

The last five or six years have not seen the predicted growth of III–V compound products, particularly in the image tube field. Photomultiplier tubes using reflective photocathodes such as the Varian Associates range[13] are the predominant commercial products. To a large extent the transmission mode image tube technology is still laboratory based, the work being carried out in UHV stainless steel chambers. The surface structure and chemical composition are studied routinely by low energy electron diffraction and Auger spectrometry, and elaborate heat cleaning procedures can be carried out.

GaAs transmission mode photocathodes in sealed-off image tubes have suffered from operational and shelf problems. The life has been particularly short in applications where it is necessary to draw high currents. The tubes require a high vacuum as the Cs(O) layer is very susceptible to contamination by gas in the tube. Such gas can change the stoichiometry of the Cs(O) layer. In addition, it seems necessary to put some caesium into the tube before the photocathode is transferred to establish an equilibrium so that the loss of caesium from the photocathode is prevented. Photoresponse decay measurements of GaAs photocathodes have been reported by Yee and Jackson.[13]

III–V photocathodes have been severely restricted in the types of image tube in which they can be incorporated. The electrostatic intensifier has been impracticable because of the necessity to produce single-crystal photocathodes on curved transparent substrates. Moreover, the use of fibre optics has been ruled out because III–V cathodes cannot be grown by epitaxy on these substrates. It would also be extremely difficult to carry out the normal heat cleaning procedure. The activation of glass-bonded GaAs transmission photocathodes was reported at the Symposium by Rodway, Holeman and Steward.[14] It is significant because a glass substrate is used in place of the usual GaP substrate. Considerable work has also gone into the determination of the optimum heat cleaning schedule for this glass-bonded cathode. It remains to be seen how successfully such devices can be produced in sealed-off tubes.

In contrast to the III–V compound cathodes considerable work has gone into the negative electron affinity properties of silicon. Williams *et al.*[15] made some of the first studies of negative electron affinity (NEA) silicon. They ignored the photoemissive properties because they were aware of the conflicting requirements whereby the layers had to be thin enough to achieve reasonable resolution and yet thick enough to have a high sensitivity notwithstanding the poor optical absorption of silicon.

Instead they concentrated on the properties of secondary emission and cold cathode emission. Because of the long diffusion length of silicon it was attractive to investigate the transmission secondary emission characteristics, and dynodes were constructed that promised to provide a TSE

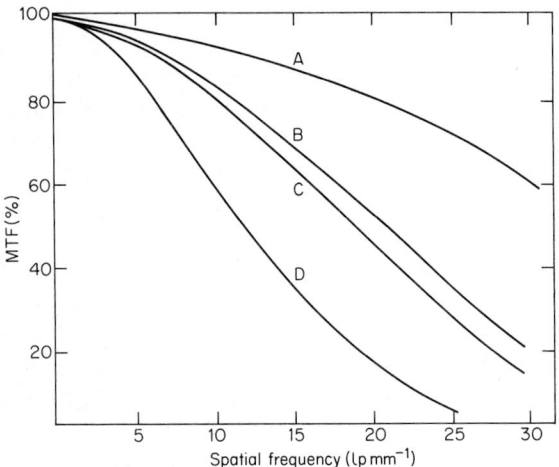

FIG. 2. Calculated MTFs of a TSEM intensifier described by Howorth et al.[17] A, electron spreading in silicon (5 μm thick); B, proximity focusing (5 kV); C, image section; D, overall MTF.

intensifier with a high gain per stage. Williams reported at that time that the dark current from silicon was very high, believed to be due to surface states, and he was hopeful that a solution to the problem would be found.

At the 1975 Symposium Howorth et al.[16] reported on further work in this field including surface physics investigations of silicon. They stated that a dark current of $10^{-12} A\,cm^{-2}$ could be achieved at room temperature by careful processing and gave figures for the spectral response of a reflective silicon photocathode having a luminous sensitivity of 1500 $\mu A\,lm^{-1}$.

At the present Symposium, Howorth et al.[17] reported further work on NEA silicon and described a TSE intensifier with a dynode gain of 1300 at 22 kV for 4–5 μm thickness. The MTF of this device is not very high, but this is largely because of the rather poor response of the electron optics in the image and proximity focus sections of the tube; Fig. 2 is reproduced from the preprints of their paper.[17] They stated, however, that the major disadvantage is still the dark current generated by surface states on the silicon surface. The dark current is about a factor of 100 too high, and satisfactory performance can only be achieved by cooling the device to approximately 0°C. It would appear that as a result of these findings all work in this area has ceased, and the device which threatened to oust the microchannel plate from many of its applications, particularly because of its better multiplication statistics and lower emission energies, now seems unlikely to pose a real threat in the near future.

Their recent work on NEA silicon transmission mode photocathodes has demonstrated sensitivities higher than the S·1 photocathode by an order of magnitude, but this is still not as high as the figures for III–V cathodes. However, the dark currents are claimed to be similar for all three types at approximately 5×10^{-11} A cm^{-2}. Reference was made to an electrostatic intensifier designed to operate at −50°C to reduce the dark current. Mention was also made of an 18 mm proximity focused microchannel plate tube. It was stated that the very narrow electron energy spread of the NEA silicon surface compensates for the finite thickness of the cathode and the MTF of this device is comparable to that of a conventional channel plate tube. It is unfortunate that no MTF curves have been published and no thickness figures given. Unlike many of the III–V compound cathodes it was claimed that the NEA silicon photocathodes are completely stable.

Cold Cathodes

The possibility of a NEA cold cathode was proposed[18] not long after the first NEA surfaces were fabricated, and the first emission from a p-n junction of GaAs was reported in 1969, by Williams and Simon.[19] However, NEA silicon looked a much more attractive material and considerable work was reported in this area by Kohn[20,21] from 1971 onwards. The silicon cold cathode was studied in a silicon vidicon by Cope et al.[22] but the results were not encouraging; peak emission current densities were orders of magnitude below the requirement, and the maximum life for stable emission was only 40 h at one third of the peak emission.

At the 1971 Symposium Deasley and Faulkner[23] described a cold cathode based on the GaAsP alloy. They conceded that silicon looked a much more promising material than GaAsP because it had a low bandgap so that one would expect a lower energy spread, but they avoided silicon because the smaller energy gap exacerbated the problem of instability of emission and the higher heat clean temperatures created problems in conventional camera tubes. They reported a conversion efficiency (ratio of emission current to bias current) of 0·1% and felt that one or two orders of magnitude improvement was possible; such an improvement in current density was considered necessary if the full benefit of the low emission energy spread was to be realized. The cathode suffered from an initial decay in emission and then remained stable for the order of 100 h.

In 1974 Howorth et al.[24] reported on the development of this cathode in a silicon vidicon. A special electron optical gun was computed to

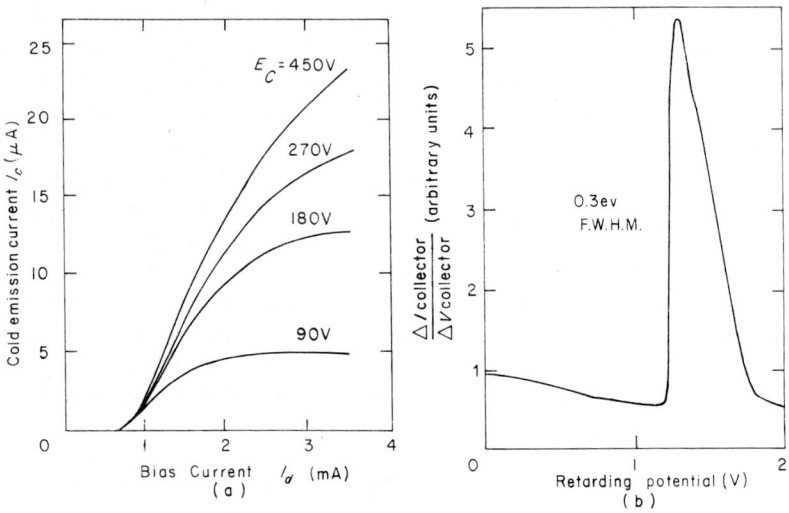

FIG. 3. (a) Variation of collected current with bias current and (b) the emitted electron energy distribution from a GaP cold cathode. (After Sukegawa et al.[25]).

provide a parallel beam so as to maintain the low beam temperature. The resolution was limited by the diode array of the target, and it was said that this was used for fabrication convenience, but a problem was encountered with caesium attack on the target. However, there was no substantial improvement in the performance of the cold cathode: the life was still quoted as of the order of 100 h. The main mechanism of decay was believed to be bombardment by ions of the residual gas (pressure of 10^{-7} Torr).

The paper presented at the current Symposium by Sukegawa et al.[25] reported on the choice of GaP as a suitable NEA surface; because of its high negative electron affinity one can expect good stability and long life.

The novel feature of the GaP-GaAlP junction structure is that the p-type GaP region consists of two parts. In the first part the acceptor concentration is low near the junction to improve electron injection efficiency and to achieve a high minority-carrier diffusion length. In the second part the acceptor concentration is high near the surface so as to improve the escape probability.

The results achieved look very promising. Figure 3 is reproduced from the paper by Sukegawa et al.[25] A maximum conversion efficiency of 0·7% was recorded for a bias current of 3 mA at a collecting potential of 450 V, and the FWHM of the energy distribution was only 0·3 eV. The most interesting fact was that the cold cathodes had been operating in excess of 2000 h. It was unfortunate that more details were not given on the operating conditions and the stability of the emission.

The cathode has been incorporated in an antimony trisulphide vidicon, but the solid state lag of the target tends to mask the effects of beam discharge lag, which in any event is improved by the dark current pedestal. It would have been interesting to see the improvement in lag with such a cathode in a lead oxide vidicon where the lag is predominantly beam discharge lag and the dark current is low. Perhaps it is too much to expect to achieve the processing of such a cathode alongside a lead oxide layer. Many low light level camera tubes, in particular the image isocon, would benefit from such a cathode.

Phosphors

There has been very little change in recent years in the methods of fabrication of the fluorescent screens in image intensifiers. The phosphors used have been predominantly P·11 and P·20 types. The former type has mainly been used for photographic recording, whereas the latter has been used for direct observation. Both types can be efficiently coupled to suitable photocathodes.

High resolution, high efficiency screens are still produced by the deposition of phosphor grains; other deposition techniques have suffered from loss of either resolution or efficiency, or both. Screen deposition is achieved in one of three ways: settling,[26] centrifuging, or cataphoresis.[27] The latter method enables the finest grain screen to be deposited, but centrifuging is by far the most common method for both types of screen.

The paper on properties of P·11 phosphors presented by Coleman[28] was of interest not because it presented any new fabrication techniques or new evaluation methods, but because of the detailed experimental study that was made of all parameters at the relatively low energy of 5 kV. It should be made clear that the measurements were part of a programme to evaluate the performance of an ultraviolet to visible image converter, and as such the screen parameters were defined for that device. The screen was a P·11 phosphor of 4 μm mean particle size that was deposited by centrifuging to a thickness from 0·8 to 1·1 mg cm^{-2}. This was coated with the usual aluminium backing by the organic film technique and on this a matt black anti-halation layer was deposited. Studies were made of screen efficiency, conversion linearity, persistence, random scintillations, backscattering and ageing.

A rather unconventional definition for the term "gain" was given in the paper. This was the ratio of the number of photons emitted to the number

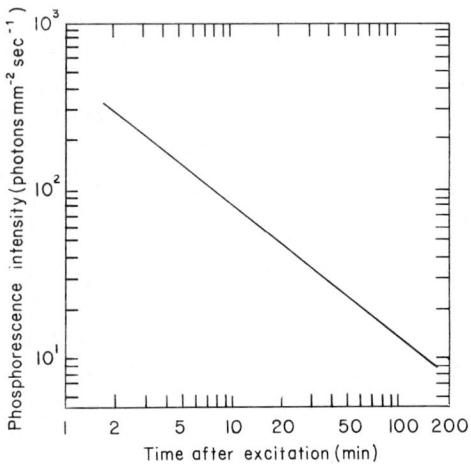

Fig. 4. Phosphorescence decay of P·11 phosphor, after Coleman.[28]

of electrons incident on the phosphor. The author prefers to use the term gain so that one is comparing like with like whether they be electrons, photons, or whatever. A better description of this quantity might have been "quantum conversion efficiency" or simply "photon yield".

A useful empirical relationship for the phosphorescence intensity as a function of time was determined from the log–log plot reproduced in Fig. 4 (taken from the Symposium preprint). It was pointed out that such long term phosphorescence can be a problem in an integrating detector, and an example was given where a ten times over-exposure can result in a 5% residual image in the succeeding image. The paper provides an interesting list of references for anyone looking further into the properties of phosphor screens.

High Speed Devices

It is interesting to see the rather novel ways in which Huston and Helbrough† have incorporated the microchannel plate intensifier into gated camera systems. Three main applications are covered in their paper. The first of these is a photographic surveillance camera operating over a wide dynamic range and using a gating system for automatic exposure control. For a given exposure time, a logic circuit computes the

† See p. 253.

gating pulse amplitude applied to the microchannel plate so that a constant exposure is achieved. The logic control signal is derived from the screen current of the intensifier prior to the photographic exposure. The operating range is 2×10^{-3} lx to 20 lx which is equivalent to 14 lens stops for an exposure range of 0·3 msec to 100 msec. Compared with a conventional photographic camera the device may suffer from fibre optic blemishes, pincushion distortion, loss of brightness and resolution at the edges, and a fall in absolute resolution; however its increased sensitivity puts the camera in a class of its own.

The second application is synchronization of the shutter action of the tube with the pulse repetition rate of a semiconductor laser. Although this technique has been used in range gating and fog penetration systems the application intended here is simple but novel, namely to read frontal detail in images with high background illumination. The illustration of reading the front number plate of a car in the glare of its head lamps is a good example.

The last example of the gated channel plate camera is a truly high speed application. Here the channel plate is used as a high speed shutter to record images with exposures of the order of 250 nsec. The special technique is that of using a multiple gating pulse which overcomes the rather limited on to off ratio of the shutter (10^4 times) by applying a second pulse to activate the tube for only the limited period (50–100 μsec) encompassing the gating pulse. By this technique a transmission ratio of 10^8 has been achieved, and it appears to be a powerful tool for recording moderately high speed events.

An interesting development in the use of an electron optical streak camera was reported by Adams, Sibbett and Bradley.† The streak camera is made to operate at a repetition rate of 140 MHz in synchronism with light pulses from a CW mode-locked dye laser. The low jitter between successive ramps (a few picoseconds) enables the integration of a very large number of sweeps, dispensing with the need for image intensifiers which are essential with single shot cameras.

The main development has obviously been in the triggering and deflection circuits. A photographic exposure of some dye laser sub-pulses with an exposure time of 0·5 sec consisted of some 7×10^7 superimposed streaks, yielding a FWHM for the sub-pulses of 13 psec compared to the 2 psec resolution that has been achieved with the single shot camera.

The authors are confident that they can remove the residual jitter to achieve picosecond resolutions. It is clear that the project is in the early days of its development, but the combination of the new trigger and

† See p. 265.

deflection circuits, coupled with the streak camera and the optical multichannel analyser promises to provide what is in essence a sampling oscilloscope with a few picoseconds resolution.

REFERENCES

1. Arndt, U. W. and Gilmore, D. J., *In* "Adv. E.E.P." Vol. 40B, p. 913 (1976).
2. Edgecumbe, J. and Garwin, E. L., *J. Appl. Phys.* **3**, 3321 (1966).
3. Driard, B., Ricodeau, M. and Rougeot, G., *In* "Seventh Symp. P.E.I.D. Preprints", p. 71 (1978).
4. Driard, B., Guyot, L. F. and Verat, M., *In* "Adv. E.E.P." Vol. 33B, p. 1031 (1972).
5. Driard, B., Rozière, G., Guyot, L. F. and Verat, M., *In* "Adv. E.E.P." Vol. 40A, p. 41 (1976).
6. Bates, C. W. Jr., *In* "Adv. E.E.P." Vol. 28A, p. 451 (1968).
7. Herstel, W., *In* "Seventh Symp. P.E.I.D. Preprints", p. 91 (1978).
8. Hounsfield, G. N., *Br. J. Radiol.* **46**, 1016 (1972).
9. Scheer, J. J. and van Laar, J., *Solid State Commun.* **3**, 189 (1965).
10. Turnbull, A. A. and Evans, G. B., *J. Phys. D.* **1**, 155 (1968).
11. Piaget, C., Polaert, R. and Richard, J. C., *In* "Adv. E.E.P." Vol. 40A, p. 377 (1976).
12. Holeman, B. R., Conder, P. C. and Skingsley, J. D., *In* "Adv. E.E.P." Vol. 40A, p. 1 (1976).
13. Yee, E. M. and Jackson, D. A., *Solid State Electron.* **15**, 245 (1971).
14. Rodway, D. C., Holeman, B. R. and Steward, J. G., *In* "Seventh Symp. P.E.I.D. Preprints", p. 101 (1978).
15. Williams, B. F., Martinelli, R. U. and Kohn E. S., *In* "Adv. E.E.P." Vol. 33A, p. 474 (1972).
16. Howorth, J. R., Sheppard, C. J. R. and Trawny, E.W.L., *In* "Adv. E.E.P." Vol. 40A, p. 387 (1976).
17. Howorth, J. R., Folkes, J. R., Hopkins, G. P. and Palmer, I. C., *In* "Seventh Symp. P.E.I.D. Preprints", p. 107 (1978).
18. Geppert, D. V., *Proc. IEEE* **54**, 61 (1966).
19. Williams, B. F. and Simon, R. E., *Appl. Phys. Lett.* **14**, 214 (1969).
20. Kohn, E. S., *Appl. Phys. Lett.* **18**, 272 (1971).
21. Kohn, E. S., *IEEE Trans. Electron Devices* **ED-20**, 321 (1973).
22. Cope, A. D., Luedicke, E. and Carroll, J. P., *RCA Rev.* **34**, 408 (1973).
23. Deasley, P. J. and Faulkner, K. R., *In* "Adv. E.E.P." Vol. 33A, p. 459 (1972).
24. Howorth, J. R., Surridge, R. K. and Palmer, I. C., *In* "Adv. E.E.P." Vol. 40A, p. 463 (1976).
25. Sukegawa, T., Kan, H., Nakamura, T., Katsuno, H. and Hagino, M., *In* "Seventh Symp. P.E.I.D. Preprints", p. 113 (1978).
26. Sadowski, M., *RCA Rev.* **95**, 112 (1949).
27. McGee, J. D., Airey, R. W. and Aslam, M., *In* "Adv. E.E.P." Vol. 22A, p. 571 (1976).
28. Coleman, C. I., *In* "Seventh Symp. P.E.I.D. Preprints", p. 571 (1976).

Stray Light in Proximity Focused Image Intensifiers

J. J. HOUTKAMP and H. MULDER

Oldelft Research Laboratories, Delft, Netherlands

Introduction

In a proximity focused image intensifier the light transmitted through the photocathode and reflected from the aluminium backing of the phosphor screen contributes considerably to the decrease in contrast of the output image. Light generated in the phosphor and transmitted through the aluminium backing toward the photocathode has a similar effect. In addition, electrons reflected with high energy by the backing will be re-accelerated towards the phosphor by the electric field between cathode and anode, thus forming another cause for loss of contrast (Fig. 1).

By applying a light absorbing layer on the aluminium backing of the phosphor the detrimental effect of the two first-mentioned phenomena will be reduced considerably. For example, a well known device is a layer of dense packed microscopic balls of aluminium obtained by evaporating pure aluminium in a poor vacuum.[1,2]

All materials to be used in anti-halation layers must meet the requirements of high transmission for electrons, and therefore a low atomic number is required.[3-5] Furthermore, they must fulfil the condition of having a low vapour pressure even at elevated temperatures, a low affinity for alkali vapours, and resistance against attack by these alkali vapours. The possibility of using a layer for preventing electron reflection seems doubtful, due to the fact that the primary electrons must be able to penetrate the layer before reaching the phosphor.

At Oldelft, where proximity focused diodes were developed for use in a system for X-ray screen photography in which green emitting ZnCdS X-ray screens are applied, investigations were made to determine the effect of different anti-halation layers on the photocurrent and thus on the stray light of the tube. The influence of these different types of backing on the reflection of electrons was also measured indirectly. The measurements of the photocurrent were carried out in a standard type of tube,

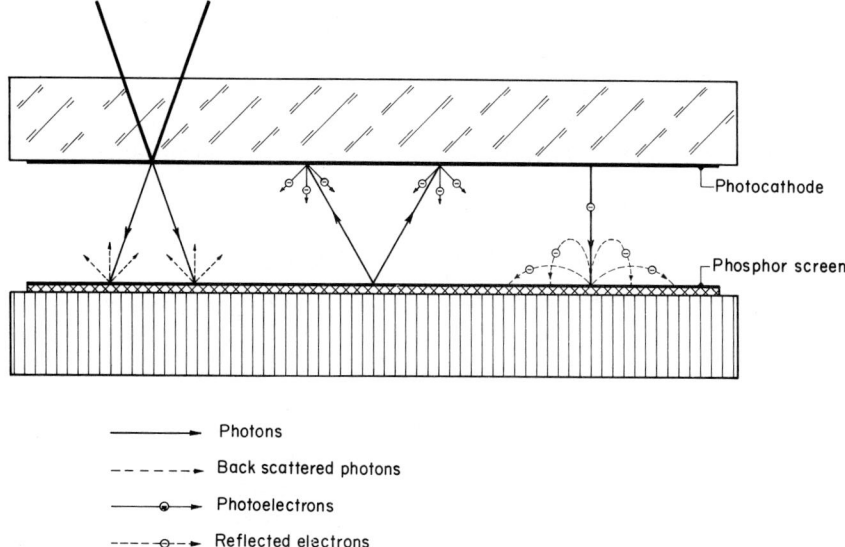

Fig. 1. Some causes of stray light in a proximity focused image intensifier.

slightly adapted for this purpose. This tube had glass input and output windows; the useful photocathode and anode dimensions were 90×90 mm^2, and the gap between cathode and anode was 3·7 mm. The phosphor was of a zinc cadmium sulfide type. Instead of the anti-halation layer commonly used, four different layers were evaporated onto the aluminium backing of the phosphor in circular areas of 15 mm diameter (Fig. 2). Two of these layers consisted of aluminium evaporated in a poor vacuum, one with a low reflection coefficient and the other with an even lower coefficient. The other two layers consisted of silicon; for one the thickness was optimized to obtain a minimum reflection for light from the above mentioned X-ray screen and for the other the reflection for blue light at 400 nm wavelength was minimized. A small part of the phosphor was omitted to permit the measurement of the enhancement of the photocathode current as a function of the applied voltage.

The Relative Influence of Different Anti-halation Layers on the Photocathode Current

Illumination of the photocathode causes a photocurrent having two components. One component is the result of the primary light, intensity a,

FIG. 2. Experimental Oldelft proximity focused image intensifier.

striking the cathode together with the light scattered back by the phosphor onto the photocathode. Let k represent the sensitivity of the photocathode, which to some extent depends on the applied electrical field. Now suppose that the transmission of the photocathode can be described by a factor p and if r represents the coefficient of reflection of the phosphor backing then this first component of the current can be expressed as

$$i_a = ak(1 + rp). \qquad (1)$$

The quantities k, r and p are related to the wavelength of the light used, while the effect on the spectral distribution of the light transmitted by the photocathode can be neglected.

The other component of the photocurrent is caused by the light, intensity b, generated in the phosphor, part of which passes through the backing of the screen towards the photocathode. We can express this as

$$i_s = ktb, \qquad (2)$$

in which tb represents the fraction of b passing through the aluminium and anti-halation layer and acting upon the photocathode. As, for our X-ray intensification systems, this light possesses almost the same spectral composition as the primary light striking the cathode, k still represents the sensitivity of the photocathode. For the light generated in the screen we can write

$$b = ag, \qquad (3)$$

in which g is the luminous gain of the tube. Substitution of Eq. (3) in Eq. (2) and adding this result to Eq. (1) results in

$$i = i_d + i_s = ak(1 + rp + tg). \qquad (4)$$

At a low voltage, when the luminous gain g is zero, the ratio between the photocurrents measured with different anti-halation layers having reflection coefficients r_a and r_b is

$$\frac{i_a}{i_b} = \frac{1 + r_a p}{1 + r_b p}. \qquad (5)$$

The ratio between the photocurrents for one anti-halation layer at two different anode voltages V_1 and V_2 is

$$\frac{i_1}{i_2} = \frac{k_1}{k_2} \frac{(1 + rp + tg_1)}{(1 + rp + tg_2)}. \qquad (6)$$

In the case of an ideal tube r and t are zero. The relative contribution to the total light output of the tube from reflection off the backing of the phosphor can be expressed as

$$R = \frac{rp}{1 + rp + tg}, \qquad (7)$$

and similarly the contribution from feedback of light generated in the phosphor is

$$T = \frac{tg}{1 + rp + tg}. \qquad (8)$$

Measurements with the Tube

Earlier measurements showed that the transmission of the normal S·20 type photocathode manufactured by Oldelft, which has an enhanced green sensitivity, is approximately 30% ($p = 0.3$). In the experimental tube the enhancement of the photocathode sensitivity as a function of the applied voltage was measured with a small parallel light beam of cross section 5 mm, impinging onto that part of the photocathode facing the anode where the phosphor was omitted. The spectral distribution of the light used corresponded to that of the light emitted by a ZnCdS X-ray screen. Next, the light beam was shifted to a spot on the cathode opposite to one of the experimental anti-halation layers and the photocurrent was measured at tube voltages varying from 1 to 10 kV. Simultaneously, the luminous gain was measured. The results for some of the layers are shown in Fig. 3.

After the completion of these measurements, the tube was taken apart. A large flat semiconductor photocell was mounted parallel to and at a distance of 3·7 mm from the phosphor and a lightbeam was passed through a 5 mm diameter hole in the centre of the cell (see Fig. 4). The reflection of the anti-halation layer with the lowest reflection coefficient

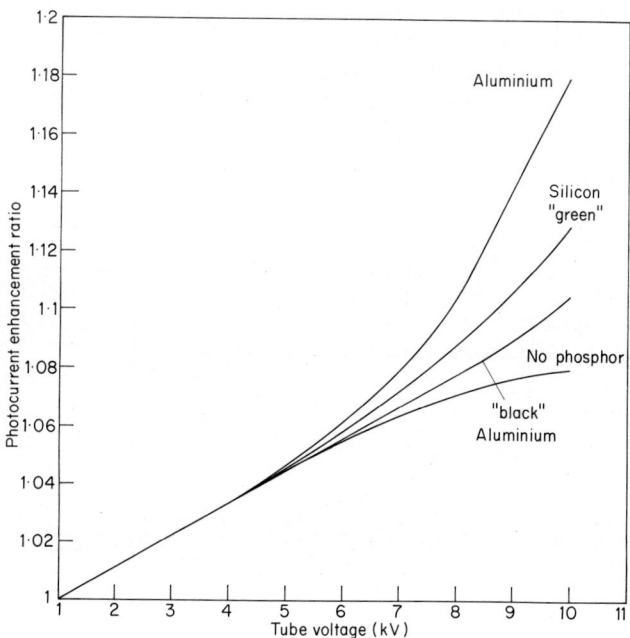

Fig. 3. Photocathode enhancement ratio for different anti-halation layers.

(aluminium evaporated in a poor vacuum) was first measured with respect to that of the uncovered aluminium backing. This ratio, which turned out to be approximately 3·5%, was subsequently used as a reference for the evaluation of the other measurements, on the assumption that for this material any change in the reflection, caused by admittance of air in the tube, was smallest and might thus be neglected.

By substituting the ratio of 3·5% and the measured values for the photocurrents at a tube voltage of 1 kV in Eq. (5), r can be calculated for each anti-halation layer. Substituting in Eq. (6) these calculated values of r and the measured values for cathode sensitivity k, photocurrent i, and gain g, (the last for tube voltages of 6 and 10 kV) results in values for the transmission t of the different anti-halation layers. Using Eqs. (7) and (8) the contribution from reflection and transmission of the layers to the stray light in the tube can be calculated. The results are shown in Table I.

When interpreting these figures one has to keep in mind that the physical characteristics of the aluminium backing will affect the light transmission and reflection. The transmission of the aluminium backing depends to a certain extent on the surface of the nitrocellulose lacquer that serves as a substrate for the aluminium evaporation. As the reproducibility of this layer is poor, the transmission of the aluminium backing will vary from sample to sample. Also the reflection of the anti-halation layer will vary to a smaller extent.

We made a second tube and found that the transmission as well as the reflection were somewhat different from those of the first tube although the layers in both tubes had been made in a similar way.

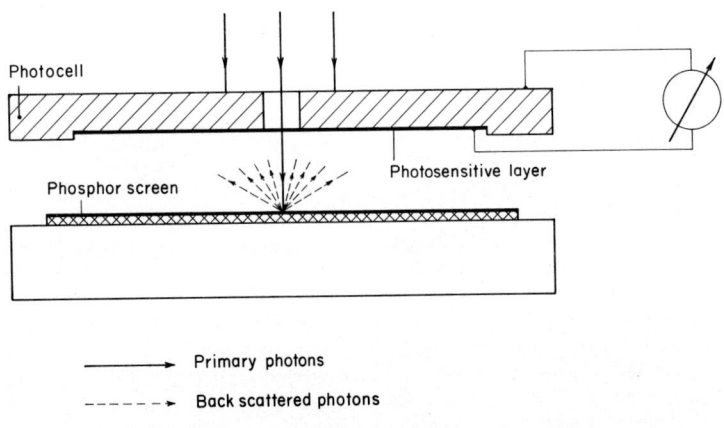

FIG. 4. Set up for measurement of reflection.

TABLE I
Performance of different backing layers

	Stray Light in % at 10 kV due to	
	Reflection R (Eq. (7))	Transmission T (Eq. (8))
Al-layer	18·5	16
Al-layer + black Al, thin	6	9
Al-layer + black Al, thick	0·9	7
Si "green"	5·5	8
Si "blue"	7	7·5

ELECTRON REFLECTION MEASUREMENTS

Secondary electrons are released from the phosphor by the impact of primary electrons. Some of the secondary electrons are elastically reflected primary electrons, which consequently have sufficient energy to excite the phosphor after having described a parabolic trajectory in the decelerating field between anode and cathode. It is a well known fact of ballistics that the maximum distance of this second impact from the original one is twice the anode-to-cathode distance. To measure the influence of the elastically reflected electrons, we have used the following set up. A fibre optic faceplate with a phosphor screen having a standard aluminium backing on one side is mounted close to a photocell. Both the photocell and the faceplate have a 2 mm diameter hole in the centre (Fig. 5). The system is placed in a vacuum chamber where electrons originating from an electron gun pass through this hole from the photocell-side to the phosphor-side. The electrons hit the target to be tested which has its flat surface parallel to and at a distance of approximately 3·7 mm from the phosphor screen on the fibre optic plate. A measurement is made of the electron current to the target which is held at approximately 10 V positive with respect to the aluminium backing in order to prevent the loss of low energy secondary electrons. Simultaneously the photocell current is measured. By exciting the phosphor on the fibre optic faceplate with an electron beam of the same intensity and energy as before, the ratio of the light produced by primary electrons to that due to reflected electrons can be determined under equivalent conditions to those in the proximity focused tube.

With 10 kV electrons striking a phosphor screen provided with an uncovered aluminium backing layer we found a value of 3·5% for the

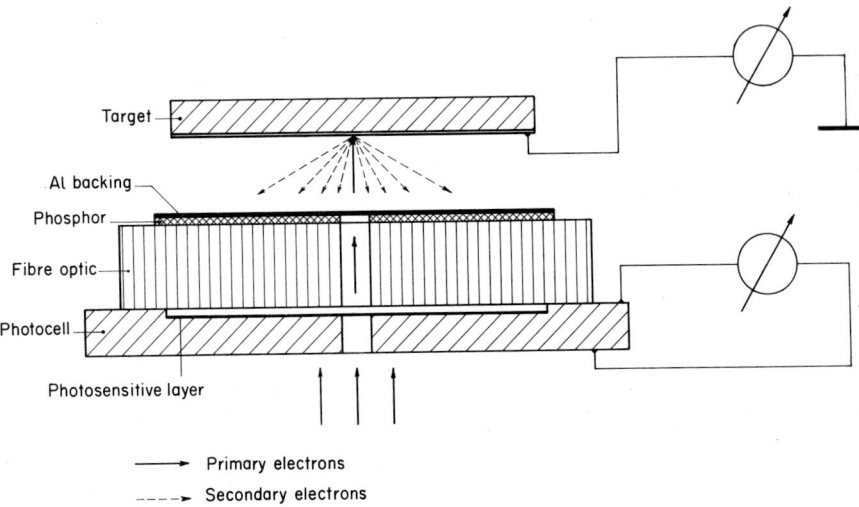

Fig. 5. Set up for measurement of electron reflection.

stray light caused by reflected electrons. Screens provided with antihalation layers showed only minor differences. However, when the energy of the primary electrons was raised to 15 kV, the stray light increased to 6·5%.

References

1. McGee, J. D., Aslam, N. and Airey, R. W., *In* "Adv. E.E.P." Vol. 22A, 578 (1966).
2. Garfield, B. R. C., Wilson, R. J. F., Goodson, J. H. and Butler, D. J., *In* "Adv. E.E.P." Vol. 40A, 11 (1976).
3. Holliday, J. E. and Sternglass, E. J., *J. Appl. Phys.* **28**, 1189 (1957).
4. Bishop, H. E., *In* "4th Intern. Congr. X-ray Opt. and Microanal." ed. by R. Castaing, P. Deschamps and J. Philbert, p. 153, Hermann, Paris (1966).
5. Heinrich, K. F. J., *In* "4th Intern. Congr. X-ray Opt. and Microanal." ed. by R. Castaing, P. Deschamps and J. Philbert, p. 159, Hermann, Paris (1966).

A Miniature Magnetic Intensifier

G. O. TOWLER

The Research and Development Establishment, British Relay Limited, Leatherhead, Surrey, England

INTRODUCTION

The shortcomings in the performance of the electrostatic image intensifier are well known (viz. pincushion distortion, loss of resolution at the edges, and fall off in brightness at the edges) but in spite of this it has gained great popularity. On the other hand the magnetic intensifier which does not suffer from these image defects has not enjoyed the same popularity largely because of power, bulk and weight problems.

PRINCIPLE OF THE MINIATURE TUBE

The miniature tube described here still depends on parallel electric and magnetic fields as its basic method of focusing. The novel feature is the creation of a sharp discontinuity in the electric field in the vicinity of the screen by interposing a mesh[1] at this point. This can be seen in Fig. 1 which is a schematic of the tube. A low voltage image section of length a occupies the majority of the focus loop, and the remainder of the loop is completed in the spacing b where there is a high potential difference.

It can be shown that if V_1 is the voltage applied between photocathode and mesh, and V_2 is the voltage applied between mesh and the screen, then the magnetic field $B \cdot$ required to focus the tube is given by the following relationship:

$$\frac{1}{B} = \frac{1}{\pi}\sqrt{\frac{e}{2m}}\left[\frac{a}{\sqrt{V_1}} + b\frac{\sqrt{V_2}-\sqrt{V_1}}{V_2-V_1}\right], \qquad (1)$$

where $\qquad E_1 = V_1/a, \qquad E_2 = V_2/b$

and the overall tube voltage $V = V_1 + V_2$.

A useful feature of this focusing condition is that, because most of the focus loop is completed in the space a and most of the electron energy is

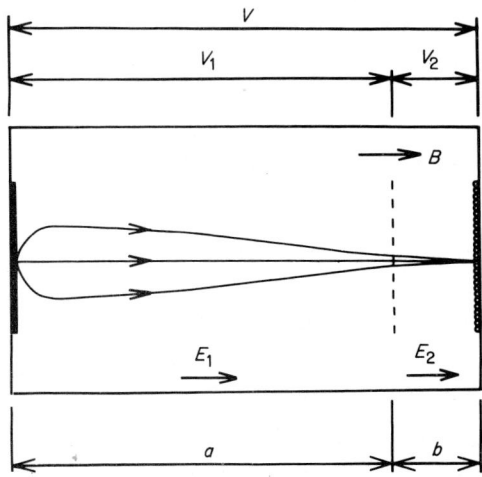

FIG. 1. Schematic of miniature magnetic intensifier.

acquired in the space b, it is possible for the focusing condition to be largely separated from the gain condition. This means that tubes can be constructed for a given magnetic field where the physical length is much smaller than for conventional tubes, and automatic gain control can be applied to a magnetic tube by varing V_2 without altering the magnetic field B. To a first approximation the tube will stay in focus; for instance for a change in V_2 from 3 kV to 10 kV, which represents a 30:1 change in gain, with B and V_1 kept constant at 10·6 mT and 1 kV respectively, the resolution is only reduced from 65 line pairs mm^{-1} to 45 line pairs mm^{-1}. For similar reasons a highly stabilized power supply is not essential for V_2, and indeed tubes have been operated in which the V_2 supply has been a conventional wrap-around Cockcroft–Walton voltage multiplier, and the supply ripple produced negligible loss of resolution even at 65 line pairs mm^{-1}.

For a typical tube $a = 24·5$ mm and $b = 3·25$ mm, i.e. $a \gg b$ then, from Eq. (1) if $V_1 \ll V_2$ it can be seen that the first term dominates and the focus condition is largely independent of V_2.

The lowest value of V_1 that could be used to produce images of reasonable resolution and contrast was 200 V, requiring a magnetic field of 6 mT. Resolution and contrast increase with increasing V_1, and although tubes have been operated with values of V_1 up to 3 kV, the typical value of V_1 which was used was 1 kV, requiring a magnetic field of 13 mT.

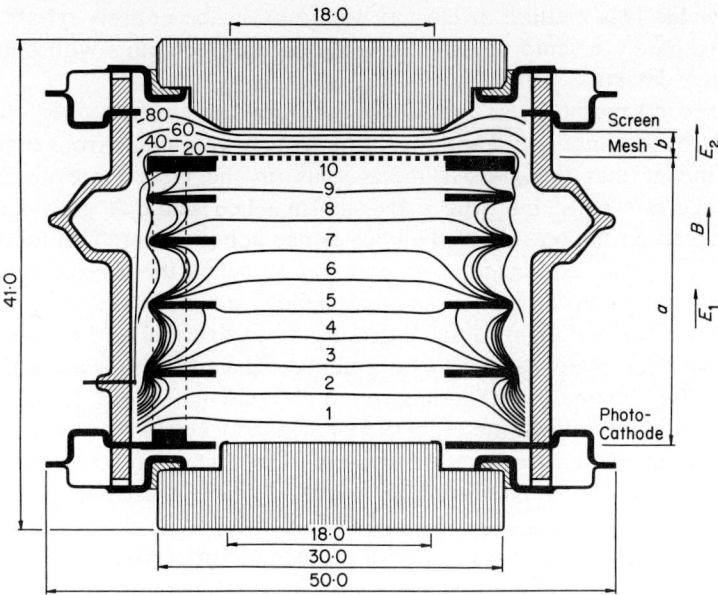

FIG. 2. Diagram of miniature magnetic intensifier, dimensions in mm.

IMAGE TUBE DESIGN AND TECHNOLOGY

In order to cover as large a range of applications as possible and to achieve a high yield it was considered necessary for the tube to be of modular construction so that multi-stage devices could easily be produced.

The final design, which was developed between December 1974 and February 1976, is shown in Fig. 2. It contains no magnetic materials to introduce perturbations in the magnetic field. The body is made of sodalime glass, and the feedthroughs for the low voltage image section are platinum pins which are spaced round the circumference of the tube close to the photocathode so as to establish a long tracking length between the mesh pin and screen flange. The faceplates are either soda-lime glass or "soft" fibre optics sealed into titanium flanges with pyroceram. Two different techniques have been used for sealing the faceplate flanges to the body flanges. The first method consisted of making these sealing flanges from OFHC copper. A titanium optic flange was brazed to the faceplate sealing flange with a silver–copper eutectic alloy. The closures were made using an annular pinch-off tool and applying a force of 16 000 lbs wt. with a hydraulic ram to effect a copper cold weld. This method of closure was found to be entirely reliable and approximately seventy tubes were sealed by this technique with only two or three defective seals.

A second method was introduced whereby the silver–copper eutectic braze was dispensed with and the entire flanges were made from titanium. This meant that the glass-to-metal seals on the body were no longer Housekeeper seals, but true expansion-matched seals. A glass annulus was sealed on the outside of the body flange in both cases. This tended to equalize stresses and acted as an end stop to define the precise length of the tube on pump down by curtailing bellows action in the flanges. This annulus contained eight radial V-grooves to enable the void between the body and faceplate flanges to be pumped. The titanium sealing flanges were closed by argon arc welding round their circumference. It was found necessary to thin down the cross section of the faceplate flanges to reduce the amount of heat required in the arc. The thinned cross section also allowed strains to be relieved in a region where there were no glass interfaces to shear.

Although both techniques avoided magnetic materials and each had certain advantages, the latter technique was eventually preferred because the outgassing time prior to processing was shorter, and there was less tendency to loss of photocathode sensitivity.

It should, however, be borne in mind that the former technique is applicable to vacuum transfer methods, which were not used on this device. Incorporating such techniques would enable tubes to be produced with large area cathodes of uniform sensitivity whilst retaining the same short physical length. These larger diameter tubes would have an improved "useful photocathode diameter" to "flange outside diameter" ratio, because the image section wall clearance and flange thickness remains substantially constant as the tube diameter increases. For example, a tube with a useful photocathode diameter of 40 mm could still be approximately 40 mm long and would have an outside flange diameter of the order of 60 mm.

It is quite feasible to carry out the copper cold weld sealing operation in vacuum and there would be no external oxidation of the copper flanges during the outgassing and photocathode processing stages, as is experienced with the conventional technique.

The annular metallization required to provide electrical contact to the photocathode and screen was sputtered nichrome. This was used because of its resistance to abrasion during substrate cleaning, and the smooth flake-free surface which it provides in the high voltage region of the screen.

The low voltage image section consisted of five stainless steel annuli

FIG. 3. An encapsulated copper flange tube and an unencapsulated titanium flange tube.

and the mesh mount assembled on three ceramic rods which were attached at 120° intervals to the photocathode body flange. This method of assembly kept all of the low voltage electrodes away from the wall of the tube and from the screen, preventing high voltage tracking.

To prevent electrical breakdown between the mesh and the screen, the edges of the screen optic were specially shaped and polished, and the mesh mount was fitted with a highly polished, gently curved anti-breakdown cap. It can be seen from Fig. 2 that this region was so designed that the electric field was at maximum over only the 18 mm diameter working area between the mesh and the screen.

The screen was a fine grain P·20 phosphor which was deposited by a modified P·11 cataphoresis technique.[2] It was aluminized in the conventional manner by evaporation on to an organic film.

The tube was pumped and activated by means of the two side ports. Figure 3 shows a copper flange tube encapsulated in silicone rubber and an unencapsulated titanium flange tube.

Tube Performance and Image Quality

Resolution

The limiting resolution increases with increasing value of V_1, but above 1 kV the improvement is marginal. Typically for $V_1 = 1$ kV the limiting

Fig. 4. An image obtained at high V_z (10 kV) from an early tube design. Image diameter is 14·25 mm and values of V_1 and B were 1 kV and 10·6 mT respectively.

resolution is 65 line pairs mm^{-1}. If V_1 is increased to 3 kV the limiting resolution saturates at approximately 80 line pairs mm^{-1}, but this requires a magnetic field of the order of 22·5 mT. The resolution was constant from the centre to the edge of the field. The limiting resolution is to a certain extent dependent on the value of V_2. This is because V_2 effects the contrast. It is highest at low V_2 (2·5–3 kV) and lowest at high V_2 (10–12 kV). The loss of contrast is the result of light induced background which is discussed in a later section. To achieve these high resolutions it is desirable to use a mesh of 2000 cycles per inch. Figure 4 illustrates the image quality that was achieved on an early tube design at $V_1 = 1$ kV and $V_2 = 10$ kV. The effect of light induced background was quite low, and produced a negligible reduction in contrast. The S-distortion was the result of an early image section design.

Distortion

The fairly high levels of S-distortion found in the early tube designs was traced to radial electric field components in the low voltage image section. Two problems were encountered, firstly there was insufficient electrostatic screening to prevent electric field penetration from the high voltage section into the low voltage image section, and secondly, because of the shape of the photocathode flanges, there was a field free region near the

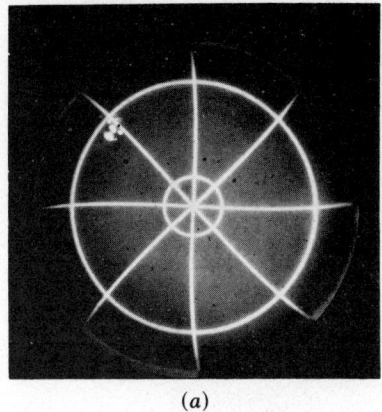 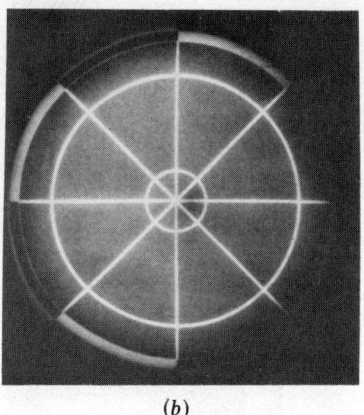

(a) (b)

FIG. 5. (a) Performance of interim tube design having a half-spacing annulus at the mesh and (b) performance of the final tube design (as shown in Fig. 2) having a half-spacing annulus at the mesh and a stepped photocathode faceplate. The larger circle occupies 75% of the useful photocathode diameter.

periphery of the faceplate which produced bulging of the equipotentials in the vicinity of the photocathode.

The first problem was overcome by using sufficient annuli with an adequate width to spacing ratio to prevent radial field breakthrough into the low voltage section (N.B. a half spacing annulus was required next to the mesh). However, some S-distortion remained as can be seen in Fig. 5(a). The second problem was overcome by the introduction of a step in the photocathode faceplate so that the photocathode substrate was coplanar with the flange that supported the low voltage image section. These modifications are shown in Fig. 2 together with equipotentials of 1% intervals of the overall voltage in the low voltage section and 20% intervals in the high voltage section. The mesh is held at 10% of the overall voltage. It can be seen that the 1% equipotentials in the low voltage focus section are all, to a good approximation, parallel to the photocathode plane over the usable 18 mm diameter. Figure 5(b) shows the improvement in S-distortion using this design. It can be seen that the S-distortion has been reduced to negligible proportions even at the periphery of the field of view. Figure 6 illustrates the general image fidelity and improved geometry of this design. The image magnification is very close to unity, a typical measured figure being 1·01.

Photocathode Sensitivity and Gain

Tubes have been produced with multialkali photocathodes. A typical tube with a white light sensitivity of 190 $\mu A\,lm^{-1}$ had a radiant sensitivity

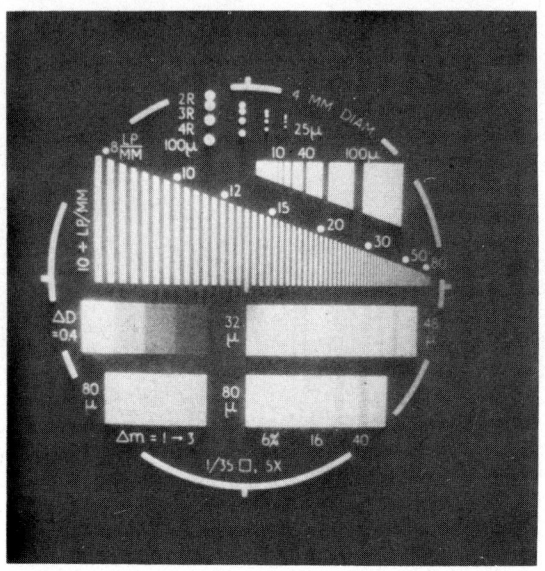

Fig. 6. Image quality obtained using the final tube design. The image diameter is 14·25 mm.

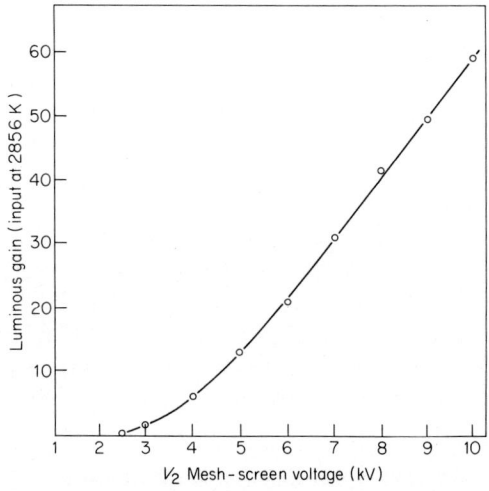

Fig. 7. Luminous gain versus mesh-screen voltage for a tube having photocathode sensitivity $190 \mu A\,lm^{-1}$.

at 450 nm of 68 mA W^{-1} which corresponds to a quantum efficiency of 18·6%. The luminous gain obtained from the tube is shown in Fig. 7. It can be seen that it is 59 at $V_1 = 1$ kV and $V_2 = 10$ kV.

Light Induced Background

It was observed for a number of tubes that, when they were operated at high values of V_2, there was some light spillage from the white areas of the image into the surrounding black areas.

The following possible sources of light induced background were investigated:

(1) Light scatter: a percentage of the input light is transmitted through the photocathode and reflected by internal surfaces so that it falls back on to the photocathode to produce a general background.
(2) Optical feedback: fluorescent light from the screen can be fed back to the photocathode.
(3) Backscattered primary electrons: some primaries may be back scattered by the mesh wires at large angles to the normal. They will be retarded by the image section electric field and will return to the mesh at considerable distances from their point of scattering.
(4) Secondary electrons generated at the mesh: these will be subjected to the same retarding field and will have an associated point spread function at the screen.
(5) Photoemission from the mesh: light transmitted through the photocathode is incident on the mesh in an out of focus plane.

The expected and observed behaviour of the light induced background for the above sources is listed below.

(1) Light scatter: this would create the same reduction in contrast at all gain levels (i.e., independent of V_2). Hence it is not the major factor in the observed effect. However, if the factor is significant, it can be decreased by blackening the anuli, mesh and screen backing.
(2) Optical feedback: this can be caused by transmission through the aluminium backing because of either too thin a layer or pin holes in the layer. It is also possible for light to be piped from the exposed edge of the screen faceplate. The remedy is to cover the edges of the faceplate with an opaque coating and to ensure that the aluminium backing thickness is optimized. The effect is most significant at high gain, and hence one would expect a decrease in contrast with increase in V_2. However, it has been observed that for a white to black edge the signal induced background extends

only over about 150 μm. This is not in keeping with the above mechanisms. Moreover, the effects of these mechanisms have been significantly reduced or eliminated.

(3) Backscattered primary electrons: scattered primaries can cause a reduction in contrast over large areas because their lateral velocity components can be high. The worst case is that where the initial longitudinal component of velocity of the backscattered electron is such that the time taken for the electron to return to the mesh enables it to complete half a cyclotron orbit. It can be shown that this can lead to a lateral displacement of the order of 14 mm. The level of such long range light induced background is relatively low and it is not a major factor. In fact it is well known that the number of scattered primaries is small compared with the number of secondaries produced.

(4) Secondary electrons: secondary electrons will behave in a similar way to the backscattered primaries, but, being of lower energy, the light induced background will be of shorter range. The higher the value of V_2 the greater will be the effect of the secondaries, which is in keeping with the observed background.

(5) Photoemission from the mesh: the amount of light induced background from this cause depends on the optical system since this determines by how much the image is defocused at the mesh plane. At high values of V_1 and low values of V_2 the background would be minimal, and changing the polarity of V_1, whilst maintaining V_2 at its normal value would not completely cut off the background image. This effect was not observed on any of the conventionally processed tubes where the antimony evaporator was introduced immediately in front of the mesh. The effect was, however, observed on one tube that was processed unconventionally where it is believed that antimony was deposited on the photocathode side of the mesh.

Enhanced Gain

It seemed probable that the most significant factor resulting in light induced background in some tubes at high values of V_2 was secondary emission. If this was the case, then it could also result in a significant contribution to the gain. A simple test was carried out on a conventional copper mesh tube that was known to exibit some light induced background at high values of V_2. The gain was measured as a function of V_1 and V_2 in such a way that $V_1 + V_2$ was held constant.

The gain was observed to rise monotonically with increase in the ratio V_2 to V_1. If $V_1 + V_2$ was held constant at 9 kV then the gain at $V_1 = 1$ kV

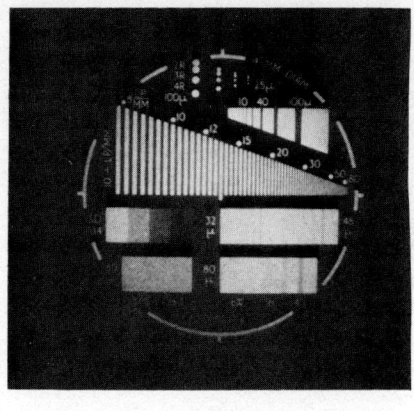 

(a) (b)

FIG. 8. Images obtained using a tube having a 10 nm kayer of MgF_2 deposited on the photocathode side of the mesh (a) with V_2 set at 3 kV and (b) with V_2 set at 5 kV.

and $V_2 = 8$ kV was approximately twice that with $V_1 = 3$ kV and $V_2 = 6$ kV. If only the energy of the photoelectrons were involved, the gain should have been constant, providing the electron transmission of the mesh remained constant. To a first order approximation the transmission will be constant because the photoelectrons are relatively "stiff" at this point. The increase in gain at $V_1 = 1$ kV, is partly the result of the secondaries having acquired more energy and partly because the secondary emission ratio is higher at a primary energy of 1 keV than at 3 keV. This was confirmed in a separate experiment to measure the secondary emission coefficient as a function of primary energy in a number of copper mesh tubes. The coefficient was found to peak at values from 1·5 to 3 for a primary energy of approximately 400 eV, whilst at 3 keV the coefficient was well below unity. This is consistent with the fact that at $V_1 = 3$ kV there is negligible light induced background. The peak values of the secondary emission coefficient of the mesh tended to be rather higher than the published figures for copper.[3] This was probably due to alkali metals deposited during the photocathode activation process.

The copper mesh was known to be a relatively poor secondary emitting surface and so tests were carried out by depositing a number of good secondary emitting materials on the photocathode side of the mesh in varying thicknesses. From the materials investigated the most promising results were obtained with MgO and MgF_2 and the optimum thickness was of the order of 10 nm. From demountable measurements the primary energy required to produce maximum secondary emission from a 10 nm MgF_2 layer was of the order of 1·3 kV. This was remarkably close to the normal operating value for V_1 of 1 kV. Figure 8 shows two images

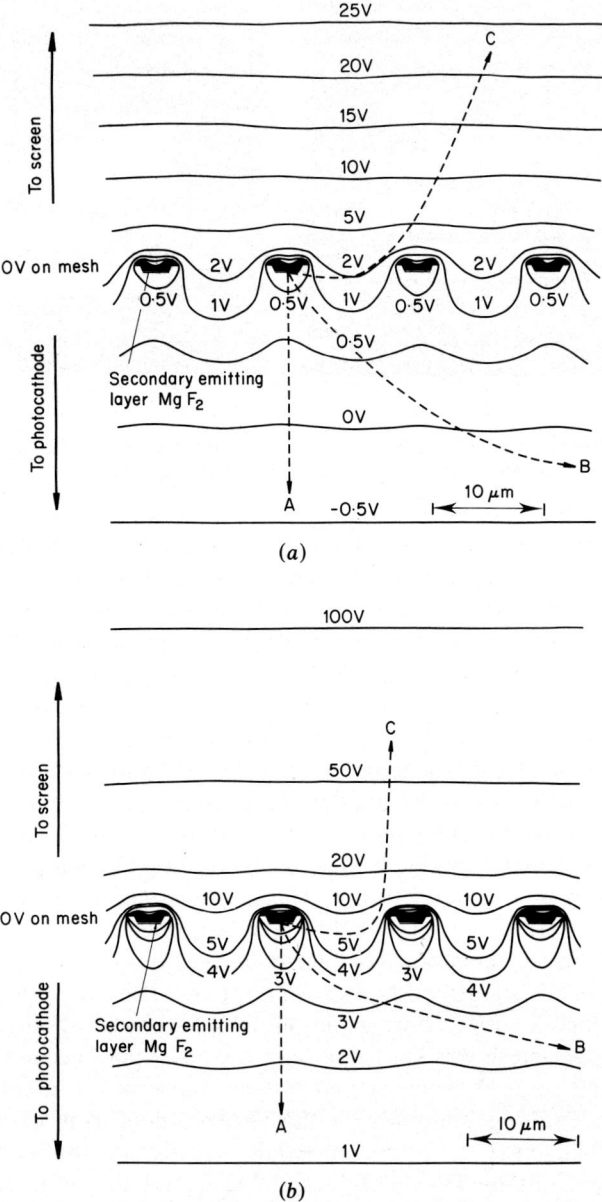

FIG. 9. Electric field distributions in the region of mesh having 2000 c in.$^{-1}$ and 45% transmission (a) with $V_1 = 1$ kV and $V_2 = 3$ kV and (b) with $V_1 = 1$ kV and $V_2 = 10$ kV. Curves A, B, C, (dashed) are typical paths of secondary electrons for various angles of emission.

obtained from a tube where the left-hand side of the mesh was coated with 10 nm of MgF_2. In Fig. 8(a) $V_2 = 3$ kV and a sharp, high contrast image resulted with no indication of the MgF_2 demarcation. However, in Fig. 8(b) where $V_2 = 5$ kV the left-hand side of the image was brighter than the uncoated side by at least a factor of 2, but the image had lost contrast because of short range light induced background. It was evident that secondary electrons were responsible for both the enhanced gain and the light induced background. In Fig. 8(a) there is negligible contribution from secondaries because the energy absorption in the aluminium backing of the screen means that for $V_2 = 3$ kV the secondaries are close to the brightness cut-off. The primaries, however, are accelerated through a voltage of 4 kV ($V_1 + V_2$) which is sufficiently removed from the toe of the gain curve to make their contribution significant.

In Fig. 8(b) the secondaries are removed from the toe of the characteristic and the secondary emission contributes significantly to the gain. The light induced background is almost certainly caused by the broad point spread function of the secondaries from the mesh. This is because when $V_2 \gg V_1$, there is field penetration through the mesh apertures into the low voltage image section in the vicinity of the mesh. The secondaries from the mesh wires no longer see an immediate retarding field, but a diverging accelerating field which enables them to move some considerable distance both laterally and longitudinally before eventually being retarded and returned to the mesh.

The effects of this process are shown in Fig. 9. Figure 9(a) shows the electric field distribution in the vicinity of the mesh for $V_1 = 1$ kV and $V_2 = 3$ kV and Figure 9(b) shows the same distribution when V_2 is increased to 10 kV. The majority of secondary electrons will have emission energies of the order of 2eV, and it can be shown that the diameter of a typical disc of confusion for the conditions of Fig. 9(b) is of the order of 200 μm. In fact a figure of 170 μm has been measured on tube images. For the conditions in Fig. 9(a) a secondary electron with 2 eV emission energy could move about 70 μm towards the photocathode and for the conditions in Fig. 9(b) this distance would be 90 μm. For lower energy secondaries the difference is much greater; for instance for 0·2 eV and with V_2 at 3 kV and 10 kV these distances are 15 μm and 40 μm respectively. However, the number of electrons emitted with such low energies is relatively small.

Although the longitudinal and lateral displacements of the secondaries are greater for the conditions in Fig. 9(b), they are not so different as to account for the noticeably different levels of light induced background in Figs. 8(a) and 8(b). It must therefore be concluded that the differences between these two images are due to the fact that in Fig. 8(a) only primaries contribute to the image (the effects of the secondaries have been removed by the energy absorption in the aluminum backing of the screen) whereas in Fig. 8(b) the secondaries pass through the screen backing and contribute significantly to the formation of the image.

Conclusions

In conclusion it may be useful to draw a comparison between the performance of these tubes and that of other conventional image tubes.

(1) Resolution: this is comparable to that of conventional magnetic tubes. It is higher than that found with proximity diode tubes and more uniform than that of electrostatic tubes.//
(2) Geometry: this is better than that of electrostatic tubes, and is not far short of that found in proximity diodes.
(3) Light induced background: this can be higher than in most tubes. It is due to secondary emission from the mesh. If high gain is not so important, the secondary emission can be supressed by coating the photocathode side of the mesh with either aluminium smoke or carbon. These materials have a secondary emission coefficient of the order of 0·5, and will also reduce optical reflection. Under these conditions a high transmission mesh should be used to retain a high equivalent quantum efficiency.

If high gain is required, the photocathode side of the mesh should be coated with a good secondary emitting material with high yield and low emission energy. The latter requirement is essential if the light induced background is to be kept low. A suitable material may yet be found in the III–V compounds. For this application the mesh transmission should be relatively low.

(4) Brightness uniformity: this is better than that found in electrostatic tubes, and can be comparable to other magnetic tubes and the proximity diode. It is dependent on the method of photocathode processing and is limited purely by the uniformity of the photocathode sensitivity.

Acknowledgments

The author would like to thank Professor J. D. McGee for his interest in the original concept and Dr. H. G. Lubszynski for useful discussions during the latter part of this development. He is grateful to the following staff of Electron Physics Ltd. for their assistance in constructing the devices: Mr G. Ensell, Mr D. Neaves, Mr R. Rainger and Mr C. Spicer. Finally he would like to thank Ralli Brothers Trading Ltd. for permission to publish this paper.

REFERENCES

1. McGee, J. D., U.S. Patent No. 3889144.
2. McGee, J. D., Airey, R. W. and Aslam, M., *In* "Adv. E. E. P." Vol. 22A, p.571 (1966).
3. Hachenberg, O. and Brauer, W., *In* "Adv. E.E.P." Vol. 11, p. 433 (1959).

Elimination of Corona and Related Problems with Astronomical Image Tubes

R. H. CROMWELL and J. R. P. ANGEL

Steward Observatory, University of Arizona, Tucson, Arizona, U.S.A.

Introduction

Detectors utilizing image intensifiers are in standard use at a large number of observatories at the present time, yet many of these intensifier systems share the problem of corona discharge or high voltage leakage on certain occasions. The problem can become very troublesome under specific environmental conditions, such as high humidity, cooled photocathode operation, or high altitude, and the effects can vary from producing a dark emission which is too high or variable to producing geometrical instabilities in the output image. With multi-stage intensifiers that operate at voltages in the range of 30 to 50 kV, corona and other high voltage leakage problems can become very severe, sometimes rendering the detector inoperable. This paper summarizes a variety of techniques we have utilized that, taken together, eliminate corona and voltage leakage problems.

Major Trouble Sources

We have found that nearly every intensifier system that exhibits corona emission or other high voltage leakage problems does so because, outside of an obvious component failure, it has one or both of the following characteristics. (1) There exists a dielectric material which insulates a component maintained at high voltage from its surroundings, but the outer surface of the dielectric is at an ill-defined or totally "floating" potential; therefore, a charge builds up on the outer surface of the dielectric until a general corona discharge, or arc, is created. (2) There exists a direct leakage path along the boundary of dissimilar dielectric materials that provides electrical conduction between the component and its surroundings. With sufficiently high voltages, of course, the leakage path need not lead directly to a ground plane to cause voltage breakdown, but only to a point where conduction can be completed by ionized gas or by surface conduction along a dielectric.

The first problem, that of floating potential dielectric surfaces, is regularly encountered in configurations where an image tube is encapsulated in an insulating material such as RTV (room temperature vulcanizing) silicone rubber, and the entire assembly is further surrounded and supported by some type of rigid dielectric material, such as plastic. Although many successful image tube systems are constructed in this way, all too often such a system is vulnerable to high humidity and other environmental conditions. The various dielectric surfaces intended to insulate the high voltage components slowly become charged and eventually discharge to a nearby ground plane, causing a visible arc or corona.

In principle, this problem can be solved by making certain that the material in which the intensifier is encapsulated has sufficient strength to hold off the full applied voltage, and then maintaining all exterior surfaces at ground potential by use of a conducting outer skin. In this manner, there are no floating potential surfaces. All voltages are intentionally dropped across a dielectric of known properties, and surface charges are conducted away harmlessly by the outer conducting layer.

We have investigated the properties of various materials in order to achieve the characteristics described above, and have determined a set of mutually compatible materials that may be used for this purpose. In selecting the various dielectric materials, it has been necessary always to check that a proper bond exists between dissimilar materials. This is to avoid the second problem mentioned earlier, that is to prevent surface conduction along the boundary between dissimilar dielectric materials.

Solution to the Problem

Figure 1 is a diagram of a fibre optic coupled three stage image intensifier comprised of three electrostatically focused single stage tubes, each with fibre optic input and output faceplates. The Figure illustrates all the various techniques we have used in eliminating high voltage breakdown problems.

In the Figure, D is a silicone rubber RTV potting compound. During encapsulation the intensifier assembly, its potting mold and the liquid RTV are all placed in a moderate vacuum in order to prevent voids in the potting caused by trapped air bubbles. A variety of RTV compounds exist, and most have similar physical and electrical characteristics. However, some compounds are not compatible with certain materials (e.g. some we have tried will not cure in the presence of silicone high voltage

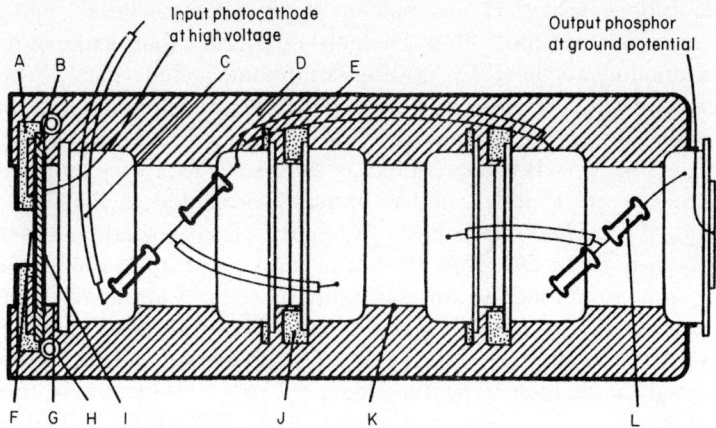

FIG. 1. Illustration of techniques used to eliminate corona and high voltage breakdown. Identified are mechanical locating and electrically insulating member A, glass window B with conducting coatings I and F with antireflecting overcoating F, silicone high voltage wire C, RTV silicone rubber D, conductive RTV layer E, conductive paint G, pneumatic pressure relief tube H, insulating couplers J, single component RTV primer K, and voltage divider network L.

wire and other materials we use during the potting procedure). Therefore, in order to minimize this variable, we have selected just one compound, RTV-511, manufactured by General Electric,† which is compatible with most materials we use.

One of the most significant problems with the silicone RTV compounds is that the silicone rubber by itself does not bond well with other materials. Therefore, any interface between the RTV and another dielectric material is likely to provide a leakage path for high voltage. A few priming compounds are commercially available that are intended to improve the bond, but we have found them inadequate. Fortunately, a very good chemical bond can be achieved with selected dielectric materials by first coating them with a single component moisture-curing RTV, such as Silastic 732 RTV Sealant/Adhesive, manufactured by Dow Corning.‡ We have found that every surface that comes into contact with the RTV-511 potting compound must first be primed with a thin coating of 732 RTV, including the silicone high voltage wire C, as well as any old, previously cured silicone potting compound. Dielectric materials that bind well to the RTV 511 once they are primed are illustrated in Fig. 1 and include the glass window B, the glass envelope of the intensifiers K, a type G-10 fibreglass/epoxy laminate used in components A and J, the

† G. E. Silicone Products, Waterford, New York 12188, U.S.A.
‡ **Dow Corning** International Ltd., Chaussée de la Hulpe 177, 1170 Brussels, **Belgium.**

silicone rubber tubing H and silicone high voltage wire C, and some epoxy materials not illustrated. Dielectric materials that we have had no success bonding to the RTV potting compound include most, if not all, plastics, especially acrylics, polyethylene, PVC, and some epoxy materials. Bonding to metals is weak, even after use of 732 RTV primer.

The glass window B is overcoated on both sides with very thin tin oxide conducting layers F and I, and is optically cemented to the fibre optic faceplate of the intensifier. The availability of such windows† was first pointed out to us by Dr. E. A. Beaver at the University of California, San Diego, and these windows are substantially more transmissive than the other conducting windows known to us. The transmission losses associated with the conducting layer of tin oxide are very slight, as they are determined by the high refractive index, ~ 2 in the visible, of the tin oxide layer. Both layers are $\ll \lambda/4$ in thickness. The reflections on either side of the inner layer are thus of opposite phase and cancel out. A 1/4 wave coating of low index is put on the outside layer, F, so the loss is no more than a few percent. The transmission through the entire window is shown in Fig. 2. The conductive coatings cover only a little more area of the glass than is required by the active area of the intensifier photocathode. The remaining uncoated portion of the glass is utilized to insulate electrically the two conductive surfaces from each other. A window thickness of 2·5 mm is sufficient for operating voltages up to 45 kV. A silver conductive paint G maintains electrical continuity between the intensifier high voltage electrode, the fibre optic faceplate, and the inner conducting layer I. The outer conducting layer F of the window is held at ground potential via contact with a conductive RTV layer E, which is painted over the entire external surface of the encapsulated intensifier, except for the input and output faceplates. Thus, the window and the potting compound serve to drop the entire potential applied to the intensifier between their respective inner and outer surfaces, doing so in a controlled, well defined manner, with no surfaces left at floating potentials. The conductive RTV provides sufficient conductivity to bleed off any troublesome charge that otherwise would build up on the outer surface, and has the additional required properties of remaining elastic and of bonding well to the silicone potting. This material is Ablebond 190–8 and is manufactured by Ablestick Adhesive Company.‡

Item A in Fig. 1 provides a structurally rigid member that is used to mechanically locate the high voltage end of the intensifier in an external

† Manufactured by the Scientific Coating Laboratory, 360 Martin Av., Santa Clara, CA, 95050, U.S.A.
‡ Ablestik Laboratories, 833W, 182nd Street, Gardena, CA90248, U.S.A.

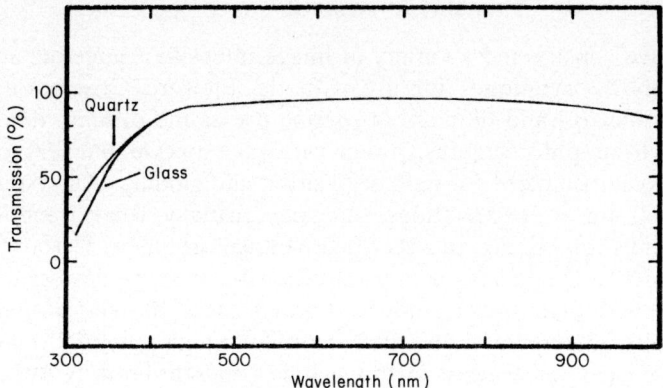

FIG. 2. Transmission versus wavelength of 2·5 mm thick window coated on both sides with a thin layer of conducting tin oxide and overcoated on one side with a $\lambda/4$ coating centred at 550 nm.

housing, not illustrated. It is made of NEMA Grade G-10 fibreglass/epoxy laminate, which has the especially desirable characteristics of high dielectric strength, high volume resistivity, good structural characteristics, good bonding (when primed) to RTV 511, and, unlike any other dielectric material satisfying these criteria, a relatively low thermal coefficient of expansion. The low thermal coefficient makes this material uniquely compatible for use as a structural member in combination with glasses and metals. There is a region between A and window B that is filled with potting compound in order to guarantee that the outer window surface is well sealed against voltage leakage.

A problem that cannot be overlooked with silicone rubber potting compounds is their relatively high coefficient of thermal expansion. Because of this property, and because cooled image intensifiers can experience temperature changes of the order of 100°C, there should be plenty of clearance between any external housing and the potted image tube. Similar considerations should be made regarding internal structures as well. An example of this in Fig. 1 is the pneumatic tube H made of silicone tubing. It serves to relieve the pressure that builds up against the glass window B when the adjacent potting expands or contracts.

Figure 1 also illustrates how we use the internally potted voltage divider network L to drop a portion of the voltage across each fibre optic coupling between stages, thus reducing the overall voltage.[1] We drop characteristically 7·5 kV across a coupling and operate each stage at 15 kV. Thus, for the three stage tube illustrated, the total operating voltage is 30 kV. The coupling assemblies, K, are made of G-10 material.

Results

We have constructed a variety of image intensifier detectors according to the above techniques for use with the Steward Observatory 90 in. (2·3 m) reflector, and they are in routine use as the primary detectors in our spectrographic cameras, direct cameras, speckle imagery cameras, and television cameras for field acquisition and guiding. All systems have performed up to expectations, showing virtually total freedom from corona emission, arcing, or other high voltage problems. Certainly, it has been easier to obtain satisfactory operation out of those devices which are operated with their photocathodes at ground potential, but proper operation can be obtained with the photocathode at high potential. Indeed, the tubes we have constructed in this manner operate entirely satisfactorily even when submersed in water.

With the addition of the multi-coated conductive window (Fig. 2) to a fibre optic intensifier, the faceplate is not only protected against high voltage breakdown, but it is more transmissive. Over the spectral region 425–900 nm, we measure an increase in transmission of 2 to 5 percent.

Application of similar window and encapsulation techniques to the Kron electronographic camera at the Mount Hopkins site of the Smithsonian Astrophysical Observatory has resulted in a reduction in spurious background emission.[2]

Acknowledgment

This work has been supported in part by the National Science Foundation.

References

1. Miller, J. S., Robinson, L. B. and Wampler, E. J., *In* "Adv. E.E.P." Vol. 40B, p. 693 (1976).
2. Chaffee, F. H., private communication.

A New Photocathode for X-ray Image Intensifiers Operating in the 1–50 keV Region

J. E. BATEMAN and R. J. APSIMON

Rutherford Laboratory, Chilton, Didcot, Oxon, England

Introduction

The X-ray image intensifier is a device extensively used in scientific research, industry and medicine. It is also, unfortunately, a difficult and expensive device to manufacture. The chief difficulty appears to lie with the initial absorption of the X-ray (usually in a scintillating layer) and one of the main costs is associated with the need to produce a large area photocathode to convert the scintillation light into an electron signal. During a study of the possibilities of fast particle detection in low density alkali halide layers (see for example Ref. 3), it occurred to the authors that such layers, while not without their problems, might offer a useful opportunity of combining the two first stages of the X-ray imaging process and enable one to pass directly from the incident X-ray to low energy electrons.

The work of Goetze, Boerio and Green[1] summarizes the techniques, properties and applications originally associated with low density films of KCl. Our interest has centred on the production of layers of ~0·5 mm thickness and ~5% relative density of CsI, a better material for X-ray detection. While it was immediately clear that the stopping power of such a layer would be low compared with conventional evaporated CsI (Na) layers (approximately 1/5), there were indications that increased spatial resolution could be anticipated (compared with conventional X-ray imaging devices).

Properties of a Low Density CsI Cathode

The CsI layers were produced essentially as described by Goetze for KCl.[1] The mounting system consisted of a 5 μm thick aluminium foil stretched and glued onto a pyrophyllite ring.

Quantum Efficiency

The detection process of an X-ray in a low density layer is rendered schematically in Fig. 1. An incident photon enters through the support foil, converts in the CsI layer giving rise to a high energy electron (photo-, Auger or Compton in origin) which dissipates the energy of the X-ray in producing free electrons in the material. Under the influence of the collecting electric field a cloud of electrons emerges from the output face of the layer to be collected into any appropriate amplifying/imaging system. Clearly the quantum efficiency (QE) of the layer can be represented as follows

$$\varepsilon = [1 - \exp(-\mu_m \rho T)](1 - P_0) \qquad (1)$$

where μ_m is the mass absorption coefficient (i.e. the molecular cross section of CsI at the photon energy concerned), ρ is the density and T the thickness of the layer. P_0 is the probability of no electron being emitted from the layer by an event. The first term is a monotonically decaying function of the incident photon energy with sudden jumps at the shell absorption edges. The second term naturally tends to increase as the photon energy rises yielding more induced secondaries in the layer.

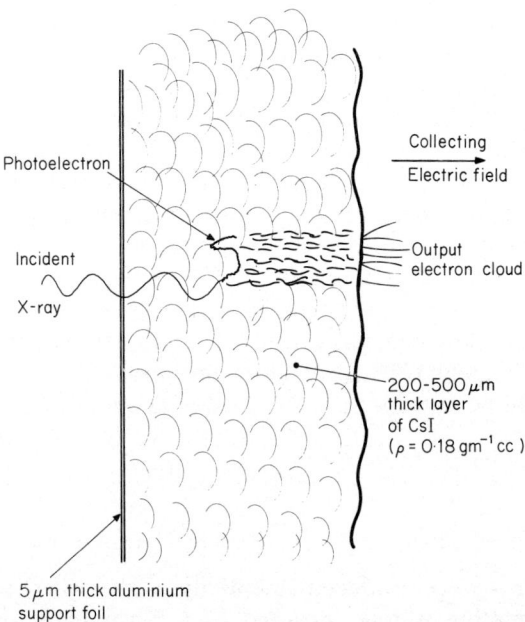

FIG. 1. Schematic diagram of an X-ray absorption event in a low density cathode.

The first measurements of the quantum efficiency were done with a simple arrangement. The secondary electrons from a cathode were collected onto a Mullard J27 channel plate multiplier (CPM) which produced a charge gain of some 10^6. An amplifier chain then supplied a pulse height analyser (PHA) which produced a pulse height distribution and counted the events above a set threshold. Quantum efficiencies were evaluated from the observed counting rates produced by a variety of X-ray sources (^{55}Fe, ^{241}Am, Rb, Mo, Ag, Ba, Tb, K shell fluorescence).

Figure 2 shows the counting rate induced by the Tb X-ray source (~44 keV) as a function of the collecting potential. The maximum field was ~3 kV cm^{-1} which was set by the poor quality of the cathode used. Subsequent cathodes have been used successfully with 3 kV across a gap of less than 1 mm. However, as Fig. 2 implies, high collecting fields are unlikely to improve the QE much further, but the mean number of secondaries ($\bar{n}$) increases rapidly making for an increase in brightness.

The spectrum of secondaries from the ^{55}Fe X-rays is shown in Fig. 3 (at maximum collecting field). The PHA spectrum consists of 256 channels, each channel calibrated at 0.9 electrons. The counting threshold is ~10 electrons and the mean approximately 30 i.e. ~5 electrons per keV of particle energy. This sensitivity is a function of the collecting field and with later cathodes has gone up by a factor of 5 to 10.

When the counting rates observed with the various sources were normalized, the QE response data of Fig. 4 were produced. The circles

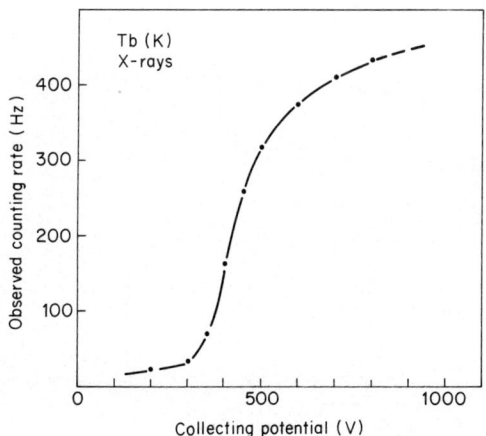

FIG. 2. Observed counting rate from a low density cathode exposed to Tb K X-rays as a function of the collecting potential.

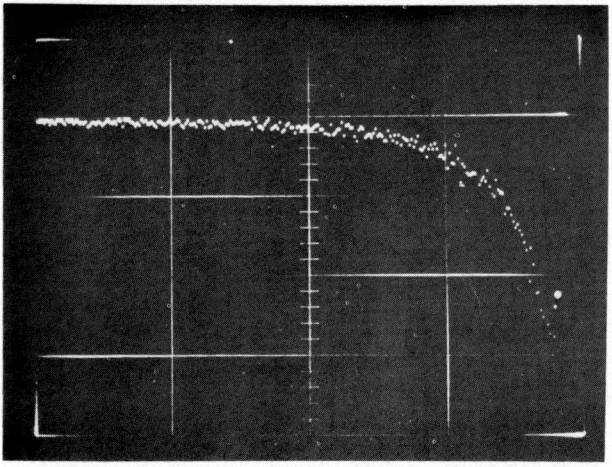

FIG. 3. Pulse height distribution observed from an X-ray detector incorporating a low density layer and a CPM. Secondary signal peak >200 electrons and mean ~30 electrons.

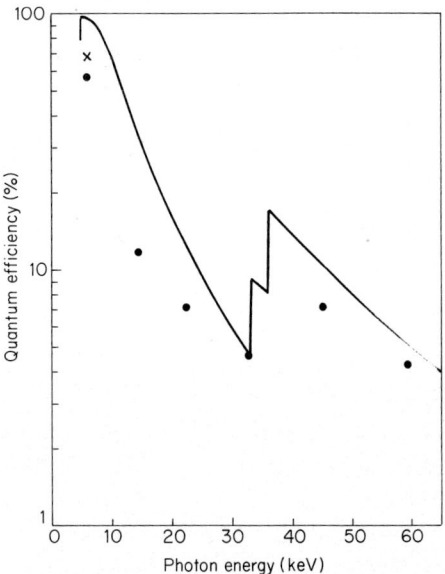

FIG. 4. Measured quantum efficiency of the X-ray detector as a function of the photon energy.

show the measured QE of a layer 0·35 mm thick and 4% dense (approximate figures). The solid curve plotted on Fig. 4 is simply the first term of Eqn. (1) with $\rho = 180$ mg cm^{-3} and $T = 0·35$ mm. Clearly this term dominates but the presence of the $(1 - P_0)$ term is indicated by the systematic upward drift of the QE relative to the absorption term as the energy deposit in the layer increases. The QE of the present cathode would appear to be very acceptable in the low energy range up to ~10 keV (where there are several applications). The cross on Fig. 4. shows the corrected efficiency for ^{55}Fe X-rays if the substrate is removed.

Spatial Resolution

While accurate measurement of the spatial resolution response functions of an X-ray image intensifier at high spatial frequencies (~20 lp mm^{-1}) is a matter of some difficulty (particularly at higher X-ray energies), it was felt that some indication of the likely spatial resolution of a low density X-ray cathode could be deduced from previously published data.[2,4] Degradation of the spatial resolution can be expected to arise from two causes: (a) the physical spread of the photoelectrons released by the X-ray in the layer, and (b) the lateral spread of the induced charged cloud as it drifts through the layer.

Information on the second effect was obtained from Goetze[4] who reported that a visible light image intensifier utilizing a KCl dynode as the gain element had 25 lp mm^{-1} resolution. While our layers are considerably thicker than those of Goetze, it was anticipated that the resolution was unlikely to be less than 10 lp mm^{-1} for a 500 μm thick layer. Thus one of the main concerns of our development work was to see how the spatial resolution varies with layer thickness.

The calculation of the limits to the spatial resolution set by the X-ray absorption processes is achieved with reasonable precision by referring to the work done by Bateman et al.[2] on xenon gas proportional counters. The geometry of the converting layer is similar (planar) and the mean atomic number of CsI is equal to that of xenon. (The elements are all neighbours in the periodic table.) Thus by simply adjusting the spatial dimensions by the ratio of the densities of the two materials (4·9 mg cm^{-3} for the proportional counter gas and 180 mg cm^{-3} for 4% CsI) one reaches a reasonable estimate of the spatial resolution of the CsI layer as a function of X-ray energy (Fig. 5). Following the $f/0·2$ curve (the spatial frequency at 20% contrast) it can be seen that a spatial resolution of better than 10 lp mm^{-1} can be expected up to 26 keV and in a band from ~38–56 keV.

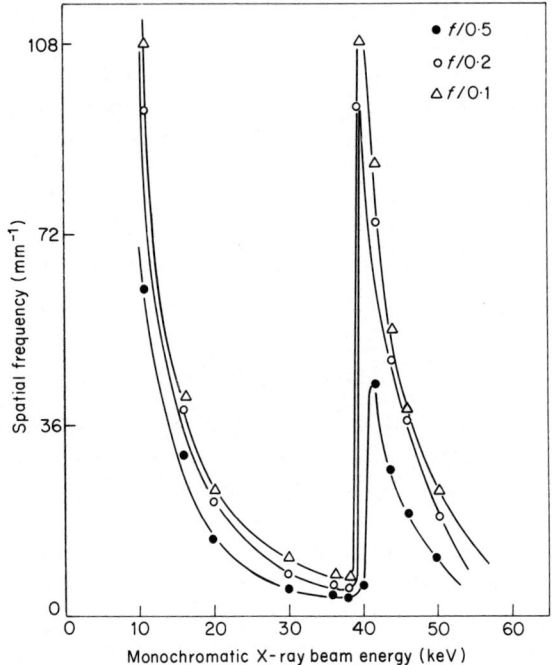

FIG. 5. Data showing the physical limits to the spatial resolution at 3 different contrasts in a xenon detector of density equal to that of a 4% CsI layer.[2]

Time Resolution

The basic secondary emission process from solid surfaces is known to be very fast ($\lesssim 10^{-11}$ sec) and the use of gold cathodes in streak cameras is established practice. The possibility of a much enhanced detection efficiency in the CsI layers (relative to a gold cathode) at X-ray energies above 1 keV led us to consider the time resolution they could yield. The major effect would appear to be the transit time fluctuations in the layer. The mean kinetic energy of emerging electrons from KCl layers is ~ 4 eV,[1] giving a transit time of $0 \cdot 8$ psec μm^{-1}. For X-ray energies of less than 8 keV, when layers of 50 μm will give useful detection efficiencies, time resolutions of ~ 40 psec FWHM can be anticipated. However as the layer is made thicker to increase the X-ray absorption at higher energies, the time resolution will degrade proportionately. Indications are that layers having a density approaching 10% are feasible so that efficiency may be maintained without sacrifice of the time resolution at higher X-ray energies.

Since X-ray streak camera images are often limited by poor quantum statistics the application of CsI layers to this technique is now under serious study.

The Test X-ray Image Intensifier

The test X-ray image intensifier (Fig. 6) is an extremely simple device consisting of a CsI cathode, a J27 channel plate multiplier (CPM) and a P·11 phosphor screen mounted closely together (for proximity focusing) in a small vacuum chamber. A small aluminium foil window at the other end allows the entry of the test X-ray beam. Test masks are supported as close as possible to the front of the cathode for imaging purposes. An ion pump keeps the pressure down to $\sim 10^{-7}$ Torr and an external resistive divider chain provides the required potentials. Typical operating potentials are 8·1 kV overall with 2·7 kV collecting potential, 2·2 kV across the CPM and 3·2 kV acceleration onto the phosphor screen.

The X-ray sources used in the tests generally had active diameters of 3 to 5 mm producing a projection spread of less than 10 μm on the cathode. One noticeable flaw is observed in some of the pictures, namely a bright spot at the bottom of the field of view. This is believed to be a fault in the CPM mounting since it is controlled only by the potential applied across the CPM. Recording of the test images was done on a Polaroid CR-9 camera with type 107C film.

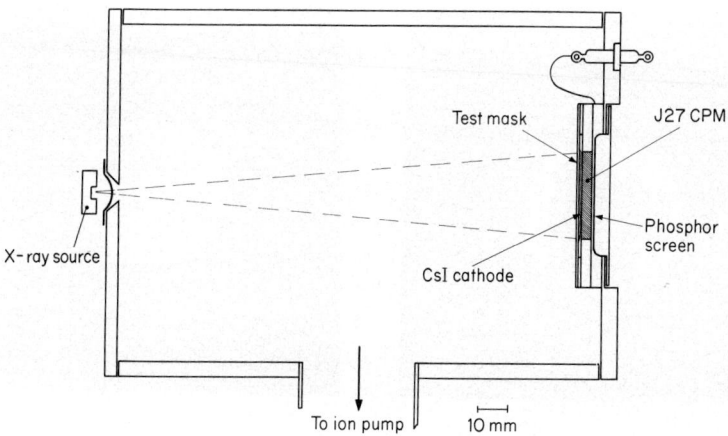

Fig. 6. Schematic sectional drawing of the test X-ray intensifier used for imaging tests on the low density cathodes.

Preliminary Tests of the Imaging Properties of the CsI X-ray Cathode

The test mask shown in Fig. 7(a) was photo-etched from 20 μm copper on a 50 μm thick substrate of Kapton film. The radial copper bars are 1 mm wide at the outside and 0·1 mm at the inside of the pattern; in fact the etching process runs out of resolution at this point leaving the bar ends ragged. The copper coating is quite adequate to produce high contrast in a beam of ^{55}Fe X-rays (5·9 keV) as Fig 7(b) shows. The tips of the radial bars appear to be reproduced faithfully indicating spatial resolution of >5 lp mm^{-1}.

The details of all exposures presented in this paper are listed in Table I. Fig 7(b) represents a 3·5 min exposure at a low EHT and an aperture of $f/11$. The photon flux at the X-ray intensifier was $8·5 \times 10^4$ cm^{-2} sec^{-1}. For higher energy X-rays a test mask consisting of a 0·5 mm thick brass watch wheel glued onto a Melinex backing was used. The tooth spacing on the gear is ~ 200 μm which provides an estimate of the spatial resolution. Fig. 8 shows the results of exposing this test object to ^{55}Fe, MoK, AgK and ^{153}Gd X-rays.

The image in ^{55}Fe X-rays (5·9 keV) (Fig. 8(a)) is almost as good as an ordinary photograph. Proceeding upwards in X-ray energy our images are severely limited by the source strength. Thus the Mo source cannot overcome the quantum mottle even after 60 min (Fig 8(b)). However, it is interesting to note that the Ag X-ray source can give a picture in 90 min with only 83·2 photons cm^{-2} sec^{-1} in the beam (Fig 8(c)).

When a strong ^{153}Gd source is used, an image of the gear wheel is obtained quickly (Fig 8(d)) but with reduced contrast. This is partly due

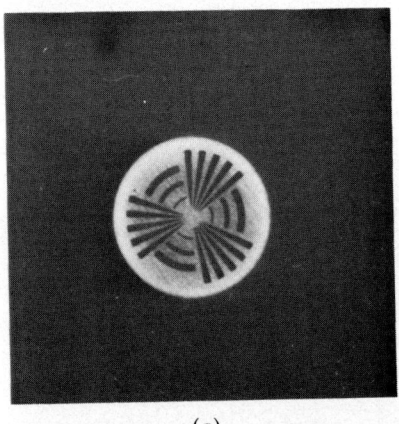
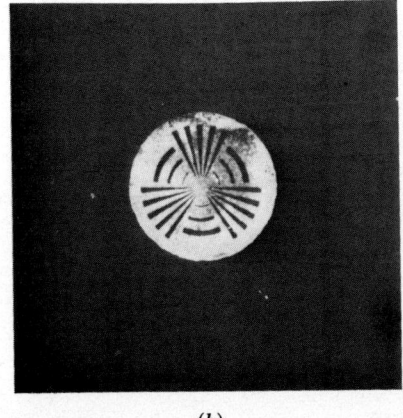

FIG. 7. (a) Copper on Kapton test mask. (Scale: ×0·8.) (b) Image of the test mask in ^{55}Fe X-rays. (See Table I for details.)

A NEW PHOTOCATHODE FOR X-RAY IMAGE INTENSIFIERS

TABLE I
Exposure data for Figs. 7 and 8. Camera aperture $f/11$.

Test object	Collection potential (kV)	Overall EHT (kV)	Exposure Time (min)	Flux at intensifier (cm^{-2} sec^{-1})	Total tissue dose for exposure (mrem)	X-ray source type	Energy (keV)
Copper mask (Fig. 7)	1·98	7·4	3·5	$8·5 \times 10^4$	49·5	^{55}Fe	5·9
Gear wheel (Fig. 8(a))	2·73	8·2	2	$8·5 \times 10^4$	28·3	^{55}Fe	5·9
Gear wheel (Fig. 8(b))	2·97	8·9	60	47.6	47.6	MoK	17.5
Gear wheel (Fig. 8(c))	2·97	8·9	90	83.2	124.8	AgK	22
Gear wheel (Fig. 8(d))	2.73	8·2	1.5	$2·2 \times 10^5$	6·9	^{153}Gd	42

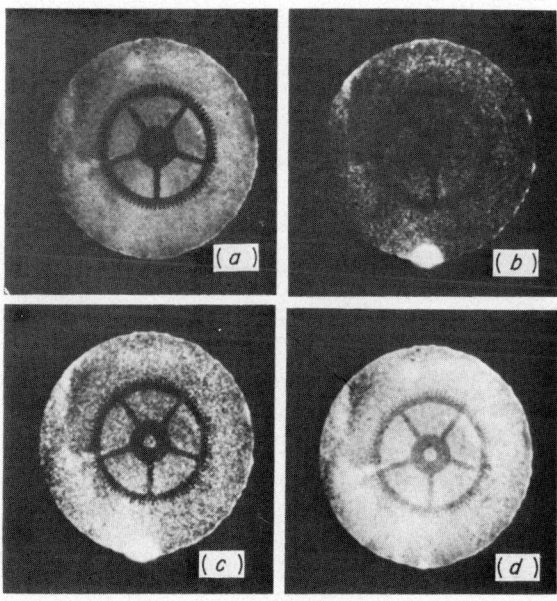

FIG. 8. Images of a watch gear wheel in various X-ray beams. Table I gives details of the exposures.

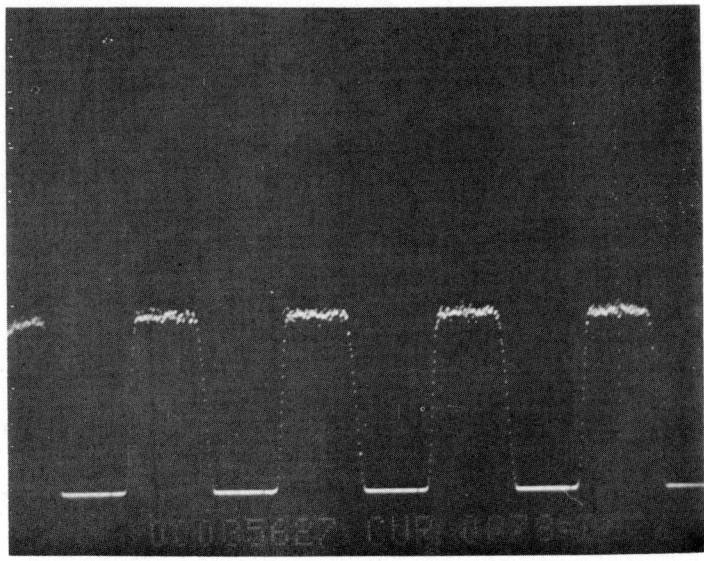

FIG. 9. Multichannel scaling spectrum given by the X-ray intensifier when exposed to a beam of ^{55}Fe X-rays chopped at 500 Hz. Each channel represents a 10 μsec counting interval.

to the partial transparency of the brass at 42 keV and the presence of 100 keV nuclear gammas in the beam. At this level of exposure the non-uniformities in the X-ray intensifier response are clearly visible as are the boundaries of the hexagonal packing structures on the CPM. Allowing for the reduced contrast, it would appear that the spatial resolution of the cathode is preserved in the 42 keV exposure as predicted by Fig. 5. However, detailed measurements with quantitative test masks are clearly required. The dose to tissue from the fluxes producing the images in Figs. 7 and 8 are given in Table I. The conversion factors used were derived from ref. 5.

A test of the time resolution of the CsI photocathode in the ~10–1000 Hz range was carried out, since this region is important for many X-ray intensifier applications. Figure 9 shows the multichannel scaling spectrum given by the X-ray intensifier when exposed to a beam of ^{55}Fe X-rays chopped at 500 Hz. The chopper waveform is faithfully represented with the counting rate in the "dark" half cycle being measured at <0·1% of that in the "light" half cycle. Tests were extended down to ~10 Hz and there was no sign of charging effects which could cause gain modulation anywhere in this frequency range.

Long term stability was investigated using the same technique (i.e. by counting the phosphor screen pulses through a discriminator into a multichannel scaler). Any change in gain shows up as a change in the measured counting rate. Several tests lasting 3 h were run and showed very similar results: a 5 to 10% droop in the counting rate followed a sharp resetting event which brought the counting rate almost back to the original level. These events occurred irregularly on a time scale of approximately 2 h. While some part of the structure of the test X-ray intensifier was clearly charging up and discharging, no significant gain change was taking place. There was no sign of the expected increase of gain as the internal field in the layer built up;[4] but perhaps this was not surprising since the predicted time scale for the current density used is several days.

The non-uniformities of the image in Fig 7(b) led us to question the quality of the cathode. However, rotating the cathode through 180° relative to the rest of the structure left the picture unchanged showing that the major blemishes arise elsewhere in the X-ray intensifier structure.

Conclusions

While the results presented are preliminary in nature, they have gone some way to indicate the potential of low density CsI cathodes for X-ray imaging. The low stopping power of the layers ($\sim 10 \text{ mg cm}^{-2}$) means that the most attractive applications lie in the soft X-ray region ($\leqslant 10 \text{ keV}$), where the high number of secondaries available ($> 10 \text{ keV}^{-1}$) also ensures good detection efficiency. In a "windowless" form the cathode should perform well down to low quantum energies ($\leqslant 10 \text{ keV}$). The ability of the cathodes to withstand exposure to the atmosphere (but not *very* high humidity) leads to great operational flexibility.

In the X-ray energy region above 10 keV the basic 4% dense cathode, while appearing to preserve good spatial resolution, has poor quantum efficiency. Work currently in progress is now indicating that our estimate of 4% for the layer density is somewhat low and that 6 to 7% is more common when argon pressures of a few Torr are used. If the secondary electrons can be successfully extracted from these layers (as appears to be the case) then enhanced performance can be anticipated in respect of the high energy X-ray detection efficiency and the time resolution of a detector of given efficiency (i.e. thickness).

Two X-ray intensifier designs are currently being pursued which utilize the cathodes. The first is a large diameter (120 mm) magnetically focused device for X-ray diffraction using Cu X-rays and aims at a spatial resolution of 12 lp mm^{-1}. The second is a small diameter (25 mm) device for X-ray topography (around 10 keV) aiming at 20–30 lp mm^{-1}.

CsI cathodes are being substituted for gold cathodes in an intensified X-ray streak camera and tests will shortly be carried out on the X-ray emissions from laser generated plasmas.

References

1. Goetze, G. W., Boerio, A. H. and Green, M., *J. Appl. Phys.* **35**, 482 (1964).
2. Bateman, J. E., Waters, M. W., and Jones, R. E., *Nucl. Instrum. & Methods* **135**, 235 (1976).
3. Faivre, J. C., Fanet, H., Gavin, A., Robert, J. P., Rouger, M. and Saudinos, J., *IEEE Trans. Nucl. Sci.* **NS-24**, 299 (1977).
4. Goetze, G. W. *In* "Adv. E.E.P." Vol. 16, p. 145 (1962).
5. International Commission on Radiological Protection, Publication NO. 21, Fig. 17. Pergamon Press, Oxford (1974).

High Resolution Phosphor Screen for X-ray Image Intensifier

H. WASHIDA and T. SONODA

Electron Tube and Device Division, Toshiba Corporation, Kawasaki, Japan

Introduction

Modern X-ray image intensifiers are required to have high resolution and high output brightness characteristics. The improvement in resolution by using CsI evaporated phosphor screens[1,2] instead of rare earth phosphors is insufficient for X-ray image intensifier photofluorography to compete with radiography using an intensifying screen. This paper describes unique input screen structures which enable a high resolution X-ray image intensifier to be constructed.

The Input Phosphor Screen

An image intensifier as shown in Fig. 1, comprises an input screen, a focusing electrode, an accelerating anode and an output screen. The input screen transforms the X-ray image to a corresponding photoelectron image. These photoelectrons are accelerated and focused onto the output screen, where they are converted to a visible light image.

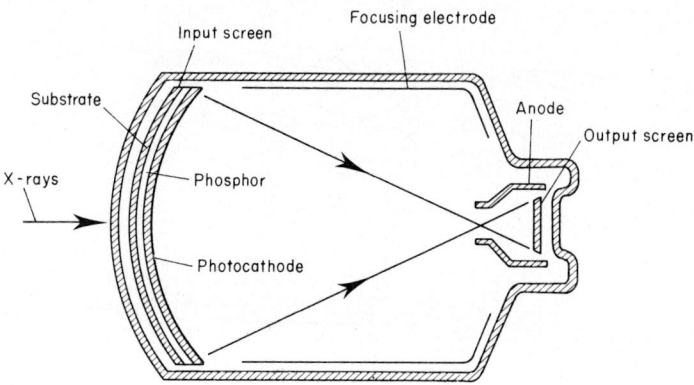

Fig. 1. Schematic illustration of an X-ray image intensifier.

In order to obtain high resolution it is important to prevent the transverse spread of luminescent light in the phosphor screen. This may be done by (1) decreasing the thickness of the phosphor layer (2) forming the phosphor in many blocks[3] and (3) making a light guide structure in the phosphor screen as shown in Fig. 2. Of these methods the light guide structure and the block structure ensure the higher X-ray absorption.[4] In general, CsI phosphor layers prepared by vacuum deposition are composed of fine crystal pillars. However, these are located so close to each other that total reflection does not occur at their boundaries and it is necessary to form a block structure. The size of the blocks should be smaller than 100 μm for high resolution.

We have developed some unique substrate structures onto which CsI phosphor layers with a block structure can be formed uniformly and reproducibly.

The Manufacturing Process

Two types of substrate structure are described below.

I: The Cracked Mosaic Pattern Substrate

An aluminium oxide mosaic pattern was formed on the concave surface of an aluminium faceplate so that the CsI phosphor layer could be grown with a block structure corresponding to the mosaic pattern of the substrate.

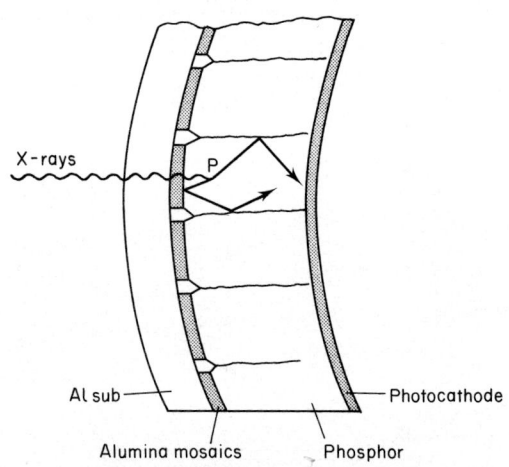

FIG. 2. Input phosphor screen with light guide structure.

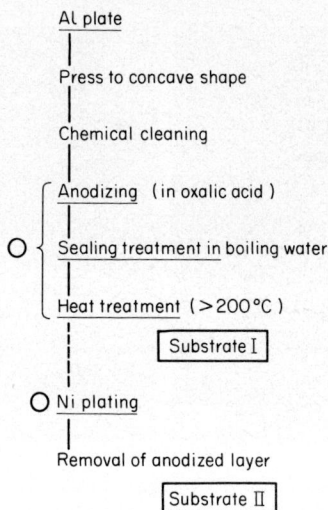

FIG. 3. Procedure for forming the two types of substrate.

The procedure is as shown in the block diagram in Fig. 3. First, the concave aluminium face plate was anodized in a solution of 3% oxalic acid to form a porous aluminium oxide surface layer of about 20 μm thickness. After sealing in boiling water this is transformed into a dense layer of boehmite (γ-Al_2O_3-H_2O) or bayerite (Al_2O_3-$3H_2O$). It is then changed back to an aluminium oxide layer by releasing the combined water by baking. This final process cracks the surface layer, creating a number of alumina areas divided by small grooves. Under proper experimental conditions, the size of these areas is 50 to 100 μm and the width of the grooves is 5 μm. A typical micrograph of the surface of this substrate is shown in Fig. 4(a).

II: The Net-like Protruded Mosaic Pattern

When Ni is deposited electrochemically on the substrate with the cracked mosaic pattern described above, selective deposition occurs on the bottom of the grooves where only a quite thin aluminium oxide layer exists. This results in the formation of a net-like protrusion on the aluminium substrate when the alumina mosaic pattern is removed. A micrograph of this substrate is shown in Fig. 4(b).

A CsI phosphor activated with NaI was evaporated in vacuum onto these substrates at a temperature of 80°C, the thickness being about 150

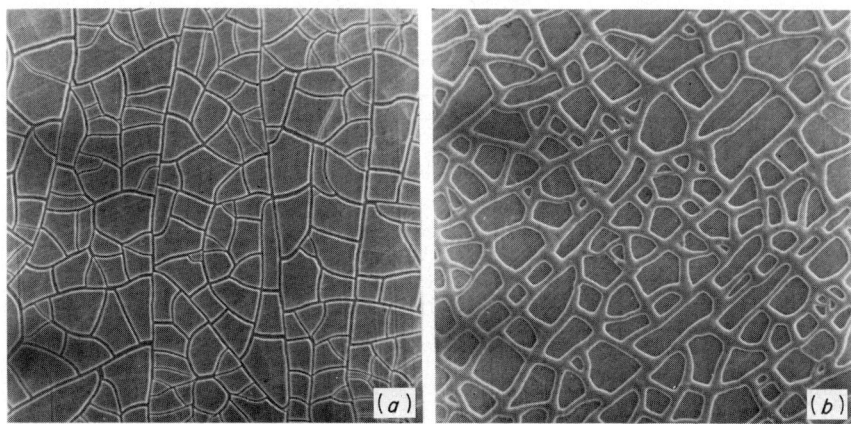

FIG. 4. Micrograph picture of (a) the cracked mosaic substrate and (b) the net-like protruded mosaic substrate. Magnification 62×.

to 200 μm. A block structure was obtained corresponding to the mosaic pattern on the two types of substrate. Using a conventional substrate the cracked structure cannot be obtained reproducibly.

STRUCTURE OF INPUT PHOSPHOR SCREEN

Cross sectional micrographs of the input phosphor screen shown in Figs. 5(a), and 5(b), and the corresponding schematic illustrations, Figs. 6(a) and 6(b) show that in each case cracks in the phosphor layer grow from the grooves or protrusions of the substrate towards the surface. Each block is composed of many pillar-like crystals several microns in diameter grown almost perpendicularly from the substrate. The difference between the substrates lies in the shape of the cracks: in the case of substrate I the cracks grow narrower towards the surface of the phosphor layer. On the other hand, in the case of substrate II the cracks become rather wider towards the surface of the phosphor layer. If the phosphor layer is made rather thicker the cracks tend to vanish in the case of the cracked mosaic substrate indicating that it is inferior to the net-like mosaic substrate. X-ray diffraction analysis shows that the preferred orientation for the growth of CsI is [200] for the first substrate and [211] for the second substrate.

MEASUREMENT OF MTF AND PHOSPHOR BRIGHTNESS

MTF and brightness characteristics of the input phosphor screen were measured using the systems illustrated in Fig. 7. A tungsten mask of thickness 50 μm on which several slits were arranged to give a spatial

HIGH RESOLUTION PHOSPHOR SCREEN FOR X-RAY IMAGE INTENSIFIER   205

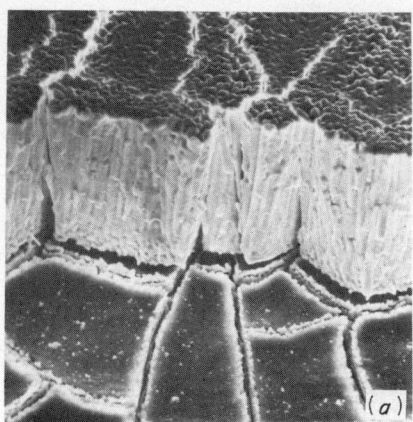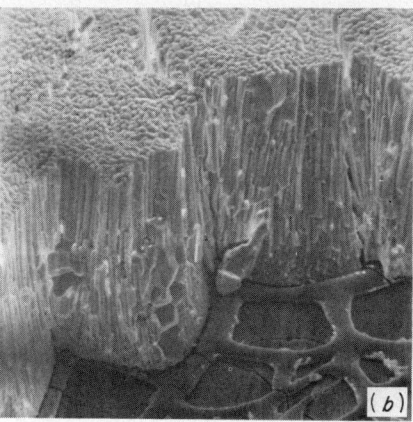

FIG. 5. Micrograph pictures of the CsI phosphor layer on (a) the cracked mosaic substrate and (b) the net-like mosaic substrate. Magnification 176×.

frequency range of 5 to 50 lp cm^{-1} was located in front of the substrate with the phosphor screen on the opposite side. It was scanned slowly. A beam of 60 kV X-rays was applied and the luminescence from the phosphor layer was collimated by a lens and detected by a photomultiplier. The measurement was carried out after the screens had undergone heat treatment in a dry nitrogen atmosphere at 250°C for an hour.

The square wave MTF characteristics are shown in Fig. 8. There is an obvious improvement over the conventional screen, especially in the range of 5 to 30 lp cm^{-1}. The net-like protruded mosaic pattern gives better results than the other probably because the size of blocks is smaller and because the electrochemically deposited nickel metal reduces the stray luminescence due to transverse scattering.

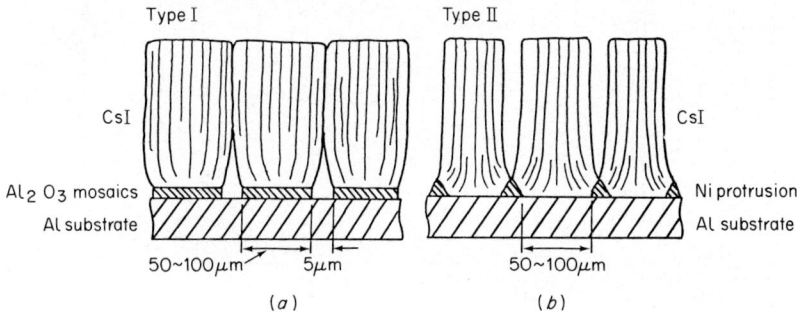

FIG. 6. Schematic illustration of the CsI phosphor layer on (a) the cracked mosaic substrate and (b) the net-like mosaic substrate.

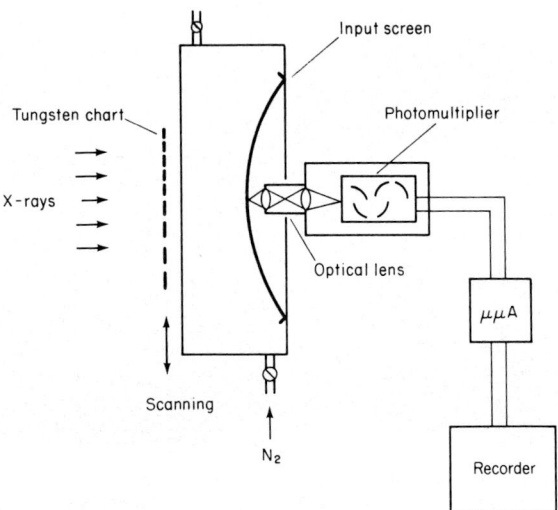

FIG. 7. Apparatus used to measure MTF and brightness characteristics.

The phosphor brightness characteristics are also considerably improved by using substrates with a mosaic structure. Results are shown in Fig. 9. The improvement is thought to be the result of increasing the external efficiency of light emission by the light guide mechanism. That is to say spreading of the output light is prevented by total reflection at the boundary of the phosphor blocks. Some decline of the brightness which occurs for the cracked mosaic substrate above 300°C may be the result of growth of the phosphor blocks, and hence a decrease in the light guide

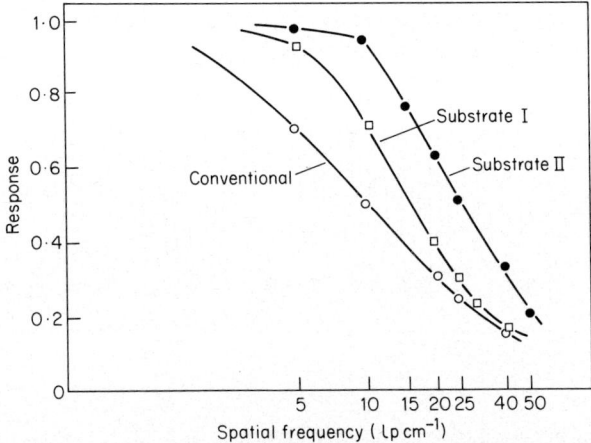

FIG. 8. Measured MTFs of phosphors on conventional and mosaic substrates.

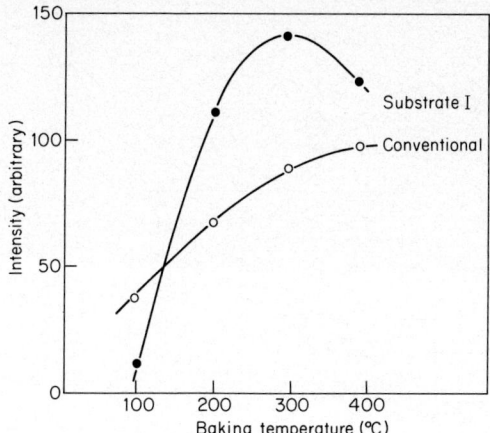

Fig. 9. Brightness characteristics of phosphors on conventional and cracked mosaic substrates.

mechanism. The two types of mosaic substrate gave almost identical results.

Application to an Image Intensifier

Intermediate preservative layers and photocathodes were deposited onto phosphor layers fabricated as described above. The preservative layer was introduced to increase the photocathode sensitivity. Nine-inch image intensifiers with these input phosphor screens showed limiting resolutions of 37 lp cm^{-1} for the cracked mosaic substrate and more than 41 lp cm^{-1} for the net-like protruded mosaic substrate.

Acknowledgment

The authors wish to acknowledge the helpful discussions and encouragement from Mr. A. Onoe, Mr M. Ishizaka and Mr N. Harao. They would like to thank Mr Y. Yamaoka for his help in the experiments.

References

1. Bates, C. W., In "Adv. E.E.P." Vol. 28A, p. 451 (1969).
2. Harao, N. and Minami, H., *Toshiba Review* **111**, 39 (1977).
3. Macleod, N. A., US Patent No. 3,041, 456.
4. Stevels, A. L. N. and Schrana-de Pauw, A. D. M., *Phillips Res. Rep.* **29**, 353 (1974).

Performance of an X-ray Television Detector for Crystallography

U. W. ARNDT and D. J. GILMORE

MRC Laboratory of Molecular Biology, Hills Road, Cambridge, England

Introduction

At previous Symposia and elsewhere we have described an X-ray Area Detector for diffraction studies.[1-3] In its present form (Fig. 1) it consists of a polycrystalline ZnS (Ag) phosphor deposited on an 80 mm fibre optic disc which is coupled to a demagnifying diode-type electrostatic image intensifier (Varian VLI–116).† The 40 mm output screen of this intensifier is, in turn, fibre optically coupled to a television camera tube (English Electric P887 Image Isocon).‡ The television camera is operated at the normal European standard of 625 singly interlaced lines and a field frequency of 50 Hz. The video signal is digitized as a 4 bit 16 grey level quantity at a sampling frequency of 6 MHz which produces 300 samples during the active line period of 50 μsec. Corresponding picture elements on odd and even lines are summed so that the digitized image consists of 300×300 picture elements. Successive digitized frames are added into a MOS shift register with a capacity of 90 000 12 bit words which acts as a frame store and which circulates once in a frame period. The summation is continued until a sufficient number of photons have been recorded for the desired statistical precision; a maximum of 256 frames can be summed before the register can overflow. The contents of the shift register can be transferred to a DEC PDP 11–40 computer, 32 words at a time, via a fast buffer which can be addressed by the computer (Fig. 2).

White noise is effectively integrated out by the summation; an analysis of the sources of the noise in the system leads to an expected detective quantum efficiency in excess of 50%.

In the present communication we discuss the reasons for past and projected alterations to the design and the measurements we have made of the performance of the detector.

---

† Varian LSE, 601 California Ave, Palo Alto, CA 94394, U.S.A.
‡ English Electric Valve Co. Ltd., Chelmsford, Essex, England.

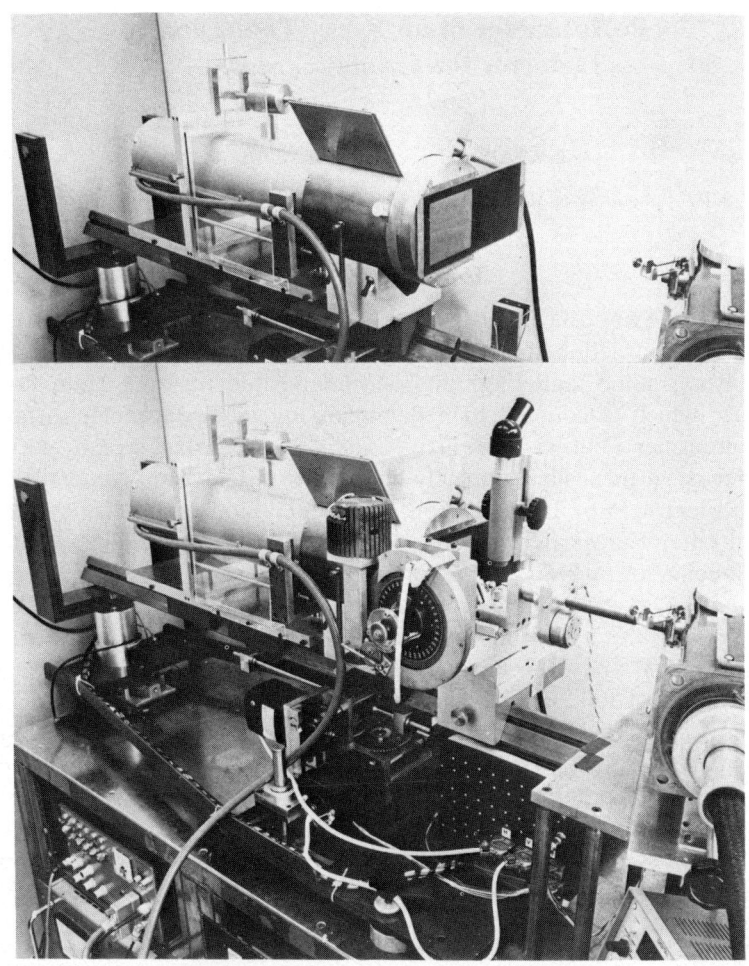

Fig. 1. The X-ray detector. The upper photograph shows the image intensifier and the camera only. The front slide carries an X-ray transparent window or a neutral density filter for the projection of optical test patterns. The camera is counterbalanced and self-aligning so that good optical contact is made between the two fibre optic faceplates. The lower photograph also shows the horizontal and vertical translations for the optical bench and the specimen rotation device. X-rays from the tube on the extreme right are crystal monochromatized and reach the camera through the horizontal beam pipe.

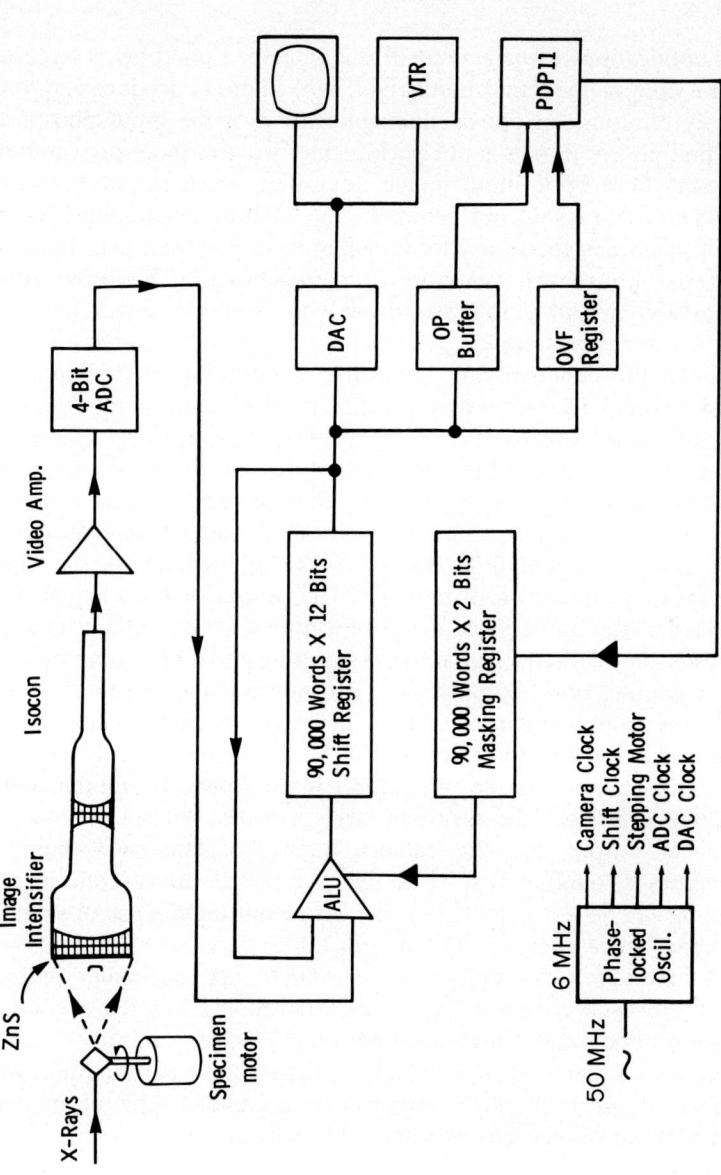

FIG. 2. Schematic diagram of detector circuits. The two-bit masking register can be directly addressed by the computer to control the four possible functions of the Arithmetic Logic Unit in different parts of the image: add, subtract (for background correction), recirculate and clear.

Components of the System

The Image Intensifier

For our applications the diameter of the detector should be as large as possible. In view of the small number of light photons produced by our 8 keV X-ray photons, an optical demagnification of the input phosphor on to the first photocathode is not permissible; we are, therefore, limited by the largest fibre optic input image intensifier which is commercially available (8 keV X-rays do not penetrate the windows of standard X-ray image intensifiers designed for harder radiation so that we are obliged to use an external phosphor). Television camera tubes with X-ray sensitive targets, equally, are only manufactured with windows which are too absorbent for our application.

The gain of the demagnifying intensifier is sufficient, in principle, to permit lens coupling its output to a low light level TV camera tube such as an isocon or an intensified silicon diode array tube (SIT or EBS). In practice we have found that the vignetting of any but the most expensive large aperture relay lens systems precludes their use and we have adopted 1:1 fibre optic coupling between the intensifier and the camera tube. Only very small potential differences may be applied across the fibre optic faceplates without risk of damage. The output of the intensifier is thus operated at the potential of the photocathode of the camera tube, so that the intensifier cathode is maintained at a negative high potential by means of a floating power supply. This procedure is inconvenient, especially with electron bombarded silicon target TV camera tubes whose photocathodes are operated at up to -12 kV; there is thus a high voltage on the intensifier photocathode. We intend, in the future, to run this input photocathode at ground; the resultant large potential difference between the intensifier output and the camera input faceplates will then be dropped across a coupling fibre optic light conduit of appropriate length This coupling can be a demagnifying one, so permitting the use of smaller diameter, and hence cheaper, TV camera tubes.

We plan ultimately to replace the Varian image intensifier with a magnetically focused X-ray intensifier incorporating a low density alkali halide photocathode and a microchannel plate electron multiplier.† This device will have a diameter of 100 mm; its light output per quantum will be sufficient to allow it to be coupled to a camera tube of modest sensitivity with an object lens of moderate aperture.

† See p. 189.

The Television Camera

We are using an isocon camera tube because at the start of our experiments it offered the best compromise between sensitivity, good signal to noise ratio and spatial resolution. In fact, a very low noise performance is important only at very high counting rates.[3] The spatial resolution of our system is primarily limited by our spatial sampling frequency and we plan to increase our array to 512×512 picture elements. Even at this sampling interval 95% of the line spread function of the detector lies within ± 0.5 pixel for a camera tube with a 50% MTF at 400 TV lines and within ± 0.75 for a tube with a 50% MTF at 250 TV lines. The latter represents a very modest resolution which is achieved by most types of camera tubes, e.g., a 16 mm SIT tube or a vidicon with two stages of external image intensification. (Reynolds and Milch[4] have shown that the line spread function is a better measure of the performance of a camera tube for X-ray diffraction applications than the modulation transfer function. The former, of course, is the Fourier Transform of the latter).

THE STABILITY OF THE DETECTOR

In order to approximate to the theoretical detective quantum efficiency of the detector a high degree of stabilization of the intensifier and TV camera tube gain is essential. We have achieved this stabilization by projecting on the lower edge of the input plane a reference patch of light from a stable radioactive light source. This patch is sampled and used to servo the gain of the system.

Equally, the black level of the television signal must have a better stability than is achieved with normal clamping circuts. A black patch outside the X-ray image is sampled in the same way as the white reference signal and used to stabilize the black level.

The resolution of the digitized image is degraded unless the image remains locked in position relative to the scanning ramps. Spatial stability of the TV image can be maintained only by locking the scan frequency to that of the mains and our 6 MHz master clock, from which the scanning ramps are derived, is, therefore, generated by an oscillator which is phase locked to the mains frequency.

SPATIAL DISTORTION

The two electron optical systems of the detector, in the image intensifier and the image section of the camera tube, necessarily introduce some spatial distortion in the image. This is further aggravated by the spillage of the magnetic focusing field of the TV tube into the image intensifier. The overall effect is one of a combination of rotation, pincushion distortion and shear of the image which together amount to about 5%. This distortion can, therefore, be corrected for in a semi-analytical way by a computer program.

Spatial Non-uniformity

The response to X-rays of the detector is not completely uniform over its surface. There is a short range non-uniformity due to unevenness of the phosphor amounting to about 1–2% which cannot readily be corrected for and which probably sets an ultimate limit to the precision of the detector. Superimposed on this is a long range non-uniformity which amounts to as much as 25% at the edges of the image and which can be corrected by interpolation from a look-up table stored in the computer and produced by calibration measurements.

Detective Quantum Efficiency

Our analysis[1,2] indicated that the overall detective quantum efficiency of the detector should be about 50%; that is, the relative variance of repeated measurements of integrated patterns should be about twice that to be expected from X-ray quantum statistics alone.

We have been able to verify that these predictions are fulfilled approximately, but we have not been able to follow the precise dependence of the relative variance on the number of summations and on the number of picture elements included in the summation. The reason is partly that our stabilization and black level clamping are still not better than about 1%. The other reason lies in the correlation of the intensity recorded in neighbouring picture elements and in successive television frames.

Spatial correlation effects exist in an area detector when the full width of the point spread function is greater than the sampling interval. Under these conditions the intensity read out from a given picture element is affected by that at neighbouring elements; accordingly intensity measurements made in the centre of a patch of uniform intensity will have a lower relative variance (which leads to a higher detective quantum efficiency) than measurements made on single element images or on images with a steeply sloping intensity profile.

Equally, temporal correlation effects lead to an artificially low variance for the sum of a small number of successive frames. The detector has a certain memory which is analogous to an integrating time constant of about 60 to 100 msec. The memory is due partly to the decay characteristics of the two phosphors of the system, the X-ray phosphor and the intensifier output phosphor, and partly to the lag of several frame periods in the isocon at very low levels of illumination. The analysis is complicated by the variation of the lag time with illumination.

The results of the memory effect are that, when gain drift effects are eliminated, the variance of repeated measurements of patterns integrated over a few time constants is actually less than the variance calculated from counting statistics. The effect leads to no disadvantage since we are not, at present, concerned with very rapidly changing patterns. After the summation of a large number of frames the memory effect becomes negligible.

Spatial Resolution

In our present instrument 300 digital samples are taken during each active line period of 50 μsec. The sampling frequency is thus 6 MHz which, by Shannon's Sampling Theorem, corresponds to an analogue bandwidth of 3 MHz, which in turn leads to a FWHM of the line spread function of 100 nsec or 0·6 picture elements. This figure is confirmed by our experimental measurements; the resolution is thus not degraded by the physical performance of the camera tube or the image intensifier or by light spreading in the phosphor.

It is worth noting that the cost of the frame store and the complexity of the high speed adders will make it difficult to increase the sampling frequency by as much as a factor of two in the immediate future, leading to a limiting resolution of the system of at best 600 TV lines. Most television camera tubes are well capable of this performance.

Linearity of Response

The television detector response is linear with integrated incident X-ray intensity over a range of several hundred to one, provided that the live TV signal is kept to well below the "knee" of the transfer curve. It is necessary to set the overall light gain of the system (by varying the image intensifier and isocon dynode chain voltages) for the range of intensities to be expected.[3]

We have verified that the response per picture element remains constant as the size of the illuminated area is decreased down to at least 3×3 picture elements.

Conclusions

The television X-ray detector is proving to be a reliable and versatile instrument with a performance which comes up to expectations. Any remaining shortcomings are due largely to the inherently lower reliability of a device whose actual circuits have undergone continuous modifications over a period of years without a complete rebuilding. These shortcomings should disappear in an adequately engineered second instrument built entirely from new components.

REFERENCES

1. Arndt, U. W. and Gilmore, D. J., *In* "Adv. E.E.P." Vol 40B, p. 913 (1976).
2. Arndt, U. W. *In* "The Rotation Method in Crystallography" ed. by U. W. Arndt and A. J. Wonacott, Ch.17, North Holland, Amsterdam (1977).
3. Arndt, U. W. and Gilmore, D. J., *Appl. Crystallogr.* **12,** 1 (1979).
4. Reynolds, G. T. and Milch, J. R., *In* "Adv. E.E.P." Vol. 40B, p. 923 (1976).

X-ray Topography with Scintillators Coupled to Image Intensifiers or Camera Tubes

Y. BEAUVAIS

Thomson-CSF, Electron Tube Division, Boulogne-Billancourt, France

and

A. MATHIOT

C.E.N.G., Grenoble, France

Introduction

X-ray imaging with monochromatic radiation is widely used for crystal studies. High intensity monochromatic X-rays are generally obtained from the characteristic radiation of metallic targets, and the corresponding photon energy does not exceed 20 keV; low energy is also required to get a convenient diffraction angle ($\theta \simeq 45°$) with common interatomic spacings ($d \simeq 1$ Å) when Bragg reflection is used.

The common need for collimating the X-ray beam, and the very low diffraction yield, result in a weak signal, so very long exposure times are needed for photographic recording.

Some very powerful X-ray sources, like that of LURE DCI at Orsay, use synchrotron radiation. These sources, when filtered to obtain monochromatic radiation, can deliver about 10^4 times more photons than conventional generators. In this case, real time imaging by means of image intensifiers or camera tubes seems possible, offering the possibilities of dynamic observation and signal processing. The difficulty encountered with image tubes is to obtain the high spatial resolution (2 to 5 μm) generally needed for topographic experiments.

The utilization of camera tubes and image intensifiers for the detection of topographic X-ray images has been the subject of papers by many authors. R. E. Green Jr[1] gives a comprehensive review of the published results and separates the detectors into two general categories:

(a) Camera tubes of the vidicon type with a window transparent to the incoming radiation, and a target modified to increase X-ray absorption.

The sensitivity of these tubes, without internal amplification, is relatively low, and the resolution is limited to between 20 and 30 μm by the thickness of the target. The use of this type of tube is thus limited to high X-ray dose rates, and the resolution is too low for the considered application.

(b) Camera tubes or image intensifiers formed by optically coupling a fluorescent screen to a light sensitive image tube can give detection efficiency and good resolution simultaneously. The high internal gains of camera tubes such as the Thomson CSF Nocticon TH 9659 (SIT), or image intensifiers with cascaded stages like the TH 9303 (see Fig. 1), or microchannel plate intensifiers permit detection of weak photon fluxes; by using thin phosphor layers, the spatial resolution is mainly limited by the tube's MTF; however, resolution can be increased by means of a magnifying optical coupling system. The performance obtainable by coupling a thin phosphor screen to a high gain image tube is described here.

FIG. 1. TH 9659 SIT camera tube and TH 9303 image intensifier with fibre optic coupling.

Resolution of the Coupled Assembly

The MTF of a phosphor screen depends on many parameters, such as the type of deposit (transparent or scattering), the thickness, the granularity, the compactness, and the nature of the interface materials (reflecting or absorbing). R. K. Swank[2] has computed the MTF of various types of phosphors and the published results are shown in Fig. 2. The MTF measured in our laboratory for a scattering layer (thin P·20 powder) under electron bombardment has been indicated on the same figure. As the excitation mechanism is not the same, results cannot be directly compared. However, for the same thickness, the practical MTF must be lower than theoretical ones because of the packing fraction which is generally about 50%.

The limiting resolution of an image tube typically corresponds to the frequency for which the MTF is about 0·05. Considering the general shape of the MTF given in Fig. 3 for Nocticon camera tubes or cascaded image intensifiers, this is easily obtained when the MTF of the image tube and that of the coupled phosphor are roughly the same, about 0·2, which occurs for a spatial frequency $f \simeq 0\cdot 4\, t^{-1}$, where t is the phosphor thickness. The corresponding thickness and limiting resolution have then been computed; as an example, the resolution is 20 lp mm^{-1} for a 20 μm thick phosphor directly coupled to a TH 9303 image intensifier. Resolution can be increased by a magnifying optical coupling system, but the phosphor thickness must be reduced in proportion to the magnification. Results are nearly the same with the TH 9659 camera tube.

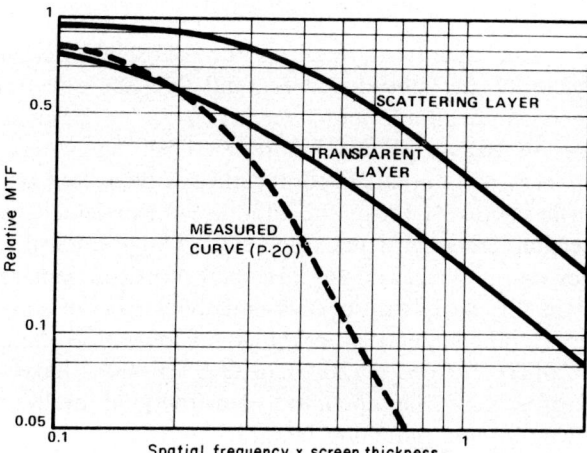

Fig. 2. MTFs of phosphor screens: computed results for scattering and transparent layers (after Swank[2]) and measured curve of P·20 screen.

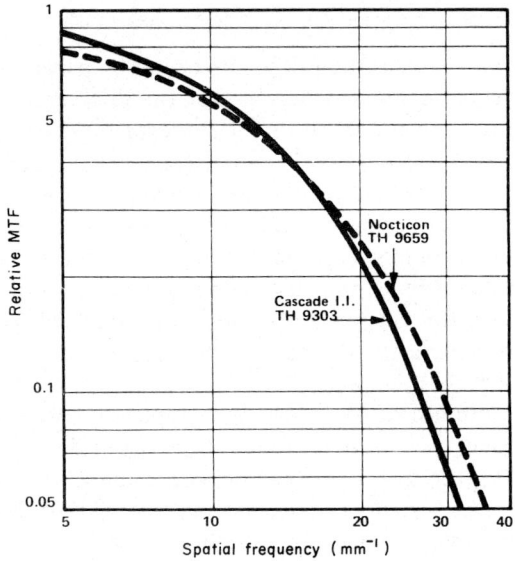

Fig. 3. Typical MTFs of TH 9303 image intensifier and TH 9659 camera tubes.

Quantum Yield

The most appropriate yield in this application is the mean number, N, of electrons released by the photocathode to which the screen is coupled: for each incoming X-ray photon,

$$N = Y_S F_{SP} Y_P, \tag{1}$$

where Y_S is the quantum yield of the phosphor screen, Y_P the maximum yield of the photocathode, and F_{SP} the spectral matching factor between photocathode and screen.

Two curves of Fig. 4 (shown in broken lines) show typical relative spectral responses of multialkali photocathodes deposited on fibre optic faceplates. The curve for the photocathode with extended red response (S·20 ER) is characteristic of the three stage image intensifier TH 9303 and the other one is typical of the TH 9659 Nocticon camera tube. The other curves of Fig. 4 are the spectral emissivity $E(\lambda)$ of some phosphor screen materials known and used for X-ray detection. Some of these materials (CsI(Tl), ZnCdS (P·20), Gd_2O_2S (P·43)), have a spectral emission centred near the maximum sensitivity of multialkali photocathodes, giving a good matching factor.

The matching factors have been computed using relative values for photocathode spectral sensitivity, and the results are given in Table I.

The quantum yield of some thin phosphor screens deposited on fibre optic faceplates have been measured with 8 keV photons by means of an image intensifier. The spectral response and the gain of the image intensifier had been calibrated before coupling to the screens; then the assembly was exposed to an 8 keV source (Cu Kα) giving a flux of about $2 \cdot 8 \times 10^4$ photons sec^{-1} (previously measured with a scintillation counter). After correction for spectral matching, the measured values of Y_S are given in Table I.

The computed values of the absorption ratios are given in the same table. They assume a 50% packing fraction for the screens which are made of agglomerated powder, and a theoretical density close to 1 for CsI(Tl) which is deposited by vacuum evaporation.

The best quantum yield is obtained with P·43 (Gd_2O_2S) which is the most absorbent material. There is some uncertainty in the values measured for CsI(T1), because of atmospheric instability.

After correction for absorption by fibre optics, the measured quantum yield corresponding to a maximum conversion efficiency of about 4%, appears to be relatively low. As the absorption decreases with increasing photon energy, the quantum yield will be approximately constant up to 25 keV for phosphor screens of the same thickness.

Noise Factor

Each step of the detection process is characterized by a quantum gain g_X which is subject to random fluctuations. The corresponding signal

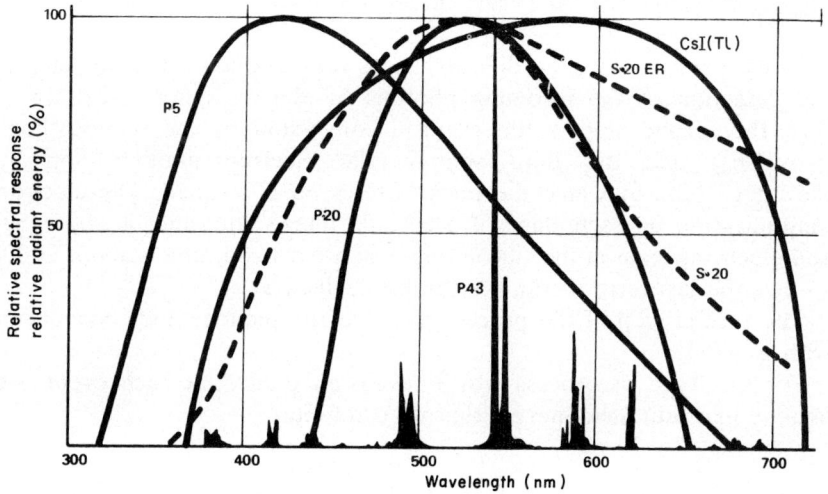

Fig. 4. Spectral emission of phosphors and photocathode spectral sensitivities.

TABLE I

Performance of fluorescent screen/photocathode combinations

Phosphor material		P·43(Gd_2O_2S)			CsI(Tl)		$CaWO_4ZnCdS$[†]	
Screen thickness (μm)		8	12	16	6	12	8	4
Spectral matching factor with	S·20	0·91			0·72		0·17	0·81
	S·20 ER	0·95			0·86		0·18	0·88
Quantum yield		71	77	86	39	69	43	31
Absorption ratio %		65	79	88	56	71	27	118

[†] with reflecting backing.

degradation can be expressed by the noise factor F, the ratio of the relative variations of the output signal to the input signal. Calculations by van Schooneveld[3] can be applied and by assuming a Poissonian fluctuation $\Delta n/n = n^{-1/2}$ for the incoming photons, the following expression can be derived:

$$F = 1 + \left(\frac{\Delta g_1}{g_1}\right) + \frac{1}{g_1}\left(\frac{\Delta g_2}{g_2}\right) + \frac{1}{g_1 g_2}\left(\frac{\Delta g_3}{g_3}\right) + \ldots \quad (2)$$

In the present case, the different steps are as follows: the first step is the detection of the incoming photons by the phosphor screen ($g_1 = Y_S$), the second step is the electron conversion by the photocathode $g_2 = Y_P F_{SP}$ and the third step is the electron multiplication by the image tube to which the phosphor screen is coupled. The electron multiplication by each stage of a cascade image intensifier is about 50, and electronic gain in the silicon target of a Nocticon tube is about 2000, so that the last term of Eq. (2) can be neglected.

The second step of the process refers to the photocathode conversion with $g_2 = 0·1$.

For this type of process which results in 0 or 1 at each event, we assume like van Schooneveld that we can write:

$$\frac{\Delta g_2}{g_2} = \frac{1 - g_2}{g_2} = 9.$$

As far as the phosphor screen is concerned, the relative fluctuation cannot be measured independently of a photocathode because of its relatively low quantum yield. We have tried a photomultiplier associated with a multichannel pulse counting system. A typical spectrum for a thin P·43 phosphor layer is shown in Fig. 5.

Calculations from this curve show that the photocathode is still mainly responsible for the dispersion, the screen influence being of the second order with $\Delta g_1/g_1$ from 0·1 to 0·2. The total noise factor is about 1·3 and will not be a limiting factor for real time imaging.

Real Time Imaging

When using image tubes for the real time observation of topographic images, the integration period is that of human eye, between 0·1 and 0·2 sec. Two cases are then possible, depending on the magnitude of the incident X-photon flux: for higher fluxes, the resolution is only limited by the MTF of the detector and corresponds to the preceding calculation; for lower fluxes, the resolution is photon noise limited. The transition between the two types of limitation appears when the number of electrons leaving the photocathode from each resolved detail is between 10^2 and 10^3 sec^{-1}. These values are consistent with the measurements made for low light level applications of the same tubes.

By referring to the preceding measurements, real time imaging appears possible with incident fluxes of over 10^6 to 10^7 photons sec^{-1} cm^{-2}; these

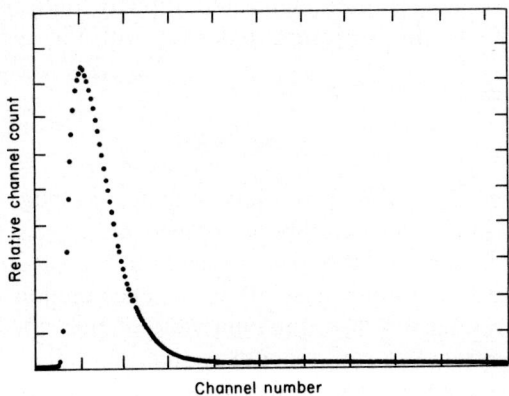

FIG. 5. Pulse height distributions of P·43 phosphor/photocathode combination.

values must be compared with the photon fluxes encountered in the topographic experiments:

(a) With a conventional generator, the maximum photon flux is about 5×10^5 photons $\text{sec}^{-1}\,\text{cm}^{-2}$ (Cu Kα, 40 kV, 20 mA). Real time imaging with a resolution better than 10 lp mm^{-1} (50 μm) is not possible, and the use of image tubes is limited to detecting image positions and setting up the experiments.

Photographic exposure times can also be reduced for recording details of about 10 μm width. However, rotating anode X-ray generators deliver about 10 times more photons and, under favorable conditions, real time imaging with direct coupling of the phosphor screen to the photocathode is possible. In this way, resolutions of 20 lp mm^{-1} (25 μm) are obtainable; optical magnification to increase real time resolution is not possible.

(b) With synchrotron radiation, fluxes of typically 10^8 photons $\text{sec}^{-1}\,\text{cm}^2$ have been measured at Orsay by the LURE group. Real time imaging is possible in these circumstances, and television experiments have been performed with a 12 μm thick CsI(Tl) layer deposited on a fibre optic faceplate directly coupled to the photocathode of a TH 9659 Nocticon tube. Resolutions of about 30 μm have been recorded. In the same experiment, the photon flux was found to be sufficiently high to permit using a magnifying coupling ($\times$2 to $\times$3) between the phosphor screen and the tube: real time imaging of details about 10 μm wide would thus be possible. The most convenient way of providing optical coupling with a minimum photon loss is to use commercially available magnifying fibre optics. The phosphor is then deposited directly onto the fibre optics. Experiments are at present underway with this type of coupling.

Conclusion

We have investigated the possibility of imaging topographic features in real time by using a thin scintillator coupled to either a high gain image intensifier or a camera tube. The camera tube must have a high gain because of the low photon fluxes that are encountered in practice, and because of the relatively low quantum yield of thin phosphors.

With conventional X-ray generators, the resolution is photon noise limited. With more powerful generators like synchrotrons, real time imaging appears possible, and the resolution is limited by the MTF of the image tube.

Higher resolution can be reached by increasing the magnification between the screen and the image tube. When doing so, the input field is reduced and thinner phosphor screens must be used, resulting in a lower yield. Each time the magnification is doubled, the minimum required photon flux is multiplied by about 8, so that the advantages of increasing the magnification are rapidly limited, so far as real time imaging is concerned.

Because image tube resolution is mainly limited by the MTF of the phosphor for image intensifiers, and by that of the target for camera tubes, improvement of photocathode resolution can be obtained by magnifying electron optics. A zooming electron optic would permit the field and the resolution of the tube to be adapted to the observed subject. Unfortunately, such tubes do not exist at present for this type of application.

References

1. Green, R. E., Jr., *In* "Adv. in X-ray Analysis" ed. by H. F. McMurdie *et al.*, Vol. 20, p. 221, Plenum, New York.
2. Swank, R. K., *Appl. Opt.* **12,** 1865 (1973).
3. van Schooneveld, C., Physics Laboratorium RVO–TNO, Report No. 42 (1965).

Image Device for Gamma Cameras Incorporating a Solid-State Localizer

H. ROÚGEOT, G. ROZIÈRE and B. DRIARD

Thomson-CSF, Electron Tubes Division
St. Egrève, France

Introduction

In 1957 H. Hanger described a gamma ray imaging system based on the use of a number of photomultipliers, closely coupled to a scintillator crystal.

In November 1975 we described an improved solution where a large field image intensifier tube eliminated most of the photomultipliers[1]. The present paper describes a new image intensifier gamma camera tube that completely eliminates the need for photomultipliers, by incorporating a silicon solid state localizer in the device. It is also worth noting that the silicon localizer tube, alone or in conjunction with a light image intensifier, may be of use in particle counting and mapping at high speeds and/or at low signal levels as in astronomy, nuclear physics, laser based equipment, etc.

General Description of the Gamma Camera Tube

The Thomson CSF THX 1427 is a two stage image converter tube designed to be coupled to a specially shaped, large diameter scintillator crystal. The rear of the tube is equipped with a plug in quadruple preamplifier delivering four voltage signals from which the location and the intensity of individual scintillations can be calculated. Figure 1 shows the camera. Its first stage is a 345 mm diameter input field image intensifier with a fast output phosphor deposited on a fibre optic coupling plate. The second stage is a silicon target localizer tube with a photocathode deposited on its input fibre optic window. Image inverter electron optics focus the input image on a silicon target electron multiplier and localizer. The output plane of the first stage and the input face of the second stage are optically coupled through an electrically insulating fibre optic block that withstands a potential difference of 45 to 50 kV.

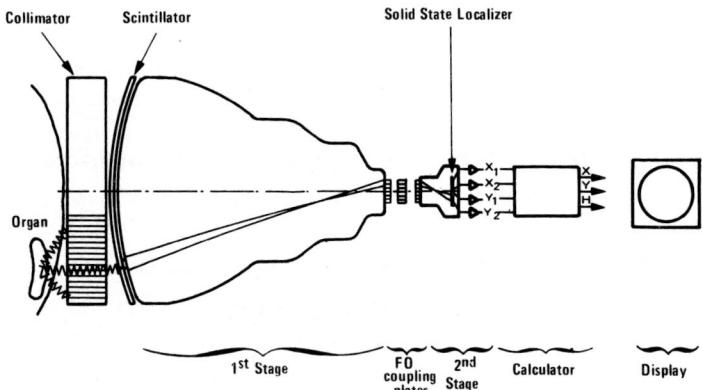

Fig. 1. Schematic diagram of scintillation camera.

When a gamma photon is absorbed by a scintillator crystal coupled to the entrance window, a burst of photoelectrons appears on the input photocathode of the first stage. These electrons are accelerated and focused by a demagnifying electron lens onto the output phosphor, where a light pulse is produced and guided through the fibre optic coupling block to the photocathode of the second stage where a burst of photoelectrons is released and in turn accelerated and focused onto the silicon target where multiplication by pair production takes place.

The silicon target is a large surface area diode with resistive electrodes completely covering both faces. These resistive coplanar electrodes work as signal dividers allowing location of the impact, as shown in Fig. 2. Four signals are collected through two pairs of parallel conducting collectors, one pair for the X direction deposited on the upper face of the target, one for the Y direction on the lower face. The quadruple preamplifier, plugged directly on the feedthrough plate of the second stage tube, delivers four localizing signals at a convenient voltage level. Figure 3 shows a photograph of the THX 1427 gamma camera tube.

The Solid State Localizer Target

The solid state localizer target is made out of a slice of n-type silicon material of $10^4 \, \Omega \, \text{cm}$ resistivity in which a diode has been formed by implanting a boron p-type layer on one side, and a phosphorus n-type layer on the other side as shown in Fig. 4. The surface resistance of these layers is adjusted to $10^4 \, \Omega \square^{-1}$ by baking. At this value the signal dispatching to the four electrodes is controlled by the surface implanted

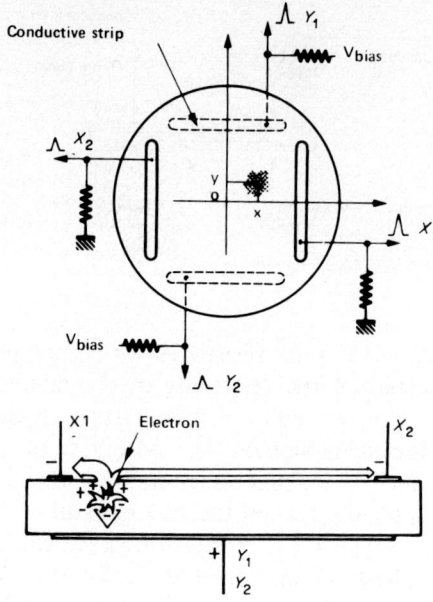

FIG. 2. Principle of solid state localizer.

FIG. 3. The scintillation camera **tube THX 1427**.

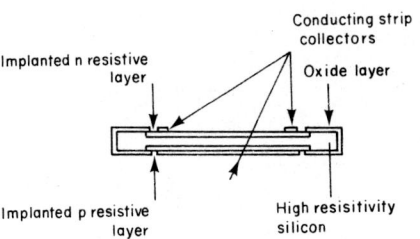

FIG. 4. The localizer target.

electrodes and not by the bulk resistance which, taking into account the thickness of the slice and the resistivity of the material, is an order of magnitude higher. The two pairs of metal strips on the resistive surface layers collect the localizing signals. The periphery of the silicon target is protected against surface leakage currents by a ring of silicon oxide. A voltage potential applied between the two pairs of conducting electrodes induces, under the p-type layer, a uniform space charge region that can extend several hundreds of micrometres across the silicon target thickness. When electrons, accelerated from the photocathode to about 15 keV, impinge on the silicon disc, they enter the diode through the thin p-type layer with practically no loss in energy and, by releasing electrons and holes in the space charge region of the diode, they are multiplied by a factor of 4000. These electrons and holes move to the resistive implanted layers where the resulting charge signal is divided into four subsignals (2 on each side) with respective amplitudes depending on their distance to the zone of impact.

Nevertheless, it must be pointed out than an important parameter of the device is the capacitance of the diode polarized in the reverse direction, given by the following formula:

$$C = K \frac{S}{\sqrt{\rho V}}, \qquad (1)$$

where: K is a constant, S the surface area of the diode, ρ the resistivity of the bulk material and V the voltage applied between the two resistive layers. This capacitance controls the jitter of the collected charges on the four conducting strip electrodes and thus, the precision in localization.

Let N_T be the number of charges by which the signal fluctuates on a collecting electrode. If we exclude the spatial fluctuation in the distribution of impinging bunches of electrons, N_T is a result of the diode leakage current noise, the Johnson noise in the surface resistive dividers and the preamplifier noise. Taking into account these three contributions, and the conclusions of Owen and Awcock[2], N_T is given by:

$$N_T = C^{1/2} \left[\frac{2}{q} I_0 R_l + \frac{4kT}{q^2}\left(1+1\cdot 4 \frac{R_s}{R_l}\right)\right]^{1/2}, \quad (2)$$

where: T is the absolute temperature, k the Boltzmann constant, q the electron charge, R_s the noise equivalent resistance of the input transistor, R_l the divider resistance of the solid state localizer and, I_0 the leakage current of the diode. It is obvious from Eq. (2) that the diode capacitance must be as low as possible. This can be achieved according to Eq. (1) by reducing the diode surface area, by choosing the highest bulk resistivity silicon material available and by applying a high reverse voltage to the diode. The diode surface area is determined by the useful image area needed which is defined by the tube demagnification and resolution. The voltage that can be applied to the device depends on the leakage current characteristic. As can be seen in Eq. (2) noise increases with leakage current. Besides reducing noise in localization, lowering the diode capacitance shortens the resolving time of the localizer which is mainly related to the product $R_l \times C$.

Typically, with a 10^4 Ω.cm resistivity material, a 12 cm^2 diode surface, 10^4 $\Omega\square^{-1}$ surface resistance, and under 90 V reverse potential applied to the diode, the capacitance C is 300 pF and the leakage current can be maintained below 15 μA.

With these values, 10^7 charges shared between the four collecting electrodes was found to resolve the location of the impact with 0·12% precision referred to the distance between the collecting strips, which means about 30 μm spatial resolution on the silicon diode. The decay time constant of the solid state localizer was found to be 3 μsec, setting the resolving time at about twice this value.

The Electronics

As indicated previously, the tube assembly THX 1427 delivers four amplified output signals (X_1, X_2, Y_1, Y_2) directly from a quadruple preamplifier.

An electronics unit (Fig. 5) was designed to shape and combine the four signals and calculate the scintillation centroid coordinates and its brightness expressed respectively as follows:

$$X = \frac{X_2 - X_1}{X_2 + X_1}; \quad Y = \frac{Y_2 - Y_1}{Y_2 + Y_1}; \quad \text{and} \quad H = X_1 + X_2 + Y_1 + Y_2.$$

A window amplitude selector incorporated in the electronics unit determines the range of H values that are to be displayed.

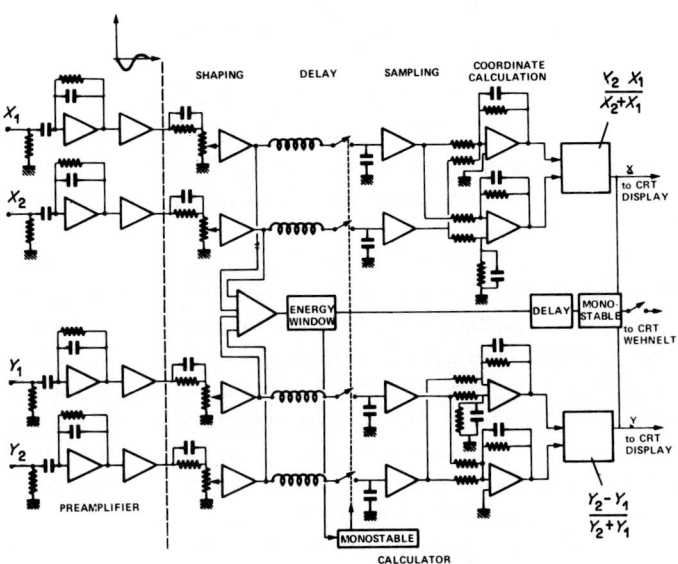

Fig. 5. Block diagram of electronics.

Performance of the Scintillation Camera Tube

A power supply sets the output phosphor of the first stage at +30 kV, keeping the input photocathode at ground potential. On the other hand the photocathode of the second stage is set at a negative (−15 kV) potential so that the tube may provide output signals at ground potential. The fibre optic block ensures the necessary insulation between the first and the second stage. The measured physical characteristics were obtained with these settings and with a NaI(Tl) monocrystal covering the hole input area irradiated by a Co^{57} (122 keV) source. The mechanical and physical characteristics of the THX 1427 tube are given in Fig. 6. The sensitivity is expressed in terms of the total voltage signal H delivered by the preamplifier for a given gamma energy absorbed in the input scintillating crystal.

Figure 7 gives the quantum conversion level at different stages of the scintillation camera tube, excluding the preamplifiers, and simultaneously the calculated relative variances at each stage which should reflect the energy resolution along the cascade. A field non-uniformity can result from crystal scintillator edge effects, and from photocathode and phosphor screen non-uniformities. To achieve good energy resolution, the uniformity must be within ±5%. Therefore non-uniformity has to be

MECHANICAL
Overall diameter   415 mm
Overall length   570 mm
Weight   27 kg

PHYSICAL
Input field diameter   345 mm
Sensitivity   3·5 mV keV^{-1}
Field uniformity   ±5%
Tube time constants   3 µs
Energy resolution with Co^{57} (122 KeV)
    with compensating filter   8.5 %
    without compensating filter 8 %
Spatial resolution (FWHM) with Co^{57}
    with compensating filter   4·2 mm
    without compensating filter 3·6 mm
Energy linearity   100 %
Geometrical distortion (integral)   <10%

FIG. 6. Performance characteristics of THX 1427 camera.

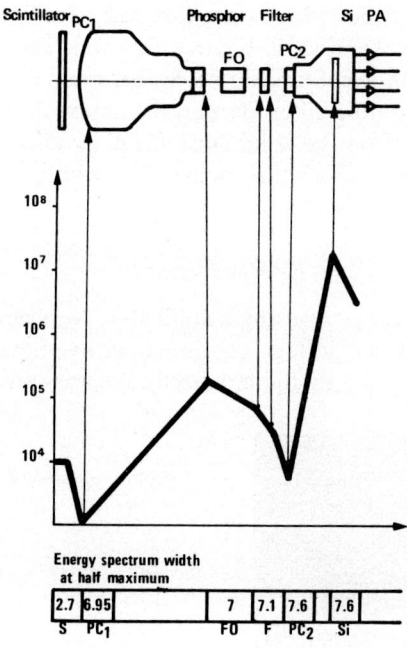

FIG. 7. Quantum levels in the camera.

corrected by a compensating filter inserted between the two stages. The time constant sets the rate at which scintillations can be counted. It depends on the output phosphor of the first stage and on the pulse response of the localizer and electronics unit. The tube time constant was found to be that of the localizer, equal to 3 μsec.

Energy Resolution

The resolution is measured with a collimated Co^{57} gamma source in contact with the input crystal. The amplitude distribution can be displayed on an amplitude selector. The energy resolution is measured as the full width half maximum (FWHM) of this peak distribution curve as shown in Fig. 8. Energy resolutions of 8% without the compensating filter, and 8·5% with the compensating filter have been achieved.

The spatial resolution is measured with the same gamma source and setting. Instead of plotting the total pulse height distribution (sum of the four localizing signals) on the amplitude selector, the position coordinate is displayed, leaving the source at a fixed position on the input crystal. Moving it 10 mm away, another count is registered. The spatial resolution is measured as the full width half maximum (FWHM) of the position distribution of one of the two peaks expressed in **millimetres** referring to the 10 mm distance measured on the input crystal. This is shown in Fig. 9. The spatial resolution was found to be in the 4 mm region both with and without the compensating filter. It may be interesting to note that this implies about 0·4 mm on the solid state localizer taking into account the demagnification factor, this value being well within its spatial resolving power.

Energy Linearity

An exclusive advantage in using a solid state localizer and multiplier in a gamma camera is its perfect energy linearity. Saturation is avoided because charges collected by the localizer collectors are supplied by a low

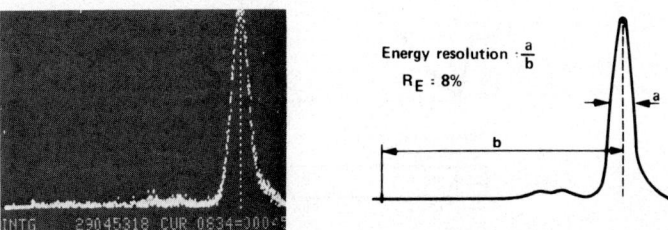

FIG. 8. Energy resolution of the camera.

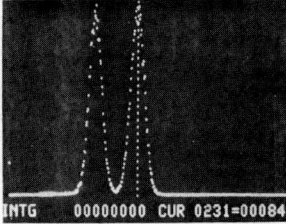 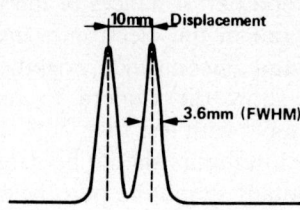

Fig. 9. Spatial resolution of the camera.

impedance source and not by an external high impedance resistance divider controlling the potentials of photomultipliers or channel plate electrodes as in present gamma cameras. This linearity is only limited by the electronic unit and can, with a properly designed electronic network, be maintained to within 1% up to energies of 600 keV.

Advantages of the THX 1427 Tube

When the THX 1427 is operated in a scintillation camera both image quality and simplicity of utilization are improved as compared to present systems. When the tube is mounted in its housing, the starting procedure and adjustments are very simple. Because the tube is delivered with its power supply, it is a simple plug-in device, gain uniformity is preset, and tube performance is stable needing no further adjustments. The tube delivers directly four low noise high level signals. No photomultiplier reading head is necessary, thus eliminating gain adjustments and drift. The image quality improvements depend on the high performance in energy resolution and linearity, spatial resolution and field uniformity. One of the most important characteristics of the tube is that the monocrystal scintillator is in close contact with the input photocathode. Therefore, almost all the photons emitted by the crystal are collected by the photocathode thus reducing fluctuations by comparison with devices in which the crystal is somewhat distant from the photocathode. On the other hand, the solid state multiplier and localizer contribution to the noise of the system is negligible, or at least far less than for a microchannel plate or a dynode multiplier. The use of high efficiency crystal coupling, and of a solid state localizer thus leads to improvements in spatial and energy resolutions. Similarly, the absence of pulse current saturation on the silicon diode assures a good energy linearity. No non-uniformity is introduced by the silicon localizer. The filter provided for compensation of field non-uniformities, however, slightly detracts from the general performances of the tube. In the future, compensation can be carried out in the electronics unit. The short temporal responses of the screen and silicon diode, together with properly designed pulse shaping, should allow the camera to reach a high count rate (exceeding 10^5 counts sec^{-1} with less than 20% counts loss). Because of the image quality obtained, tumours should be detected earlier than with other systems and more details should be perceived with gamma cameras incorporating this new tube. The solid state localizer tube should also be a valuable new tool for research in nuclear physics, astronomy and industry.

References

1. Driard, B., Verat, M. and Rozière, G., *IEEE Trans. Nucl. Sci.* **NS23,** 502 (1976).
2. Owen, R. B. and Awcock, M. L., *IEEE Trans. Nucl. Sci.* **NS15,** 290 (1968).

The Image Intensifier as a Convolution Processing Device

R. J. GELUK

Oldelft Research Laboratories, Delft, The Netherlands

INTRODUCTION

Although the digital computer has become a very successful system for picture processing, there are still applications that involve an extremely large number of processing steps or a very high processing speed and consequently the use of a digital computer may not be possible. Although clock rates of digital computers have become very high and speed can be increased by connection of several systems in parallel, the computer is fundamentally a serial data handling device. Optical systems on the other hand have the fundamental property of parallel data handling and could become a favourable alternative.

We can distinguish between *coherent* optical processing, making use of both the amplitude and phase of the light waves (interference), and *incoherent* optical processing, based on the light intensity distributions only. Although coherent optical processing systems seem to have more applications, in practice there are many practical problems still to be solved in this field. The two systems which will be described here are of the incoherent type and are very tolerant to mechanical inaccuracies. The principal component in both systems is an image intensifier tube of the inverting type.

The first system to be described is meant for two-dimensional spatial filtering of optical images. The input to this system is a transparent photograph, the output is a real time television signal enabling virtually instantaneous results. The second system to be described is a major component in a system for analogue reconstruction of cross sectional images of the human body using X-ray transmission values. This technique named Transverse Analogue Tomography is thus an analogue version of Computed Tomography as introduced by E.M.I.[1] The image intensifier is used for one-dimensional filtering of the X-ray transmission data.

Principle of Operation

Figure 1 shows an image intensifier of the inverter type in a magnetic deflection yoke. The deflection yoke generates two orthogonal magnetic fields. Any light distribution $f(x, y)$ on the photocathode will be imaged electron optically on the output phosphor of the tube with a lateral displacement (x', y') proportional to the currents in the deflection coils. A transparency with a transmission distribution $h(x, y)$ is placed in the image plane on the output phosphor (in Fig. 1 fibre optic coupling is assumed). A photomultiplier placed behind the transparency responds to the total light flux transmitted through the mask. In the literature this is called "area integration" as the photomultiplier responds simultaneously to all light passing through the two-dimensional mask.

If we neglect conversion constants, the photomultiplier output current $i(x', y')$ can be expressed as:

$$i(x', y') = \iint_{xy} f(x - x', y - y')\, h(x, y)\, dx\, dy. \qquad (1)$$

This integral is known as the convolution integral and is said to be the convolved version of f with the convolution function h. In optical terminology $h(x, y)$ is called the point spread function and in electronic

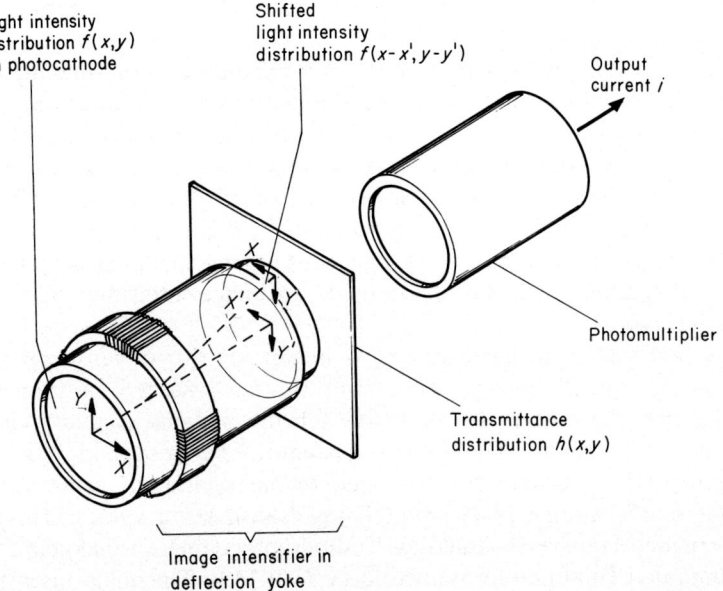

Fig. 1. Lateral displacement of $f(x, y)$ using an image intensifier placed in a deflection yoke.

terminology it is referred to as the impulse response. All values of the convolution integral can be obtained by generating the appropriate values of the deflection (x', y'). The deflection can be done at high speed in any scanning pattern. It is interesting to note here that, as it lacks area integration, a digital computer system would have to perform a complete scan in order to generate only a single value of the convolution integral. To generate all values of the convolution integral, a digital computer has to do a *double scanning* operation. On the other hand a coherent optical convolution system needs no scanning at all and is, in principle, the most elegant convolution system one can imagine.

As the image on the output phosphor of the image intensifier can change its position at a high rate, it is desirable to have a fast-decaying phosphor in order to avoid smearing. When using a phosphor of the yttrium silicate type, decay is as short as 100 n sec and a system based on such a tube can generate signals of more than 5 MHz bandwidth.

Although a convolution system as described above looks very favourable, there are two limitations that may influence its applicability. The first limitation lies in the point spread function, $t(x, y)$ of the image intensifier tube itself. As a result of $t(x, y)$ the corrected point spread function $S(x, y)$ of the total system becomes:

$$S(x, y) = h(x, y) * t(x, y), \qquad (2)$$

where $*$ denotes convolution. The absolute number of useful picture elements that can be separated in $S(x, y)$ is a figure of merit for the convolution system. In electronic processing, where signals are a function of time rather than space, the corresponding figure of merit is known as the time-bandwidth product, which is the maximum number of entities that can be processed simultaneously. A practical electrostatically focused image intensifier with 300 resolvable line pairs over its field diameter will have a "time–bandwidth" product of 9×10^4. The second limitation lies in the electron shot noise in the tube, which limits the signal to noise ratio that may be obtained. The photocurrent of the tube cannot be increased at will, and care must be taken in the choice of the deflection speed. The signal to noise ratio also depends on the convolution function $h(x, y)$ and consequently the allowable deflection speed depends on the application.

Application to Two-dimensional Spatial Filtering

It is well known that blurred images can be improved by convolution with a suitable "deconvolution" function. The functions needed for the proper deconvolution however, have both positive and negative values. It will be clear that negative transmission values are not physically possible in the mask shown in Fig. 1 and a different construction of the optical system has to be found. As shown in Fig. 2 we can use two separate masks, one that contains only positive values of the deconvolution function as its transmittance distribution and one that contains only negative values of the deconvolution function as its transmittance distribution. Although both masks transmit only positive light fluxes, two separate photomultipliers will generate electrical signals that may be combined in a differential amplifier. Thus the electrical output signal contains both

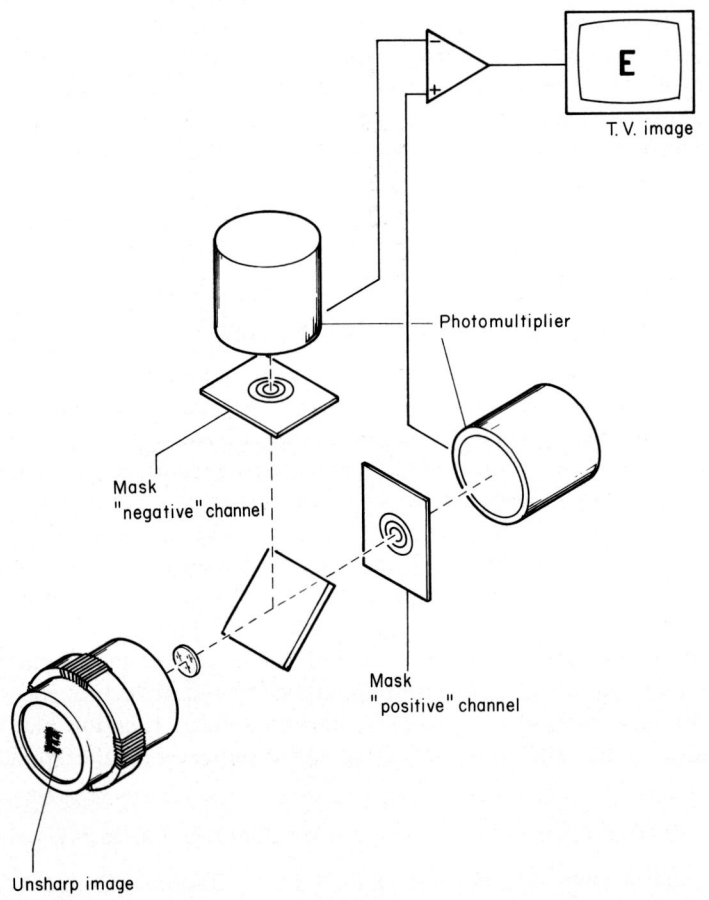

FIG. 2. System for spatial filtering of an image. The filtered image is displayed on a television monitor.

positive and negative contributions from the image. The necessary tandem optics and the beam splitter cause some attenuation of the light fluxes to each photomultiplier, but, when using a lens with a high numerical aperture, the effect on shot noise is negligible.

To observe the spatially-filtered image the photomultiplier differential current must be converted into a light intensity distribution. This may be done by a television display which is facilitated by applying a standard TV scanning pattern to produce the co-ordinate shift (x', y').

Figure 3 shows photographs taken from a television monitor screen which was modulated by the photomultiplier currents of the system as described. In Fig. 3(a) the mask contained a small hole in the positive channel and no transmission in the negative channel. By this means the image is scanned with a small spot, resulting in an unfiltered image on the television monitor (the "original"). Figure 3(c) shows the filtered image using a circular transmission pattern in the negative channel and a small hole in the positive channel. Figure 3(b) shows in a perspective view, the convolution function $h(x, y)$ that was used for the filtering process. This is the point spread function of the system and is obtained when a point light source is imaged on the image intensifier. Figures 4(a), (b) and (c) show corresponding results for an image that was linearly blurred by motion.

Application to X-ray Cross Sectional Imaging

Cross sectional X-ray imaging as introduced for medical purposes by E.M.I. has become known as Computed Tomography (CT). It is a technique that uses X-ray transmission values of tissue layers. This way a linear image called a "profile" or "view" is obtained for each direction of transmission. In order to obtain enough spatial information to reconstruct an image of the tissue layer, a large number of profiles is taken in different directions. CT scanners use discrete one dimensional element arrays for the detection of the transmission values. The system developed at Oldelft, however, uses an existing television fluoroscopy system as the detector. As seen in Fig. 5, only a strip of the detector area is used in the reconstruction mode. A digital computer is normally used to transform the measured transmission values, but the Oldelft system uses only electro-optical methods. Contrary to the case of the digital computer, the analogue system is not limited by the same fundamental relation between number of picture elements, processing speed and cost and can therefore be tailored to the high bandwidth at the output of the existing fluoroscopic system. As for most computer systems, the analogue system uses the principle of "filtered backprojection". This will be explained first.

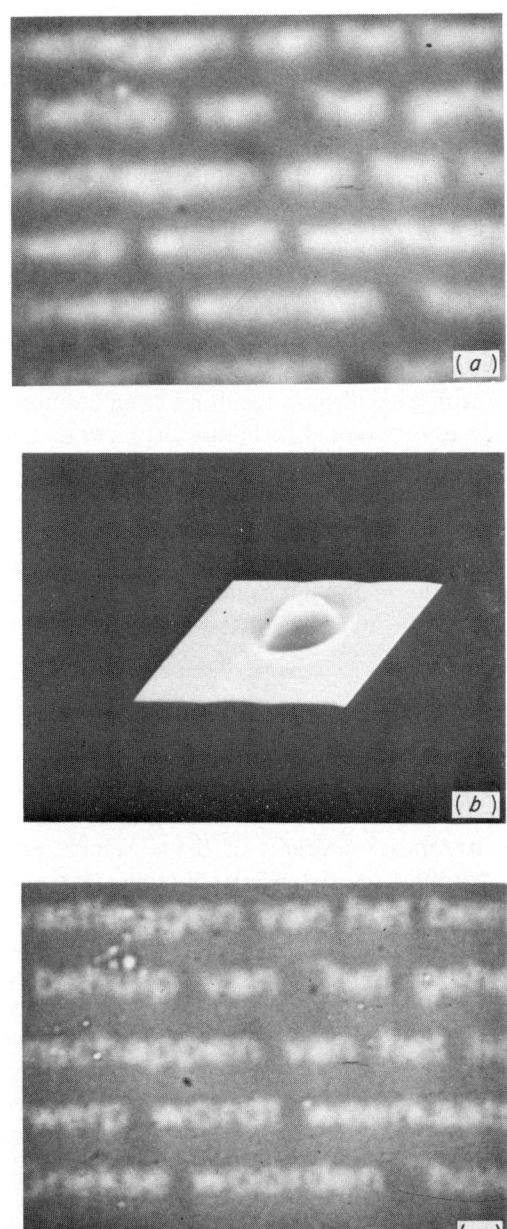

FIG. 3. Spatial filtering of unsharp text using the system of Fig. 2: (a) the input image, (b) the point spread function of the convolution system and (c) the output image.

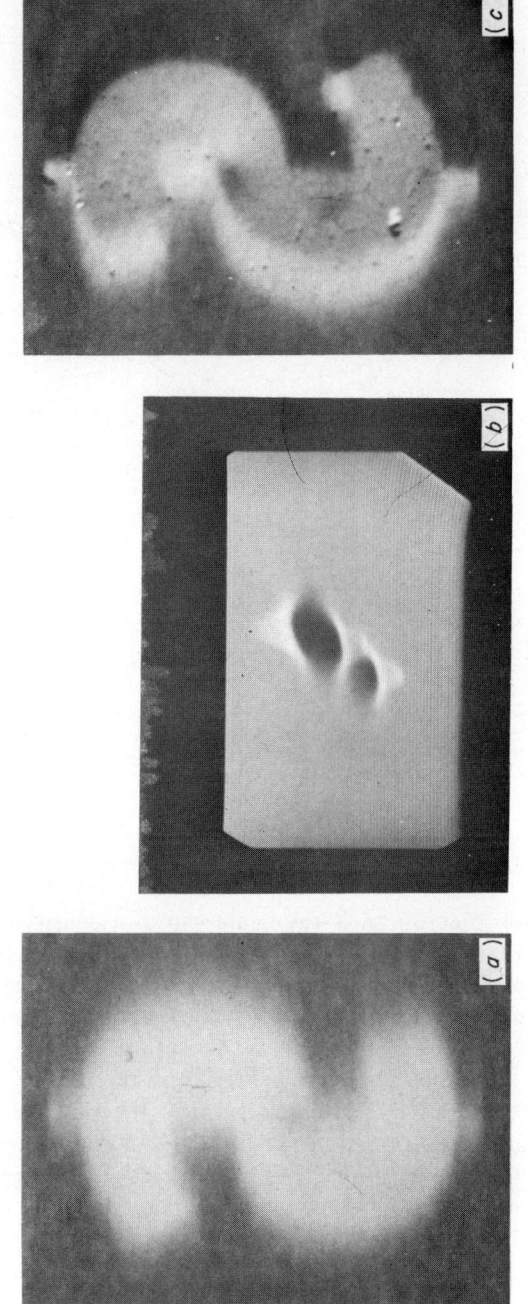

FIG. 4. Spatial filtering of a linearly blurred picture: (a) the input image, (b) the point spread function of the convolution system and (c) the output image.

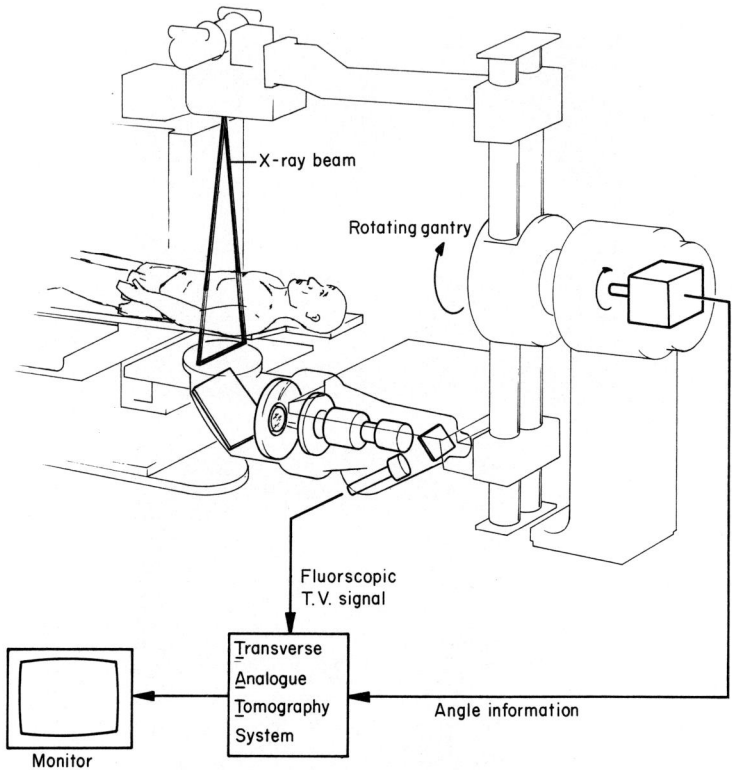

FIG. 5. Simulator for therapy planning in combination with Transverse Analogue Tomography system.

BACK PROJECTION

A single small object in an X-ray beam causes a density change in the profile as shown in Fig. 6(a). In a plane of reconstruction we may "backproject" this density response as shown in Fig. 6(b). This is repeated in a virtually continuous process while rotating the beam of X-rays over 180 or 360 degrees. The density responses are continuously summed resulting in an "unfiltered backprojection". All the density responses of the small object will have a common point in the reconstruction plane. This point, where the summation reaches its peak value, corresponds to the position of the original object. At other points the summed contribution decreases in proportion to the distance from the centre of the object. All other "small objects" will be imaged in the same way with a $1/r$ point spread function, where r is the distance from the true position.

Imaging with a $1/r$ point spread function is far from ideal and may be regarded as a first order approximation to reconstruction only. Methods to improve the imaging technique will be described below. The summation may be performed by a digital computer system in a random access memory or photographic film can be used. At Oldelft experiments have been carried out using photographic film but the need for film processing and the inability to sum negative contributions make the photographic system rather impractical. We now use a charge storage tube for summation of backprojection patterns. Complete systems containing such a charge storage tube are available as a scan converter.

A Posteriori Filtering of a Backprojected Image.

The simple backprojected image is a representation of the original object with $1/r$ point spreading. Experiments using photographic film for the backprojection have been carried out and yielded the image seen in Fig. 7(a). It shows a section of a polyester car bumper which was X-rayed and backprojected via a television screen on a rotating photographic camera. In an attempt to improve the picture quality, it has been two-dimensionally spatially filtered using the system described above. Figure 7(b) shows the result, which is clearly of a higher quality. As described in the previous paragraph the procedure has several shortcomings and has now been abandoned.

Filtered Backprojection

It can be shown that, by suitable processing of the profile signal *prior* to backprojection, the $1/r$ point spreading can be avoided. Most CT scanners use this principle of simple one-dimensional filtering as it can be

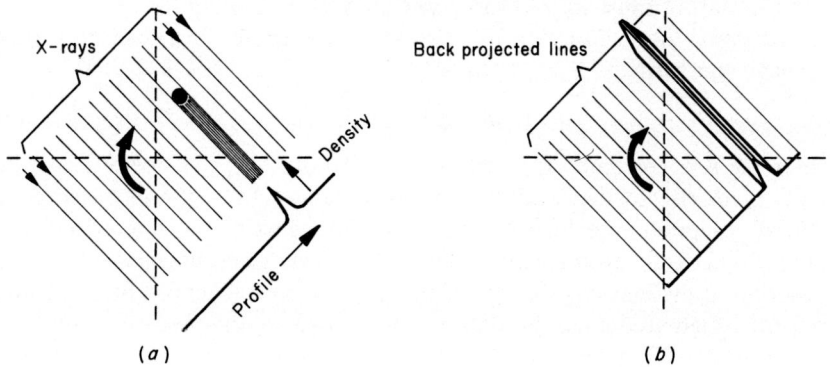

FIG. 6. Principle of backprojection: (a) original density profile; (b) density profile backprojected.

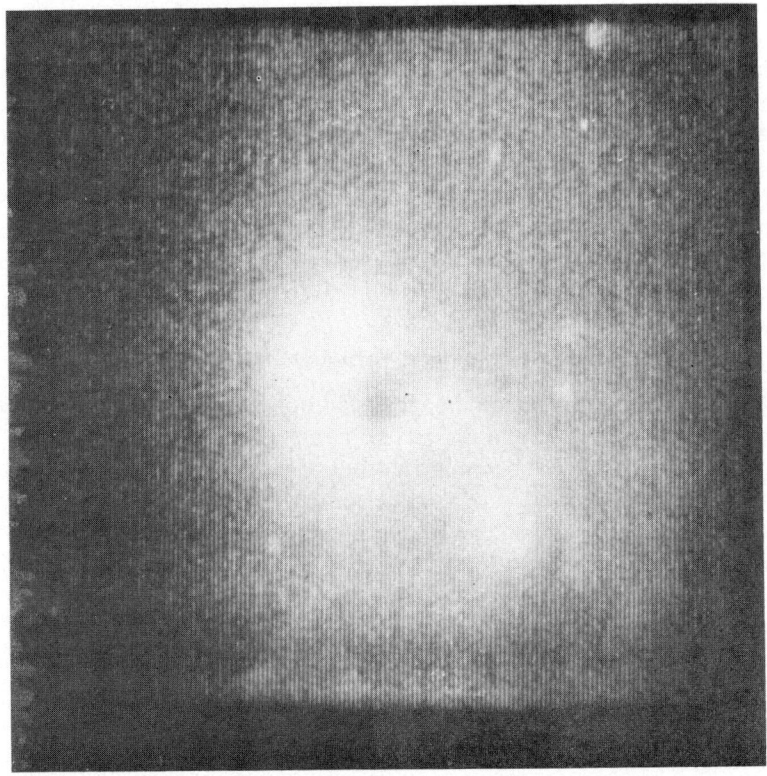

FIG. 7. Two dimensional filtering of backprojection image: unfiltered backprojected cross section of a car bumper.

done simultaneously with the acquisition of profile data and thus reduces the computing time after completion of X-ray scanning.

The convolution function to operate on the profile $p(x)$ can be shown to have the form:

$$C(x) = a\delta(x) - (1/x^2), \qquad (3)$$

where a is a constant. It is interesting to note that the positive part of $C(x)$ gives rise to the unfiltered backprojection. In a system based on a digital computer, the high frequency components of the filtered profiles $p(x) * C(x)$ will cause moiré ("aliasing") effects when backprojected in a random-access memory. To avoid this, a version of $C(x)$ can be used with rigorous high frequency limitation, giving rise to an oscillating convolution function. In an analogue reconstruction system moiré effects do not occur and the function of Eq. (3) can be used. When using an image intensifier for convolution, the separate masks for the positive and the

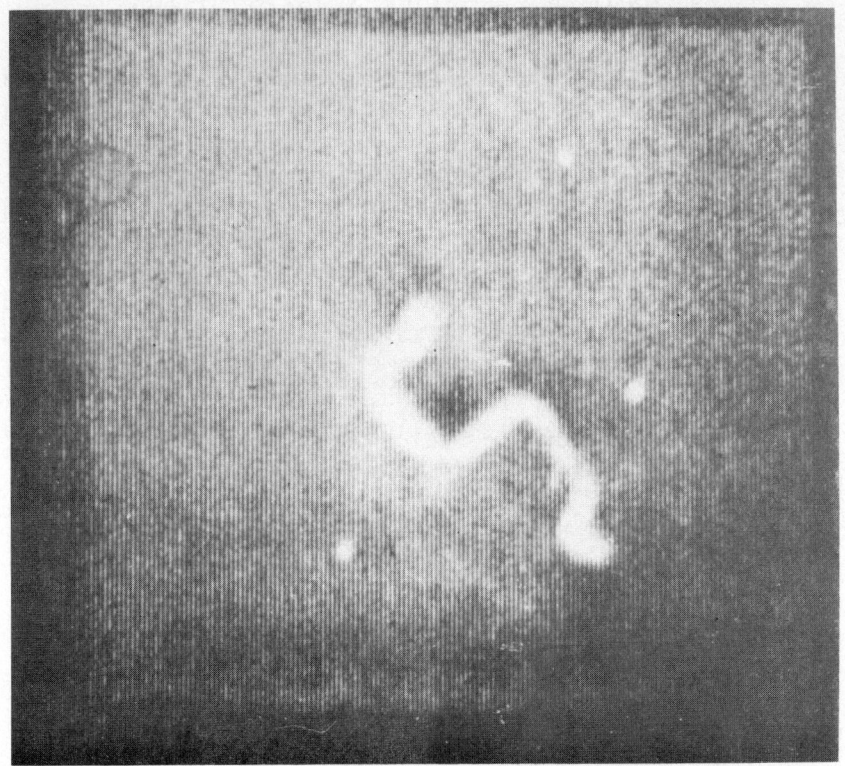

FIG. 8. Two dimensional filtering of backprojection: image of cross section of car bumper after spatial filtering using the system of Fig. 2.

negative parts of $C(x)$ have completely different shapes. However, if we realize, that

$$p(x) * C(x) = \frac{dp(x)}{dx} * \int C(x)\,dx = \frac{dp(x)}{dx} * \frac{1}{x}, \qquad (4)$$

it becomes obvious that we can use masks of equal shapes for both the positive and negative channels. As $p(x)$ is available as a television signal, differentiation is a simple electronic operation.

The convolution described here is not only more practical because of the symmetry of the masks to be used, but it has the extra advantage (as will be clear after reading the next section) that blemishes in the electro-optical system will be imaged in a smeared way and consequently will have a less harmful effect.

THE ANALOGUE CONVOLUTION SYSTEM AS APPLIED IN TRANSVERSE ANALOGUE TOMOGRAPHY

As shown in Fig. 5 the reconstruction system accepts a standard television signal as profile information. The X-ray beam is collimated to a narrow "fan beam" and the television lines containing the profile information will be selected by the TAT system for filtering and backprojection. As the scanning lines are parallel to the profile, the relevant profile signal is read out in a relatively short time (a few milliseconds) leaving the rest of the frame time for convolution and backprojection. This time is needed for the backprojection, as each convolved profile is backprojected over an area and to address an area in the storage tube at least several milliseconds are needed.

In the TAT system the incoming video signal is first logarithmically amplified in order to obtain a signal proportional to densities rather than

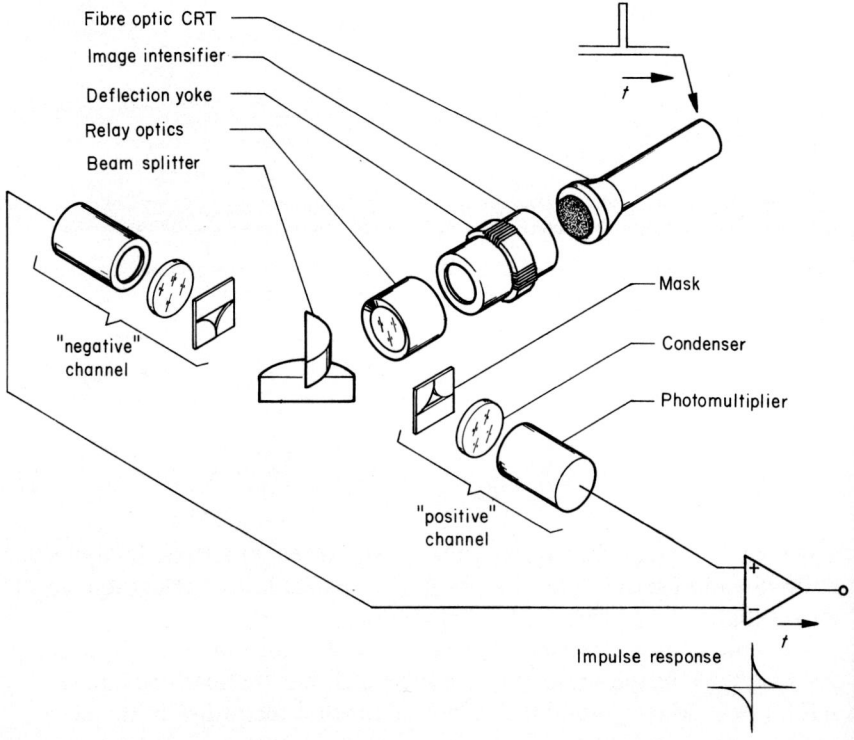

FIG. 9. Electro-optical system for filtered backprojection.

transmission values. It is then differentiated because the convolution process will be done according to Eq. (4). After differentiation the relevant lines are written into the phosphor of a small fibre optic CRT which is coupled to a 25 mm image intensifier (Fig. 9). The phosphor of the CRT is chosen to have a persistence long enough to allow for a relatively slow convolution process (18 msec). A magnetic field deflects the profile image over its total length and two convolution masks are used for the positive and negative parts of the convolution function. As the convolution is only one-dimensional the masks can be made to have a variable opening representing the $1/x$ function; the beams are therefore expanded in a direction perpendicular to the profile. In Fig. 9 it can be seen that a special beamsplitter with cylindrical mirror surfaces is used to direct the light beams to the masks. Condensers concentrate the light flux onto the photocathode of a small photomultiplier.

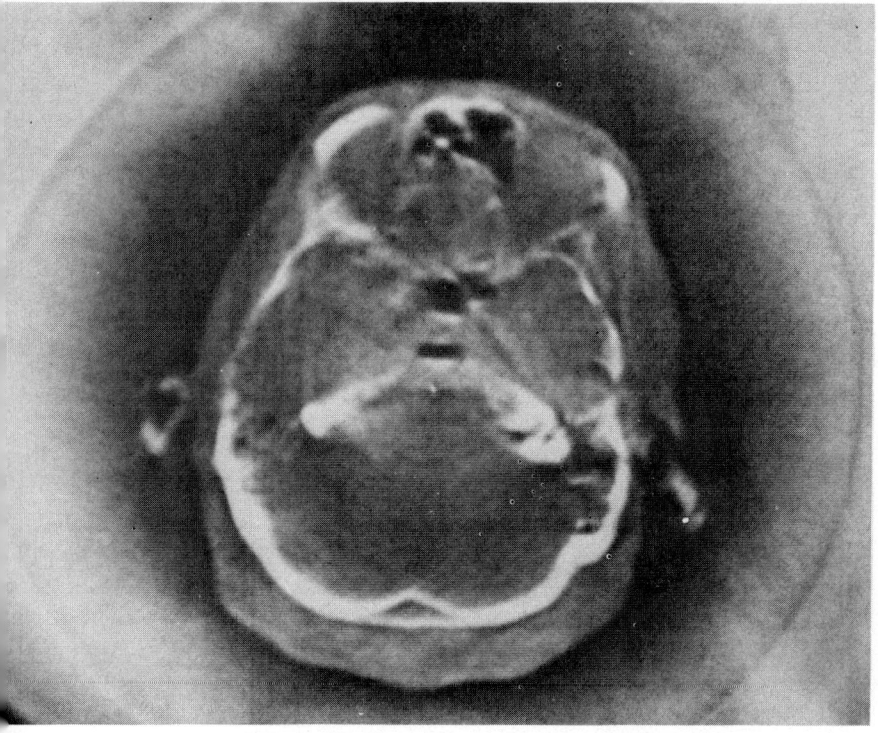

FIG. 10. Cross sections of a human skull using TAT technique.

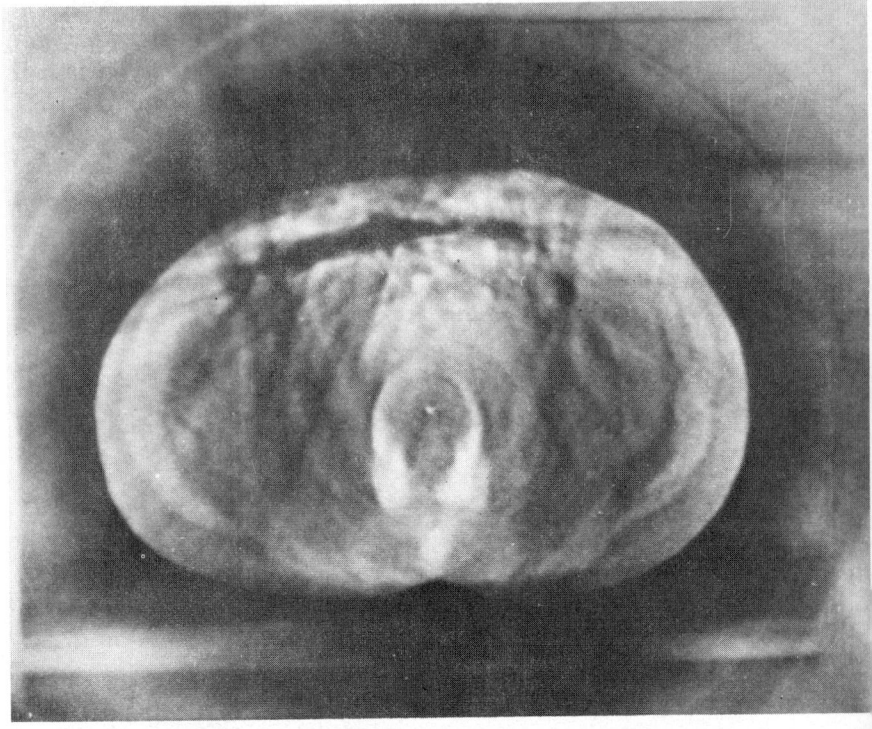

Fig. 11. Cross section of a human abdomen using TAT technique.

The system of Fig. 9 not only performs a convolution but also offers a time-base expansion from a line time of 64 μsec to a frame time of 20 msec. Moreover, the optics combine the contributions of several TV lines into the light flux directed to the masks. The output signal of the convolution system is suitable for intensity modulation of a backprojection pattern. As mentioned above, a scan-converter system is used for the integration and storage of the summed backprojections. In the storage tube of the scan converter an electron beam writes a line raster in which the lines conform to the X-rays passing through the object plane. The angular position of the line raster is made to correspond to the direction in which the object is being X-rayed.

The system described here can be used as an accessory to a simulator system for treatment planning. By adding the Transverse Analogue Tomography equipment to the existing system, which already contains an X-ray apparatus, a T fluoroscopic camera and an accurate gantry, cross sections of the body can be simply reconstructed. The time needed for acquisition of data and reconstruction is completely determined by the time the gantry needs to complete a rotation (60 sec). The reconstructions shown in Figs. 10 and 11 are built up from 1500 views with 500 lines per view.

REFERENCE

1. Hounsfield, G. N., *Br. J. Radiol.* **46,** 1016 (1972).

Photography with Gated Microchannel Plate Intensifiers

A. E. HUSTON and K. HELBROUGH

John Hadland (P. I.) Ltd., Bovingdon, Herts., England.

Introduction

Microchannel plate intensifier tubes first became readily available in the early 1970s, and, although their prime purpose has been as aids to viewing under very low light conditions, several workers have applied them as aids to photography.[1-4] These applications include low light situations such as are encountered in surveillance operations, and a variety of high speed (submicrosecond) photographic uses. This paper discusses developments at John Hadland (P. I.) Ltd. in methods of gating and control of inverter type microchannel plate intensifier tubes for photographic applications, and includes results with two operational camera designs.

The Channel Plate as a Gating Electrode

Microchannel plate intensifier tubes are intended to be used as viewing aids and are therefore normally operated in "DC mode". However, they may fairly readily be gated, this being particularly so with the so-called "wafer" design, by applying a pulse of a few hundred volts amplitude between the photocathode and the input connection to the channel plate. Most of the work at John Hadland (P. I.) Ltd. has been concentrated on the inverter type of intensifier tube, a design which would require a pulse amplitude of several kilovolts between cathode and channel plate to achieve satisfactory gating. To avoid this, the channel plate itself has been pressed into service as a gating point, and has proved quite effective. Some advantages of channel plate gating are:

(1) Variation of channel plate potential does not affect the tube focus, and therefore the pulse shape is not critical.
(2) The gain/channel plate voltage characteristic is non-linear, and has the effect that the transmitted light pulse rises and falls more rapidly than the applied electrical pulse.

The disadvantages are:

(1) The channel plate is inherently of high capacity, necessitating high current capability in the electrical pulse generator.

(2) Some 800 to 1000 V pulse amplitude is required to switch from "off" to "on", and this is also a demand on the pulse generator.

A pulse generator has been developed[5] which is capable of applying the required 800 to 1000 V pulse to the channel plate of the Ni-Tec R–6340 25 mm diameter tube,† with rise and fall times of about 40 nsec (the capacitance in this tube is ~ 85 pF). The generator makes use of two chains of transistors, one to generate the leading edge of the pulse, and the other to generate the trailing edge. The transistors are not allowed to avalanche, and a special feedback circuit reduces the rise and fall times and inhibits false triggering of the transistor chains. Tube gain may conveniently be varied by adjustment of the gating pulse amplitude, and this, in turn, is readily achieved by variation of the EHT potential applied to the transistor chains.

When the non-linearity of the gain/channel plate voltage characteristic is taken into account, as mentioned above, the rise and fall times of the transmitted light pulse are ~ 25 nsec, a figure quite adequate for the intended applications.

GATING AND CONTROL SYSTEM FOR A SURVEILLANCE CAMERA

The high gain and gating capability of microchannel plate intensifier tubes may be exploited in a camera for surveillance purposes. Such a camera, to be effective, needs intensified viewing, and an instrument has been developed which combines this facility with efficient photography.

Lens coupling from the intensifier screen to the film has been avoided since this entails a loss both of light transmission and resolution and photography is carried out by direct contact of the film on the fibre optic output plate. To permit the screen also to be viewed for observation purposes a special mechanism[6] has been developed and this is shown diagrammatically in Fig. 1. The intensifier screen (1) is viewed through a magnifying binocular optical system, including the two mirrors (2) and (3). When an exposure is required, the button is depressed and a motor drive causes the shutter (4) to roll along the rack (5), until it comes to rest with the mirror (3) lying horizontally and the compartment below effectively light sealed. During the period of shutter movement, the film loop (6) is driven forward so that the film is pressed into contact with the fibre optic faceplate of the tube. In this position a gating pulse is applied to the tube electrodes to determine the exposure.

† Ni-Tec Corp., 7426 Linder Ave., Skokie, IL 60076, U.S.A.

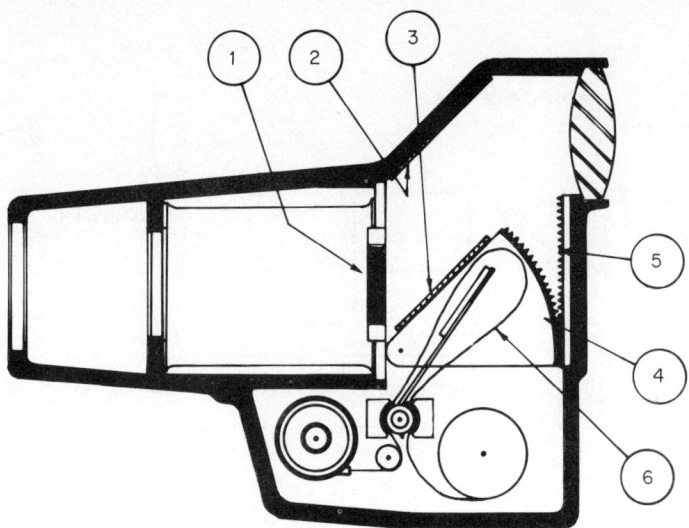

Fig. 1. Mechanism of surveillance camera.

As a viewer, the tube operates under DC conditions and automatic brightness control circuits compensate for variations of scene brightness to provide a comfortable viewing level for the dark adapted eye. On pressing the exposure button, the viewing facility is interrupted and film is brought into contact with the fibre optic faceplate of the tube. At the same time, an automatic photographic exposure control (APEC) becomes operative. This incorporates a "memory" which recalls the light level immediately prior to exposure and computes the voltage to be applied to the channel plate to give the correct exposure on the film, taking into account the exposure duration previously selected by the operator.[7] The way in which this is achieved is shown diagrammatically in Fig. 2. The tube screen current is monitored continuously, providing a control signal during the viewing phase which is memorized when the exposure button is depressed. This signal is fed to the control logic, as is also information relating to the exposure duration which the operator has selected. The logic circuits control the voltage supply fed to the gating pulse generator, which in turn controls the amplitude of pulse fed to the channel plate.

Measurements taken from the APEC system are reproduced in Fig. 3. Examination of the curves shows that, for a range of light levels from about 2×10^{-3} lx to 20 lx and for exposure durations from 0·3 msec to 100 msec, the photographic gain of the tube is adjusted so that the same amount of light always falls on the film, and thus the same density of

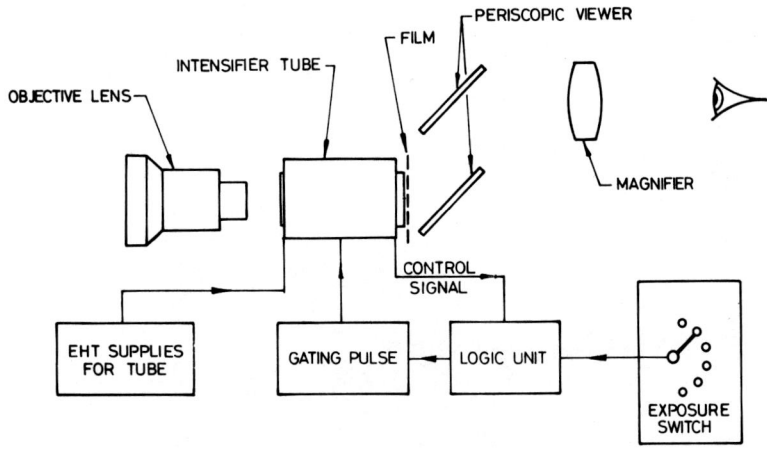

FIG. 2. Scheme of Automatic Photographic Exposure Control (APEC).

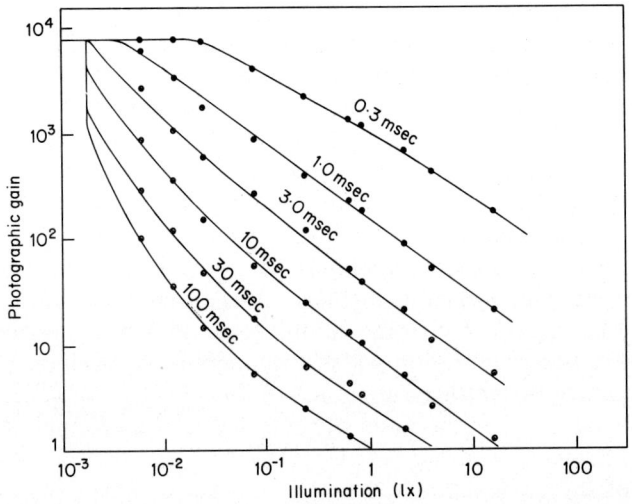

FIG. 3. APEC characteristics. Photographic gain vs. illumination for exposures from 0·3 msec to 100 msec.

record is obtained. Below 2×10^{-3} lx, the control fails for the three longest exposures and an over-ride switch is incorporated in the camera to enable manual control to be exercised under these conditions.

Photographic Gain

The term "photographic gain" has been used above and is the ordinate axis in Fig. 3. We define this simply as ratio between the light required to produce a fixed density on the film, with and without the intensifier tube. The data in Fig. 3 were compiled using tungsten lighting and HP5 film processed in Teknol $(1+9)$ at 20°C for 6 min. These choices are of course arbitrary, but approximate to realistic conditions. It is noteworthy that the measured photographic gain of the tube is an order of magnitude less than the manufacturer's quoted gain, which of course refers to viewing rather than photographic conditions.

RESULTS WITH THE SURVEILLANCE CAMERA

The completed surveillance camera is shown in Fig. 4, and Fig. 5, is a series of photographs obtained with it, using the control system described above. The APEC control was effective throughout, and the operator had only to set the exposure duration to the required value, view, and press the exposure button. That printable records may be obtained over a wide range of illumination is clearly demonstrated.

FIG. 4. The surveillance camera.

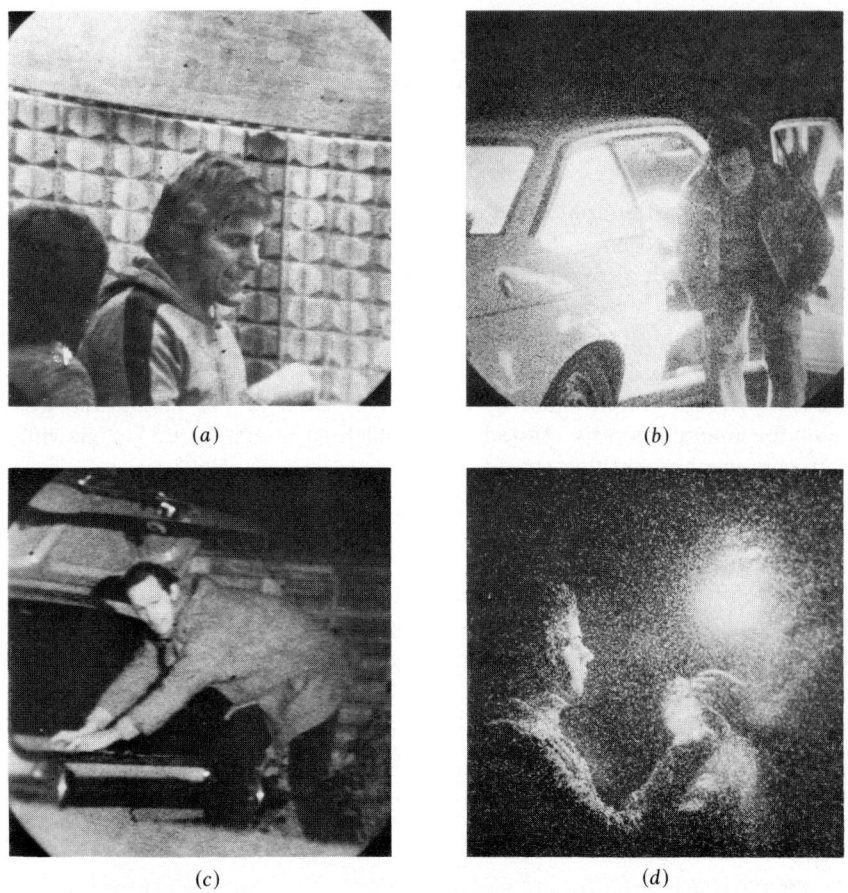

FIG. 5. Records taken with surveillance camera. The stated conditions are, respectively, illumination, distance, and exposure. (a) 1 lx, 15 m, 10 msec; (b) 3×10^{-2} lx + interior light, 17 m, 30 msec; (c) 5×10^{-2} lx, 20 m, 10 msec; (d) cigarette glow, 13 m, 30 msec.

GATING AT HIGH REPETITION RATES

Channel plate gating is capable of operating at surprisingly high repetition rates. It has been found possible to synchronize the shutter action of the intensifier tube at the pulse repetition rate of semiconductor laser diodes, thus assembling a system similar in principle to the familiar range gating technique described by many workers.[8] In this case, however, the purpose is not to determine range but to permit photography in conditions which are normally impossible, for example to reveal frontal detail in a brightly back-lit subject.

Experiments have been conducted using an array of four laser diodes, producing light pulses of ~200 nsec duration at a frequency of ~500 kHz, with a mean power of 50 mW. The intensifier gating pulse width was set at 250 nsec.

Examples of the results obtainable with this technique are shown in Fig. 6: in 6(a) the bright light in the field of view makes it extremely difficult to see the figure in the foreground, but in 6(b) with the laser illumination and the intensifier tube switched on for a total duration of 1 msec, a dramatic improvement was obtained. Similarly, in Fig. 6(c), the headlights prevent the registration plate of the car from being photographed until in 6(d) the laser and intensifier are used in synchronism. It is

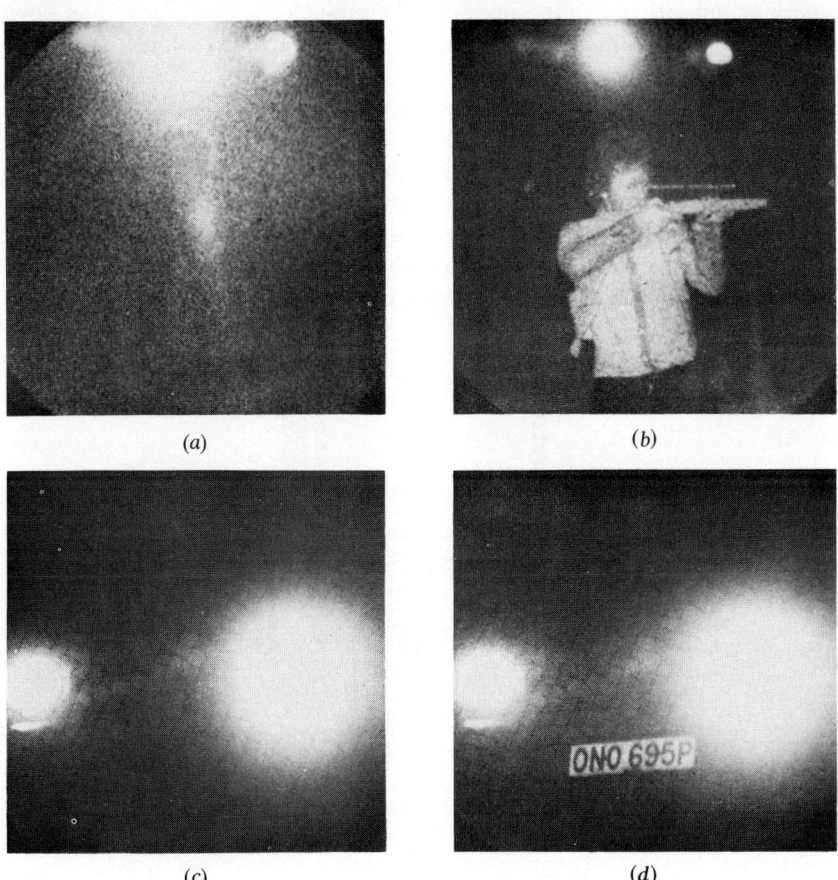

FIG. 6. Records with synchronized illumination. (a) and (c) without laser; (b) and (d) with laser.

noticeable, especially in Fig. 6(b), how effective is the reduction in background, including the bright light, due to the intensifier tube being "on" for only a fraction of the time during the 1 msec exposure.

A Camera For Photography of Projectiles

Military proving ranges make use of single shot photography of projectiles in flight to ascertain damage, condition of driving band, to study sabot separation, etc. Often the information required is of a purely qualitative nature, and the high standard of definition that would be needed to make quantitative measurements is unnecessary. A camera which simplifies the process of obtaining a suitable result is highly advantageous to the range operators, and such an instrument has been designed incorporating a microchannel plate intensifier tube.

The exposure duration required depends on the size and velocity of the projectile, and for normal range work, durations from 100 nsec to 10 μsec are needed. There is no difficulty in providing such exposures by the channel plate gating method, and the camera may be triggered by the approach of the projectile, using an optical detector ("sky screen"). It is, however, necessary under range conditions for the camera to be "live" for several seconds during the approach and passage of the projectile, and if channel plate gating alone is used, there is sufficient breakthrough of

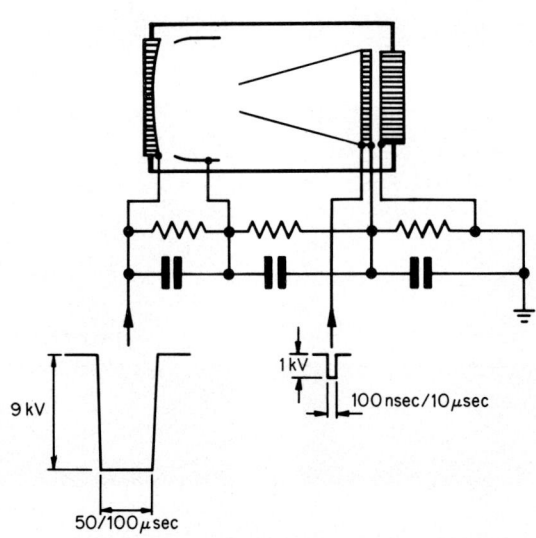

Fig. 7. Scheme of projectile camera.

FIG. 8. The projectile camera.

light to fog the film. The ratio of light transmission in the "on" state to that in the "off" state is of the order of 10^4 times for channel plate gating, so that for an exposure time of, say, 500 nsec a "live" time of 5 msec would cause fogging of the film. To combat this, the system shown in diagram form in Fig. 7 has been devised. A multiple gating method is used: a broad pulse of amplitude equal to the full working potential of the tube is generated, and triggered from the sky screen a short distance (usually < 1 m) up-range. This makes the tube live for a period of 50 to 100 μsec while the projectile is passing, and a suitable built-in delay circuit triggers the channel plate pulse at the correct time to secure a synchronized picture. Very high transmission ratios ($>10^8$ times) have been measured with this system.

A camera has been built using the Mullard XX1330 tube† which has a 50 mm diameter cathode and a 40 mm diameter screen. Recording is by direct contact, using Polaroid, on the fibre optic faceplate of the tube.

Figure 8 is a photograph of the completed camera, and Fig. 9 is a representative result showing a bullet in flight. It is also possible to apply more than one pulse to the tube during the passage of the projectile, and this technique has been used to obtain the result shown in Fig. 10, which has been taken by applying two 250 nsec pulses to the channel plate with

† Mullard Ltd., Torrington Place, London WCI E7HD, England

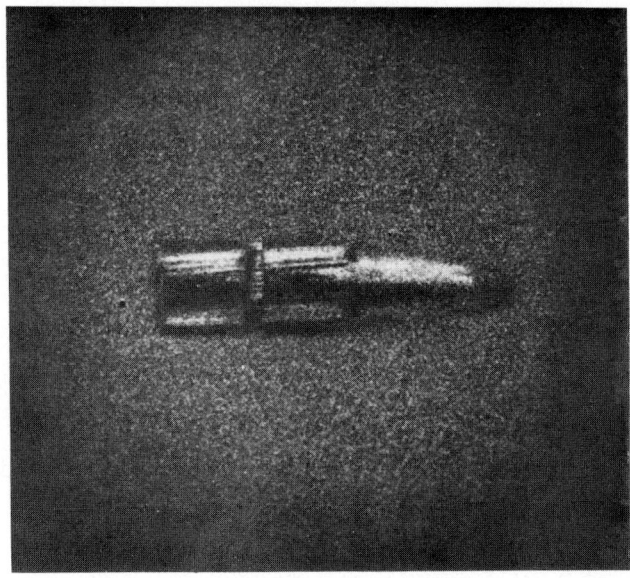

Fig. 9. Bullet in flight; velocity 600 msec^{-1}; exposure 200 nsec; aperture $f/4$; illumination, overcast daylight.

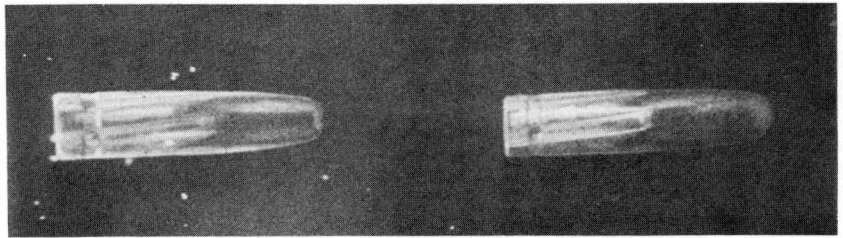

Fig. 10. Double pulsed operation; illumination 1-J flash.

an interval of 100 μsec between them. It demonstrates that acceptable records of projectiles can be obtained in a simple manner without the necessity for powerful synchronized flash illumination.

Conclusion

The microchannel plate intensifier tube is shown to be a very useful device for certain specialized photographic applications, such as surveillance and projectile photography. Synchronized operation with lasers pulsed at high repetition rates is also possible.

Acknowledgments

The authors acknowledge the valuable contributions made by their colleagues, S. Cross, B. Speyer, D. J. Bowley, J. R. Beeley and S. Bellis. The work described in the paper was mostly carried out under M. O. D. Contracts.

References

1. Eschard. G., Graf, J. and Polaert, R., "Proc. 9th International Congress on High-speed Photography", p. 493. Denver (1970).
2. Graf, J., *Acta Electron.* **15,** 357 (1972).
3. Lieber, A. J., *Rev. Sci. Instrum.* **43,** 104 (1972).
4. Grover, C. G., and King, W. L., *Proc. SPIE* **108,** 39 (1977).
5. Harris, R.B.A., British Patent No. 1470065.
6. Haynes, K.A.F., British Patent Application No. 45295/77.
7. Huston, A. E. Helbrough, K., and Beeley, J. R., British Patent Application No. 35396/78.
8. Gillespie, L. F., *J. Opt. Soc. Am.* **56,** 883 (1966).

Electron Optical Picosecond Streak Camera Operating at 140 MHz and 165 MHz Repetition Rates

M. C. ADAMS, W. SIBBETT and D. J. BRADLEY

Blackett Laboratory, Imperial College, London, England

Introduction

Single shot electron optical streak cameras have already been demonstrated to have picosecond and subpicosecond time resolution when used in conjunction with pulsed laser systems.[1] It is, however, rather inconvenient to employ such cameras in experiments where CW mode locked lasers are preferred.[2,3] For this reason, the deflection circuits of a Photochron I streak tube have been redesigned to provide streak operation at repetition rates of about 140 MHz and 165 MHz and jitter < 5 psec.

In this paper the repetitively operating Synchroscan streak camera is described. The details of its evaluation using both synchronously and passively mode locked CW dye lasers are also presented.

The Synchroscan Camera

The principle of operation of the Synchroscan camera is outlined schematically in Fig. 1. The linear voltage ramp that is normally applied to the streak plates of the single shot camera is replaced by a continuous sinusoid for fast repetitive streak operation. When light from some periodic luminous phenomenon is synchronized to the period or to an integral number of periods of the sinusoidal waveform then many individual streak images can be superimposed on the image tube phosphor screen. Since only the central half amplitude of the sine wave is generally used, the departure from linearity is ≤ 5%. Thus at a frequency of 140 MHz this implies a useful linear streak of about 1 nsec extent. Provided the jitter between the arrival of the optical pulses and the application of the streaking voltage cycles is insignificant, then the temporal resolution capability of the camera in this mode of operation should equal that of its single shot counterpart.

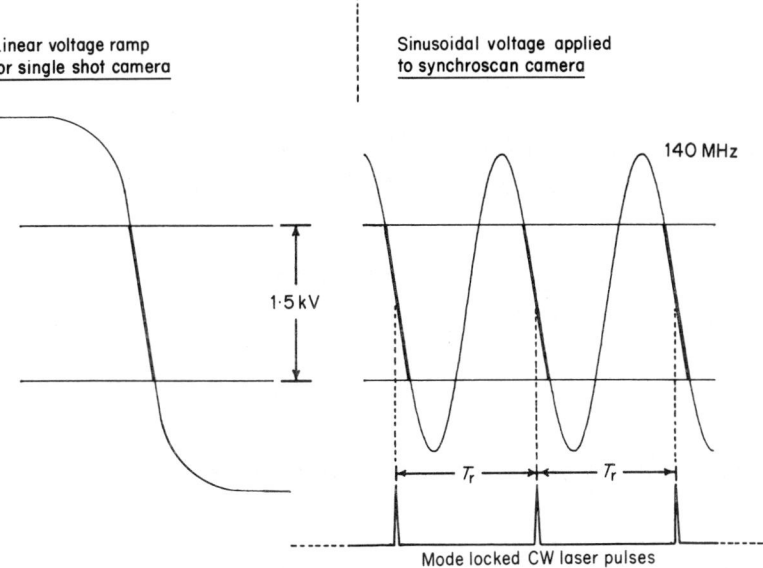

Fig. 1. Principle of operation of Synchroscan camera.

There are several advantages associated with a synchronously operated picosecond streak camera. These result from the fact that a very large number of low intensity streaks are accumulated to provide a recordable time resolved image. The photoelectron currents are therefore sufficiently low that space charge dependent increases in time dispersion can be avoided without the requirement for additional image intensification.[4,5] Furthermore, the integration of successive streaks effectively enhances the sensitivity and enables very low intensity luminous events to be investigated. The continuous nature of the output screen image can be readily photographed with relatively long exposure times of the order of seconds or alternatively recorded and processed using an optical multichannel analyser.

THE EXPERIMENTAL ARRANGEMENTS

Synchroscan with Synchronously Mode Locked CW Dye Laser

The first experimental configuration that was used for the operation and evaluation of the Synchroscan camera is shown in the schematic of Fig. 2.

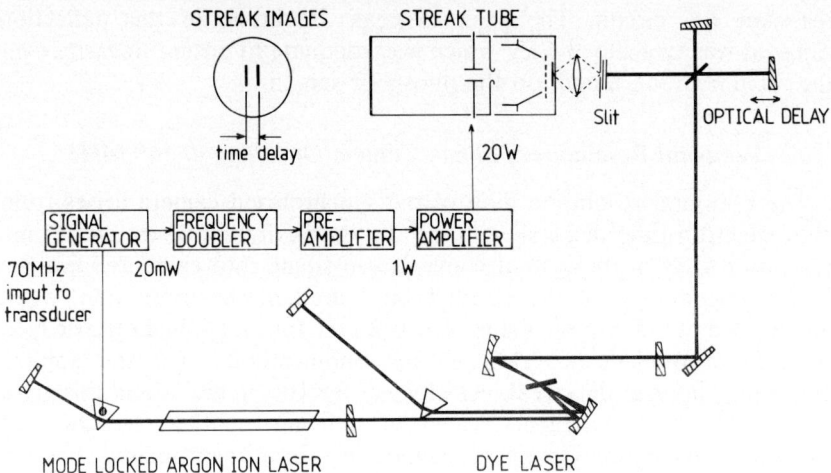

FIG. 2. Experimental configuration of Synchroscan camera and synchronously mode locked CW dye laser.

The CW RH6G dye laser was synchronously pumped by a mode locked argon ion laser (Spectra Physics Model 164).† The acousto-optic mode locker in the argon laser was driven by approximately 1 W of RF power at 70 MHz. The average power available at the 514·5 nm excitation wavelength was typically 800 mW and pulse durations were normally ~90–150 psec. When the dye laser cavity length was optimized and tuned to an accuracy of about 1 μm and the argon ion laser stably mode locked for minimum pulse widths then the dye laser pulses were ~1–2 psec long.[6] These pulses had peak powers of about 200–300 W and were tunable over the range 560–630 nm.

Two separate methods were investigated for deriving the synchronized trigger signals for the camera. In one approach (shown in Fig. 2) about 2% of the signal generator output was used. Slightly better performance was obtained in an alternative arrangement where about 50% of the output intensity of the dye laser pulses was directed on to a fast photodiode (TIXL 56). These electrical pulses ($\gtrsim$ 50 mV, 250 psec FWHM) were supplied to a tunnel diode which was biased for monostable operation so that a sine wave output voltage was produced at the cavity frequency of the dye laser.

In both cases the trigger sinusoid was frequency doubled and then amplified in a three stage transistor amplifier. This voltage was fed to the deflection plates of the streak tube which were incorporated into a high Q

† Spectra Physics, 7 Stuart Rd, Chelmsford, MA. 01824, U.S.A.

resonant *LC* circuit. The peak to peak amplitude of this deflection sinusoid was typically 3·5 kV which was adequate to ensure linearity over the 50 mm streak length on the phosphor screen.

Temporal Resolution of Streak Camera Operating at 140 MHz

The temporal resolution limit of the synchronized camera arises from photoelectron time dispersion t_1 and the streak limited or technical time resolution t_2 as in the case of conventional single shot cameras.[3] For the S·11 photocathode of the Photochron I used in the evaluation, when illuminated at 600 nm the value of t_1 is 2 psec for an applied electric field strength of 6·6 kV cm^{-1} close to the photocathode. For the applied deflection sinusoid the streak velocity was 5×10^9 cm sec^{-1}, and this gives $t_2 = 4$ psec when combined with the dynamic spatial resolution of 5 lp mm^{-1}. Assuming Gaussian profiles, an overall instrumental function for the camera of 4·5 psec is obtained.

The durations of the pulses from the dye laser were independently measured from intensity autocorrelation traces obtained by second harmonic generation. Under optimum operating conditions autocorrelation traces similar to that shown in Fig. 3 were recorded.

Time calibration of the streak camera records was obtained from the optical delay arrangement illustrated in Fig. 2. The intensity profiles of the streak images were recorded using an Optical Multichannel Analyser; (PAR 1205D)† which was optically coupled to the streak tube via an

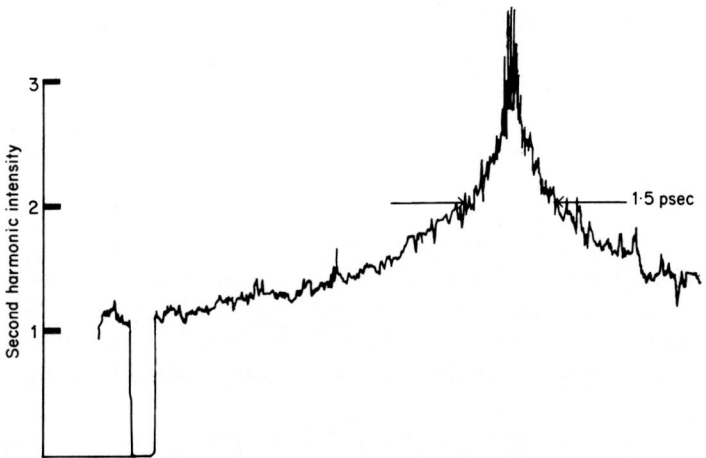

FIG. 3. Autocorrelation trace of pulses produced by synchronously mode locked dye laser.

† Princeton Applied Research, P.O. Box 2565, Princeton, NJ 08540, U.S.A.

$f/1\cdot 5$, 80 mm lens. A storage oscilloscope (Tektronix type 605) was used for the real time display of the streak profiles and permanent records were made with a chart recorder.

When the pulses from the synchronously mode locked dye laser were streaked using the Synchroscan camera, the recorded intensity half-widths significantly exceeded the expected value of ~5 psec. The shortest recorded streak durations were 13 psec as indicated in Fig. 4. This result could be explained by one or more of the following possibilities. There may be jitter of about 10 psec between the argon ion laser pulses and those of the dye laser as already suggested for two mode locked CW dye lasers operating in tandem.[7] Alternatively, the laser pulses may not have been particularly reproducible in duration and shape during the recording period of the streak records. This irreproducibility has already been observed when the output of this type of CW laser was investigated using a single shot streak camera.[8] A third explanation could be that there was jitter of about 10 psec between the generation of the streaking sinusoid and the incident laser test pulses. Furthermore, some feedback from the RF amplifiers to the acousto-optic transducer could also contribute to some overall jitter in the combined diagnostic.

In order to establish the reason for the disparity between the predicted and demonstrated performance of the Synchroscan the evaluation involving the passively mode locked CW laser was undertaken as described in the next Section. In any case, the resolution capability of the camera operating at 140 MHz (~15 psec) was adequate for investigating and

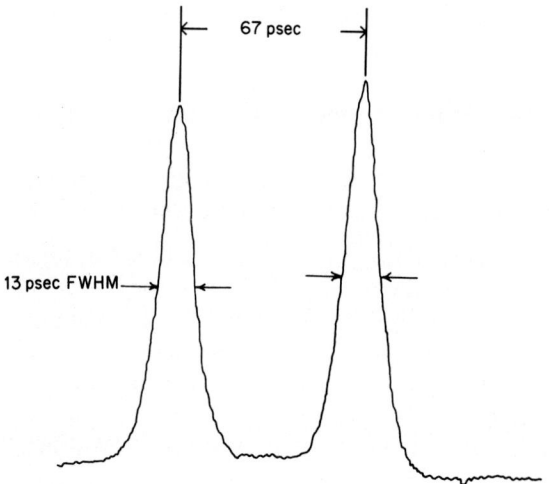

FIG. 4. Intensity profiles of streaked dye laser pulses for camera operating at 140 MHz.

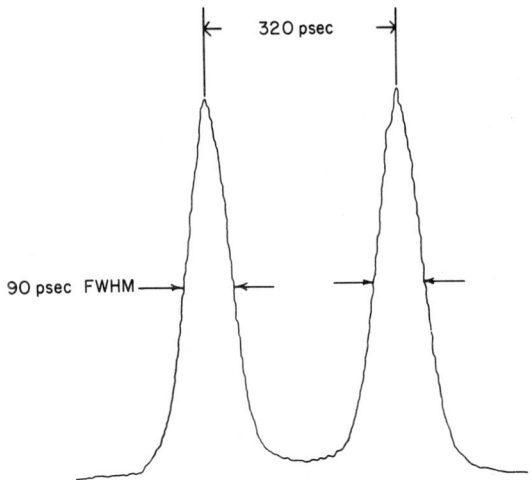

FIG. 5. Intensity profiles of streaked argon ion laser pulses.

optimizing the mode locking of the argon ion laser. It was found that when a high quality oscillator (Wandel and Goltermann-model LM 568)† was substituted for that supplied by Spectra Physics, then pulses as short as 90 psec could be reliably generated (see Fig. 5). In practice, this camera proved to be a very convenient and powerful diagnostic for monitoring these pulses with appreciably better time resolution than is currently available with photodiode and sampling oscilloscope combinations. The fluorescence lifetimes of several dyes have also been measured using this camera system.[9] In these observations it was clearly evident that the repetitive streak operation provides the necessarily high detection sensitivity that is required to record weak optical signals.

Synchroscan with Passively Mode Locked CW Dye Laser

By using the experimental arrangement outlined in Fig. 6 which incorporated a passively mode locked CW dye laser and the Synchroscan camera then the uncertainties associated with the temporal character of the synchronously mode locked CW laser pulses and any possible interference between the acousto-optic mode locker and streaking sinusoid could be eliminated.

The dye laser was similar to that described elsewhere.[10] During the evaluation of the camera the laser cavity frequency was 82·5 MHz and

† Wandel & Goltermann, Postfach 45, D-7412 Eningen, W. Germany

the wavelength of the output was 605 nm. Under these conditions subpicosecond pulses were reliably produced and an autocorrelation trace of a typical second harmonic profile is shown in Fig. 7.

Temporal Resolution of Streak Camera Operating at 165 MHz

About 50% of the intensity of the ultra-short laser pulses was directed on to a photodiode which provided the electrical trigger signal for the tunnel diode oscillator. The transistor amplifier that had been used at 140 Mhz was retuned for operation at 165 MHz. This retuning was at the expense of a reduced sinusoidal amplitude which in turn decreased the streak writing from 5×10^9 cm sec^{-1} to $4 \cdot 3 \times 10^9$ cm sec^{-1}. Consequently, the theoretical instrumental resolution was now 5 psec.

From the streak profiles that are reproduced in Fig. 8 it can be seen that the recorded half-width was 5·2 psec. Since this is just greater than the theoretically predicted value, it indicates that the repetitively operating streak camera has a very low jitter, probably appreciably less than 5 psec. This result would therefore indicate that the synchronously mode locked dye laser was setting the limit to the previously demonstrated resolution of the Synchroscan of 13 psec.

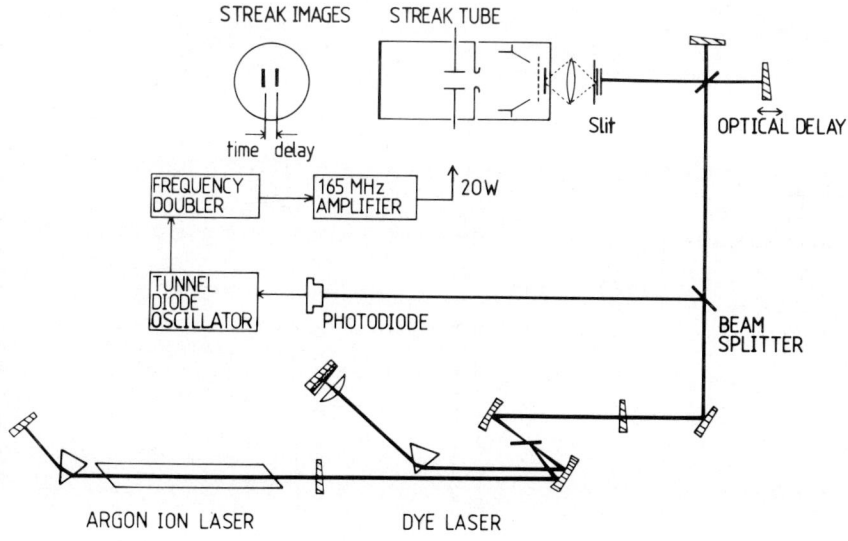

FIG. 6. Experimental configuration of Synchroscan camera and passively mode locked dye laser.

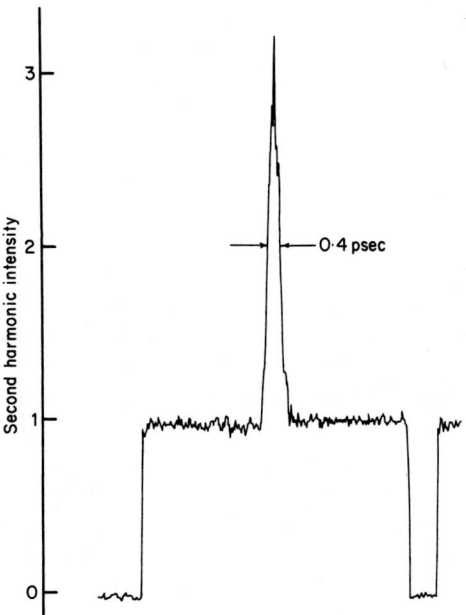

FIG. 7. Autocorrelation trace of pulses produced by passively mode locked dye laser.

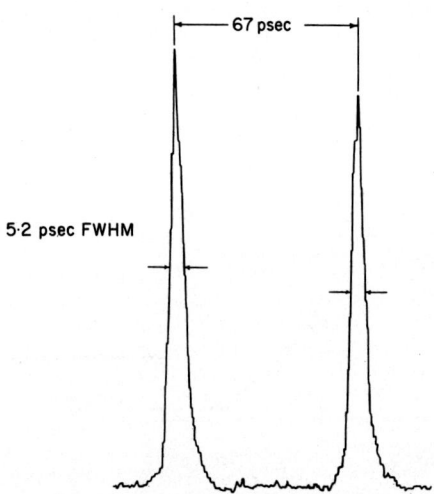

FIG. 8. Intensity profiles of streaked dye laser pulses for camera operating at 165 MHz.

Conclusions

The work described in this paper illustrates that the Synchroscan streak camera is an important diagnostic tool which is particularly well suited to the investigation of repetitive picosecond luminous phenomena. It is believed that the temporal resolution can be further improved to ~1 psec by employing a Photochron II streak tube with improved deflector geometry.

This technique which has the advantage of ultra-high detection sensitivity should therefore have wide application in the picosecond spectroscopy of fluorescence, luminescence, etc. at intensity levels which may be significantly lower than that which is required for measurements involving the single shot streak cameras.

Acknowledgments

The authors wish to thank S. F. Bryant and J. P. Ryan for many helpful discussions. Financial support from the Science Research Council and the Paul Instrument Fund is also gratefully acknowledged.

References

1. Bradley, D. J. and Sibbett, W., *Appl. Phys. Lett.* **27**, 382 (1975).
2. Ruddock, I. S., Sibbett, W. and Bradley, D. J., *Opt. Commun.*, **18**, 26 (1976).
3. Bradley, D. J. Topics in Applied Physics Vol. 18, 'Ultrashort Light Pulses', ed. by S. L. Shapiro, pp17–122, Springer-Verlag, New York (1977).
4. Bradley, D. J., Liddy, B., Roddie, A. G., Sibbett, W. and Sleat, W. E., *In* "Adv. E. E. P." Vol 33B, p. 1145 (1972).
5. Kalibjian, R., To be published in Proc. 13th International High Speed Photography Congress, Tokyo, (1978).
6. Bradley, D. J. and Ryan, J. P., Paper J6 at 10th Int. Quant. Elect. Congr. Atlanta, Georgia, (1978).
7. Heritage, J. P. and Jain, R. K., *Appl. Phys. Lett.* **32**, 41 (1978).
8. Ryan, J. P., PhD Thesis, Imperial College, London (1978).
9. Adams, M. C., Sibbett, W. and Bradley, D. J., *Opt. Commun.* **26**, 273 (1978).
10. Ruddock, I. S. and Bradley, D. J., *Appl. Phys. Lett.* **29**, 296 (1976).

A Review of Astronomical Applications

J. RING

Astronomy Group, Blackett Laboratory, Imperial College, University of London, England

It is no accident that the seven Symposia on "Photo-Electronic Image Devices" which have taken place at Imperial College, London since 1958, have recorded remarkable developments in astronomical image detectors. One of the principal objectives of Professor J. D. McGee in setting up his research group at the College was to develop an ideal detector for use in astronomy and the Symposia which he instituted, although covering the whole field of photoelectronic technology, have always strongly emphasized astronomical applications.

The published proceedings of the Symposia form a valuable and concise record of the improvement to astronomical detectors over the past two decades and it is in this perspective that we can best consider the present "state of the art" as exemplified by the contributors to the 7th Symposium.

It is, perhaps, obvious why detector technologists have been attracted to astronomical observation; practically all our information about the extra-terrestrial universe comes from an analysis of the photons collected by telescopes and the efficiency of the process depends strongly on the quality of detecting devices. In 1958 the photographic plate was used by almost all astronomers, although some observers were beginning to use single-point photoelectric detectors such as the photomultiplier tube. Having pushed telescope design to its limit (there is still only one telescope larger than the 5m Hale instrument) astronomers were wasting as much as 99·9% of the photons, collected at great expense, in inefficient detectors. Low quantum efficiency, threshold intensity levels, non-linearity and limited storage were the more obvious defects of the photographic technique; there were other, more subtle, difficulties. The photomultiplier tube overcame most of these problems at a stroke, but could not compete with the one great asset of the photographic plate—its spatial multiplicity. If the observation being conducted required information from more than about one hundred spatial or spectral elements then, despite its superiority in all other respects and a quantum efficiency advantage of order 100 times, the fact that the photomultiplier had to obtain its information sequentially rather than simultaneously was a severe limitation.

It was clear that what was needed was an imaging device which retained the advantages of the photomultiplier tube, and it is interesting to note that at the 1st Symposium the problem was well understood and the types of device which could satisfy the needs of astronomers were clearly outlined. Baum opened the proceedings with a clear statement of the astronomical requirement and with a warning that a simple improvement of speed was of limited use—it should be accompanied by a greater storage capacity. Lallemand described his electronic camera and reported preliminary astronomical trials; Zacharov and Dowden, members of McGee's group at Imperial College, described a tube with a thin mica window which was clearly the forerunner of the Spectracon to be described by McGee himself at the 3rd Symposium. There were several papers on the use of "cascaded" image tubes (without the benefits of fibre optics) and a description of a true "cascade" tube with transmission secondary electron multiplication. McGee outlined a scheme for a channeled image intensifier, the precursor of the channel plate systems currently being investigated. Finally there were papers on charge integration and storage tubes, but without the benefits of diode arrays.

What has happened in the twenty years between the First and Seventh Symposia? Initially most of the time seems to have been spent on technological problems such as preventing cathodes from dying, increasing resolution, reducing background, improving efficiency and so on. Later on much effort went into obtaining a better understanding of the more important characteristics of image detecting systems. For example, although it was realized quite quickly that nuclear emulsions were less than ideal for astronomical electronography, it was not until the 6th Symposium that Worswick presented a clear analysis of the complete system including the microdensitometer, which showed how difficult it was to achieve photometric accuracies better than 1 or 2%. Throughout the period, all kinds of tubes were being tried in astronomy, but despite efforts by several authors results were often quoted in a way which made it difficult to intercompare them.

Slowly the field of devices narrowed, although new components such as fibre optic plates, channel plates and diode arrays were incorporated. At the present time the systems finding favour with astronomers seem to be either electronographic cameras or commercial image tubes feeding a photographic plate or a rapid scanning detector such as a television camera or solid state array. Both types of system can yield a DQE of at least half that of the primary photocathode, can be linear over a wide range and can have high storage capacity. Each suffers from limitations, but these are fortunately different so that the types of device are complementary. Looming over the horizon at the present time is the prospect of large arrays of diodes, with sufficiently low readout noise to allow them to be used as analogue detectors without intensification, but that story is told elsewhere in this volume. Let us consider the present state of the art in electronography and in the use of commercially available image tubes.

Several papers at the 7th Symposium showed that electronographic cameras are now in almost routine use at a number of observatories. Wlérick† describes a variety of applications of the 81 mm version of the Lallemand camera, which yields high resolution (70 lp mm^{-1}) and very low background (<10 electrons pixel^{-1} h^{-1}). The spatial multiplicity of the camera, with 3×10^7 pixels of a size well matched to the prime focus of the largest reflectors, makes it an almost ideal instrument for the study with moderate photometric accuracy of faint extended objects. The camera has been used mainly at the Cassegrain focus of the 1·93 m telescope at Haute Provence where it is now available as a "common-user" instrument; three similar cameras are being constructed for use with the 3·6 m C. F. H. telescope in Hawaii. Wlérick and his colleagues have shown the considerable merits of the electronographic method in studies of radio-galaxies, clusters of galaxies and quasars.

At the Royal Greenwich Observatory, McMullan‡ and his colleagues have continued to develop large field (80 mm) electronographic cameras and have supplied 40 mm cameras to observatories in Israel, South Africa, Australia and Chile. The cameras are of the mica window type and numerous improvements are reported which simplify the routine use of the device. Operation of the camera is automatic, with discs of emulsion, previously mounted on nylon rings, being loaded in less than one minute. Here too, the electronographic method has been applied mainly to the study of the morphology of galaxies and nebulae.

Griboval§ reports progress towards the ambitious aim of constructing a 200 mm diameter electronographic camera at the University of Texas. His Mk II experimental laboratory model, with a 50 mm field, has been tested on the 0·76 m telescope at the McDonald Observatory. The electronographs of galaxies which were obtained were limited by poor seeing and high sky brightness but the camera performed well.

If the cameras mentioned so far go a long way towards refuting early fears that electronographic systems were too complicated and fragile for

† See p. 295.
‡ See p. 315.
§ See p. 305.

use at observatories, the work of Carruthers† must surely finally lay that ghost. He describes the development of a 123 mm camera with 10 μm resolution for use in Shuttle/Spacelab missions, at the focus of a 1 m telescope which will yield an angular resolution of 0·2 arcsec over a useful field of half a degree. The camera will be sensitive in the ultraviolet below 200 nm and will be focused either by a permanent magnet or by a superconducting solenoid. The intense magnetic field generated by the latter (1 tesla) allows the use of curved photocathodes, which match the focal surface of a telescope or spectrograph, without loss of electronic resolution. Laboratory tests have confirmed the expected resolution whilst disclosing problems of radial non-uniformity in magnetic fields. Attempts are being made to extend the wavelength range of the camera to 300 nm. If the remaining technical problems can be overcome it is difficult to imagine any other imaging detector rivalling Carruthers' system for space surveys from vehicles which permit the recovery of data stored on film.

Now that electronographic cameras are coming into routine use, more attention is being paid to the reduction of the resulting data. Hardwick et al.‡ describe their analysis of electronographs of the elliptical galaxy NGC 3379, obtained with a Spectracon at the Newtonian focus of the 74 in. telescope of the Helwan Institute in Egypt. (Despite the spread of the cameras described by Wlérick and McMullan, there are still many astronomers who do not have ready access to them and the Spectracon continues to be put to excellent use.) After scanning the image with a microdensitometer using a 50 μm square aperture (equivalent to 1·1 arcsec), Hardwick et al. derive isophotal contour maps which are used to select an area for detailed study. A computer rejects extraneous objects such as stars and image defects and then derives luminosity profiles around the galaxy. By averaging pixels within successive annuli a measure is obtained of the excess brightness of the galaxy above sky background, which extends to the outermost regions of the object. In this way, the photometric accuracy of the electronographic method can be greatly improved until the errors are comparable to those of photoelectric photometry.

Astronomical applications of commercial image tubes with a phosphor output now seem to be concentrated in two rather different areas. The phosphor may be photographed, in which case the principal gain is that of increased speed over the unintensified plate, or the individual photoelectron events may be detected, when several other important advantages accrue.

† See p. 283.
‡ See p. 329.

The contribution by Craine et al.† describes an excellent system of the first kind. It employs an ITT single stage, magnetically focused tube with a 146 mm diameter photocathode sensitive to 900 nm and an output phosphor deposited on a fibre optic faceplate. 300 photons are emitted from the phosphor for each incident photoelectron and the gain in photographic speed is about 40 when IIa-D emulsion is used. The plate limits the resolution to about 25 lp mm^{-1} but the peak DQE of the complete system is of order 7·5%. The fibre optic faceplate produces some discrete shears (<30 μm) in the image. Although this camera has fewer pixels than the electronographic ones and is limited in its linearity and storage capacity by the plate, it clearly lends itself well to survey type observations and similar systems are being installed on most large telescopes. With the Steward Observatory system a plate limit of 21^m is achieved in two minutes at the Cassegrain focus of a 2·3 m reflector. The authors describe how the camera has been used in a near infrared sky survey and in a study of the variation of intensity and polarization across M82. The polarimetric camera has also been used to discover new BL Lacertae objects.

When an image intensifier is coupled to a rapid scanning detector such as a TV camera or CCD array the limitations of the photographic plate are removed completely. Each photoelectron scintillation is directly registered and the resultant imaging, photoelectron counting system can have high resolution, is linear and can have almost unlimited digital storage; in addition the image is available on-line. Photoelectron counting systems are now extremely popular amongst astronomers and their use is rapidly being extended to all major telescopes, although their cost and complexity do not yet allow them to be universally available.

The most widely adopted photoelectron counting system is that developed by Boksenberg et al.‡ at University College, London. In their contribution to the 7th Symposium they give a lucid account of the way in which the system achieves the properties mentioned above and go on to describe a ruggedized and improved version which is being developed for use with the Space Telescope. There are however still some annoying limitations even in the photoelectron counting systems. The number of pixels may be limited by the TV detector or by the available computer memory, and the need to detect each scintillation limits the maximum flux to much less than one scintillation per pixel per frame.

An alternative form of photoelectron counting system has been developed by Stapinski et al.§. The image intensification is achieved by no

† See p. 339.
‡ See p. 355.
§ See p. 389.

less than six electrostatically focused, fibre optically coupled Varo tubes. The output phosphor is scanned by a Reticon, dual, 1024 diode array. Event centring logic is used to improve the resolution and the system has given good results on the 1·9 m telescope at Mt. Stromlo in a variety of spectroscopic observations. It is intended to extend the detector to two dimensional operations by using a CCD to scan the phosphor.

A similar system was described by Chaffee,† who made an interesting comparison of its performance with that of a Kron electronographic camera. He uses three fibre optically coupled image tubes and a dual Reticon array. The performance of the detector was established by obtaining spectra of ζ Oph using an echelle spectrograph on the 1·5 m reflector at Mt. Hopkins. It is very interesting to note that both detectors achieved statistical uncertainties only 1·6 times those expected from Poisson statistics. When exposed to give 2% photometric precision the performances of both detectors were indistinguishable. At lower signal to noise ratios the Kron camera yields too low a density to be measured with conventional microdensitometers; however when a large range of spectrum must be covered the Kron camera is superior on account of its 6×10^4 pixels as compared to 10^3 for the photoelectron counting system. (The author does not make a comparison at higher signal to noise ratios but presumably the electronographic system is limited both by the difficulty of measuring high densities and by variations in emulsion sensitivity which cannot be corrected for.)

The expense and complexity of the first photoelectron counting systems is partly due to their use of cascade intensifiers. Attempts are being made to simplify the device by using microchannel plate tubes to provide sufficient gain for photoelectron detection. Two papers report the development in France of systems based on such an intensifier fibre optically coupled to a television camera. Rosier *et al.*‡ give a detailed description of the characteristics of their channel plate tube, mentioning the restrictions on photon gain imposed by optical feedback and the efforts being made to remove this limitation. They give pulse height distributions for photoelectron scintillations with various channel plate voltages. Boulesteix§ describes the astronomical tests of a similar system using a Thomson -CSF TH9503 microchannel plate tube coupled by fibre optics to a SIT camera. The images comprise 256×256 pixels with event centring logic and are stored in a computer. The dynamic range extends from 2 to 3000 events pixel^{-1} h^{-1}. Images of HII regions through

---

† See p. 415.
‡ See p. 369.
§ See p. 379.

interference filters and interferometers and of the spectrum of the spiral galaxy M51 demonstrate the capability of the system which has a gain in speed (for the same signal to noise ratio) of four over a two stage cascade image tube and of 25 over a 103 a-E plate.

It is, at first sight, surprising that these devices work so well. There are many statements in the literature that the channel plate tube can only be used efficiently in the saturated mode, because of the negative exponential pulse height distribution at lower gain. Yet Rosier, working at low gain, claims a DQE of 3·4% with a cathode quantum efficiency of 8·6%. Much of this reduction must occur in the loss of photoelectrons at the entrance to the channel plate and in zero yield on the first collision inside it. Further reflection suggests that the earlier literature may well be misleading. The negative exponential shape of the pulse height distribution, whilst increasing the noise of an analogue detector, is of little importance when pulse counting, provided that a sufficiently large fraction of the photoelectrons yield detectable pulses. Simple calculations suggest that the distribution curve must reach a peak well above zero electron gain and there is one observation[1] which appears to confirm that this peak occurs at a gain of 10^4 when the mean gain is of order 10^5. A substantial fraction of the incident photoelectrons will lie at gains higher than this peak. If this is indeed the case, or if the optical feedback at high gain can be suppressed, then astronomers can look forward to a very simple photoelectron counting system consisting of a channel plate tube coupled to a CCD array.

The dynamic range limitations of the photoelectron counting systems may be removed, whilst retaining the quantum noise limited performance, by using analogue detectors with sufficient gain to overcome readout noise. All that is necessary is to ensure that the pulse height spread of the intensification system does not increase the noise unduly, by unequally weighting individual photoelectrons. The contribution to the 7th Symposium by Hege et al.† reports tests of five different systems all based on this principle. A variety of image tubes and optical couplers was used to feed a Reticon dual array. Detailed laboratory comparisons were made and several of the systems were tested on the 2·3 m telescope of the Steward Observatory. Performances near to the theoretical limit were achieved at fluxes as low as $0·1 \text{ pixel}^{-1} \text{ sec}^{-1}$ and with a dynamic range of 10^5.

The intensified television camera is regularly used at many observatories with direct display of the image on a monitor. Such a system, which is much less expensive and complicated than the photoelectron

† See p. 397.

counting devices, can nevertheless be extremely useful in finding and guiding on faint objects. Angel et al.† describe the Steward Observatory system which employs a three stage electrostatic image tube coupled to a vidicon with a 10 sec lag. When used on the 2·3 m telescope the camera detects all the stars visible on the E plates of the Palomar Sky Survey over a field of 4 arcmin.

For reasons of convenience, papers on purely solid state detectors are published in a separate section and are not reviewed here. It is clear however that they will have important applications in astronomy, particularly when high signal to noise ratios are required; in such circumstances sufficient photons must be collected in each pixel that their shot noise exceeds the readout noise of the detector.

What is the astronomer who is not a detector technologist to make of the plethora of devices described at the 7th Symposium? First of all he should be grateful that he has such a wide choice—it has only been provided by thousands of man-years of effort. He should note that he can record objects comparable in brightness to the night sky using cascaded intensifiers with a television or photographic output stage. He can obtain photometric surveys of the useful field of paraboloid reflectors with electronographic cameras which are linear, extremely sensitive and capable of accuracies of 1 to 2%. For fainter objects he can study a smaller area of sky or of a spectrum with imaging photoelectron counting systems whose photometric accuracy is limited only by the statistical noise of the number of photoelectrons he can afford to collect.

Finally, he can look forward to simpler, more convenient ways of obtaining the results described above, but not to any further dramatic improvements in performance. The ideal astronomical detectors envisaged at the first few Symposia are not yet here, but present systems are so close to that performance that the law of diminishing returns must soon begin to apply.

Reference

1. Chalmeton, V., *Acta Electron.* **14,** 99 (1971).

† See p. 347.

High Resolution Large Format Electronographic Cameras for Space Astronomy

G. R. CARRUTHERS

E. O. Hulburt Center for Space Research, Naval Research Laboratory, Washington, D. C., U.S.A.

INTRODUCTION

NASA is presently studying possible UV/optical astronomical telescopes for use in Shuttle/Spacelab missions. In such missions, the instrumentation remains attached to pallets in the payload bay of the Shuttle during its 7 to 21 day stay in orbit, following which the instruments return to earth with the Shuttle. A particular advantage of this type of mission is that it allows the use of film recording imaging devices, such as electronographic cameras, instead of electronic readout detectors (with transmission of data to the ground by telemetry) such as must be used with the unmanned Space Telescope (ST).

Specific facilities under study include STARLAB, which is based on a 1 m aperture telescope with 0·5 deg diameter useful field and 0·2 arcsec resolution, and a Spacelab Wide Angle Telescope (SWAT), having about 0·8 m aperture, 4 to 6 deg field of view, and 1 to 2 arcsec resolution. (For comparison, the Wide Field Camera on ST will have a 3 arcmin field of view and 0·1 arcsec resolution.) Both of these Spacelab telescopes will require imaging detectors having both very large image format and very high resolution, in order to make full use of their performance capabilities.

Large format electronographic cameras are ideally suited to these applications. These devices are capable of recording a much larger number of picture elements per frame (10^8 or more) than the largest available low light level electronic readout image sensors (typically limited to at most about 10^6 pixels per frame). The electronographic camera is also simpler to implement in space missions, because of its lack of complex electronics, and little or no requirement for telemetry or on-board data handling and processing. In comparison to image intensifier/film combinations, it offers the advantages of linearity of response, higher resolution, and wider dynamic range.

Development of a large format electronographic camera meeting the requirements of STARLAB and SWAT is particularly straightforward for the far ultraviolet (below 200 nm) wavelength range, where alkali halide photocathodes can be used.[1,2] However, even in the longer wavelength ranges, such as the middle ultraviolet (200–300 nm) and near visible (300–10 00 nm), the potential advantages of a large format electronographic camera justifies a vigorous effort in its development.

Far–UV Large Format Camera

We are presently developing, under NASA sponsorship, a large format, magnetically focused electronographic camera (Fig. 1) capable of recording images 123 mm in diameter (on 5 in. roll film) with resolution better than 10 μm over the entire format. This format and resolution are directly applicable to the present concept of STARLAB. The camera is sensitive in the wavelength range below 200 nm, by the use of a semi-transparent CsI photocathode.

Figure 2 shows the components of the camera with the permanent magnet focusing assembly, and Fig. 3 is a photograph of the complete camera assembly. The spacing between the photocathode and film plane

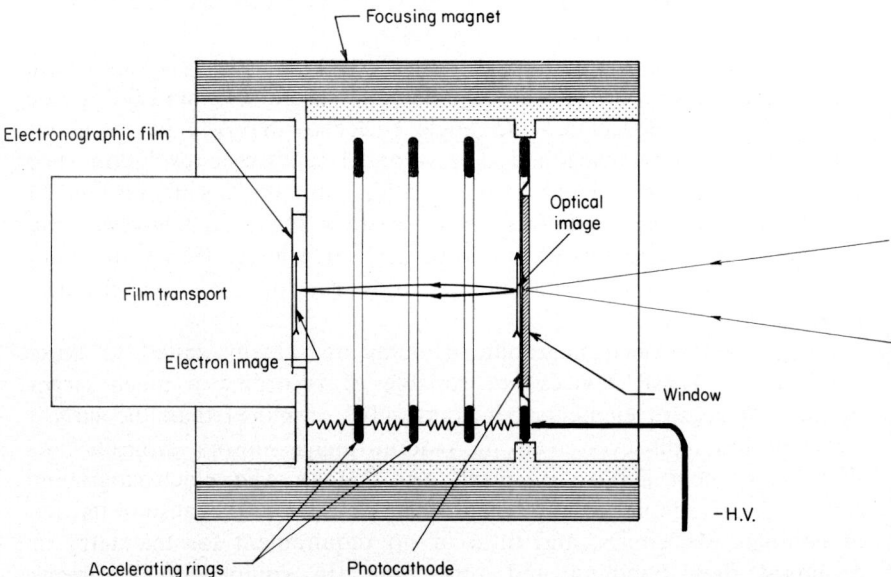

Fig. 1. Diagram of a magnetically focused, large format far-ultraviolet electronographic camera.

FIG. 2. Components of large format far-ultraviolet electronographic camera and permanent focusing magnet.

FIG. 3. Photograph of assembled far-UV large format electronographic camera.

for the initial tests was about 125 mm, which corresponded to three loop focusing with the permanent magnet and at about 18 kV operating voltage.

In addition to tests using the permanent magnet, the camera was used with a superconducting solenoid (Fig. 4) to provide the focusing field. The superconducting solenoid and liquid helium dewar assembly† has a 25 cm diameter bore which is at room temperature. It allowed operation at much higher field strengths (up to 10 000 G as against 350 G for the permanent magnet) and in addition provided much better field uniformity, even at low field strengths, than did the permanent magnet.

The use of very intense magnetic fields permits the use of non-flat photocathodes with flat recording emulsions without an unacceptably large variation in electron optical focus over the image format.[3] This condition of "infinite depth of focus" results from the radius of gyration of the photoelectrons being kept to less than the maximum acceptable image radius. The radius of gyration is given by

$$R = 26 \cdot 92 \frac{\sqrt{V_t}}{B} \text{ mm},$$

where V_t eV is the transverse component of the photoelectron emission energy, and B G is the magnetic field strength. Taking a typical value

FIG. 4. Superconducting solenoid, with 25 cm diameter room temperature bore, used for testing large-format electronographic camera.

† Manufactured by American Magnetics Corp., 250 S. Jackson, Carterville, IL. 62918, U.S.A.

of 2 eV for V_t,[4] R is less than 10 μm for field strengths in excess of 4000 G (or less than 5 μm above 8000 G). The possibility of using non-flat photocathodes is of importance for several proposed Spacelab telescopes and auxiliary instruments. For example the focal surface of a Ritchey-Chretien telescope, as studied for STARLAB, is concave (outward from the detector), as is the focal surface of a Rowland spectrograph. The focal surface of an all reflecting Schmidt optical system for use in SWAT is convex outward from the detector. However, in this latter case, the use of intense magnetic fields is not essential as the recording film can also be given a convex curvature (by the use of a dome-shaped platen) which matches the curvature of the photocathode, hence allowing uniform focusing at low field strengths. (It is not generally practical, on the other hand, to produce a concave recording surface in a camera using roll film.)

In order to test the resolution capabilities of the camera a semitransparent CsI photocathode was deposited onto a flat ultraviolet grade fused silica substrate on which was a series of USAF 1951 test patterns. These patterns were prepared by Metrigraphics† using thin film metal deposition onto the substrate; the bar spacings were as fine as 228 lp mm^{-1}. The faceplate of the camera was illuminated uniformly by a H_2 microwave discharge lamp, in front of which was placed an interference filter with central wavelength of 160 nm (30 nm bandwidth).

For testing the camera with curved photocathodes, an ultraviolet grade fused silica substrate was prepared with concave and convex radii of 500 mm. As it was not feasible to have the USAF 1951 patterns prepared on a curved substrate, we made a somewhat cruder test pattern by evaporating nickel onto the substrate through nickel meshes of various spacings, held onto the substrate with tape during the evaporations. Since it was difficult to obtain adequate contact of the shadowing mesh with the substrate at all points on the concave photocathode surface, the resulting test pattern was less than satisfactory in some areas of the format. An additional resolution test pattern was therefore provided by scribing the evaporated CsI coating with a needle, giving typical line widths of 10 to 20 μm. Artifacts, such as localized defects in the nickel coating, also provided resolution test patterns.

Flat CaF_2 windows 125 mm in diameter were also obtained for use as photocathode substrates. As we found it would be very expensive to have the USAF 1951 test patterns produced on this substrate in the same way as on the flat silica substrate, we elected instead to produce our own USAF 1951 patterns by evaporating nickel through a thin nickel foil

† Metrigraphics Division, Dynamics Research Corporation, 50 Concord St., Wilmington, MA. 01887, U.S.A.

1951 test pattern obtained from Buckbee-Mears.† This foil mask was held against the CaF_2 window during evaporations by a magnet placed on the front side of the window.

Some tests were also made with a mesh-based semitransparent CsI photocathode[4] formed on a Buckbee-Mears nickel mesh having 60 lines mm^{-1} spacing; the resolution achievable was limited by the mesh spacing (and Nyquist criterion) to about 30 lp mm^{-1}. Nevertheless, this still provided a capability of recording 7000 pixels across the format, or 4×10^7 pixels over its entire area, and is of particular interest for spectrographic or imaging applications in the extreme ultraviolet (below 105 nm) where conventional photocathode substrate materials such as LiF and CaF_2 do not transmit. Even in the wavelength range above 105 nm CsI photocathodes deposited on mesh substrates have considerably higher quantum efficiencies than do conventional semitransparent photocathodes.

The focusing field of the permanent magnet was found to be significantly non-uniform in the radial direction, varying from 355 G along the centre line to 370 G (axial component) at a position 60 mm out from the centre line (corresponding to the edge of the image format). There was therefore significant variation in resolution at a fixed operating voltage (or focus voltage for best resolution) from the centre to the edge of the image. With best focus at the centre of the image, the central resolution was in excess of 120 lp mm^{-1} but there was a fall-off about 25 lp mm^{-1} at the edge of the image. However if the voltage was adjusted for best focus at an intermediate zone, it was possible to maintain better than 40 lp mm^{-1} over the entire format.

Fortuitously this field strength variation was found to compensate almost exactly for the curvature of the convex outward photocathode substrate (500 mm radius). As a result, the convex photocathode gave much more uniform resolution over the 123 mm format when imaging onto flat film, than did the flat photocathode!

Using the superconducting solenoid uniform resolution of better than 120 lp mm^{-1} was obtained over the entire format with the flat CsI on silica photocathode (see Figs. 5 and 6), even at field strengths as low as 350 G. The expected centre to edge variation was observed, however, with the curved photocathode. In our initial attempts to operate at very high field strengths, we found that at field strengths above 1200 to 1500 G the camera would go into high voltage discharge (although the residual gas pressure for all runs was of the order of 2×10^{-6} Torr). This problem was solved by reducing the photocathode/film spacing to 40 mm which corresponded to single loop focusing at 18 kV and 350 G. In this

† Buckbee–Mears Corp., 245E 6th Street, St Paul, MN 55101, U.S.A.

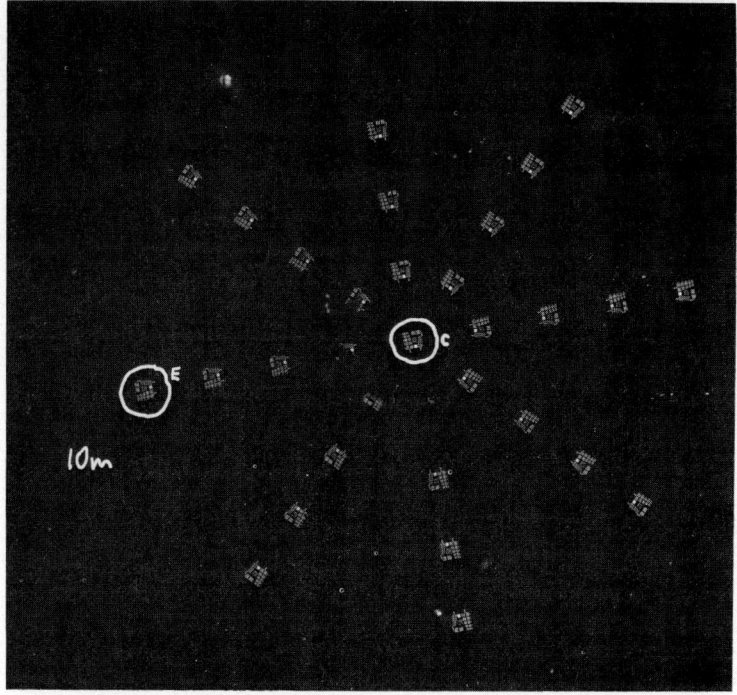

FIG. 5. Full-frame image (×0·75) of USAF 1951 test pattern array, obtained with large format electronographic camera and flat photocathode substrate.

configuration the camera was successfully operated at field strengths of up to 8000 G.

In accordance with what we expected theoretically, it was found that with the curved photocathode and using the superconducting solenoid there was no marked improvement in centre to edge focus uniformity with increasing field strength until a field strength of 4000 G was reached. With this field strength the difference between the photocathode/film path lengths at the centre and at the edge (about 4 mm) was just equal to one loop out of the theoretical ten loops executed by the photoelectron over the nominal 40 mm spacing. Hence the image was sharply focused (5 μm or better resolution) at the centre and at the edge of the format, while the worst resolution in the intermediate zone was still better than 20 μm. At 8000 G (20 loop focus) there was theoretically an intermediate zone of sharp focus in addition to the centre and edge; however, the variation in image sharpness across the format was almost undiscernible with the resolution better than 10 μm everywhere in the image.

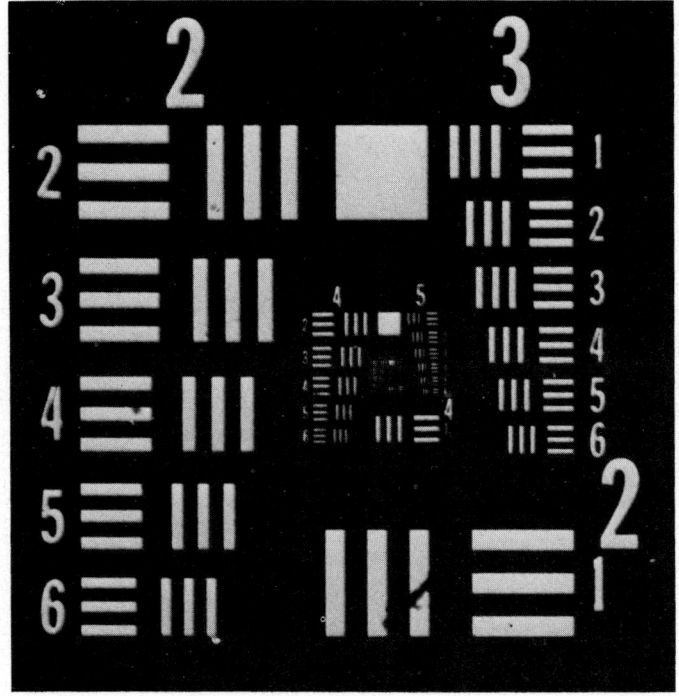

FIG. 6. Enlargement of the central USAF 1951 test pattern image from Fig. 5. On the original negative, bars with spacings corresponding to 120 lp mm^{-1} or better are clearly resolved.

For the proposed application of large-format electronographic cameras to the SWAT instrument, an image format of 200 mm diameter is required. This does not present any particular problems in itself, since it is readily achieved by simply scaling up the present camera. Nine inch roll film is a standard aerial film size, and windows of CaF_2 of the required diameter can be obtained. However, a special problem in faceplate design is encountered with the electronographic cameras for the SWAT application owing to the fast focal ratio ($f/2 \cdot 5 - f/3 \cdot 5$) of the beam converging to the image point. Because of the variation of refractive index with wavelength, particularly in the far ultraviolet below 160 nm, chromatic aberration produced in the faceplate can cause unacceptably large image diameters unless the thickness of the faceplate is kept very small (1 to 3 mm depending on the range of spectral coverage).[5] For faceplates 125 mm and larger in diameter, of materials such as LiF and CaF_2, this can be difficult to achieve especially if a convex photocathode is required (as in the all-reflecting Schmidt optical system).

We have been investigating a specific method for producing thin, curved photocathode substrates from thin, flat LiF and CaF_2 windows obtained from Harshaw Chemical Co.† In this method, the window is placed between two fused silica mandrels having matching concave and convex radii equal to the desired final substrate curvature. The assembly is then subjected to a combination of applied force and heating, causing the window to curve until it is in nearly uniform contact with the mandrels over its faces. With LiF windows 75 mm in diameter and 2 mm thick, it was found possible to produce a curvature of 800 mm radius in this manner using force alone; however for a CaF_2 window of the same size and using the same amount of force, heating to about 600°C was required.

This experiment is now being extended to CaF_2 windows 125 mm in diameter and 3 mm thick. In initial attempts it was found that the curvature produced was not uniform (the window took on a "panelled" appearance, with preferential bending along certain directions in the crystal structure) and breakage problems were encountered. However it seems likely that this technique will be useful for providing a large fraction of the required curvature, with the final inner and outer surfaces being best produced by conventional grinding and polishing.

EXTENSION TO LONGER WAVELENGTHS

We are presently attempting to extend the wavelength range capability of the large format camera toward longer wavelengths by incorporation of the appropriate photocathode materials and means for protection of the photocathodes from emulsion outgassing. Our initial efforts are towards the utilization of semitransparent caesium telluride photocathodes, which are sensitive to wavelengths as long as 310 nm. In related device developments incorporating opaque Cs_2Te photocathodes[6] we have found that these seemed to be considerably less susceptible to degradation by small leaks and outgassing than were Cs_3Sb photocathodes processed in the same devices. Therefore, it was felt that an electronographic camera using a Cs_2Te photocathode should be easier to develop and operate than one using a photocathode sensitive in the visible.

The planned mid-ultraviolet sensitive electronographic camera will utilize the barrier membrane method of Griboval‡ for photocathode protection (see Fig. 7). At present we are developing processing techniques for large diameter semi-transparent Cs_2Te photocathodes, and simultaneously are investigating various barrier membrane materials

† Harshaw Chemical Co., 1933 E 97th Street, Cleveland, OH 44106, U.S.A.
‡ See p. 305.

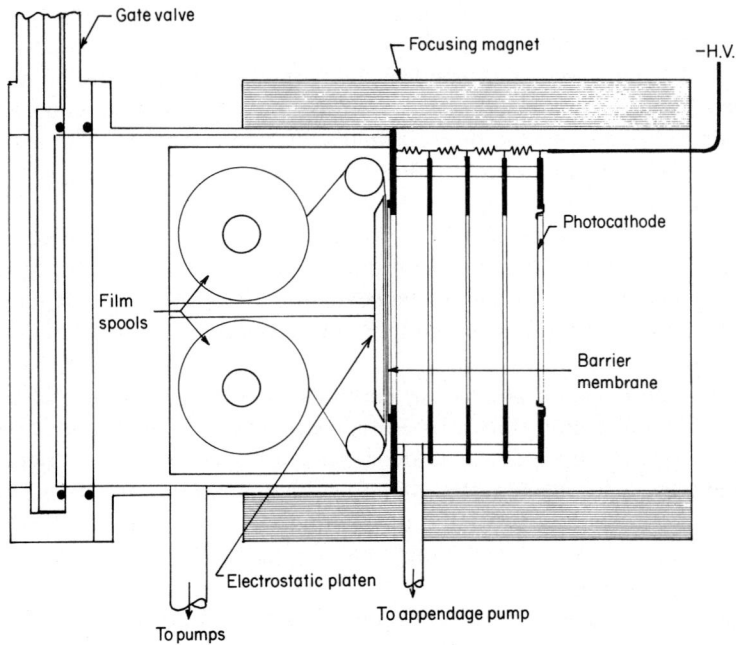

Fig. 7. Diagram of a large format electronographic camera which will incorporate semitransparent Cs_2Te photocathode, for sensitivity in the mid-ultraviolet up to 310 nm.

using the far-ultraviolet large format camera. A list of commercially available candidate membrane materials (with their upper temperature limits) follows: (a) 2μm Mylar (used in NRL far ultraviolet electronographic Schmidt cameras[1] to protect the electronographic film from visible light exposure), 100°C; (b) 3-μm Teflon, 200°C; and (c) 7·5 μm Kapton polymide (used in the Griboval camera), 300°C. The last of these is the most desirable from the standpoints of its physical properties, temperature resistance, ease of handling, etc., but has the major disadvantage of requiring very high electron energies (above 35 kV) for useful transmission. We are therefore investigating methods of making thinner (2 μm) polyimide films from a commercially available polyimide varnish.

For camera applications in which a flat film plane is used, we would employ the electrostatic hold down technique of Griboval. However, if a convex film plate is used, as would likely be the case in the SWAT application, this is not required since a convex pressure plate of the appropriate curvature behind the film would ensure adequate contact of the film with the barrier membrane. This pressure plate would, of course, have to be retracted a suitable distance to advance the film between exposures.

We are also looking into methods for sealing large diameter fused silica photocathode substrates to the tube bodies. A particularly promising method, if bakeout temperatures above 200°C do not prove to be necessary, is the use of a metal spring ring seal, with a soft coating of Teflon, polyimide, or soft metal such as lead or indium. This approach provides the important practical advantages of demountability and reuseability.

Once the camera achieves routine and reliable operation with Cs_2Te photocathodes, it would be logical to proceed, step by step, towards longer wavelengths by incorporation of photocathodes having progressively greater red sensitivity.

Acknowledgments

The development of large format electronographic cameras is supported by the National Aeronautics and Space Administration under DPR's W-14, 157 and T-7902F. I thank Mr Joseph Kervitsky, Dr Harry Heckathorn, and Dr Karl Henize for their assistance and for useful discussions.

References

1. Carruthers, G. R., In "Electrography and Astronomical Applications", ed. by G. L. Chincarini, P. J. Griboval and H. J. Smith, p. 93. University of Austin, Texas (1974).
2. Carruthers, G. R., In "Astronomical Applications of Image Detectors with Linear Response" (IAU Symposium No. 40), ed. by M. Duchesne and G. Lelièvre, p. 6-1. Paris–Meudon Observatory (1976).
3. Picat, J. P., Astron. & Astrophys. **11,** 257 (1971).
4. Carruthers, G. R., Appl. Opt. **14,** 1667 (1975).
5. Schroeder, D. J., "Optical Study of the Deep-sky Ultraviolet Survey Telescope". Report submitted to NASA–JSC June 1976.
6. Carruthers, G. R., Kervitsky, J., Hicks, G. T. and Opal, C. B., Proc S.P.I.E. **78,** 95 (1976).

André Lallemand
1904–1978

Cet article est dédié à la mémoire d'André Lallemand (1904–1978).

Etudes Extragalactiques avec la Caméra Electronique "Grand Champ".

G. WLÉRICK, G. LELIÈVRE, B. SERVAN, L. RENARD et B. LEFÈVRE.

Observatoire de Paris, Meudon, France

Introduction

La Caméra Electronique Lallemand à focalisation magnétique, développée au cours des dernières années,[1-3] est opérationnelle. Son diamètre utile est de 81 mm. Par rapport à 1976, la pression dans le tube a été diminuée par un facteur 2 et la valeur actuelle lue sur la jauge est de l'ordre de 10^{-9} Torr. La densité optique correspondant à l'émission parasite est située dans la gamme de 0,02 à 0,06 pour des poses d'une heure sur film Kodak Industrex A (Definix), lorsq'on utilise une tension de 25 kV. Avec une résolution de 70 pl mm^{-1} sur l'ensemble du champ, le tube présente 30×10^6 éléments-images (en anglais: pixel) et la densité optique de 0,02 correspond à 4 électrons parasites par élément-image et par heure. En supprimant avec un masque l'arrivée de la lumière sur deux petites zones à la périphérie de la photocathode, on peut connaître effectivement la valeur de l'émission parasite pour chaque cliché astronomique.

L'instrument a été essayé pendant quatre missions à l'Observatoire de Haute-Provence (C.N.R.S.), en 1976, 1977 et 1978. Il sera mis, dans cet Observatoire, à la disposition de la communauté astronomique, à partir d'Avril 1979. D'autre part, un ensemble de trois caméras est en construction pour le foyer Cassegrain du télescope C.F.H. de 3,6 mètres, à Hawaii; sa mise en service est prévue en 1980.

Observations

Elles ont été effectuées à l'O.H.P., au foyer Cassegrain du télescope de 1,93 m.

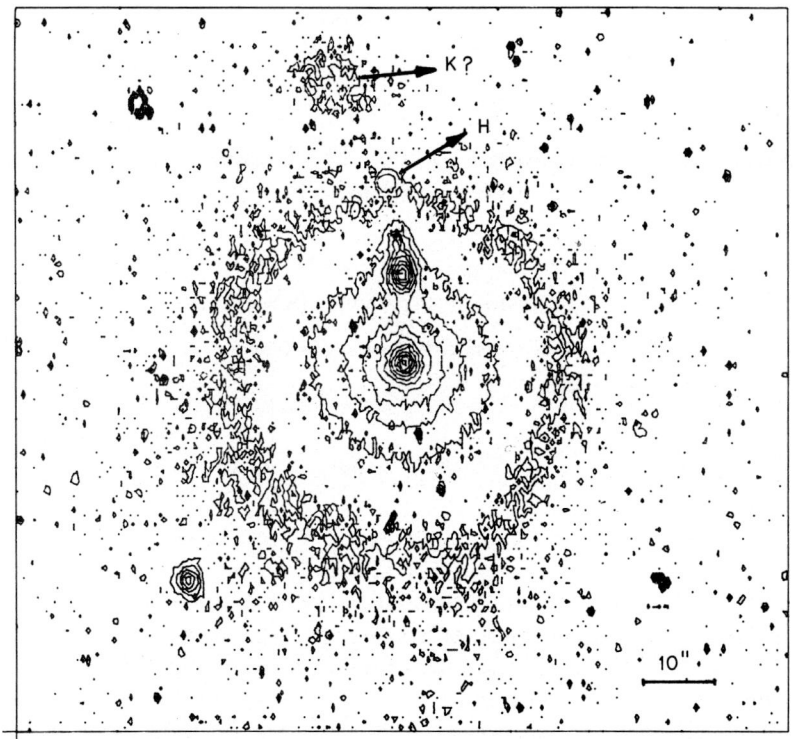

FIG. 1. Isophotes de la région centrale de la radiogalaxie Virgo A; cliché PbY 10, pose 195 min, couleur U, émulsion Ilford L4, 19 avril 1977.

Jet de la Radiogalaxie Virgo A

Nous avons observé cet objet en U, B et V en Avril 1977 et en lumière polarisée, en couleur V, en Mars 1978. La Fig. 1 montre les isophotes correspondant à un cliché pris en ultraviolet. L'intensité du jet à été obtenue en soustrayant du signal le flux de la galaxie mesuré au point situé symétriquement par rapport au noyau. La Fig. 2 présente les résultats : l'intensité trouvée pour la condensation H est nettement plus importante que dans les mesures publiées.[4] De plus, au-delà de H, apparaît une zone large, notée K? Un autre cliché UV à longue pose serait nécessaire pour confirmer sa réalité.

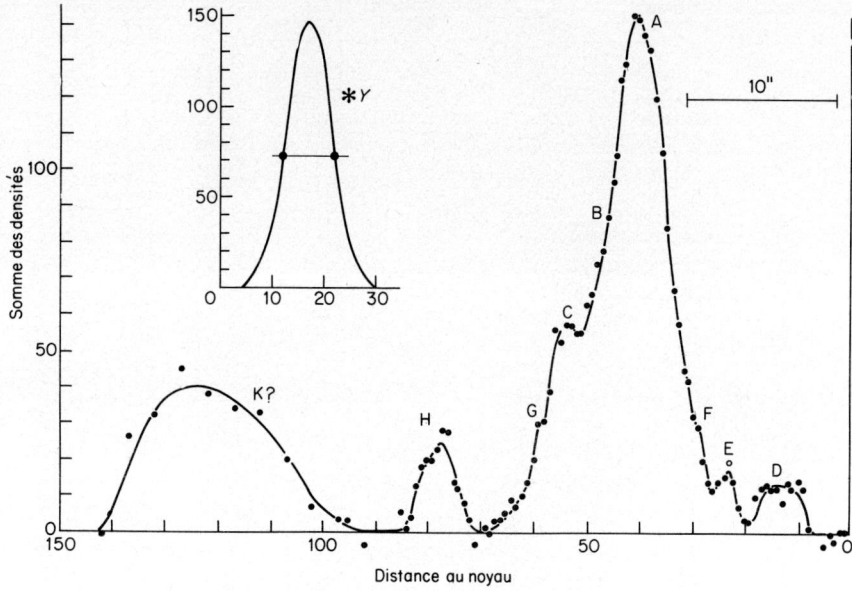

Fig. 2. Intensité en ultraviolet des condensations du jet de Virgo A (c.f. la Fig. 1).

Champs de Galaxies Compactes

Pour étalonner photométriquement des galaxies compactes observées avec un télescope de Schmidt, un astronome suédois A. Kinnander a pris en U, B et V, des clichés de champs contenant des galaxies sélectionnées par Richter. La Fig. 3 reproduit un de ces champs. On remarque la finesse de l'image de certaines galaxies spirales lointaines.

Amas Compacts de Galaxies "Compactes"

Avec J. Roland, nous avons observé les amas Shakbazian 78 et 82 en B et V (Figs. 4(a) et 4(b)). Il apparaît que l'objet n° 3 de Shakbazian 82 n'est pas une galaxie. De même, les astres 1, 2 et 3 de Shakbazian 78 ont un aspect stellaire.

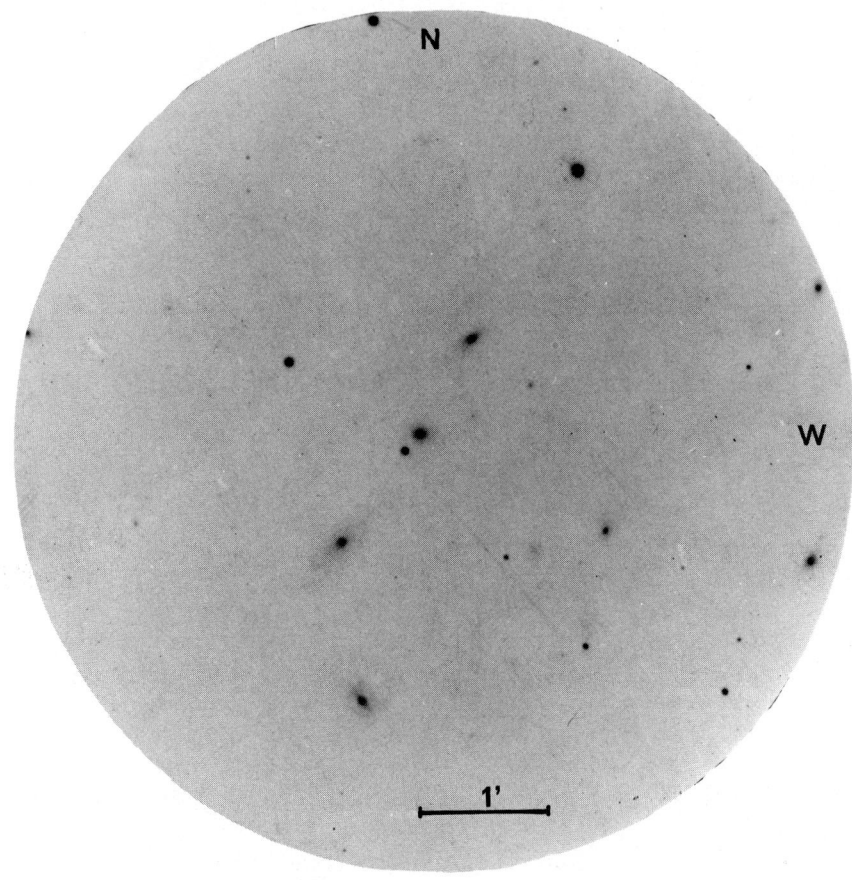

FIG. 3. Champ de galaxies Kinnander 2, pose 60 min, filtre V, émulsion Ilford G5, 20 avril 1977. L'électronographie permet une utilisation intéressante d'un foyer Cassegrain peu ouvert (f/15), en particulier pour la photométrie d'astres présentant de forts gradients d'éclat.

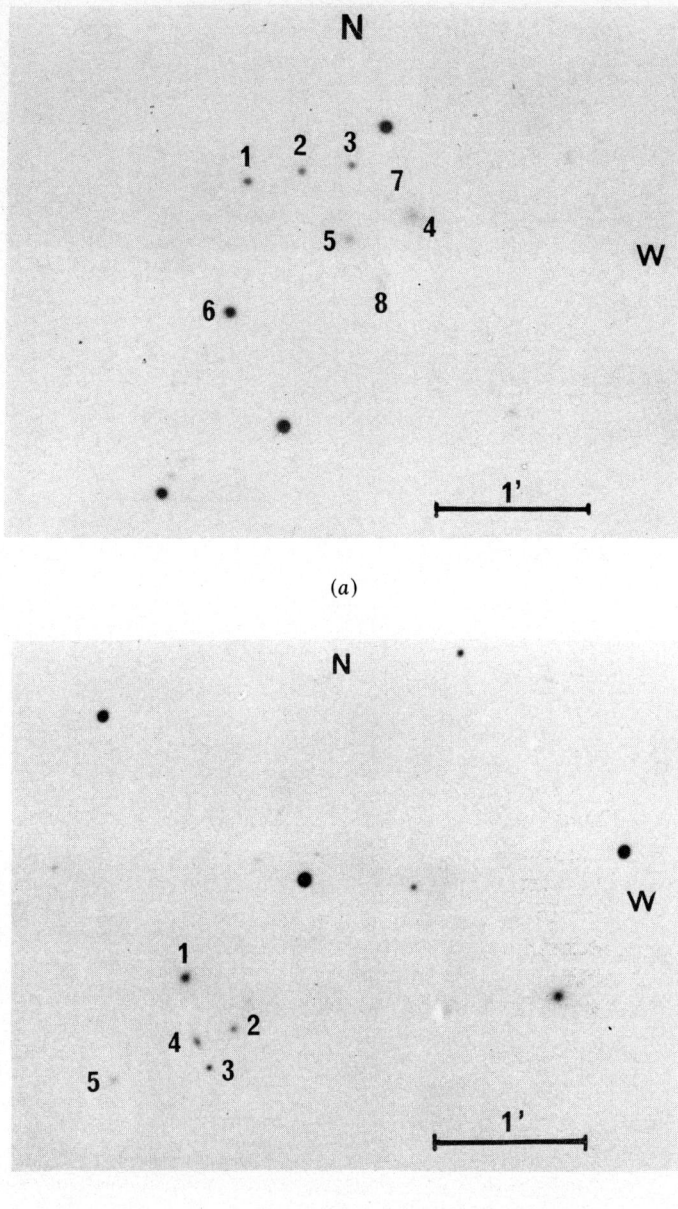

Fig. 4. Amas compacts de galaxies "compactes". (a) Shakbazian 78, filtre B, 90 min, émulsion Industrex. Noter l'aspect stellaire des objets 1, 2 et 3. (b) Shakbazian 82, filtre V, 100 min, émulsion Industrex. Noter l'aspect stellaire de l'objet 3.

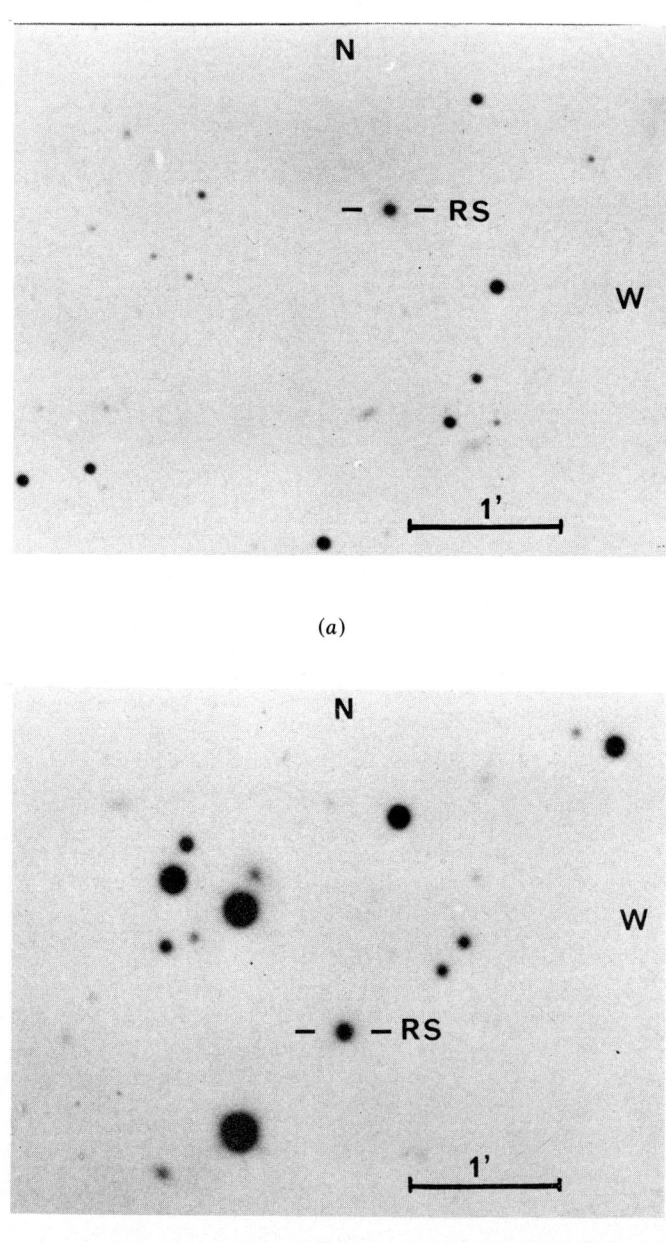

Fig. 5. Radiosources de type galaxie N: (a) IZw 187, filtre B, pose 102 min, émulsion Industrex; (b) 3C 390.3, filtre V, pose 80 min, émulsion Industrex. L'éclat de galaxies associées à ces sources est faible par rapport à celui d'une galaxie elliptique géante.

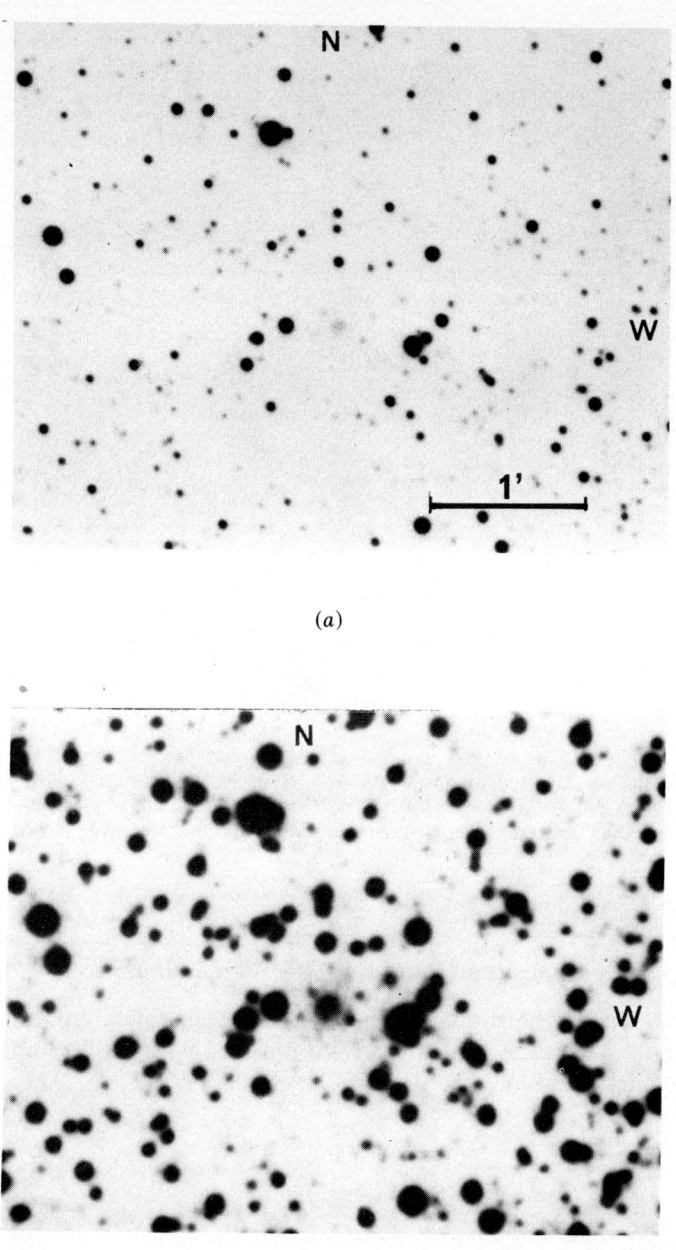

FIG. 6. Radiogalaxie Cygnus A: (a) cliché électronographique filtre bleu, pose 60 min, émulsion Industrex A; (b) cliché du "Palomar Sky Survey", sans filtre, 15 min

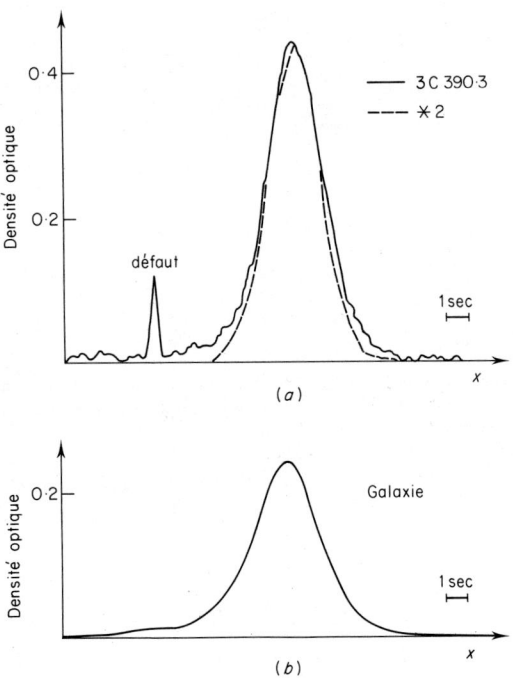

Fig. 7. Galaxie N 3C 390,3. Cliché PbY 61, couleur V, 9 Juillet 1978, pose 120 min, émulsion Ilford L 4. (a) coupe diamètrale de l'objet et d'une étoile présentant un flux comparable; (b) résultat de la séparation de la galaxie et du noyau ponctuel.

Quasars à Grand Décalage Vers le Rouge

Pour $z > 2,5$, certains quasars présentent un très faible flux en UV et leur mesure, dans cette couleur, devient difficile. Nous avons pu prendre, en ultraviolet, de bons clichés de PKS 0830+115 ($z = 2,97$) et PKS 1004+141 ($z = 2,71$).

Radiogalaxies

En Juillet 1978, nous avons obtenu des clichés de certaines radiogalaxies dont le décalage vers le rouge z est de l'ordre de 0,06. Les objets étudiés ont été : la galaxie de Zwicky, IZw 187, de type BL Lacertae (Fig 5(a)), la galaxie N 3C 390,3 qui est une source X intense (Fig. 5(b)) et la radiogalaxie Cygnus A qui est une des sources radio les plus puissantes (Fig. 6(a)). La longueur focale équivalente du télescope, au foyer Cassegrain, $F = 28$ m, permet d'enregistrer les champs stellaires avec une très bonne résolution spatiale, comme le montre la comparaison avec une reproduction du champ de Cygnus A extraite du "Palomar Sky Survey", pour lequel la longueur focale est de 3,07 m (Fig. 6(b)).

Pour les galaxies N et les galaxies de Zwicky, les clichés doivent permettre de déterminer les flux respectifs de la galaxie et du noyau actif qui se trouve en son centre.

Des mesures préliminaires ont été effectuées pour 3C 390,3. La Fig. 7(a) montre une coupe diamétrale de l'objet et, pour comparaison, la coupe relative à une étoile; on remarque le faible éclat de la galaxie au-delà des pieds de l'image de l'étoile. La séparation de la galaxie et du noyau a été effectuée, en admettant la symétrie circulaire de l'image de la galaxie. Le résultat est présenté en Fig. 7(b). A. Sandage[5] a fait l'hypothèse que les galaxies N sont formés d'une galaxie elliptique géante et d'un noyau ponctuel. Le résultat trouvé ici correspond à un éclat trois fois plus faible que celui d'une elliptique géante ($\Delta V = 1,2$).

Spectroscopie de Type Prisme Objectif

Munie d'un réseau de diffraction, la caméra a permis de prendre, avec une dispersion de 360 Å mm,$^{-1}$ les spectres des astres d'un champ de 10 min d'arc avec pose de 30 min sans filtre, la magnitude limite étant de l'ordre de 17. Avec filtre bleu, bonnes images et ciel bien noir, une magnitude bien supérieure serait atteinte.

Conclusion

La Caméra Électronique Lallemand à grand champ est un récepteur très puissant. Associée à un grand télescope, elle doit permettre de reculer les limites de l'Univers accessibles à l'observation.

References

1. Lallemand, A., Servan, B. et Renard, L., *C. R. Acad. Sci.* **270,** 385 (1970).
2. Lallemand, A., Servan, B. et Renard, L., *Dans* "Electrography and Astronomical Applications" ed. G. L. Chincarini, P. J. Griboval and H. J. Smith, p. 1, University of Texas, Austin (1974).
3. Lallemand, A., Servan, B. et Renard L., Colloque N° 40 de l'UAI, éd par M. Duchesne et G. Lelièvre, P. 1–1. Observatoire Paris–Meudon (1976).
4. Schmidt, G. D., Peterson, B. M. et Beaver, E. A., *Astrophys. J.* **220,** L 31 (1978).
5. Sandage, A., *Astrophys. J.* **180,** 689, (1973).

The U.T. Electronographic Camera: Present Status, Astronomical Performance and Future Developments

P. J. GRIBOVAL

Department of Astronomy, The University of Texas at Austin, Texas, U.S.A.

INTRODUCTION

The electronographic camera is a unique instrument capable of recording *simultaneously* and *accurately* millions of information elements. Its high detective quantum efficiency, extended spectral range and low noise make it an invaluable detector for astronomy as well as for many other fields of physics research. More precise information is obtained in less time than with photographic emulsions.

During the past decades, much effort has been made to produce electronographic cameras of various designs and some excellent astronomical results have been obtained with those instruments.

The electronographic camera developed at The University of Texas during the past twelve years represents another approach. The experimental laboratory model which was built some years ago, and the modified version presented here, provide data for designing and building a large field (~20 cm diameter) electronographic camera for both ground based and space telescopes.

THE MARK II CAMERA

The Mark II electronographic camera (Fig. 1) is a modified version of the first model,[1] having the same photocathode diameter and internal structure; only the major changes and new operating methods will be described here.

In the U.T. camera, photoelectrons are accelerated and focused by highly corrected electric and magnetic fields which ensure freedom from image distortion and uniform high resolution over the entire 5 cm diameter field. The photocathode substrate and the front window are two optically flat silica discs. A specially shaped silica cup, internally coated

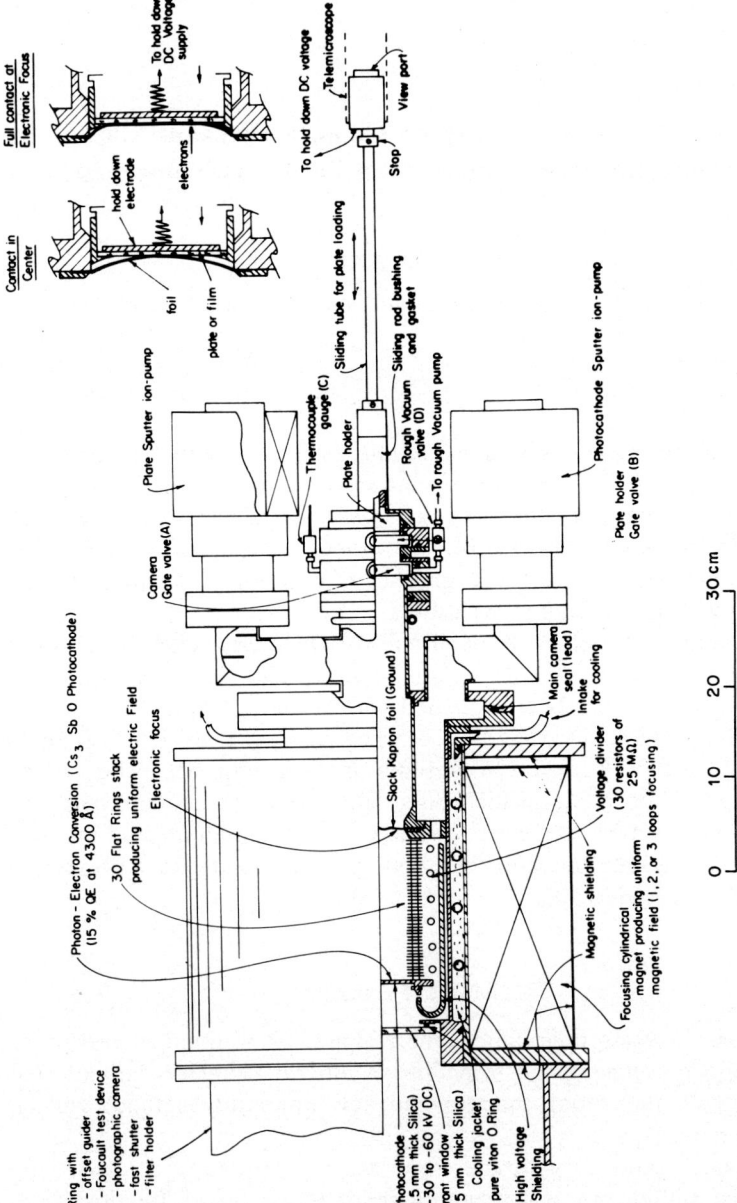

FIG. 1. Sketch of the Mark II electronographic camera. High voltage feedthrough and lead not shown.

with conductive and semiconductive layers, entirely surrounds the electrode stack which is composed of 30 flat rings and by suppressing field emission enables the camera to operate at a voltage of up to 60 kV. In addition, a high voltage rating 10 MΩ resistor is placed near the camera in series with the high voltage lead to reduce spark energy. Thanks to these two features the high voltage can be raised quickly to maximum value with very little chance of sparking; if a spark occurs at 60 kV, it will be weak and will produce a negligible rise in pressure, and so will be harmless to the photocathode.

The aluminium oxide foil used in the earlier model has proved to be unreliable and has been replaced by a 7·5 μm thick Kapton foil which is cut from large sheets of factory produced material, aluminized on both sides. This new foil is almost unbreakable and has an average leak rate for water vapour and air of 2×10^{-10} Torr l sec^{-1} cm^{-2}. The foil is mounted slackly at the stack base where the electronic focus occurs, and is sealed with two 175 μm thick flat washers of the same material. Such foils are easy to prepare, cheap and very reliable as none have failed since we began using them. They represent a good solution to the problem of protecting the photocathode from the emulsion outgassing. Foil slackness ensures an intimate contact between the foil and the emulsion over the entire field when the 1·5 kV DC voltage is applied at the back of the film or plate. Thus, in spite of its greater thickness, the Kapton foil enables, with either film or a glass plate, the same uniform high resolution to be obtained as did the aluminium oxide.[1]

After a 48 h camera bake-out, at a temperature of 260–270°C, the outgassing rate is of the order of 2×10^{-10} Torr l sec^{-1} and the photocathode and plate chamber pressures are the order of 2×10^{-10} Torr. No extended study of photocathode life has been made since camera improvements necessitate opening it between scheduled operations at the telescope. The longest sensitivity survey covered a 6 week period during which the photocathode maintained its sensitivity within 2% in both red and violet. The sensitivity was unaffected even when both ion pumps were turned off for a 6 day period while the camera was submitted to mechanical stress (vibrations) and heat up to 50°C.

The $Cs_3Sb(O)$ photocathode we are presently using is processed in a separate chamber attached at the front window end of the camera. The two chambers are separated, during processing, by a plug which prevents caesium contamination of the camera. The photocathode is formed by alternating evaporations of thin layers of caesium and antimony with two or three partial oxidations of the multilayers thus built. This technique produces uniform, high efficiency, robust photocathodes. Typical quantum efficiencies are 15% at 430 nm and 2 to 3% at 630 nm.

After being tested, the photocathode is pushed into the camera at the top of the stack, and the camera is sealed with a prebaked pure Viton O-ring attached, with an acrylic sealant, to the front window.

A 100 μm wide sharp metallic line is evaporated onto the edge of the photocathode, in such a way as to be imaged on the picture, and can be used to check the accuracy of the electronic focus during camera operation.

One of the major problems occurring with cameras using separation foil is the dust or foreign particles introduced by the emulsion into the camera and sticking to the foil, thus producing undesirable shadows in the picture. Such difficulties occurred when loading the plate holder by hand. The new technique we are using prevents this happening and significantly shortens the camera loading time. The round piece of film or plate, carefully cleaned by a blast of nitrogen, is seated in a holding ring and loaded into a loading/outgassing chamber which holds 20 such rings; the chamber is pumped out by a small 2-stage rough vacuum pump. The plate or film is picked up, in the vacuum, by the plate holder and loaded into the camera using a light 2-stage rough vacuum pump for pumping-out the dead volume between the two gate valves. Since emulsion outgassing is considerably reduced, loading can be achieved in less than 45 sec. With the plate or film in the camera, the plate chamber pressure comes down to 5×10^{-6} Torr a few minutes after being loaded. There is neither a glow discharge when applying the hold-down voltage nor blackening of the film when separated from the foil and returned to atmospheric pressure. The two identical plate holders permit fast operation, one being used on the camera, while the other is pumped out in the outgassing chamber located near the camera. Hundreds of films have been loaded so far without trouble, and the pictures are free of pinholes.

Astronomical Performance of the Mark II Camera

The Mark II electronographic camera has been attached (Fig. 2) to the $f/13\cdot5$ Cassegrain focus of the Boller and Chivens 0·76 m telescope at McDonald Observatory. The 10 m focal length of this medium size telescope gives a scale of 55 μm per arcsec. The telescope image scale is well matched with the camera resolution to give a good balance between adequate sensitivity and image resolution when good seeing (~1 arcsec) occurs. During the observing runs the seeing was mostly poor to average and the sky brightness was high; both conditions have severely limited the overall performance.

The weight of the camera (~140 kg) has not affected the telescope

FIG. 2. The electronographic camera attached to the Cassegrain focus of the 0·76 m telescope, with the rig for filters and guiding (top) and the plate holder (bottom).

pointing and guiding. After a few observing nights, during which we got accustomed to the camera/telescope operation, the job became so easy that the observer was capable of serving both instruments, i.e., loading and unloading the camera, and setting and guiding the telescope without difficulty or loss of time. The camera worked flawlessly and was trouble free.

Since the magnetic field and high voltage supplies were kept on at all times during the night, the exposure began when the hold-down voltage was turned on to contact the foil. Most of the pictures were obtained at 45 or 50 kV with 2-loop focusing; however, other voltages from 35 up to 60 kV can be selected and the magnetic field matched for 1-, 2- or 3-loop focusing according to a setting table. The loading method proved efficient and fast. Working at elevations of up to 2,000 m and sometimes with a relative humidity of up to 95%, there is no corona. At 50 kV and with a 15% Q.E. photocathode, no background was detectable after a 2·5 h exposure.

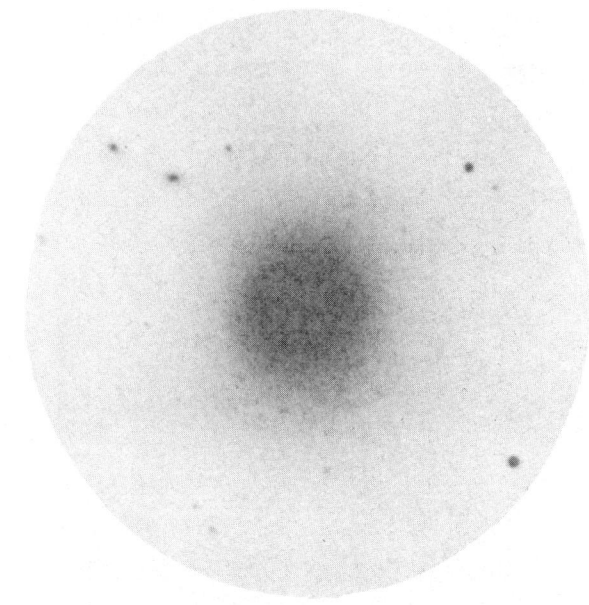

Fig. 3. A 4·5 times enlargement of an electronograph of NGC 4486 (M87) obtained at the 0·76 m telescope. 60 min exposure through an B filter at 45 kV on Kodak Electron Image Film. Average seeing during exposure ≈3 arcsec. The electronograph recorded globular clusters down to 22 mag (electronograph 1315).

No physical damage occurred when carrying the camera along the 10 000 km we drove in the many trips between the observatory and the laboratory. The only way to damage the instrument is to break the front window or to open the gate valve when no plate holder is attached to it.

Kodak Electron Image Film and Kodak NTB2 and NTB3 nuclear research emulsions have been used for these observations. Films are hung freely in the Kodak D19 developer by clips attached to an automatic processing machine which produces uniform developing for accurate photometry; eight films can be processed at the same time. These 10 μm thick Estar base films are clean, free of defects and easy to handle since they are coated on both sides by a 1 μm thick layer of hard gelatin.

Four among the hundreds of pictures obtained in various conditions are shown in Figs. 3 and 4. If observing conditions had been better fainter magnitudes would have been reached in less exposure time. Nonetheless, these pictures reveal the same faintest details as those obtained, on photographic emulsion, at the prime foci of the world's largest telescopes,[2] thus clearly demonstrating that, without complicated handling, or

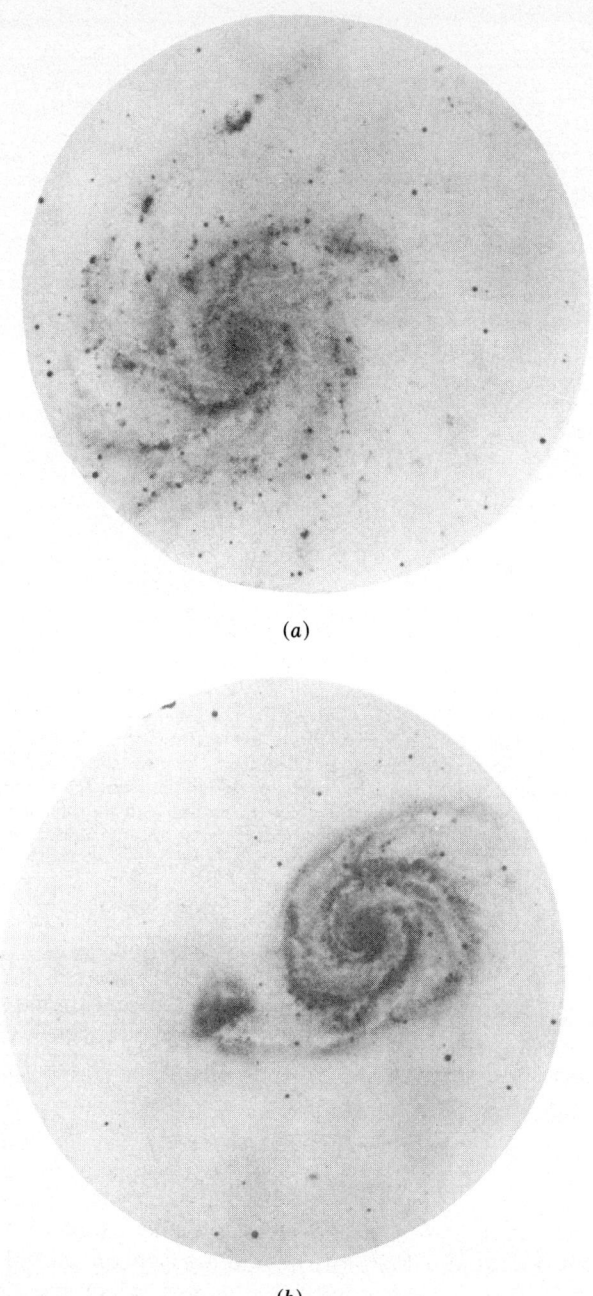

FIG. 4. Enlargements of whole electronographs obtained at the 0·76 m telescope through B filter at 45 kV on Kodak Electron Image Film. (a) NGC 5457 (M101): 60 min exposure; average seeing ≈3 arcsec. (Electronograph 1318). (b) NGC 5194–95 (M51): 45 min exposure: average seeing ≈4 arcsec. (Electronograph 1338).

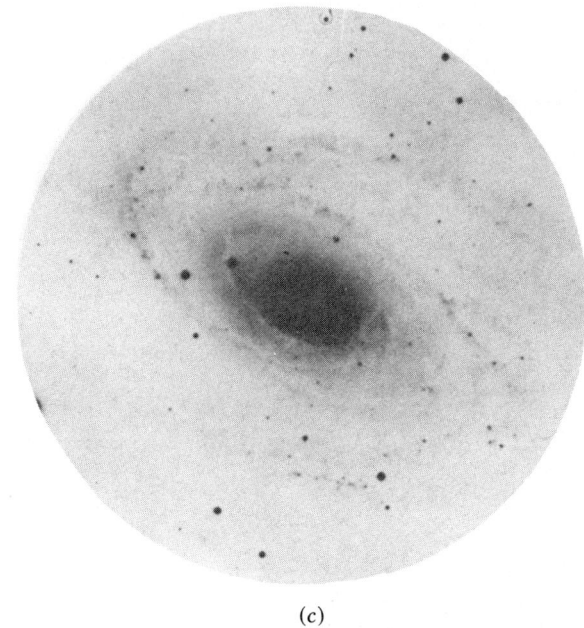

(c)

Fig. 4.(c) NGC 3031 (M81): 70 min exposure; average seeing ≈4 arcsec. (Electronograph 1344). The seeing has limited the overall resolution. Compare the diameter of the faint stars with the 80 μm wide line (at the edge) recorded by the film, which also indicates sharp electronic focus. Note also the very close contact of the foil with the film over the entire field, the absence of pinholes or local photocathode defects, and the uniformity of the sky background (average density ≈1).

the need of specialists, an electronographic camera can increase substantially the power of a telescope, with no price to pay in return for its use. Dead time between exposures was due to the telescope setting, not to camera loading.

Camera Improvements

To render it faster, even more convenient and simpler to use, the camera and its operating conditions have already been or will be modified. The two 30 l sec^{-1} triode sputter ion pumps, which have proved to be too powerful, have been replaced by two 5 l sec^{-1} diode pumps, permitting a reduction in size and a saving of nearly 25 kg in weight. A $K_2CsSb(O)$ photocathode, which has more than twice the sensitivity of the present $Cs_3Sb(O)$ will be used and will give less dark current and a slightly extended spectral range. A roll film magazine giving 20 exposures of 34×40 mm^2 on a 35 mm unperforated film will be used conjointly with the existing plate holder. The expected factor of four gain in speed of loading will allow, for example, four "deep" exposures of the same object through a blue filter to be made in an hour. By rotating the camera along its axis by a quarter of a turn between exposures, we will be able to reduce local photocathode or emulsion defects and, thus, to detect much fainter details in the object's structure.

The Proposed 20 cm Mark III Camera

There is no apparent difficulty in scaling up the Mark II camera and producing, subject to technical and financial support, a model having a 20 cm plate/film diameter to fit larger ground based or space telescope fields.

No scientific instrument can be 100% reliable, especially when operated by people of different skills; consequently, servicing is a part of the electronographic camera problem and should be solved if one wants to broaden the camera's use among laboratories and observatories.

The photocathode will be processed in a separate chamber, *not attached to the camera* but instead to a special compact loading chamber, which will permit loading the photocathode and then the foil into the camera from the *plate side*. This new technique will allow servicing of the camera *on the spot*, eliminating the need to bring it back to the laboratory for photocathode replacement. The camera will use either a film magazine containing 20 round films or plates of 20 cm diameter, or a roll film magazine giving a 14×14 cm^2 image. Both will be introduced sideways so that the size of the gate valve can be reduced.

Conclusion

We have demonstrated once more that an electronographic camera, thanks to its sensitivity and resolution, increases the power of a telescope. The camera resolution and the telescope scale should be well matched in order to obtain maximum information in the shortest possible time. Good seeing and low sky brightness are essential for reaching very faint magnitudes. In these conditions a telescope of moderate size can give much more information in a shorter time, especially of faint detail, than a much larger one with conventional photography. Thus, since we have proved that an electronographic camera is no longer an instrument that is difficult to use, a wider field of research is open to numerous medium size telescopes the world over, permitting them to contribute more efficiently to astronomical research. Then will come the problem of extracting *faster* and *more accurately* the information produced by the billions of pixels contained in large electronographs especially those taken at the focus of a large field, diffraction limited space telescope, thus giving a new and surely exciting view of the Universe.

References

1. Griboval, P., *In* "Adv. E.E.P." Vol. 40B, p. 613 (1976).
2. Sandage, A., "The Hubble Atlas of Galaxies", Carnegie Institution, Washington, D.C. (1961).

Operational Experience with the RGO Electronographic Cameras

D. McMULLAN and J. R. POWELL

Royal Greenwich Observatory, Herstmonceux, Sussex, England

Introduction

The development of the RGO electronographic cameras commenced in 1971 and, as was stated at the 5th Symposium,[1] the aim was to produce a large field camera of high efficiency that would be reliable and easy to use by astronomers. This aim has been met and cameras of the standard type are at present in use at Herstmonceux (since 1974) and at a number of other observatories in the Northern and Southern Hemispheres: the Wise Observatory, Israel (1975), the South African Astronomical Observatory (1976), the Anglo-Australian Observatory (1977) and on the Danish 1·5 m telescope at the European Southern Observatory, La Silla, Chile (1978). Two sizes of image tube are used giving respectively 40 mm and 85 mm diameter electronographic images but currently the larger cameras are available only at Herstmonceux and La Silla.

The designs of the image tubes and cameras are substantially as were proposed in 1971,[1] but many unforeseen technological problems had to be solved during the initial development period.[2-4] Since the first successful operation in 1974 numerous improvements have been introduced. Developments are continuing, at present at rather a slow rate, but information about these should be sought in other recent publications.[5,6]

Reports of the astronomical application of the cameras have been given in a number of astronomical journals and some examples of their earlier use have been summarized in Ref. 7. Recent examples of electronographs are given at the end of the present paper after a description of the standard type of camera and a discussion of operational experience.

Description of the Camera

The camera incorporates an image tube of the Lenard window type with a mica window either 40 mm or 85 mm in diameter which is protected from atmospheric pressure by a vacuum lock through which the electronographic film, generally nuclear research emulsion on a polyester base, is inserted and pressed into contact with the mica. The loading and unloading of the film through the lock is carried out by an automatic electropneumatic system which incorporates a number of safety interlocks to safeguard the tube against operator error or faults in the control system.

40 mm Image Tube

Figure 1 shows a cross section of the 40 mm camera with the image tube in position in its solenoid. Focusing is by approximately parallel magnetic and electric fields. Because of the divergence of the magnetic field in the photocathode region there is ~10% demagnification and the photocathode diameter is 44 mm. The vacuum envelope, 128 mm in diameter, is of fused silica and the S·20 photocathode can be formed directly on the faceplate although in most of the tubes at present operating a separate photocathode plate made of silica is used because of manufacturing difficulties.[6] Two fiducial marks are provided at the edge of the photocathode as references for maps of photocathode sensitivity. The electrode assembly is built up from titanium annuli spaced by soda-lime glass cylinders 20 mm long. Since these are slightly conducting they form closing surfaces of uniform potential gradient between the electrodes; they are also coated with chromium oxide which reduces the secondary emission coefficient of the glass surface. The cylindrical internal surface of the envelope is similarly coated. Metal oxide glaze resistors (15×200 MΩ) form the potential divider and are mounted directly on the electrode structure.

The 40 kV high voltage connection to the photocathode is made by bringing the high voltage cable through a glass tube, the insulation being provided by the tube vacuum and the silica envelope. There are no exposed surfaces at high potential and the tube can be operated under the most humid conditions and at observatories at the highest altitudes.

The mica window, 4 μm thick and 40 mm diameter, is stretched tight and sealed to a titanium mount; a thin evaporated aluminium layer renders it conducting and on this there is a black film of finely divided aluminium to reduce the optical reflectance. The window is protected from atmospheric pressure by the pneumatically operated gate valve.

A 2 l sec^{-1} ion appendage pump is connected to the image tube vacuum space but is turned on only when the tube is operated. Its function is to pump those residual gases which do not react with the barium getter or the photocathode and which in time would build up to an unacceptably

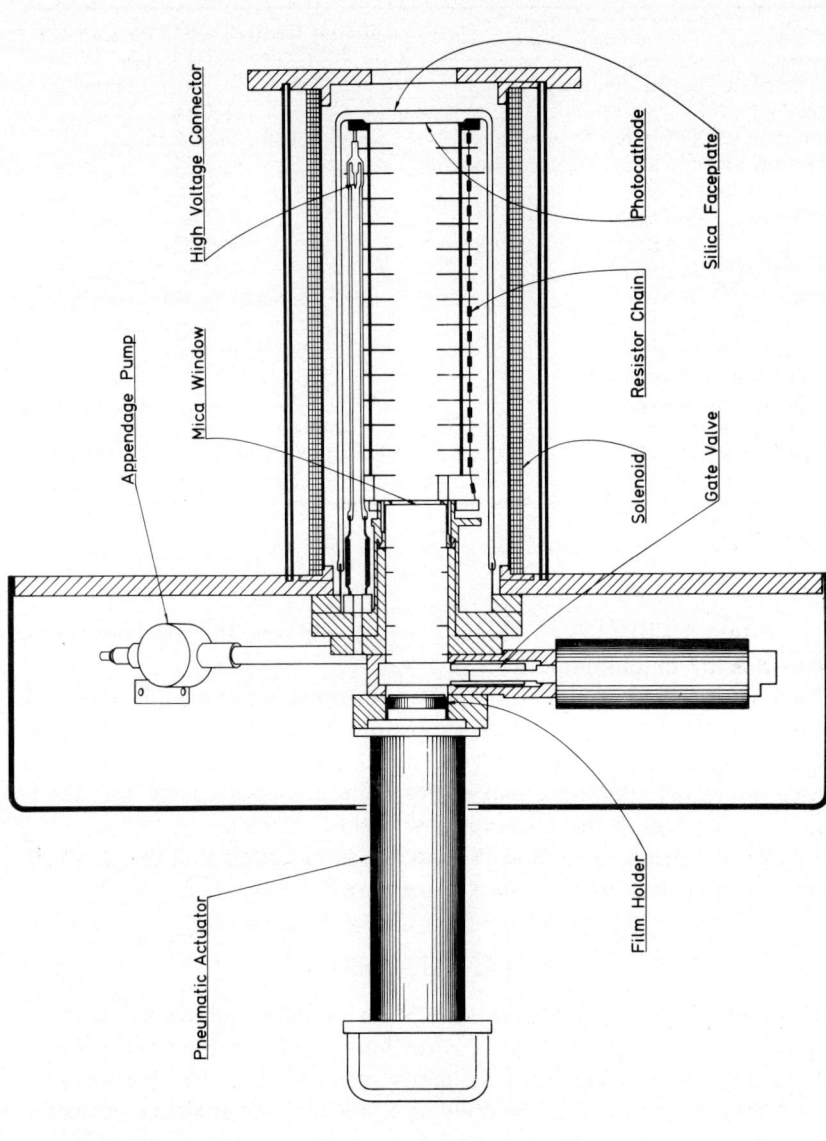

FIG. 1. Cross section of the 40 mm camera with the image tube mounted in the solenoid and the film applicator in position. Control system components are not shown.

Table I
Electronographic camera parameters

	40 mm Camera	85 mm Camera
Photocathode: useful diameter	44 mm	93 mm
Type	S·20	
Luminous eff. (2854 K)	180–250 μA lm^{-1}	
Quantum eff. 445 nm	20–23%	
545 nm	12–18%	
644 nm	4–7%	
793 nm	0·3–0·8%	
Accelerating potential	40 kV	
† Dark current 25°C	700 electrons cm^{-2} sec^{-1}	
10°C	70 electrons cm^{-2} sec^{-1}	
Resolution (limiting)	50 lp mm^{-1}	
Solenoid power dissipation (single loop focusing)	18 W	60 W
Film loading time	45 sec	120 sec
Dimensions W×H×L	67×28×89 cm^3	76×47×86 cm^3
Weight	66 kg	113 kg

† Typical

high pressure ($>10^{-6}$ Torr); helium which diffuses through the silica envelope is the main component.

The image tube is of demountable construction with gold wire and indium seals. If there is a loss of photocathode sensitivity (e.g., through exposure to too high a light level), or in the unlikely event of mica window breakage, the tube can easily be reprocessed using the same components including the mica window if this is intact.

The characteristics of the image tube are summarized in Table I which gives the parameters of both sizes of camera.

85 mm Image Tube

The construction of the 85 mm tube is very similar, the main difference being that the envelope is made of Pyrex glass (190 mm diameter) instead of silica, because of cost. Pyrex is slightly conducting so it is necessary to insulate the envelope and this is done by a silicone rubber sleeve provided with a silica window.[5] Electrodes and a potential divider are embedded in the rubber to produce a uniform potential gradient along the inner surface of the sleeve corresponding to that inside the image tube. Table I gives a summary of the tube characteristics.

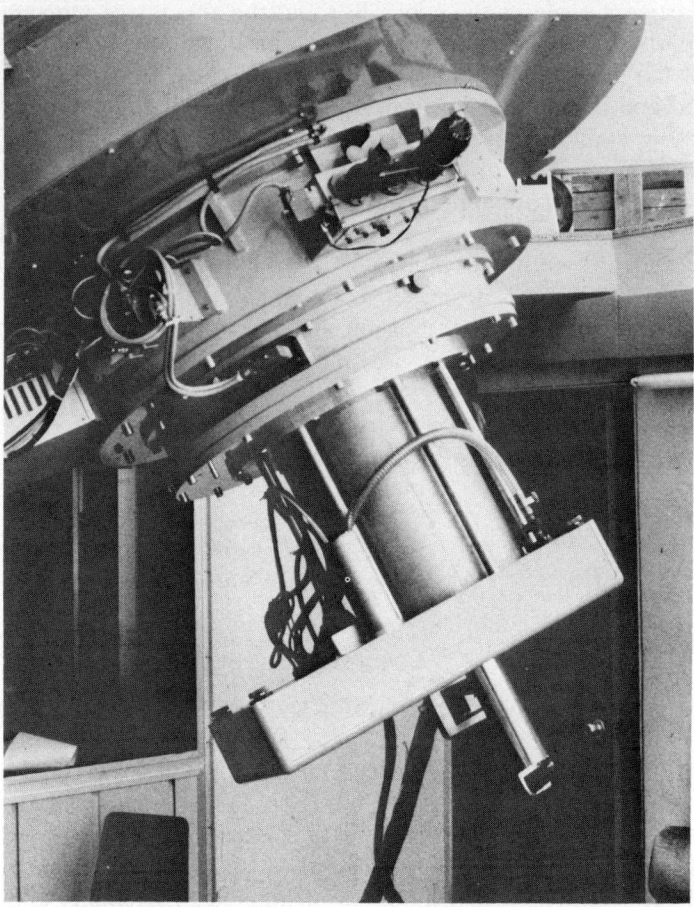

Fig. 2. 85 mm camera mounted at the Ritchey Chretien focus of the Danish 1·5 m telescope at the European Southern Observatory, La Silla, Chile. Filter wheels, shutter and autoguider are incorporated in the telescope instrument adaptor to which the camera is bolted.

Camera Assembly

The part of the camera assembly that is mounted on the telescope is shown in cross section in Fig. 1 (40 mm version) and in the photograph in Fig. 2 which is of the 85 mm camera on the 1·5 m Danish telescope at the European Southern Observatory, Chile. The rectangular box at the rear of the camera encloses the gate valve and the appendage pump and also some parts of the pneumatic control system (not shown in Fig. 1). At the rear of the box there is an aperture through which the film applicator is inserted and locked into position by twisting in a bayonet mount.

The film applicator consists of the film holder on which the electronographic film, mounted on a nylon ring, is placed and retained by detents, and a pneumatically operated piston actuator. After the vacuum lock has been pumped and the gate valve opened the piston rod pushes the film holder forward until the film contacts the mica window. Air at low pressure (~15 Torr) is admitted to the space behind the film to ensure good contact between the emulsion and the mica. The thickness of the film base (generally polyester) must not exceed about 75 μm if adequate contact is to be achieved.

The solenoid is uncooled because single loop electron focusing is employed and the power dissipation is only 18 W for the 40 mm camera and 60 W for the 85 mm camera. The heat is conducted away by the telescope structure.

Control System

The operation of the film applicator and vacuum lock is controlled by an electropneumatic control system, the sequence of operations being as follows: the film applicator is locked into the position shown in Fig. 1; the pressure in the space around the film is reduced to about 1 Torr by a mechanical pump; the gate valve is opened; the film holder is pushed forward to the mica window; and air at ~15 Torr pressure is admitted behind the film. The aim has been to achieve high reliability and so the timing is done by motor driven cam operated microswitches. As already mentioned, part of the system is inside the camera box; the remainder, including the power supplies and a silent mechanical pump for evacuating the vacuum lock, is enclosed in a control cubicle (Fig. 3) which can stand on the observing floor or can be carried by a large telescope. The cubicle is joined to the camera assembly by up to 30 m of multiway cable and tubing.

The various operations, e.g., opening of gate valve, movement of the film applicator etc., are monitored by microswitches and vacuum switches and the sequence continues only when the earlier operations have been properly completed. A number of electrical and pneumatic interlocks are incorporated to minimize the possibility of damage to the mica window resulting from failure of the control system or operator error.

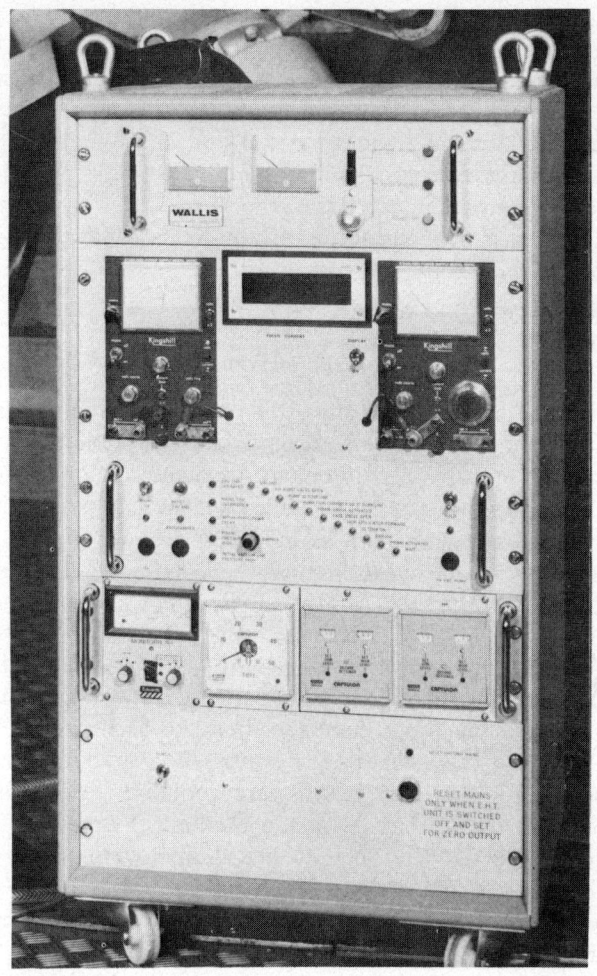

Fig. 3. Camera control cubicle containing (from the top): the high voltage, solenoid current and control system power supplies; the control system timing module; vacuum gauges and relays for controlling the 15 Torr air pressure for film contacting; and the mechanical vacuum pump.

Operation at the Telescope

For the astronomer, operation of the camera is very simple. The electronographic film, generally nuclear track emulsion on 50 μm polyester base, has to be cut into discs, 46 mm or 95 mm diameter for the two sizes of camera, which are then stuck to nylon rings using precut annular gaskets of double sided pressure sensitive adhesive tape. Mounting of the films is generally done in advance so that the astronomer has only to deal with the easily handled rings. To load the film, a ring is placed on the film holder, the applicator is inserted, and the "Load" button on the camera is pressed. The film is brought automatically to its working position in less than 1 min for the 40 mm camera and the exposure can then be made. Unloading takes about 20 sec.

Operational Experience

As mentioned at the beginning of the paper, the cameras are in operation at a number of overseas observatories. Reliability has on the whole been high in spite of maintenance being carried out mainly by local staff who have not had specialist training on the equipment. Most of the difficulties have been with tube manufacture and reprocessing at Herstmonceux and these have arisen mainly because of changes aimed at improving the performance of the tubes, for example processing the photocathode on the tube faceplate.[6] Even so the availability of the cameras at most observatories has been high (>90%) although tube performance has sometimes not been as good as we would have liked.

Reprocessing of the tubes has been found to be necessary on average about every 18 months, generally because of a gradual loss of photocathode sensitivity. Vacuum leaks were probably responsible in most cases and recent improvements in the method of preparing the indium seals are expected to increase tube life. Light overloads will obviously cause impairment of the photocathode but they may occur without the astronomer being aware and so it is very difficult to collect reliable data on the subject.

Breakage of the mica window has occurred on occasion due to various causes such as operator error when setting up a camera, faults in the control system, and the film or film ring becoming detached from the film holder during the load/unload cycle and being caught up in the gate valve mechanism. Improvements have been made to the cameras over the years with a view to preventing these occurrences; examples of these are a locking mechanism to prevent the film applicator being removed once the load cycle has started, means for safely resetting the camera to the gate valve "closed" position in the event of a fault or power failure, better

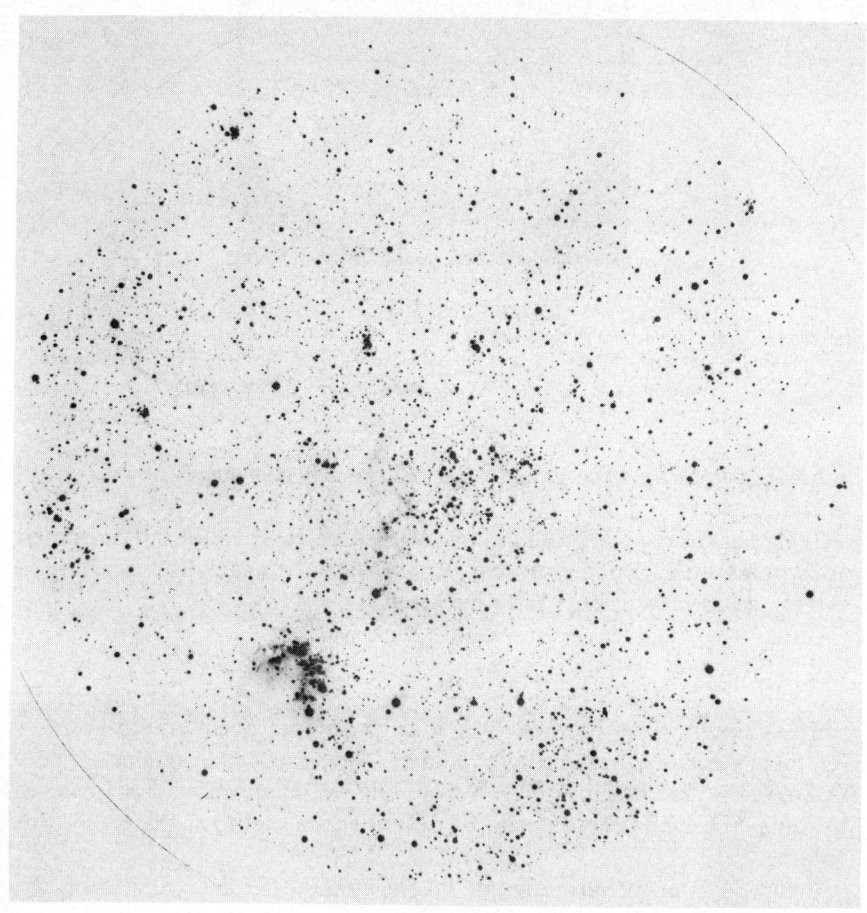

FIG. 4. Electronograph taken with the 85 mm camera on the Danish 1·5 m telescope (f/8·7 Ritchey Chretien) at ESO of the open cluster NGC 2081 and surrounding field in the Large Magellanic Cloud. 'V' filter, $\frac{1}{2}$ h on Ilford G5, seeing 2–3 arcsec. Electronograph reproduced 1·3 times full size, image scale 12·8 arcsec mm^{-1}.

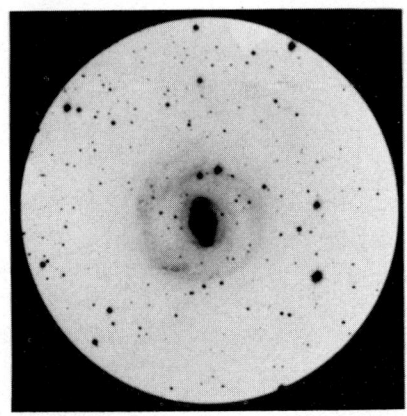

Fig. 5. Electronograph of NGC 2217, an SBa(r) galaxy, taken with the 40 mm camera on the Danish 1·5 m telescope. A dark sky blue filter ($\lambda_{max} = 490$ nm, $\Delta\lambda = 80$ nm) was used to give maximum contrast against sky background. Exposure $\frac{1}{2}$ h on Ilford G5. Electronograph reproduced 1·3 times full size, image scale 12·8 arcsec mm^{-1}.

adhesive for sticking the films to the rings and tighter dimensional control of the rings.

Transport of the camera and image tubes overseas by air has presented no problems. A sprung crate has been designed to carry the image tube and no cases of breakage have occurred.

Astronomical Results

The cameras have been used by a large number of different observers for their various observing programmes. Three recent examples of electronographs are shown in Figs. 4 to 6, and the application of the camera to time resolved photometry is described in another paper in this volume.†

Figure 4 is an electronograph taken with the 85 mm camera on the 1·5 mm Danish telescope at ESO, Chile, of an open cluster (NGC 2081) and surrounding field in the Large Magellanic Cloud. The emulsion used was Ilford G5 and it can be seen that there are only a few blemishes (some being on the photocathode) thus illustrating a recent improvement in the quality of these emulsions.

Another example taken with the same telescope but using the 40 mm camera is shown in Fig. 5. It is of the spiral barred galaxy NGC 2217, and

† See p. 109.

is one of a series of electronographs taken for a study of the morphology of galaxies.

Finally Fig. 6 shows the application of the 40 mm camera to the study of polarization in extended objects.[8] The two images on the left are of the nebula Eta Carinae; they have been recorded in orthogonal polarization with the Durham University polarimeter[9] and the 40 mm camera on the 3·9 m Anglo-Australian telescope at Siding Spring. The images have been processed on a contour drawing television system developed at Herstmonceux. The drawing on the right shows the derived planes of polarization. Analysis of the results indicates that the nebula medium contains small solid particles, and that the central source of energy is probably a double star; it also helps to establish the relationship of this object to other cataclysmic variables.

CONCLUSION

The standard type of RGO electronographic camera is now being used at a number of observatories in both Northern and Southern Hemispheres. Although the camera performance is reasonably good there are several parameters needing improvement, in particular, photocathode uniformity, reduction of scattered light and better resolution through improved contact between emulsion and mica window. The ways of achieving these improvements are for the most part well known but their introduction is unfortunately very slow because of the limited resources available and the necessity of giving priority to the production of the standard type of tube.

ACKNOWLEDGMENTS

The electronographs shown in Figs. 4 and 5 were taken during a joint observing run with Copenhagen University Observatory on the Danish 1·5 m telescope at ESO, Chile, and our thanks are due to Professor K. Gyldenkerne and Dr R. Florentin Nielsen of CUO for making this possible, and to NATO for a research grant (No. 1137). The electronograph in Fig. 6 was taken by the Durham University Astronomy Group and we thank them and Dr Bingham of RGO for making it available. We also wish to thank Messrs N. A. Curtis, W. E. Matthews, P. Terry, D. J. Bonnick, E. Wilson and D. Mayhew of the RGO for their assistance in the construction, testing and application of the cameras. The paper is published by kind permission of the Director of the Royal Greenwich Observatory.

REFERENCES

1. McMullan, D., Powell, J. R. and Curtis, N. A., *In* "Adv. E.E.P." Vol. 33A, p. 37 (1972).
2. McMullan, D., *In* "Instrumentation for Large Optical Telescopes" ed. by A. Reisz and S. Laustsen, p. 433, ESO/CERN, Geneva (1972).

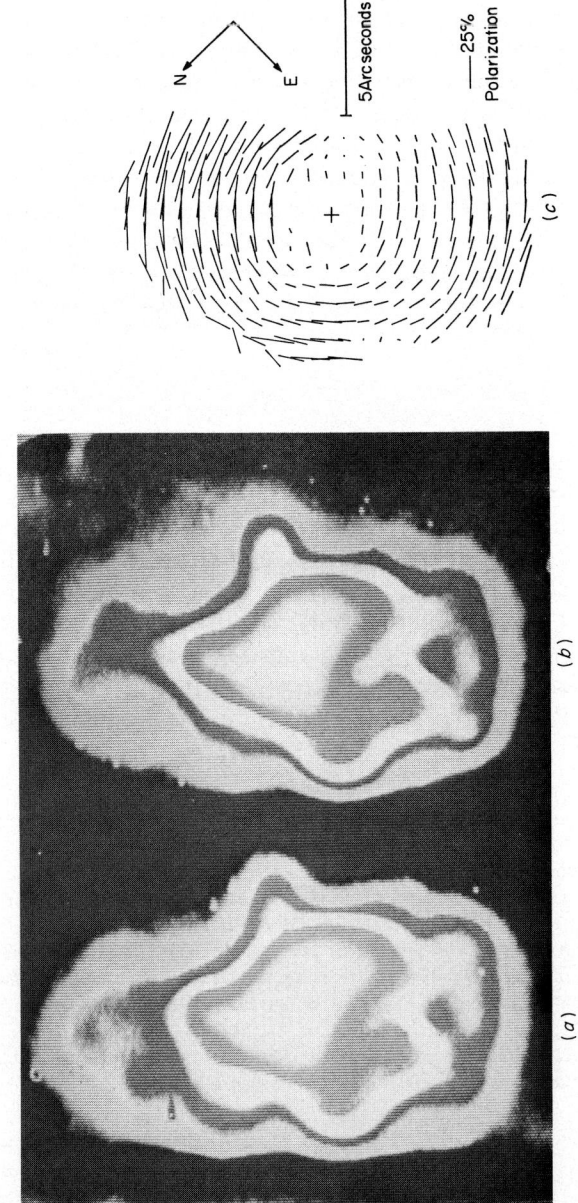

Fig. 6. (a) Two images with orthogonal polarization of the nebula Eta Carinae taken on the Anglo-Australian 3·9 m telescope with the Durham Polarimeter and the 40 mm electronographic camera. Images processed by television contouring system. (b) Derived planes of polarization.

3. McMullan, D., Hartley, K. F. and Powell, J. R., *In* "Electrography and Astronomical Applications" ed. by G. L. Chincarini, P. J. Griboval and H. J. Smith, p. 37, University of Texas, Austin (1974).
4. McMullan, D., Powell, J. R. and Curtis, N. A., *In* "Adv. E.E.P." Vol. 40B, p. 829 (1976).
5. McMullan, D. and Powell, J. R., *In* "Astronomical Applications of Image Detectors with Linear Response" (Proceedings IAU Colloquium No. 40) ed. by M. Duchesne and G. Lelièvre, p. 5-1. Paris-Meudon Observatory (1976).
6. Powell, J. R. and McMullan, D., *In* "7th Symp. PEID Preprints", p. 225 (1978).
7. McMullan, D. and Wehinger, P. A., *Endeavour* **1** (New Series), 32 (1977).
8. Warren-Smith, R. F., Scarrott, S. M., Murdin, P. G. and Bingham, R. G., *Mon. Not. R. Astron. Soc.* **187,** 761 (1979).
9. Bingham, R. G., McMullan, D., Pallister, W. S., White, C., Axon, D. J. and Scarrott, S. M., *Nature* **259,** 463 (1976).

Electronographic Photometry of NGC 3379

M. A. R. HARDWICK, A. B. HARRISON and B. L. MORGAN

Astronomy Group, Blackett Laboratory, Imperial College, University of London, England

Introduction

In this paper photometric data are presented for the elliptical galaxy NGC 3379. The results were derived from electronographic exposures of the galaxy which were scanned with an automated microdensitometer. The surface brightness profiles and integrated magnitudes in the standard V and B system are compared with those obtained photoelectrically by other observers. The $(B-V)$ colour profile reveals a reddened nucleus.

The Elliptical Galaxy NGC 3379

NGC 3379 is an elliptical galaxy in the Leo cluster, designated as E0 by Hubble[1] and E1 by de Vaucouleurs.[2] In the Morgan[3] system, which takes into account integrated colours as well as morphology, it is classified as kED1, the k denoting spectral dominance of K-giants and the ED1 indicating that it is an early-type elliptical.

Photoelectric observations of NGC 3379 have been made by Burkhead and Kalinowski,[4] Miller and Prendergast[5] and de Vaucouleurs.[6] Photographic observations have been made by Hubble,[7] Redman and Shirley[8] and Fish.[9] These studies have sought to determine accurate luminosity profiles and to derive radial colour gradients.

Radial velocity studies indicate that NGC 3379 belongs to a group including at least six other galaxies. The companion galaxy NGC 3384 is a member of this group and its close proximity to NGC 3379 has led Burkhead and Kalinowski to search for observational evidence of a tidal interaction between the two. They report an anomalously high U surface brightness between the two galaxies and the rapid fall-off in the $(B+V)$ profile which they observe near their detection limit, might also indicate some physical interaction.

Assuming a Hubble constant of $50 \text{ km sec}^{-1} \text{ Mpc}^{-1}$, NGC 3379 is at a distance of 14·9 Mpc. The irregular galaxy to the South-East is at a distance of 24·1 Mpc and is therefore not likely to be a member of the same group.

OBSERVATIONS

The observations described here were obtained using a Spectracon camera at the Newtonian focus of the Helwan Institute's 74 in. telescope at Kottamia, Egypt. The Spectracon camera, which was designed and built at Imperial College, is shown mounted on the telescope in Fig. 1. A guiding eyepiece can be scanned over a circular field of radius 75 mm around the Spectracon. A "flip-in" mirror enables the object field to be located and accurately positioned on the photocathode. The Newtonian focal ratio is $f/4·9$, corresponding to a plate scale of $22·3 \, \text{arcsec mm}^{-1}$. This is readily resolved by the Spectracon and gives a blackening rate adequate for the surface photometry of the outer regions of galaxies.

The Spectracon had an S·11 photocathode and the images were recorded on Ilford G-5 nuclear emulsion. Standard B and V filters were employed. Table I lists details of the electronographs on which the results are based.

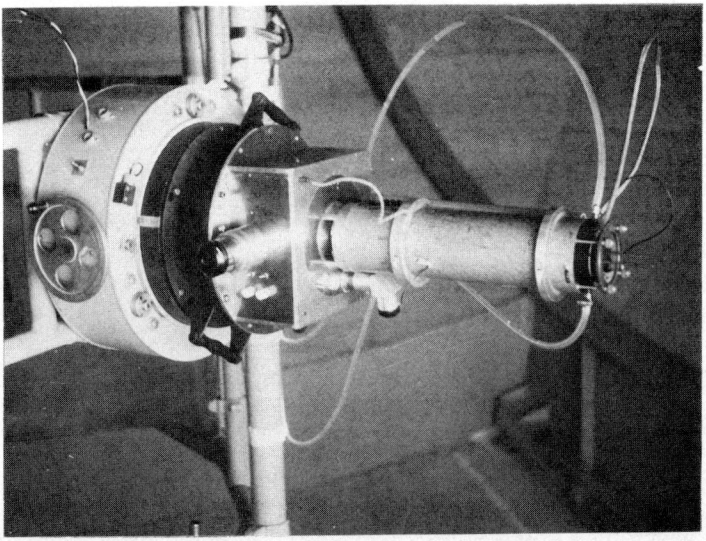

FIG. 1. The Spectracon camera mounted at the Newtonian focus of the 74 in. telescope of the Helwan Institute, Kottamia, Egypt.

TABLE I
Observational data

Plate number	Filter	Exposure	Date
K 22e	B	40 min	13/14 March 1978
K 22f	V	20 min	13/14 March 1978

Seeing: 4 arcsec (visual estimate)
Emulsion: Ilford G-5
Spectracon: AS-9 (S·11 photocathode)

Figure 2 shows a reproduction of a Palomar Schmidt survey plate containing the field.

DATA REDUCTION

Preparation

Two-dimensional scans of the electronographs were made using an automated Joyce–Loebl microdensitometer which measures the density at each sample point and records the result digitally on magnetic tape for subsequent computer processing. The microdensitometer scanning aperture, which defines the pixel size, was a square of side 50 μm. This

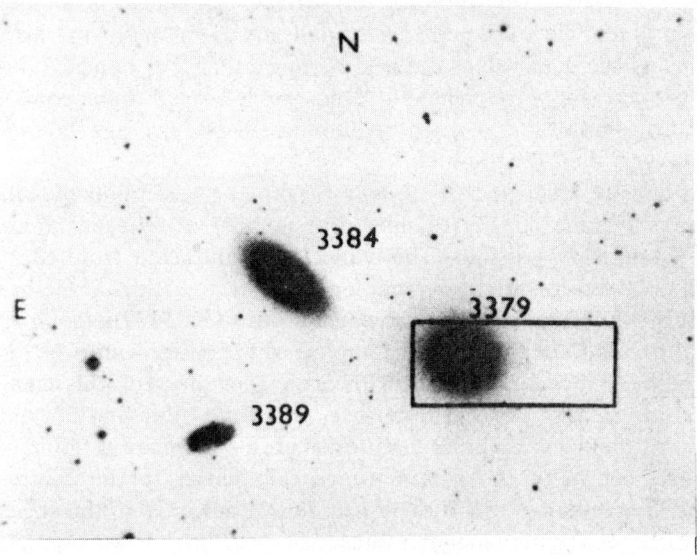

FIG. 2. Reproduction from the Palomar sky survey chart of the area around NGC 3379. The superimposed rectangle represents the image area of the photocathode.

corresponds to 1·1 arcsec in the image, about one quarter of the diameter of the seeing disc during these observations.

The scans were displayed as isophotal contour maps and a rectangular area around the galaxy and extending over most of the photocathode was selected for further analysis. A number of extraneous objects, such as foreground stars, were rejected. Rectangular zones enclosing them were discarded by specifying the elements in the data array corresponding to vertices of the rectangles. Pixels which fell within these "forbidden" areas were labelled and ignored in subsequent calculations. The data values of the picture points were not replaced by average values as this would have biased later statistical analysis.

Luminosity Profiles

Luminosity profiles were obtained in the following manner. The pixel with the largest measured density was assumed to be the centre of the object and the distance of each pixel from this point was calculated. This measure was used to ascribe each pixel to an annulus. The computer stored the number of corresponding array elements for each annulus, the sum of the data values and the sum of the squares of the data values. From these sums the average, mean error and standard deviation of density readings in the individual annuli were computed. Since NGC 3379 is an E0 or E1 galaxy, the variation of surface brightness as a function of position angle was expected to be small and elements of the array in an annulus whose data values differed by more than 2·6 standard deviations from the average were rejected. This enabled small areas contaminated by dust or emulsion defects to be eliminated. The average value for each annulus was then recomputed.

The density level due to the sky brightness was found by taking the weighted average density reading of a number of adjacent annuli at as large a radius as possible. This value was subtracted from the average values obtained for the individual inner annuli to yield surface brightness as a function of radius. Since the image of NGC 3379 is comparable in size to the photocathode the adopted sky density value is inevitably contaminated by galaxy light. However, the effect of this small over-estimation in the sky brightness is important only in the outermost regions of the galaxy. The B luminosity profile is somewhat more affected since the centre of the galaxy image falls closer to the centre of the photocathode than in the V exposure, thus requiring that the sky estimate be made at a slightly smaller radius.

The contribution to the total relative luminosity from an annulus was found by multiplying the average value for that annulus by its area. The luminosity within a chosen radius was determined by summing the contributions from the enclosed annuli.

Calibration

No standard stars were observed for direct calibration; instead the photoelectric integrated magnitude determinations made by other observers[4,10,11] were employed. A preliminary estimate was made of the integrated magnitude within a particular radius. This was used in conjunction with the relative luminosity measured from the electronograph within the same radius to calibrate the results. Using this calibration the average difference between the electronographic and photoelectric measures was calculated. The preliminary estimate was then revised so that the average difference became zero. For the exposure taken with the V filter the mean error between the electronographic and photoelectric results is 0·02 magnitudes, and for the B filter it is 0·03 magnitudes. These errors are of a similar order or are less than those quoted for the photoelectric results. The differences between the electronographic and photoelectric measures are shown in Figs. 3 and 4.

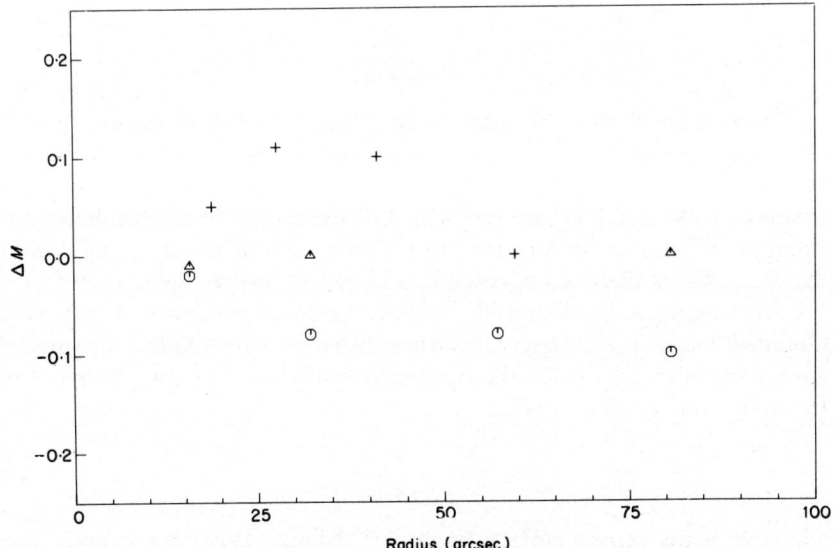

FIG. 3. Differences between photoelectric and electronographic results for integrated V magnitudes of NGC 3379 plotted as a function of radius. ○, Burkhead and Kalinowski; △, de Vaucouleurs; +, Sandage.

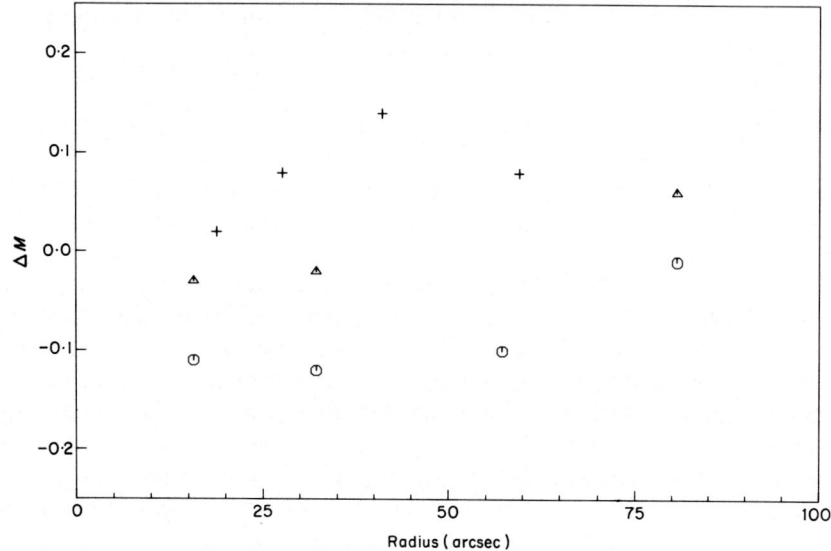

FIG. 4. Differences between photoelectric and electronographic results for integrated B magnitudes of NGC 3379 plotted as a function of radius. ○, Burkhead and Kalinowski; △ de Vaucoulers; +, Sandage.

Results

Figures 5 and 6 show the distribution of surface brightness in V and B respectively. The crosses indicate observations, the curves are fits of the various luminosity laws to the data and are discussed below. The V and B profiles are in good agreement with the observations of Burkhead and Kalinowski[4] and of Miller and Prendergast[5] except at large radii where the B profile is likely to be slightly affected by galaxy light.

Early attempts by Reynolds[12,13] and later Hubble[7] to fit empirical luminosity laws to the observed surface brightness distribution of galaxies gave acceptable results for the data then available. The law suggested by Reynolds and Hubble was:

$$S = \frac{S_0}{(1+r/\beta)^2}, \qquad (1)$$

where S_0 is the central surface brightness. More recent observations have shown that this law falls off too slowly at large radii. Oemler[14] in an attempt to overcome this difficulty has proposed a modified version of Hubble's law which introduces an exponential cut-off factor. The law

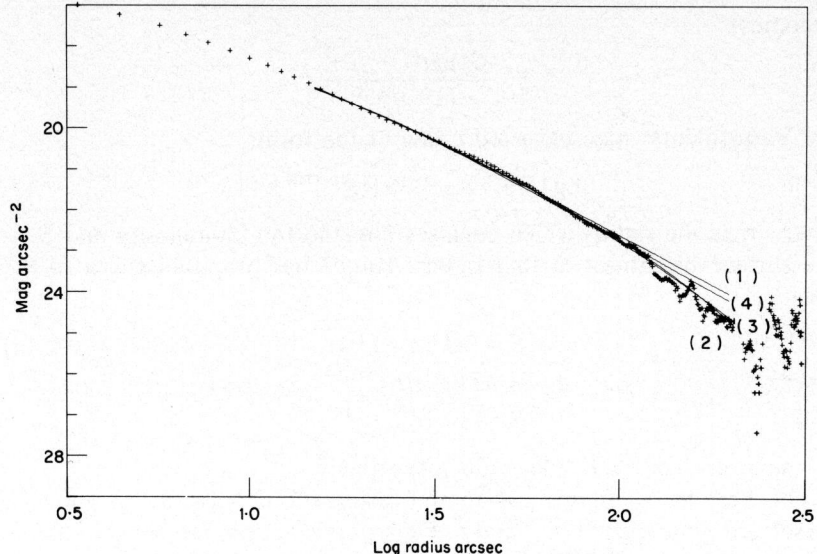

FIG. 5. The V surface brightness of NGC 3379 as a function of radius. Crosses are observed values, curves are the fits of luminosity laws: (1) Hubble; (2) Oemler; (3) de Vaucouleurs; (4) King

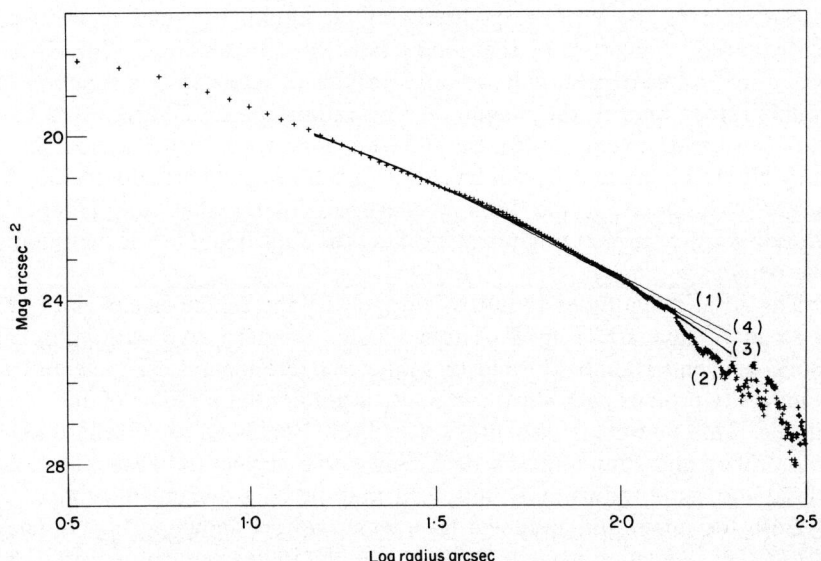

FIG. 6. The B surface brightness of NGC 3379 as a function of radius. Crosses are observed values, curves are the fits of luminosity laws: (1) Hubble; (2) Oemler; (3) de Vaucouleurs; (4) King.

becomes:

$$S = \frac{S_0 \exp(-r/a)^2}{(1+r/\beta)^2}.$$ (2)

De Vaucouleurs[2] has suggested a law of the form:

$$\text{Log}(S/S_e) = -3\cdot 33\,[(r/r_e)^{1/4} - 1]$$ (3)

where r_e is the radius which encloses half the total luminosity and S_e is the surface brightness at that radius. King[15] has proposed a law of the form:

$$S = k(1/A - 1/B)^2,$$ (4)

where $\quad A = [1 + (r/r_c)^2]^{1/2}$

and $\quad B = [1 + (r_t/r_c)^2]^{1/2};$

r_c and r_t are core and tidal radii respectively.

The best fits of these laws to the observations between 15 and 200 arcsec are plotted in Figs. 5 and 6. Data points at radii less than 15 arcsec were not used in the calculation because of the poor seeing at the time of the observations; those beyond 200 arcsec were excluded because of the possible errors introduced by incorrect estimate of the sky brightness. These best fits were determined by a computer program which varied the parameters of the law being fitted until a minimum value of reduced chi-squared, a measure of the goodness of fit, was obtained. This fitting was done in linear space, i.e., to observations of intensity as a function of radius rather than in the magnitude–log radius space of Figs. 5 and 6.

It can be seen that despite the differences of form of the various laws they all closely match the observations in the range 15 to 100 arcsec. At larger radii none of the laws is markedly better than the others in describing the surface brightness profile, although Hubble's is marginally the worst.

The total magnitudes found by integrating the luminosity profiles are 9·35 in V and 10·27 in B. These results compare well with those of Burkhead and Kalinowski and of Miller and Prendergast. The B and V luminosity profiles each show a change of gradient at a radius of about 50 arcsec. This feature is also present in both Burkhead and Kalinowski's and Miller and Prendergast's data. Since this gradient change occurs at about the same radius in B and V, it may be of a dynamical origin.

Both the integrated and local $(B-V)$ curves are shown in Fig. 7. They display a reddened nucleus with a rapidly increasing colour gradient within a radius of 20 arcsec, rising to a maximum value of 1·3 at the centre. The integrated $(B-V)$ profile remains close to a value of 0·92 from 120 to 200 arcsec. The local $(B-V)$ profile which is a more

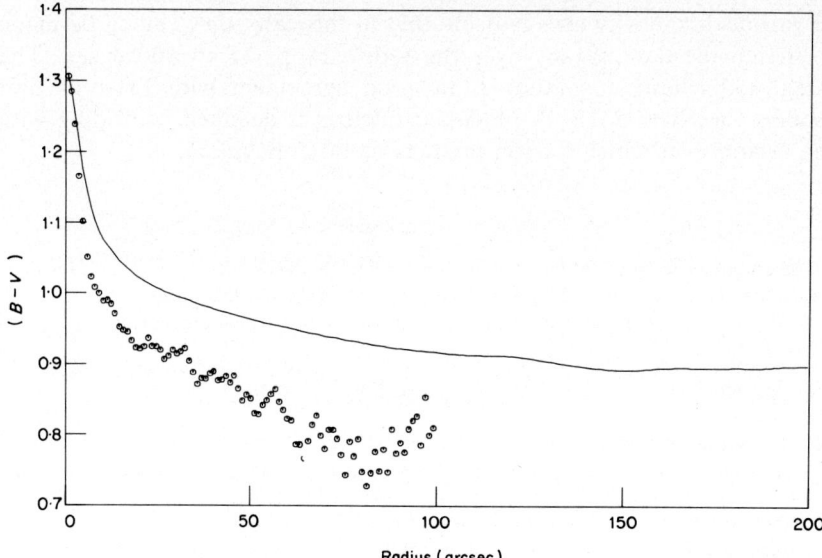

FIG. 7. Integrated and local $(B-V)$ colour index as a function of radius: ○, local; —, integrated.

sensitive indicator of colour variations at larger radii, falls to a value of 0·75 at 80 arcsec. Values at greater radii are probably influenced by errors in the B sky brightness level.

Larson[16] has described a number of simplified models of spherical galaxies that reveal a rapidly increasing metallicity gradient in nuclear regions. For the model that provides the best overall fit to the observed photometric properties of NGC 3379, he predicts a dynamical boundary at a radius of 1·5 kpc or 21 arcsec separating "Nuclear" and "Halo" regions. The model suggests that stars in the halo region were formed during the initial evolution of the galaxy and are therefore metal-poor. In the nuclear region the continuing inflow of gas enriched with the debris of supernovae leads to the formation of metal-rich stars. Outside the nuclear region the model shows a constant metallicity and in a later paper[17] Larson and Tinsley predict a value of 0·95 for $(B-V)$ in the outer region of the model.

Conclusions

Integrated magnitudes and surface brightness and colour profiles have been derived from electronographs of the galaxy NGC 3379. They are found to agree well with the photoelectric results of other observers. Fits of various luminosity laws indicate that in this case, they can all be made to match the data closely over the radius range 15 to 100 arcsec. The integrated colour magnitude is in good agreement with Larson's best models for NGC 3379. A reddened nucleus is detected coinciding with the region over which Larson predicts metal enrichment.

ACKNOWLEDGMENTS

The authors wish to thank the Director and staff of the Helwan Institute, Cairo, for their invaluable assistance in obtaining the observations described in this paper.

A. B. H. was in grateful receipt of a Science Research Council Studentship.

REFERENCES

1. Hubble, E., *Astrophys. J.* **64,** 321 (1962).
2. de Vaucouleurs, G. and de Vaucouleurs, A., *In* "Reference Catalogue of Bright Galaxies". Published by the University of Texas Press (1964).
3. Morgan, W. W., *Publ. Astron. Soc. Pac.* **71,** 394 (1959).
4. Burkhead, M. S. and Kalinowski, J. K., *Astron. J.* **79,** 835 (1974).
5. Miller, R. H. and Prendergast, K. H., *Astrophys. J.* **136,** 713 (1962).
6. de Vaucouleurs, G., *Ann. Astrophys.* **11,** 247 (1948).
7. Hubble, E., *Astrophys. J.* **71,** 231 (1930).
8. Redman, R.'O. and Shirley, E. G., *Mon. Not. R. Astron. Soc.* **98,** 613 (1938).
9. Fish, R. A., *Astrophys. J.* **139,** 284 (1964).
10. de Vaucouleurs, G. and de Vaucouleurs, A., *Mem. R. Astron. Soc.* **77,** 1 (1972).
11. Sandage, A. R., *Astrophys. J.* **183,** 711 (1973).
12. Reynolds, J. H., *Mon. Not. R. Astron. Soc.* **74,** 132 (1913).
13. Reynolds, J. H., *Mon. Not. R. Astron. Soc.* **80,** 746 (1920).
14. Oemler, A., *Astrophys. J.* **209,** 693 (1976).
15. King, I., *Astron. J.* **67,** 471 (1962).
16. Larson, R. B., *Mon. Not. R. Astron. Soc.* **166,** 585 (1974).
17. Larson, R. B. and Tinsley, B. M., *Astrophys. J.* **192,** 293 (1974).

Laboratory Tests and Astronomical Application of an Image Intensifier with a 146 mm Diameter Photocathode

E. R. CRAINE and R. H. CROMWELL

Steward Observatory, University of Arizona, Tucson, Arizona, U.S.A.

Introduction

An extensive programme of testing and application of a large format image intensifier has been underway at Steward Observatory since 1973. In this paper we briefly describe this image intensifier, outline the results of our evaluation of the device and comment on some of the astronomical applications for which this tube is being used.

Instrument Description

The instrument is a single stage, magnetically focused ITT model F-4094 image intensifier.† Similar tubes have been described in Ref. 1. The image tube has a 146 mm diameter photocathode of the multialkali extended red type with a spectral response which declines rapidly at 900 nm. The output phosphor is deposited on a fibre optic faceplate. Photographs are taken by directly contacting the photographic emulsion to the faceplate; normally a D-sensitized emulsion is used for this purpose.

A photograph of the tube/camera assembly (Fig. 1) indicates the major components of the system. The sub-assemblies consist of: (1) a rectangular unit containing filter slides, a dark slide, and a test pattern projector (used in determining a proper focus setting of the solenoid); (2) a cylindrical unit which houses the image tube, its solenoid, and a solenoid cooling jacket; (3) a plate holder (or film holder); and (4) an interchangeable eyepiece unit.

The plate holder is composed of a pressure plate plunger assembly which upon retraction of the plate holder darkslide permits the photo-

---

† ITT Electro-Optical Product Division, 7635 Plantation Road, Roanoke, VA 24019, U.S.A.

Fig. 1. The 146 mm image tube camera: tube and solenoid housing A, cooling valves B, filter carrier box C, resolution projector slide D, step wedge projector receptacle E, plate holder F, plate holder dark slide G, plate holder plunger H and eyepiece unit I.

graphic emulsion to be pressed forward against the fibre optic output faceplate. The camera was originally used exclusively with 180 mm diameter photographic plates which must be circular in order to fit inside the focus solenoid. Therefore, a special circular plate cutter unit was built which has proved to be as simple and reliable to use as a standard linear plate cutter. With the application of the image intensifier to all sky survey photography (see below) the large quantity of photographs obtained made the use of film more desirable than plates. Conversion to film was accomplished by slight modification of existing plate holders and construction of a circular film cutter.[2]

The assembled camera system has a total weight of about 52 kg. The cylindrical assembly is 33 cm long by 30 cm diameter and the filter carrier box is 30 cm square by 10 cm deep. The camera can be mounted at the Cassegrain focus of the 2·3 m reflector or at the Newtonian prime focus of the 0·5 m Baker Reflector-Corrector, both at the Steward Observatory field station at Kitt Peak.

Laboratory Tests

Measurements have been made of a number of the performance characteristics of the intensifier using methods described in detail in earlier references;[3-5] the results of these tests are summarized here.

The photocathode quantum efficiency remains above 10% over the spectral region 320–550 nm and then slowly decreases with increasing wavelength to about 1% at 850 nm and rapidly decreases thereafter, being about 0·1% at 900 nm. Large scale uniformity of response across the field remains better than ±2% over the useful wavelength range of the photocathode, becoming only slightly worse (±4%) at 900 nm. Some local blemishes, shadings, and fibre optic bundle patterns amounting to a few percent variation are apparent in the tube, but these are similar to other intensifiers of selected quality.[3]

The gain of the intensifier operated at 15 kV is about 40, where gain is defined as the photographic blackening rate of an intensified Eastman Kodak IIa-D emulsion compared to a directly exposed IIa-D emulsion. The number of phosphor screen photons emitted per photoelectron is 300. Measurement of the detective quantum efficiency (DQE) using a IIa-D emulsion to photograph the image tube output shows the combined system to have a peak DQE of 7·5% when the input radiation is at 425 nm.

Geometry and resolution are quite uniform throughout the field. Incremental magnification (local magnification of a small image) remains within 1% of unity to a 60 mm radius, and S-distortion is about 100 μm at this radius. Limiting resolution at the output phosphor is better than 80 lp mm^{-1} over most of the field. When a IIa-D emulsion is used to photograph the output, the limiting resolution of the image tube plus emulsion is about 25 lp mm^{-1} when averaged over the entire field. Discrete displacements of an image are produced at certain points in the field due to shear in the fibre optic faceplate. The faceplate in our tube was especially selected for minimum shear. The largest displacements in this tube are 29 μm, and the average frequency of occurrence of an identifiable shear (one producing a displacement of >6 μm) is about 20 cm^{-2}. These numbers are not significantly different than those found in selected faceplates of the smaller standard 40 mm tube size.

Astronomical Applications

The 146 mm image tube has been used in a variety of photographic programmes at Steward Observatory which take advantage of the tube's large field and large number of resolution elements (~10^7 with IIa-D

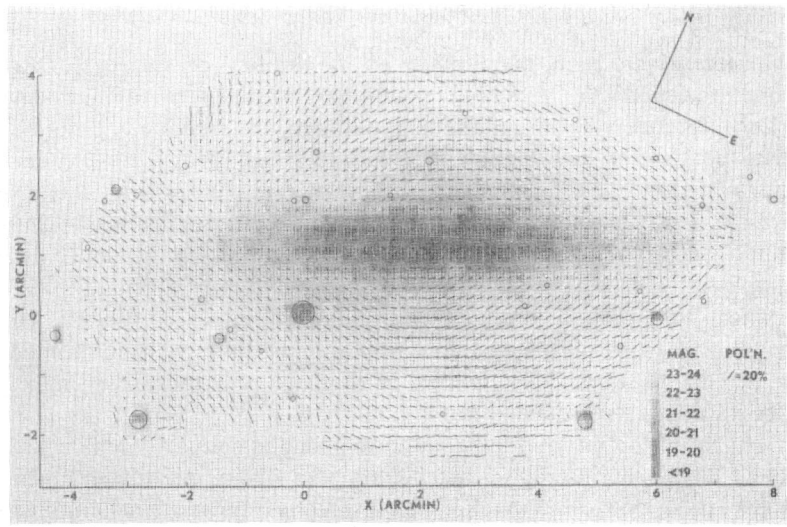

FIG. 2. Polarization and isophotal maps of M82. The blank outer region indicates that the galactic light was fainter than 25·0 mag arcsec^{-2}, except at the NW edge, where this isophote extends beyond the boundary of the traced rectangle. Coordinates are measured from the star BD+70° 587.

emulsion). At the Cassegrain focus of the 2·3 m reflector it is possible to photograph a field 24 arcmin in diameter at a scale of 9·85 arcsec mm^{-1}. Unfiltered exposures on IIa-D emulsion typically achieve a plate limit of about 21 mag in 2 min. In the following sections we discuss several programmes which have been conducted with the 146 mm intensifier.

Intensity and Polarization Map of M82

Figure 2 shows the results obtained in a study of the intensity and polarization in the continuum of the galaxy M82, derived from four successive photographs taken through 45° rotations of a polarizing filter.[6] A wavelength range where the galaxy's emission line contribution is small (500–590 nm half-power points) was obtained with appropriate filters. The exposure time for each plate was 12 min, bringing the night sky density to 0·5.

The Figure displays the absolute intensity and electric field polarization vector of elemental square areas of 1 mm (9·85 arcsec) side. The blank outer region indicates the galactic light was fainter than 25·0 mag arcsec^{-2}. No polarization data are given in this region because of the unacceptably high noise level.

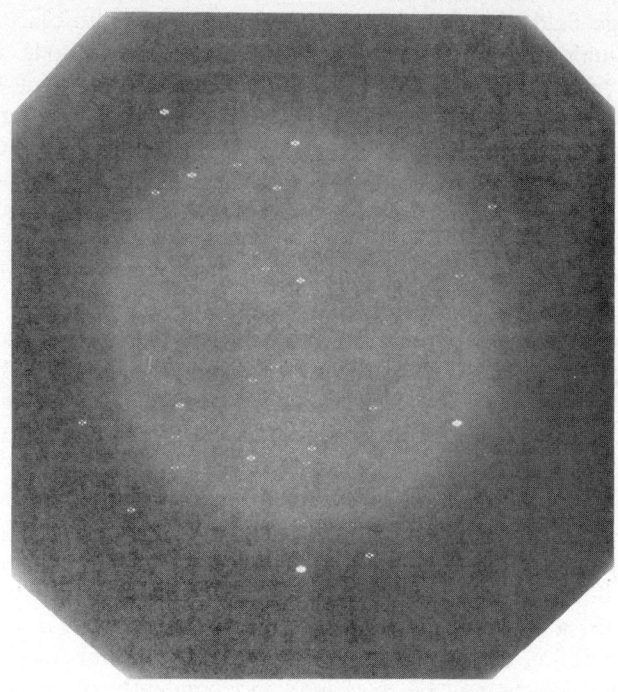

FIG. 3. A 146 mm image intensifier photograph of a star field through the calcite plate shows multiple polarized images.

Observational Studies of BL Lacertae Objects

One of the first extensive survey programs to utilize this image tube was a search for new BL Lacertae objects by means of photographic polarimetry. Although the programme has more recently been conducted using a smaller (40 mm) image tube, the original work was done with the 146 mm intensifier. Since BL Lacertae objects seem to be characterized by their high degree of linear polarization (5–35%), it was felt that this trait offered a new method of quick survey of candidate objects. Previous discoveries were primarily the result of very time-consuming object by object optical spectroscopy. A second characteristic of known BL Lacertae objects is that most have rather flat radio spectra; for this project we had available a list of such radio sources, but their positions were poorly known. In view of the inaccuracies of the radio positions, optical spectroscopy was ruled out as a search technique since it would require far too much time with a large telescope.

The large field of the image intensifiers allowed us to place a 75 mm square double calcite plate in front of the photocathode to obtain multiple, polarized images of every object in the field of the radio source.[7] An example of one of these photographs appears in Fig. 3, which clearly shows the four images of each object which arise from the calcite plate. By measuring these images with an iris astrophotometer it is possible to determine which objects are polarized as well as the degree of their linear polarization.

This technique has been found very satisfactory in that it is time-efficient and reliable; it is particularly useful as a supplement to photoelectric polarimetry programs. Preliminary analysis of the data in this programme has yielded several new and particularly interesting BL Lacertae objects.[8] It is interesting that several suspected BL Lacertae objects which were derived from this photographic work have subsequently been confirmed.[9]

A second application of the large format image intensifier to the problems of BL Lacertae objects is the study of underlying structure and galaxy population in the vicinity of the objects. For this purpose we have obtained deep, high resolution photographs of several BL Lacertae objects using a variety of red filters. The speed of the intensifier and its ease of operation allows the observer to obtain a number of very good photographs which may be examined independently or co-added after scanning by a PDS microdensitometer. By use of these photographs we have discovered underlying structure associated with BL Lacertae objects, and we have compiled information on the galaxy content of regions surrounding the objects.[10]

An Optical Infrared Photographic Sky Survey

Since 1977 the 146 mm image intensifier has been used in an extensive photographic sky survey[2] which yields pairs of photographs (visual, 520–630 nm; optical infrared, 800–900 nm) of fields covering the entire northern sky. The intensifier is mounted at the Newtonian prime focus of a highly modified Baker Reflector-Corrector telescope of 0·5 m aperture. A corrector plate and field lens provide a wide (4·55 deg diameter) and flat field which permits the use of the intensifier. This arrangement provides a scale of 110 arcsec mm^{-1} on the photographs. The survey is unusual in that it provides photographs of the sky in hydrogen emission-free bands as well as being the first high resolution photographic survey conducted with an image intensifier. The survey is also unusual in that the photographs are recorded on film rather than on plates.

In an effort to expedite the survey, exposure times are short (15 min in

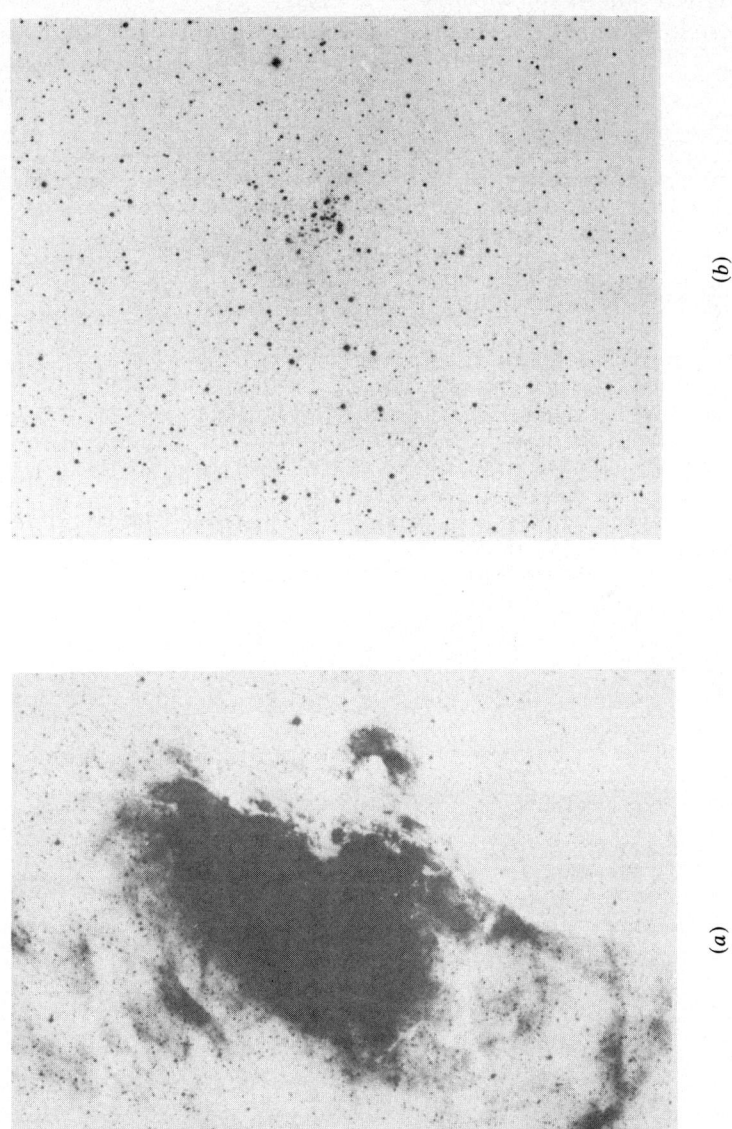

FIG. 4. (a) M16 from the Palomar Sky Survey red print; north left, west up. (b) The same field in the optical infrared obtained with the 146 mm image tube camera.

the optical infrared giving a limit of about 15·5 mag; 5 min in the visual, limit about 18 mag). With exposures matched in this way, we are cataloguing red objects for which $V - I \geq 3$ mag.

An example of the effect obtained by the near infrared photography may be seen in Fig. 4 showing M16 in the visual and the optical infrared.

Acknowledgments

Testing and application of the 146 mm image intensifier is supported by research grants from the National Science Foundation, the National Aeronautics and Space Administration and the National Geographic Society.

References

1. Ceckowski, D. H., *Proc. S.P.I.E.* **42,** 25 (1973).
2. Craine, E. R., *Astron. J.* **83,** 1598 (1978).
3. Cromwell, R. H. and Dyvig, R. R., "Laboratory Evaluation of Eleven Image Intensifiers," Optical Sciences Center Tech. Report No. 81 (1973).
4. Cromwell, R. H. and Dyvig, R. R., *In* "Adv. E.E.P." Vol. 33B, p. 677 (1972).
5. Cromwell, R. H. and Smith, G. H., *Proc. S.P.I.E.* **42,** 155 (1974).
6. Schmidt, G. D., Angel, J. R. P. and Cromwell, R. H., *Astrophys. J.* **206,** 888 (1976).
7. Serkowski, K., *Acta Astron.* **10,** 227 (1960).
8. Craine, E. R., Duerr, R. and Tapia, S., *In* "Proc. of the Pittsburgh Conf. on BL Lac Objects.," University of Pittsburg (1978).
9. Miller, H. R., private communication (1978).
10. Craine, E. R., Tapia, S. and Tarenghi, M., *Nature* **258,** 56 (1975).

An Intensified Storage Vidicon Camera for Finding and Guiding at the Telescope

J. R. P. ANGEL, R. H. CROMWELL, and J. MAGNER

Steward Observatory, University of Arizona, Tucson, Arizona, U.S.A.

Introduction

It is a great convenience at astronomical telescopes to use a television viewing system which can see faint objects directly. Such systems are used routinely at major observatories to find and guide stars at the slit of a spectrograph. Most systems currently in use employ an electrostatic image intensifier fibre optically coupled to a vidicon with an integrating target (SEC or cooled SIT vidicons). Integrations of several seconds are made to obtain an image of adequate signal to noise ratio, and then the target is read out. The image is stored in a scan converter and is continuously read out to give a normal television video signal for display on standard monitors. In some systems very deep images are obtained by digitally adding several integrations to improve the signal to noise ratio.[1]

Such systems, while very effective, are rather expensive. At Steward we have felt the need for an inexpensive TV system with integration that could be made available for smaller telescopes. In this paper we describe such a system which has been developed over the last year or so and which has a current component cost of under $4000. The basic idea is to use a special vidicon which has the property of extremely long lag (~10 sec), even while being scanned at normal TV rates. This removes the need for a scan converter, timing circuits and special camera electronics; a standard CCTV camera can be used. The tube we use is a Teltron 9300 ST/fo,† which is like a standard vidicon except for the target material. Unlike the SEC and SIT tubes, this vidicon has no gain provided by an internal electron imaging stage, and therefore requires cascaded external intensifiers. We use a three stage electrostatic image tube, fibre optically coupled to the vidicon. The gain at full voltage of the

† Teltron Inc., 2 Riga Lane, P.O. Box 416, Douglasville, Pa. 19518, U.S.A.

selected 25 mm Varo intensifier tubes used is about 40 per stage. That is to say, a primary photoelectron from the front photocathode yields 40 electrons from the second cathode, 1600 from the third, and 2×10^5 charge carriers in the vidicon target. The gain is sufficient that in a lag free vidicon individual photoelectrons would appear as flashes, and the sensitivity would be limited by the ability of the eye and brain to integrate. With the long lag vidicon, the light output from the photoelectron is spread over several seconds. Photoelectrons are not apparent individually, but for diffuse illumination, such as the sky, appear as a mottled background which changes on the time scale of a few seconds, set by the target decay time. A star near the sky limit appears as a brighter patch against the sky background, and the faint limit is set by the photoelectron statistics of the background.

Details of the Camera

A first version of the system has been in operation at the 90 in. telescope for nearly two years. It uses an external 45 kV high voltage power supply with the resistor divider chain potted in with the intensifier. The camera electronics are from a very inexpensive Ampex camera, rehoused in a box attached to the image tube. A second version has now just been completed, and is being duplicated to use at several telescopes, including the multimirror telescope on Mount Hopkins. The new version is more rugged, convenient and stable, and will now be described in detail. This camera is a self-contained unit incorporating the intensifier, vidicon, electronics and all power supplies. Its output is 60 frames $\sec^{-1}$ video, and this is displayed with a standard TV monitor in the telescope control room.

The intensifiers are 25 mm electrostatically focused diodes (Varo model 8585†). We have chosen to assemble and pot our own selected premium tubes, rather than buy a ready made three stage intensifier. Higher gain is achieved by using all three tubes selected for high gain. In addition, we arrange the tubes to operate with the final phosphor at ground potential, eliminating the losses associated with the thick fibre boule needed if the final phosphor is at high potential. The grounded back phosphor also eliminates electrical interference from the high voltage power supply, since the vidicon target does not then pick up the high frequency oscillation of the supply. The power supply for the intensifier, consisting of a high frequency oscillator and voltage multiplier made by Varo is potted in with the intensifier. Details of the vacuum encapsulation and

†Varo Electron Devices, 2203 Walnut St., P.O. Box 1437, Garland, TX75040, U.S.A.

construction of the image tube are given in the accompanying paper by Cromwell and Angel.†

The 9300 ST storage vidicon used in our system was developed by Teltron to give an integrated display in a radar system. The target of antimony trioxide is very resistive, so that only a small fraction of the stored image is read out with each scan. The tube has a lifetime of 10 000 h, and is thus good for several years of normal use at the telescope. It currently costs $350. The quantum efficiency is high for the green emission of the P.20 phosphor of the image tubes. The target has the desirable property that for bright scenes the integration time is short, so focusing can be easily achieved with immediate feedback. Also telescope motions can be followed in real time by the images of bright field stars. For faint scenes the characteristic decay time is 5–10 sec, depending slightly on operating conditions. Bright stars leave a faint after-image which can persist for minutes; however it is possible to erase the target nearly completely by flooding it with light for a few seconds while raising the beam current.

A problem to which we have given some consideration is how to mount the heavy intensified camera tube assembly together with the sweep and focus coils. In particular the focus coil is a load which must not bear on the vidicon neck, or the glass envelope could easily be smashed if the assembly were shocked. Our solution has been to use the focus coil of a conventional wire wound sweep focus assembly, but with printed circuit (PC) type sweep coils, Penntran Corp. CYL–103–PC.‡ The latter are very light, and are fixed with RTV silicone rubber to the vidicon without risk of damage. There is then a large clearance between the sweep coil (OD 1·25 in.) and the focus coil (ID 1·62 in.). The focus coil can easily be hard mounted with no risk of straining the vidicon.

While it is possible to operate the intensified vidicon with any normal CCTV camera, we include here details of the design we are now making for several new cameras. The starting point for the design was a camera described by Townslee.[2] Substantial modifications were made to meet the requirements of the PC sweep coils and additional circuitry for the erase. The only PC card that remains "untouched" is the video processor.

The deflection coil presented a small problem when operated at the standard 15,750 Hz horizontal rate. The DC resistance is $\sim 9\,\Omega$ and the inductance is $\sim 150\,\mu\text{H}$. These low values necessitate a current drive scheme, but the inductance is large enough to present a problem on retrace. The current necessary to fully deflect the beam with 300V on g-2

---

† See p. 183.
‡ Penntran Corp., P.O. Box 508, Belleforte, PA. 16823, U.S.A.

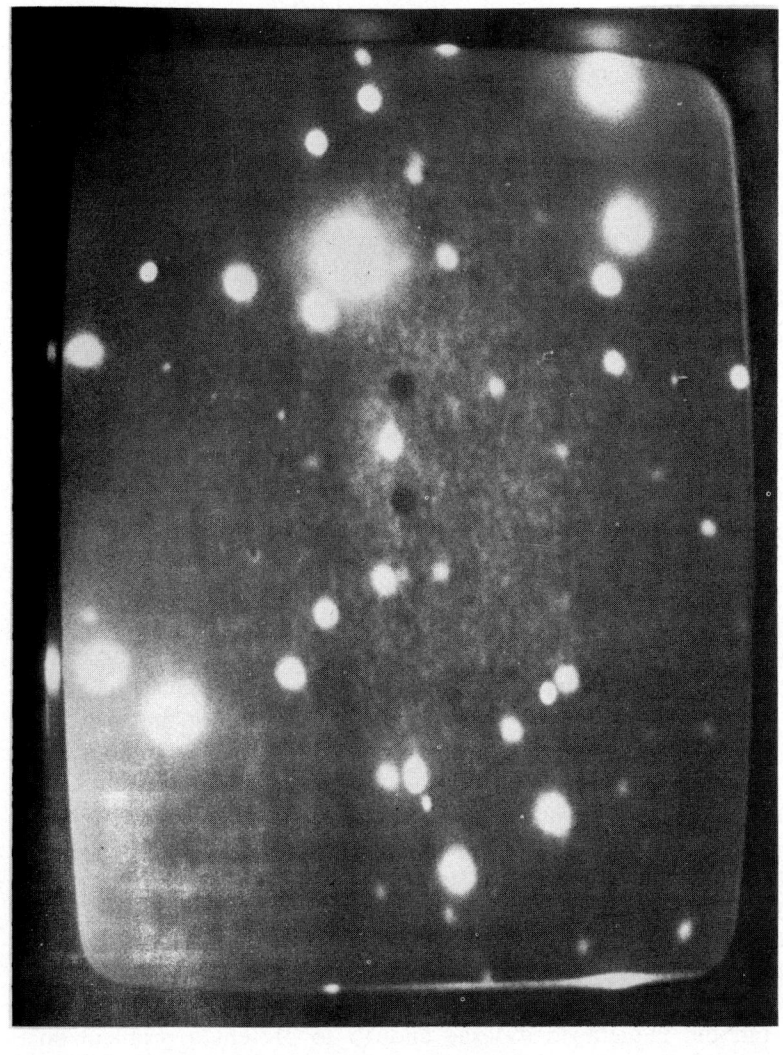

FIG. 1. Photograph of intensified storage vidicon monitor. The entire television display shows a 2·6×2·2 arc minute region of the open cluster NGC 2168 (M35). Two 5 arcsec spectrograph apertures appear as dark circles near middle of field. Seeing is about 2 arcsec. Faintest stars are near $V = 21$. The present second-grade vidicon has a few bright target defects such as the one located between the spectrograph apertures, but at the telescope these are easily differentiated from star images.

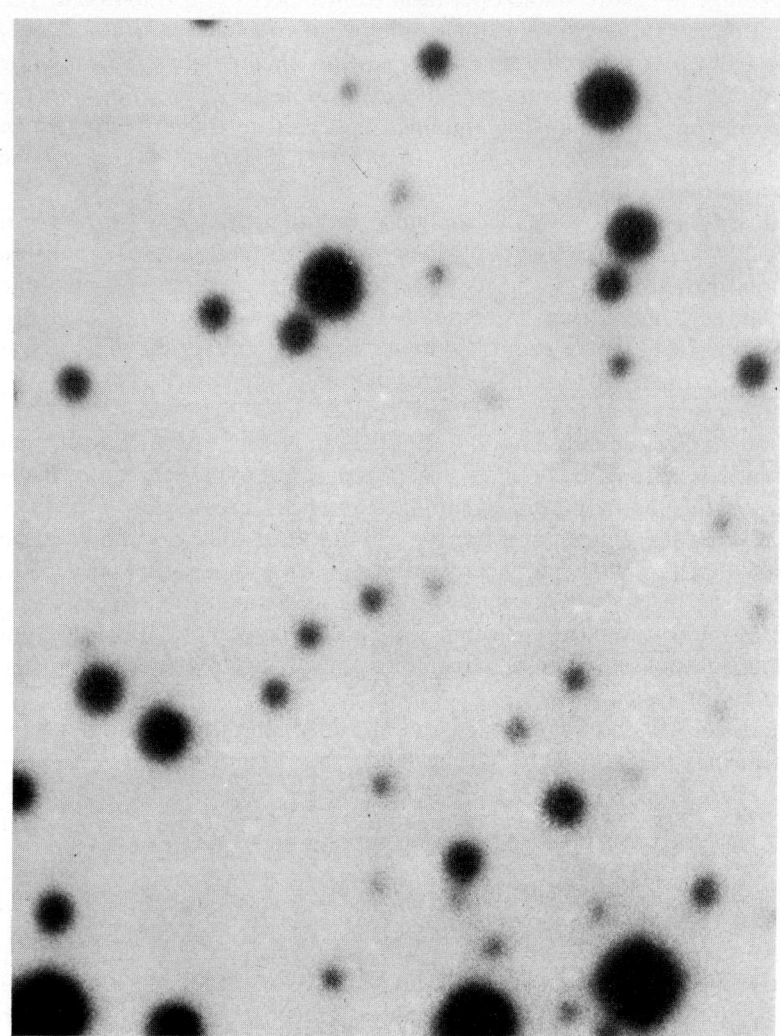

FIG. 2. The Palomar Sky Survey E print of the field shown in Fig. 1 at the same scale.

is approximately ±250 mA. A push-pull output amplifier is used with the negative rail at −22 V, and the positive rail at +6 V. The high negative rail allows for the retrace ($L\, di/dt$). Care is needed in adjusting the bias of the output stage to ensure stable operation from −30 to +50°C ambient. The vertical deflection circuit presented no special problem.

The vidicon is protected against deflection circuit failure by sampling the voltage developed across the current sense resistors in series with the deflection coils. This signal is amplified and used to trigger retriggerable one-shots upon retrace. A failure in V or H deflection will cause the vidicon to be blanked within $1\frac{1}{2}$ lines or $1\frac{1}{2}$ fields.

The sync generator was completely redesigned around a National MM5320 integrated circuit that takes care of all basic housekeeping. In addition to the MM5320 and its buffering there is circuitry to control the erase lamps, beam current increase, and blanking which are necessary for the 9300 vidicon. A new regulated high voltage supply for the vidicon was designed so that all controls were mounted on the PC cards. The transformers were also mounted on a PC card.

The complete camera circuit is divided among six PC cards such that each card is a functional unit (sweep generator, sync generator, video processor, high and low voltage supplies, and transformers). All electronic parts, except the vidicon and the focus and sweep coils, are mounted on these six cards. With spare cards available, fast repair can be accomplished by "black box" replacement. An additional advantage is that modification for applications other than acquisition can be easily accomplished. (It would be a simple change to convert to slow scan or to use a silicon target vidicon).

Full details of the circuit design and printed circuit boards and mechanical drawings of the camera will be made available on request.

Operation and Performance at the Steward 2·3 m Telescope

The camera is used to view the reflecting slit or apertures of spectrographs at the $f/9$ Cassegrain focus. After reflection by the slit mirror, tilted at 15° to the telescope axis, light from the star field strikes a folding flat mirror, and then is collimated by a 216 mm focal length projector lens. At the exit pupil of 25 mm diameter is located a photographic zoom lens of 85–210 mm focal length, which forms an image of the slit on the camera tube. The plate scale at the camera can be varied from 10 to 20 arcsec mm^{-1} without vignetting the pupil. Normally the system is operated at maximum magnification (10 arcsec mm^{-1}), so that stellar images are well resolved and can be accurately located on the slit.

A photograph of the TV monitor image obtained with the latest system is shown in Fig. 1. For comparison we show the same field from the E print of the Palomar Sky Survey in Fig. 2. The area displayed is a region of the open cluster NGC 2168 (M35). Two spectrograph apertures 5 arcsec in diameter appear as dark circles near the middle of the field. Because seeing conditions produced about 2 arcsec star images, the threshold images are not as faint as those obtained under better seeing, but stars near $V=21$ are seen. Observers who use the system are in general agreement that the TV in good seeing conditions brings up all the stars visible on an E plate. A few bright target defects of the present second grade vidicon appear to mimic star images in the figure, but at the telescope these are immediately differentiated because of their invariant character. A premium grade vidicon with very few defects may be installed in applications where such defects would be a problem.

In operation it is found that the long persistence, which cannot be switched off, is not a practical difficulty. This is because of the property of the target to respond immediately to bright objects ($V \lesssim 18$).

Acknowledgment

This work has been supported in part by the National Science Foundation under grant AST 75 17845.

References

1. Wampler, E. J., *Science and Technology* **28**, 337 (1972).
2. Townslee, A. C., *Ham Radio*, **11**, 10 (1978).

Image Photon Counting Detectors for Spaceborne Applications

A. BOKSENBERG and C. I. COLEMAN

Department of Physics and Astronomy, University College, University of London, England

Introduction

The University College London Image Photon Counting System (illustrated schematically in Figs. 1 and 2) is now well proved. Its most important features are summarized below. The construction and operation of the UCL system have been described by Boksenberg et al.[1-4] Here, we review the present status of the ground-based Image Photon Counting System (IPCS), and explore the development of such systems for spaceborne applications.

The UCL Image Photon Counting System

The IPCS has been in routine observing use since 1973. A common-user version is installed permanently at the Anglo-Australian Telescope (AAT), and another is nearing completion for the UK Northern Hemisphere Observatory. The UCL system has recently been expanded, by the addition of a larger data memory, to full two-dimensional operation and is now able to record in 2×10^5 pixels simultaneously. Extensive use of this has been made for two-dimensional (long-slit) spectroscopy of extragalactic objects at the ESO 3·6 m telescope and the Hale 5 m telescope. A complete Faint Object Camera (FOC) embodying a range of filters and masks, and having optional spectrographic and objective dispersion facilities, has also been developed at UCL, and the IPCS/FOC combination has been used at the Cassegrain focus of the Hale telescope. Examples of direct images obtained with the IPCS/FOC are given in Fig. 3. Time-resolved direct observations of the Vela pulsar field have been made at the AAT, resulting in identification of the pulsar.

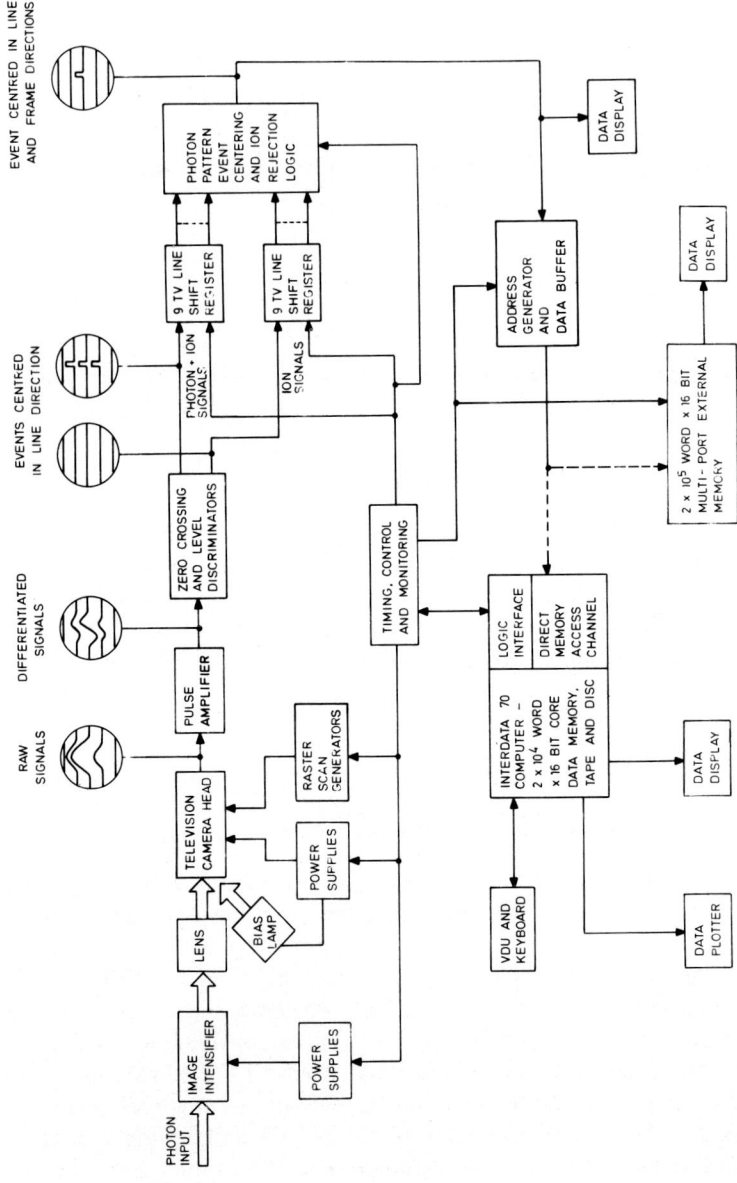

FIG. 1. Block diagram of UCL Image Photon Counting System.

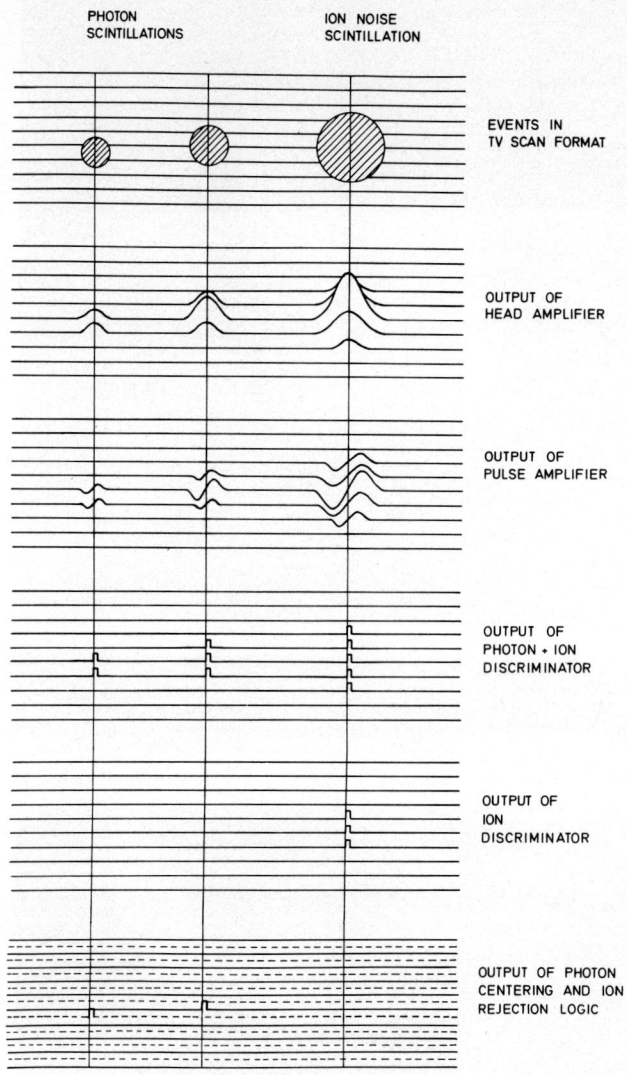

FIG. 2. UCL Image Photon Counting System: illustration of electronic processing of video signals.

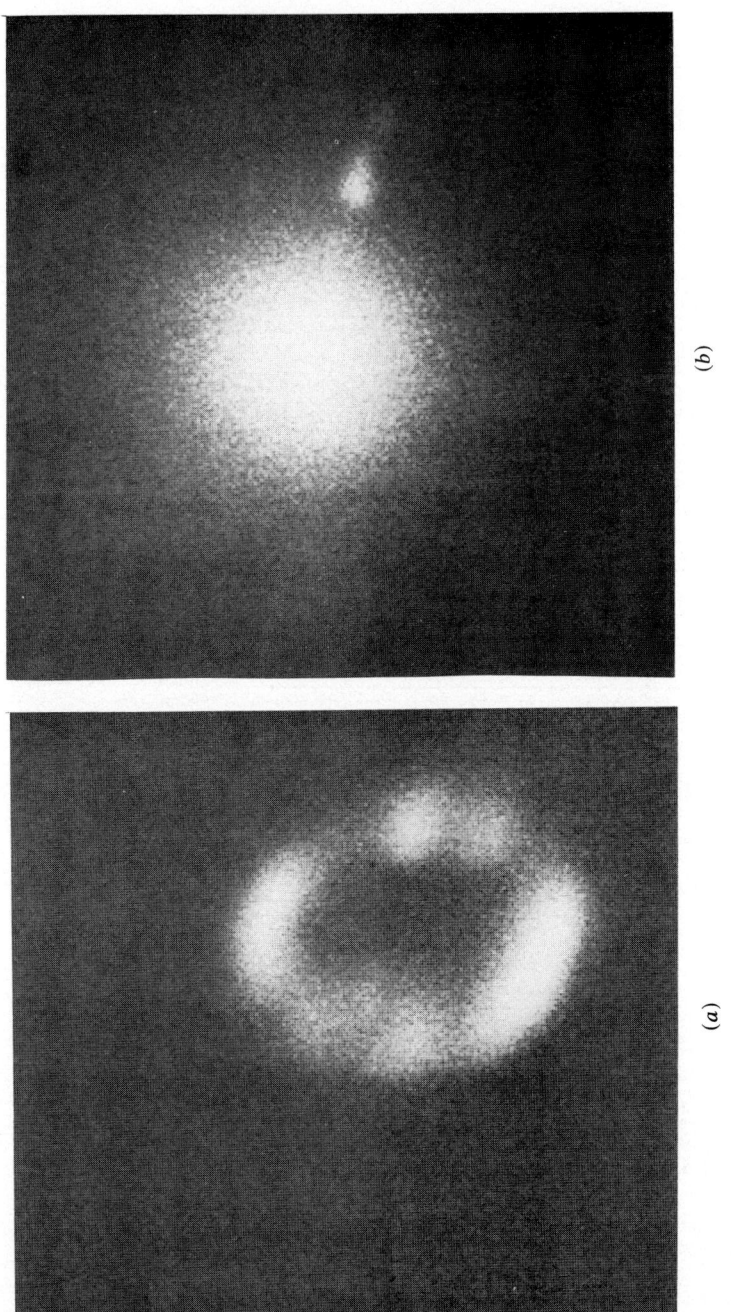

FIG. 3. Examples of observations obtained with the IPCS/FOC at the 5 m telescope: (a) the ring galaxy VII Zwicky 466, and (b) the giant elliptical galaxy M87. The large dynamic range present in the original data cannot be reproduced here.

IPCS Performance Summary

Because of the digital (photoevent/no photoevent) mode of operation, the system performance is far less sensitive than in the analogue case to detailed aspects of the detector's construction; the performance of an IPCS is generally limited only by the efficiency of the primary photocathode and by the resolution of the first intensifier stage. An important feature of the system, essential to the operation of a true IPCS, is the centroiding function; it is carried out by a special hard-wired digital signal processor, and performs three distinct tasks:

(a) Location of the geometric centre of each event and elimination of multiple counting of a single event. The latter is necessary because the finite size of a scintillation means that it will be detected on perhaps two or three adjacent television scan lines. Centroiding thus ensures maximum DQE by according equal statistical weight to each counted event.[5] The other important advantage, compared with analogue image detection, is that recording the position of the centre of each scintillation results in markedly improved spatial resolution. The mechanism of this improvement is illustrated in Fig. 4, and the effects on spectral arc line resolution and on system MTF are shown in Figs. 5 and 6.

(b) Rejection of low-amplitude amplifier noise pulses, and of high-amplitude pulses due to ion noise scintillations in the intensifier, as illustrated in Fig. 2.

(c) Encoding the positions of the centroided photon events and passing the data to the computer where the appropriate memory addresses are incremented.

The separation of the detection and storage functions of an IPCS results in several advantages:

(i) There is no low level threshold for very weak exposures of only a few photoelectrons per pixel, so that very faint images can be recorded.

(ii) Storage capacity is effectively unlimited, being completely independent of the detector and depending only upon the size of the computer memory (and, ultimately, upon magnetic tape resources). As presently configured, the system can count photoevents in up to 2×10^5 pixels, and the data acceptance format is completely flexible, i.e., data can be accepted from any specified area of the image format (e.g., full two-dimensional imaging over a large area, or high-resolution imaging over a long thin rectangle for spectroscopy); it is also possible to sum over adjacent groups of pixels.

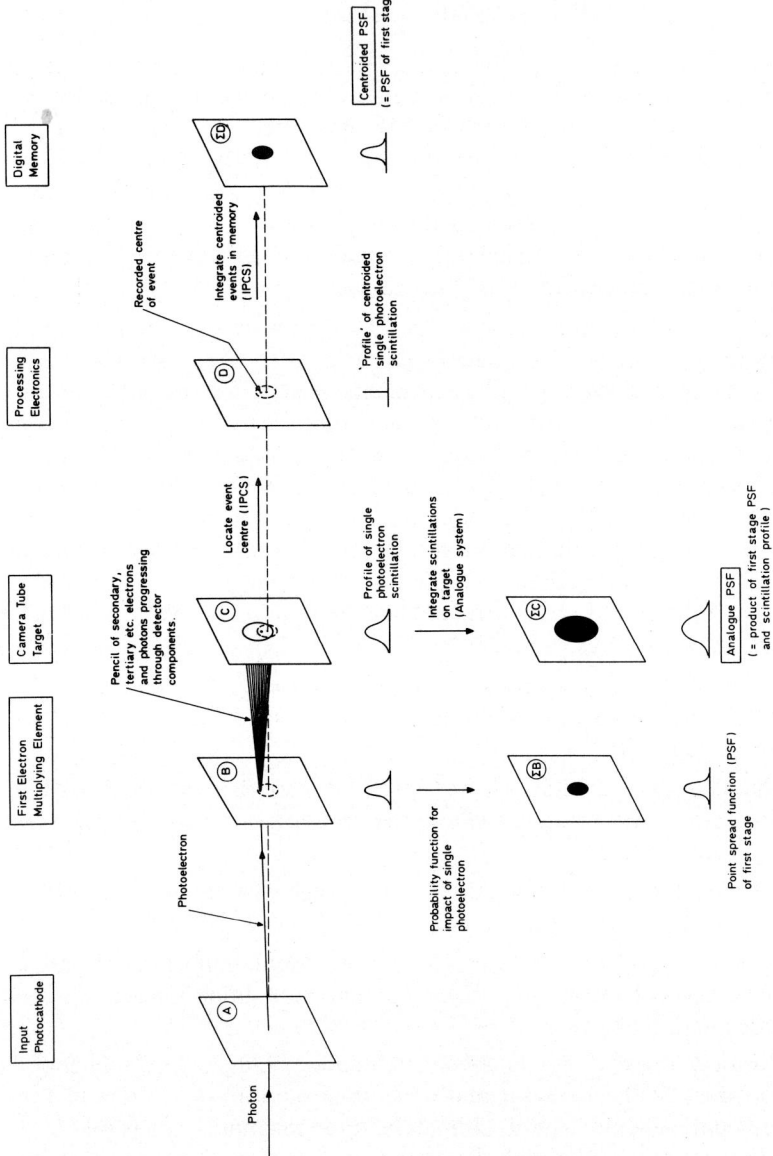

FIG. 4. Mechanism of resolution improvement in a centroiding IPCS, compared with an analogue detector.

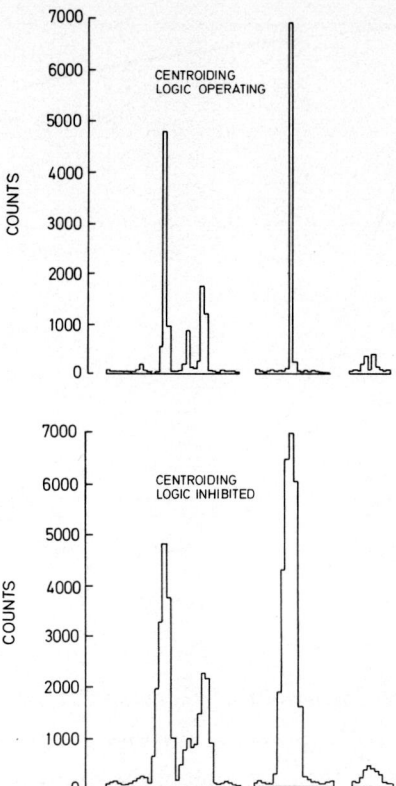

FIG. 5. Exposures to arc spectra with the UCL IPCS. The centroided line-spread function is clearly less than one pixel in width (25 μm in this case, but 15 μm is now used routinely), whilst the non-centroided line-spread function extends over several pixels.

(iii) Rapid time varying effects can be accommodated, with a resolution of a few milliseconds or less, depending on the image format.

(iv) Real time display of the accumulating image on a monitor results in high operating efficiency; for example, in astronomical observations, considerable time is saved by terminating exposures as soon as the required image signal to noise ratio has been achieved.

(v) The exposure linearity and system stability allow accurate photometric calibration and background subtraction. The latter is particularly useful in astronomical spectroscopy of faint objects, where the signal can be extracted from a background containing bright night sky emission lines; subtraction and calibration may also be carried out in real time.

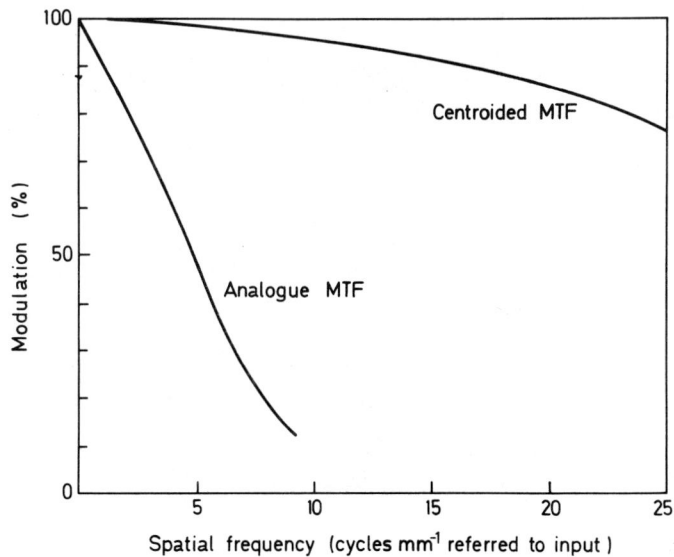

Fig. 6. Comparison of analogue and centroided MTFs for IPCS (blue light).

Development of a Spaceborne System

The development of a spaceborne IPCS/FOC is being funded by the European Space Agency (ESA) as one of the focal plane instruments for the Space Telescope. This instrument extends the spectral coverage into the ultraviolet and will be capable of 0·1 arcsec resolution and a limiting magnitude of $m_v \sim 29$ for point objects (S/N~3 in a 3 h exposure). Incidentally, it is interesting to note that comparison with the projected performance of a cooled UV-sensitized CCD shows that for faint objects (implying long exposures) an IPCS will have superior signal to noise throughout the ultraviolet (from 112 nm) and in much of the visible, and that the CCD wins only in the red and near infrared because of the high long-wavelength quantum efficiency of the CCD. The results of the model calculations[6] are presented in Fig. 7. Note that at progressively lower light levels, the IPCS performance improves relative to the CCD; this is because of the IPCS's very low inherent dark current, and its complete lack of low signal threshold.

The UCL IPCS as it stands is not space-worthy, principally because it is not ruggedized, its power consumption is too great, and its spectral coverage extends only into the near ultraviolet (~320 nm). Following a general philosophy of employing only well developed intensifiers and

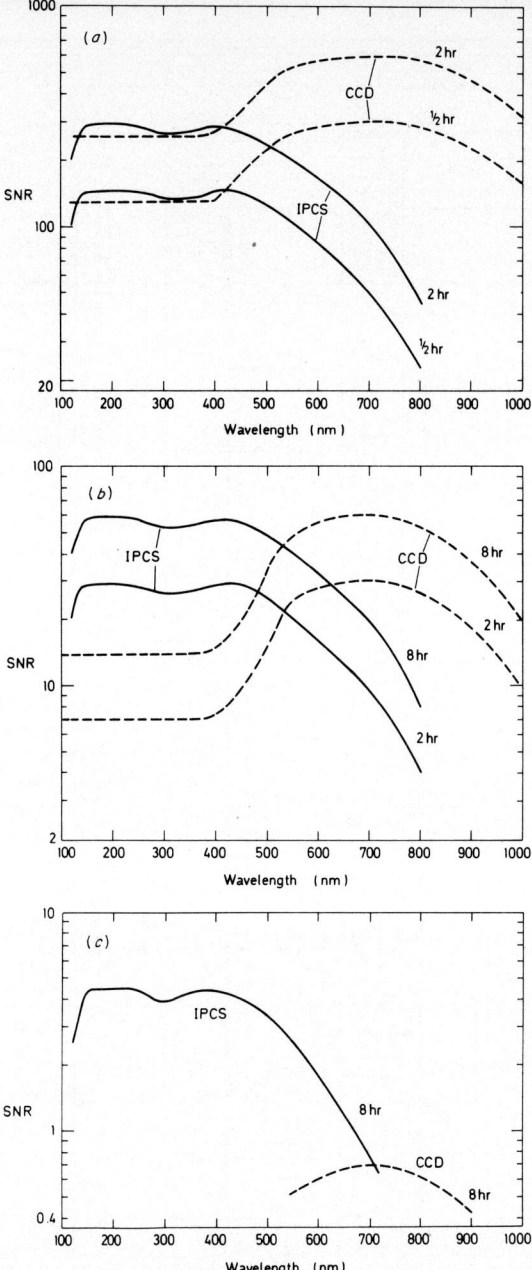

FIG. 7. Comparison of calculated signal to noise ratio obtainable with IPCS or CCD in Space Telescope FOC, for stars of (a) 20 mag, (b) 25 mag, (c) 30 mag for various exposure times. Assumptions include: 100 nm spectral bandwidth, 23 mag arcsec^{-2} sky background (independent of wavelength).

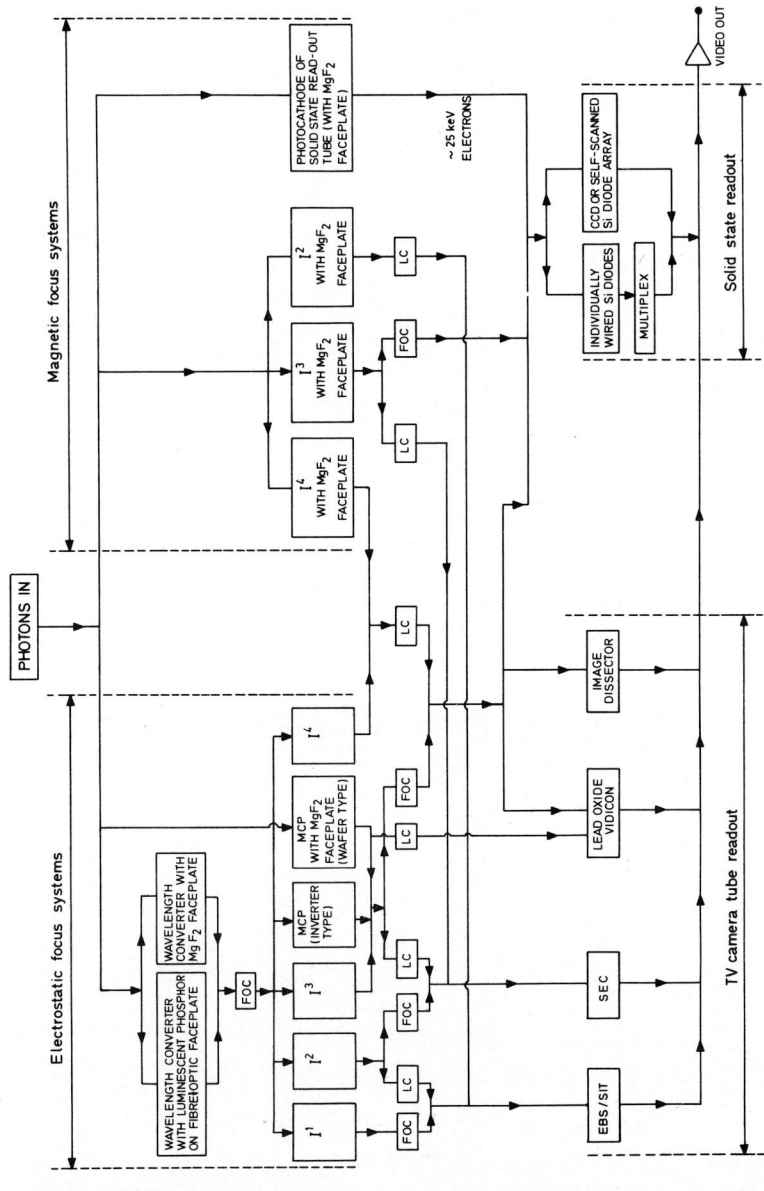

FIG. 8. Some of the more feasible alternative IPCS detector head configurations considered for use in the Space Telescope Faint Object Camera (key: I^1, I^2, I^3, I^4 = 1,2,3,4 stage intensifiers, MCP = microchannel plate intensifier, FOC = fibre optic coupling, LC = lens coupling).

camera tubes, several other possible detector head configurations (Fig. 8) have been evaluated against the following requirements:

(i) For efficient photon detection it is necessary to have adequate gain for the events to have a favourably peaked amplitude distribution (for high S/N), and for their diameters (FWHM) to be no more than two or three scanned lines in extent as in the present UCL system. The latter ensures that an appreciable fraction of the signal appears on a small number (1 to 3) of TV lines (see Fig. 9).

(ii) High and spatially uniform sensitivity in UV and visible. As fibre optic faceplates are not available for the far UV, it is necessary to use either a magnetically or a proximity focused first stage with a plane MgF_2 entrance window.

(iii) High resolution after centroiding ($>512^2$ pixels).

(iv) Large dynamic range. This implies a high frame rate and the possibility of long exposure times, and also requires small event diameters, in order to minimize count rate non-linearity at high signal levels caused by spatially or temporally overlapping events. The detailed operation of the centroiding logic is also important here.[7] The response of an IPCS may be written as:

$$c \approx \eta p \exp(-np)$$

where c is the output count rate, p is the rate of photoelectron production at the photocathode, η is the photoelectron detection efficiency and n is ~ 3 (depending upon event diameters and centroiding logic); c and p are expressed in terms of counts per pixel per scanned frame period. Thus, for a mean primary photoelectron rate of 0·05 per pixel per frame (or $\sim 2\cdot 5$ per pixel per second for the 20 msec frame time appropriate for 512^2 pixels, and correspondingly higher rates for smaller formats), the non-linearity may amount to $\sim 15\%$. Moderate non-linearity occurring at high signal levels may, of course, be accurately calibrated.

(v) Stable response and good image geometry.

(vi) Low inherent noise (may require photocathode cooling).

(vii) High reliability.

(viii) Tolerance of hostile environment (launch vibration, particle radiation, thermal, magnetic).

(ix) Compliance with specified limits on mass, size and power consumption.

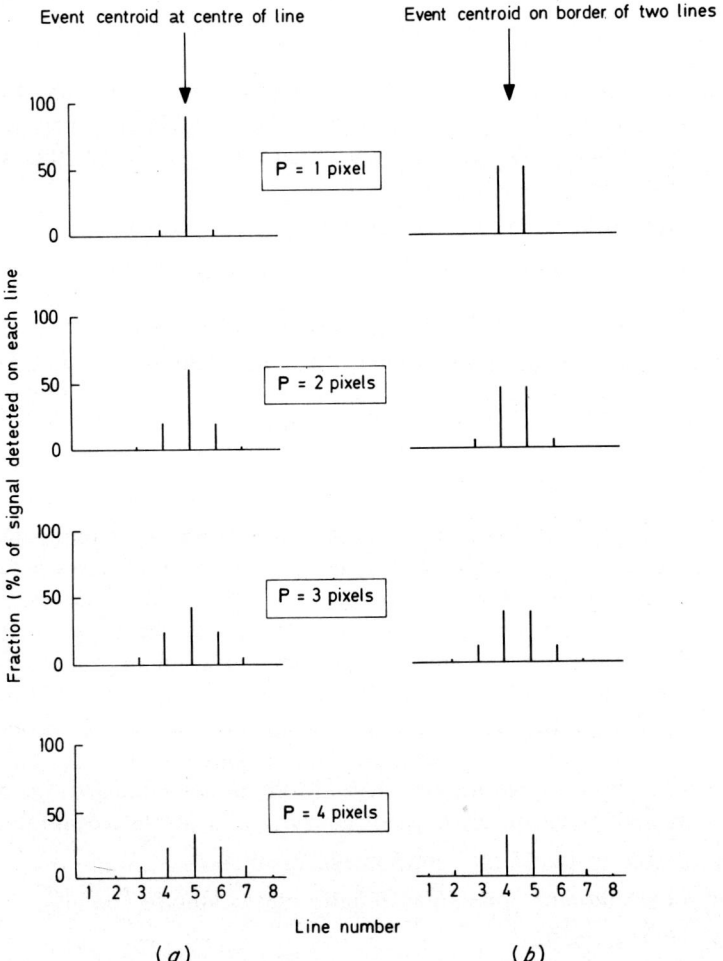

FIG. 9. Fraction of total photoevent signal read out on successive scan lines for (top to bottom): photoevent sizes P = 1,2,3,4 pixels, for the two extreme cases of (a) event centroid at centre of line and (b) event centroid on border of two lines. Gaussian event profiles assumed.

Magnetically focused intensifiers give better performance (particularly centroided resolution) than those using proximity focusing, so the former have been selected despite the penalties paid in terms of size, mass, and stability requirements for EHT supplies. ESA have supported a programme for the ruggedization of EMI intensifiers, and for the incorporation of UV-transparent MgF_2 windows. Because of the power limitations, it is necessary to use a permanent magnet† rather than a focusing coil.

The 4-stage intensifier/PbO vidicon combination, as presently used, has undoubtedly the best photometric performance of all the systems evaluated. There are problems in ruggedizing the PbO vidicon, but this could be substituted by, for instance, an FPS vidicon; however, this approach is not currently being pursued. Other solutions have been investigated in which there is less gain in the intensifier, and more gain in the television camera tube. Studies have shown that a 2-stage intensifier/EBS (or SIT) camera tube combination should have marginally adequate gain and S/N properties. An advantage is that EBS tubes are already well ruggedized; potential disadvantages, besides the possible inadequacy of S/N, include the effects of the structured EBS target.

A breadboarded version of a 2-stage intensifier (in focusing coil)/EBS detector head has just been set up at UCL, under ESA contract. Photon events are clearly visible on a TV monitor; with the system still not optimized, the event signal to noise ratio is ~ 10, and event sizes are 2 to 3 pixels. While this performance is promising, it is now planned that a 3-stage intensifier will be substituted in order to provide a good margin of safety in system gain.

References

1. Boksenberg, A., *In* "Astronomical Use of Television-Type Image Sensors", NASA SP–256, p.77 (1970).
2. Boksenberg, A., *In* "Auxiliary Instrumentation for Large Optical Telescopes", ed. A. Reisz and S. Laustsen, p. 295, ESO/CERN Geneva (1972).
3. Boksenberg, A. and Burgess, D. E., *In* "Adv. E.E.P." Vol. 33B, p. 835 (1972).
4. Boksenberg, A., *In* "Optical Telescopes of the Future", ed. R. West, p. 497, ESO/CERN, Geneva (1978).
5. Coleman, C. I., *Photogr. Sci & Eng.* **21,** 49 (1977).
6. Coleman, C. I., *In* "Proceedings of the 8th Symposium of IMEKO Technical Committee on Photon Detectors" ed. J. Schanda, p. 21, Imeka, Budapest (1978).
7. Fort, B., Boksenberg, A. and Coleman, C. I., *In* "Astronomical Applications of Image Detectors with Linear Response" (IAU Colloquium No. 40), ed. M. Duchesne and G. Lelièvre, p. 15–1. Paris-Meudon Observatory (1976).

† See p. 89.

An Image Tube with a Curved Microchannel Plate and its Use in a Photon Counting Imaging System

J. C. ROSIER and R. POLAERT

Laboratoires d'Electronique & de Physique Appliquée, 3 avenue Descartes, 94450 Limiel Brévannes, France

and

T. N'GUYEN-TRONG and B. SIDORUK

Laboratoire d'Astronomie Spatiale, Traverse du Siphon, 13012 Marseille, France

Introduction

Photon counting is a new versatile technique which appears to be quite suitable for the imaging of faint light sources in the field of astronomy. The performance of the imaging system must allow the detection and the two-dimensional localization of single photons.[1,2]

For this purpose, LEP has designed, with the support of CNES, new double proximity image intensifiers for direct fibre optic coupling to a pick-up tube.

The first part of this paper gives a description and the characteristics of wafer tubes incorporating a curved microchannel plate. The second describes a photon counting imaging system using such a wafer tube and gives the preliminary results obtained at LAS (Laboratoire d'Astronomie Spatiale, Marseille).

Image Intensifier with Curved Microchannel Plate

The advantage of the curved microchannel plate (CMCP), described earlier,[3] consists in the reduction of the backward flow of ions through the channels. The CMCP can therefore be operated at a very high electron gain with low internal noise.[4] The most interesting point is the ability of the CMCP to work in the charge saturated mode: this occurs when the electron gain reaches a sufficiently high value that the total charge delivered approaches the available wall charge. As a result of the

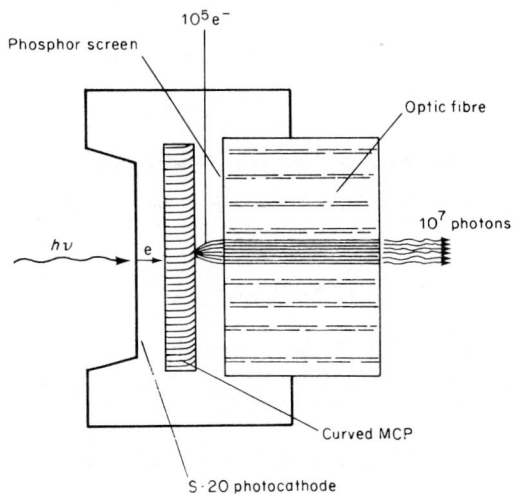

FIG. 1. Double proximity curved MCP image intensifier.

charge saturation, the CMCP shows a well peaked pulse height distribution curve; typically the variance of the electronic gain falls to 0·13 (instead of 1 with the conventional MCP) for a CMCP with 16 μm channels, giving a 5×10^5 electron gain at 1800 V.

Such a low value of the variance is interesting for two reasons: it produces a valley between the noise and the signal in the pulse height distribution curve where the threshold of the discriminator can be set; and the scintillations on the screen have a uniform diameter which enables a more accurate localization of the events by the centroiding electronic circuit.

A schematic drawing of a high gain wafer tube is shown in Fig. 1. The tube has a S·20 type photocathode, a curved microchannel plate, and a phosphor screen deposited on a fibre optic output window. The phosphor screen is specially selected for its light opacity (reciprocal of its transmission), because the first wafer tubes indicated a photon gain limitation due to internal optical feedback.[2,5,6] In principle, the light which is emitted from the powder screen cannot reach the photocathode owing to the aluminium backing layer, but because of pinholes the mean opacity of the aluminium is limited depending on the size and the density of pinholes.

A special technological effort has been undertaken, with support from ESTEC,[7] in order to improve the opacity of the backing layer to the 10^6 that is needed for avoiding optical reaction in the wafer tube when it is operated at very high photon gain, in the range of 10^6 photons per photon.

TABLE I

Characteristics of two wafer tubes

	Tube No. 1	Tube No. 2
Useful diameter (mm)	18	18
Photocathode: type	S·20	S·20
sensitivity (μA lm^{-1})	90	100
Quantum yield at $\lambda = 546$ nm (%)	8·6	9·5
Phosphor screen type (on fibre optic plate)	P·20	P·11
Screen opacity (mean)	5×10^5	5×10^5
Microchannel diameter (μm)	12·5	16
Length to diameter ratio	80	100
Electron gain at 1200 V	$2·5 \times 10^3$	–
1400 V	7×10^4	$1·8 \times 10^4$
1600 V	$1·8 \times 10^5$	$1·7 \times 10^5$
1800 V	–	5×10^5
Plate resistance (MΩ)	240	750
Nominal operating voltages for 2×10^5 luminous power gain (W/W):		
V_1 between photocathode and plate (V)	200	350
V_2 across the plate (V)	1320	1630
V_3 between plate and screen (V)	6500	6000
Spatial limiting resolution at 4% contrast (lp mm^{-1})	20	20
Light spot diameter on the screen (μm)	50–60	50–60
Background counting rate at room temperature (counts cm^{-2} sec^{-1})	280	40
Maximum counting rate (Ref. 2) (counts cm^{-2} sec^{-1})	4×10^5	2×10^5

Table I gives the characteristics of two wafer tubes which have been built with different kinds of screen and plates having different channel diameters.

For a luminous power gain of 2×10^5, at the voltages indicated in Table I, the microchannel plate does not operate in the saturated mode and consequently the pulse height distribution is nearly exponentially shaped, as can be seen on Fig. 2(a) for the tube No. 2. The saturated mode can be obtained with this tube when it is operated at a luminous gain near 10^6 as shown on Fig. 2(b). However, at that gain, the noise level becomes too high for using the tube in a counting system.

Both the P·11 and P·20 phosphors have been tried and they have similar efficiencies when coupled with a Plumbicon tube. Due to the high charge density of the electron avalanches delivered by the channels, it

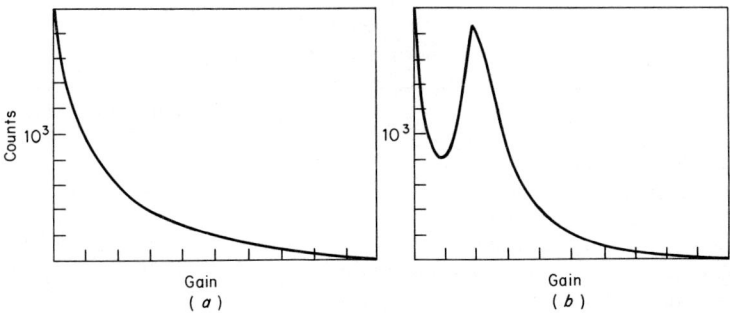

FIG. 2. Light pulse height distribution taken from the screen of a 16μm curved MCP image intensifier at various plate voltages. (a) $\bar{G} = 2 \times 10^5$ W/W; $V_2 = 1630$ V; $V_3 = 6$kV. (b) $\bar{G} = 10^6$ W/W; $V_2 = 1800$ V; $V_3 = 6$kV.

appears that the decay time of these phosphors is much shorter than usually mentioned in the data handbooks: at 10% of the maximum response, the decay time is about 0.2 μs for the P·11 and 30 μs for the P·20.

The pulse height distribution is measured by collecting the electron pulses from the anode of the wafer tube through a charge preamplifier followed by a multichannel analyser. In the particular case of the P·11 screen, which is very fast under the present conditions, it is possible to record the light pulse amplitudes by using a fast photomultiplier placed in front of the screen; its anode signal is then integrated and shaped before entering the multichannel analyser. As expected, the pulse height distribution curves measured in both ways are similar.

With both tubes, the mean diameter of the bright spots that are generated at the screen by the individual photoelectrons is around 50–60 μm due to the spreading of the electron avalanche after its exit from the microchannel.[2-8] It must be emphasized that the spatial resolution of 20 lp mm^{-1} that has been measured on these tubes at 10^4 photon gain, does not correspond to the limit which could be obtained from such high resolution MCPs; in fact, the resolution was limited by the photocathode to plate gap and by the plate to screen gap which have been chosen to be rather large in order to prevent electrical breakdown. At full gain, the measurement of the spatial limiting resolution by looking at a bar pattern on the screen is no longer feasible as the number of scintillations per pixel per second is limited by the recovery time of the channels. Such a measurement would need a much longer integration time than that of the eye.

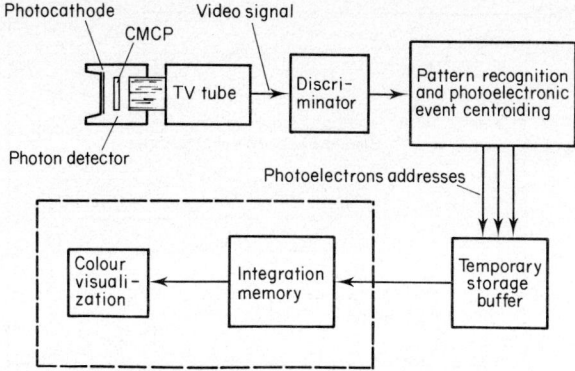

Fig. 3. Schematic drawing of the photon counting imaging system.

Tube life has been investigated at the nominal operating voltages and the plate gain appears quite stable when compared with the photocathode stability.[9] The aging of the tubes has shown that the cathode sensitivity decay depends on the total amount of charge delivered to the screen. A 40% loss in cathode sensitivity is observed when the amount of charge is 1.6×10^{-3} C cm^{-2}; this corresponds to 10^{11} counts cm^{-2} at 10^5 electron gain.

Photon Counting Imaging System

An experimental photon counting imaging system has been set up at LAS, incorporating an image intensifier with a curved microchannel plate (Fig. 3). The curved microchannel plate multiplies each photoelectron coming from the photocathode and generates, on the phosphor screen, a bright scintillation which is guided through optical fibres to the Plumbicon target. This pick-up tube delivers a video signal which, after rejection of low amplitude noise, is transformed into a binary signal.

The binary image of a photoelectron is distributed over a few adjacent pixels. A centroiding electronic circuit calculates the coordinates of the central pixel and sends its address to the integrating memory. The available image format is presently 256×256 pixels with a dynamic range of 16 bits. A display unit is used for the visualization of the data stored in the memory.

Wafer tube No. 1 has been used in this photon counting system. The operating conditions are: scanning area 12.6×9.8 mm^2; integration time 100 sec; and pixel size 50×37.5 μm^2.

In Fig. 4, is plotted the number of counts per pixel per second versus

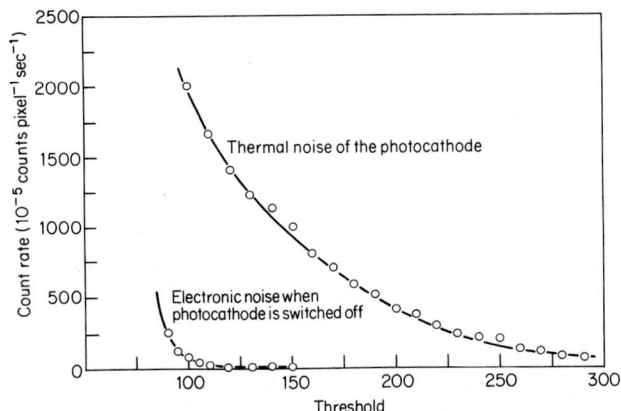

Fig. 4. Count rate versus the electrical threshold (arbitrary units).

the electrical threshold of the discriminator. In the same Figure is the electronic noise (including the plate noise and the electrical readout noise), which is that measured when the image intensifier is switched off by applying a reverse voltage on the photocathode. When the threshold level is set at 130 a.u. (Fig. 4), the electronic noise completely disappears. For this value, the dark counting rate, which is mainly due to the thermoelectronic emission of the photocathode, is 70 counts cm^{-2} sec^{-1} ($1\cdot 35 \times 10^{-2}$ counts pixel^{-1} sec^{-1}) at room temperature.

Under those conditions the total quantum efficiency, defined by the ratio between the number of counts measured after rejection of the low amplitude pulses and the number of photons which impinge onto the photocathode, is 3·4%. By comparison with the 8·6% quantum efficiency of the photocathode at the same wavelength (546 nm) one may estimate that the counting efficiency of the whole electronic system following the photocathode is about 40%.

Astronomical Use

Our equipment has been used on the 193 cm telescope of the Observatoire de Haute Provence, France, with the Pellet-Deharveng spectrometer.

In the three following Figures, the wavelength range is spread over 256 pixels, from 654 nm at the left to 675 nm at the right.

Figure 5 depicts the spectrum of the "Dumb-bell" nebula NGC 6853: one can see the forbidden lines [N II] ($\lambda = 654\cdot 8$ and $658\cdot 4$ nm) and [S II] ($\lambda = 671\cdot 6$ and $673\cdot 1$ nm), in addition to the H$_\alpha$ hydrogen line and the He I ($\lambda = 667\cdot 8$ nm).

AN IMAGE TUBE WITH A CURVED MICROCHANNEL PLATE

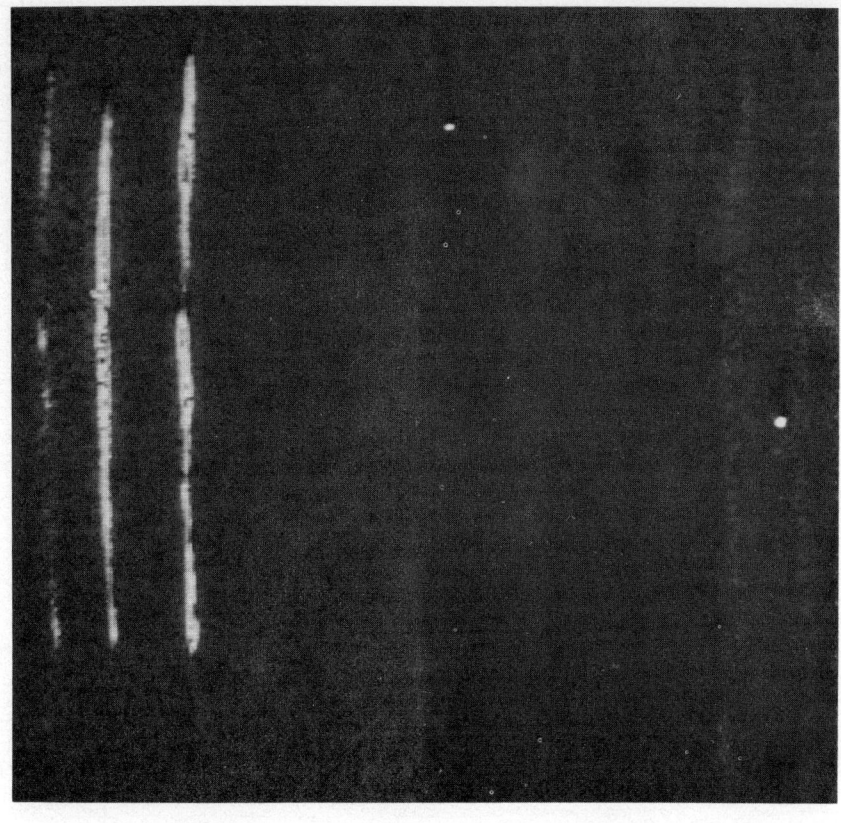

[N II]	H$_\alpha$[N II]	He I	[S II]	[S II]
(654·8 nm)	(658·4 nm)	(667·8 nm)	(671·6 nm)	(673·1 nm)
	(656·3 nm)			

FIG. 5. Spectrum of the "Dumb-bell" nebula NGC 6853.

Figure 6 shows the spectrum of NGC 6888: besides the very bright H$_\alpha$ line, are the forbidden lines [N II] and [S II] and the rather low He I.

Figure 7 shows the spectrum of the spiral galaxy NGC 6643: the horizontal line corresponds to the continuum emitted from the nucleus of the galaxy. The red shift of the H$_\alpha$ line corresponds to a velocity of recession of the galaxy of 1440 ± 50 km sec^{-1}, and the inclination of this H$_\alpha$ line yields the angular velocity of the arms.

CONCLUSION

The high performance of these new wafer tubes has been obtained because of the outstanding properties of the curved microchannel plates

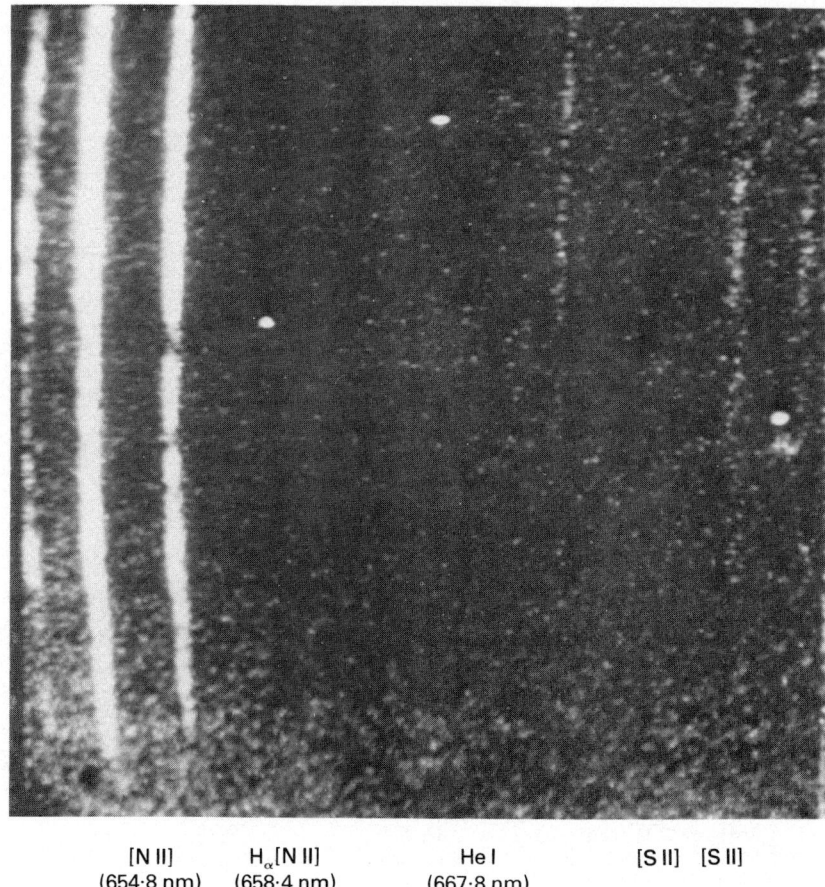

| [N II] | H$_\alpha$[N II] | He I | [S II] [S II] |
| (654·8 nm) | (658·4 nm) | (667·8 nm) | |

FIG. 6. Spectrum of NGC 6888.

and the improved light opacity of the screen backing. A better counting efficiency would be obtained by operating the plate in the saturated mode, but this will depend on reducing the optical feedback still further.

At present we are building at the request of CNES, a high gain, high spatial resolution wafer tube having a MgF_2 input window allowing the tube to be sensitive from 120 nm to 800 nm.

Because of the well known advantages of wafer image intensifiers, namely small size, low weight, supply voltage lower than 8000 V, no focusing adjustment, no image distortion and insensitivity to external magnetic fields, this kind of image intensifier is very promising for photon counting imaging applications.

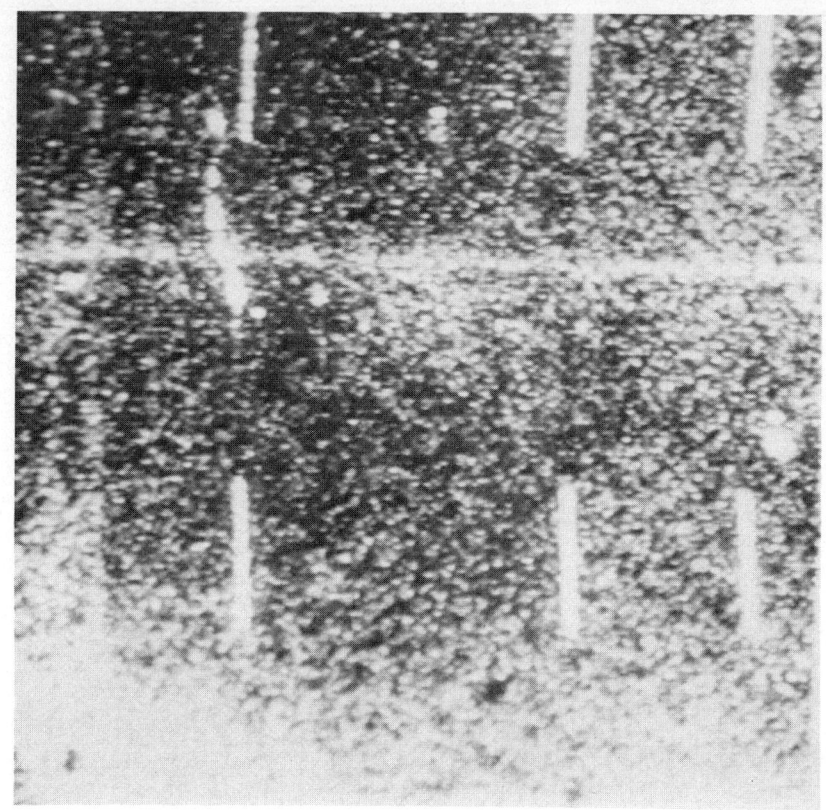

H_α

Fig. 7. Spectrum of the spiral galaxy NGC 6643.

Acknowledgments

This work was carried out at LEP with the support of CNES. The authors wish to thank Mr Bricard of CNES for his support throughout this work, and Messrs Jean, Duchenois and Dietz, Mrs Goutelle and Miss Desvaux for their assistance in preparing and measuring the tubes.

Finally, the authors are grateful to Mr Marie and Mrs Fouassier for their advice and comments while preparing this paper.

References

1. Macau, J. P., Jamar, J. and Gardier, S., ESTEC contract No. 2321/74 PP
2. Rosier, J. C. and Fouassier, M., ESTEC contract No. 2659/76/NL/AK.

3. Boutot, J. P., Eschard, G., Polaert, R. and Duchenois, V., *In* "Adv. E. E. P." Vol. 40 A, p. 103 (1976).
4. Audier, M., Delmotte, J. C. and Boutot, J. P., 5*th J. Opt. Spatiale*, p. 33 (1975).
5. Polaert, R., Rosier, J. C. and Barat, C., 5*th J. Opt. Spatiale*, p. 17 (1975).
6. Rosier, J. C. and Polaert, R., CNES contract No. 0588/76.
7. Rosier, J. C. and Duchenois, V., ESTEC contract No. 3307/77/NL/AK.
8. Vibrans, G. E., Techn. Rep. No. 308, Lincoln Lab. (1953).
9. Sandel, B. R., Broadfoot, A. L. and Shemansky, D. E., *Appl. Opt.* **16,** 1435 (1977).

First Observations of Faint Extended Emission Sources with an Image Photon Counting System

J. BOULESTEIX

Observatoire de Marseille, 2, Place Le Verrier, 13004 Marseille, France

INTRODUCTION

One of the requirements of modern astrophysics is the detection of very faint fluxes with optimum sensitivity, spatial resolution and spectral resolution. The detection of very faint radiation is nevertheless fundamentally limited by the quantified nature of the radiation itself, the photon. So, the aim is to individually detect at time t each photon in two dimensions. (This problem was just approximated by the counting of traces in electronography, where however the temporal information is destroyed).

This is the purpose of the photon counting system previously conceived by Boksenberg[1,2] which is based on the physical distinction of two functions:

(1) the individual detection of each photon and the precise computation of its position and the assignment of an identical weight to each event;

(2) the acquisition of the events with real time integration or preservation of the temporal information.

Such a photon counting system is but the logical result of the evolution of astrophysical receptors toward higher sensitivity. Its advantages, compared with usual receptors, are not only due to its detection capabilities, but also to the way in which it furnishes the data: the image can be displayed as it is being acquired and is stored digitally, thus permitting computer processing of the data without the need for a photodensitometer or calibration plates.

INSTRUMENTATION

The detector used by the Marseille Observatory is the same as that of Labeyrie[3] and the Institut d'Astrophysique et Geophysique (INAG) at the Observatoire de Haute Provence. It consists of a microchannel plate electrostatically focused intensifier (TH 9304),[†] coupled by fibre optics to an SIT camera (TH 9655).[†] The diameter of the S·20 photocathode is 25 mm but the silicon target is only 18 mm wide. The scanning electronics are conventional. The detector is cooled to $-15°C$, the photocathode end being in a dry nitrogen chamber.

The electronics used have been developed by the Laboratoire d'Astronomie Spatiale (LAS) and consist of the following units.

(1) Special detection logic circuits built by Cenalmor[4] including a comparator and a logic matrix to compute the exact location of the centres of the events. Such a system allows one to multiply the resolution by a factor of three. For the present, with a 256×256 grid, the size of each pixel corresponds to 55 μm on the photocathode.

(2) Circuits for the acquisition of the addresses of the centered events with real time integration and display[5] that allows the astronomer to see the image building up and to monitor control acquisition. The images (256×256) are stored on floppy disks, while connection to a computer offers a complete interactive treatment of the data.[6]

The 193 cm telescope of the Observatoire de Haute Provence (OHP) has been used at the Cassegrain focus ($f/15$) with a focal reducer ($f/3$) giving a scale of 36·7 arcsec mm^{-1} and allowing observations with Fabry-Perot interferometers.

Working Characteristics

The relative sensitivity is the gain in exposure time on the same object, for a particular signal to noise ratio. This "gain" is 3·5 to 4 with respect to a two stage magnetically focussed image tube (RCA 33063) as shown by spectrographic observations of red airglow lines. With respect to an Eastman Kodak 103a-E photographic plate, the "gain" is about 25 for an integration time between 5 min and 1 h. The detection limit of extended sources with a focal reducer at the 193 cm telescope is 3×10^{-7} erg cm^{-2} sec^{-1} sterad^{-1}.

The resolution is actually limited by the size of the electronic pixels (55 μm for 256×256). For a MTF of 50%, it corresponds to 9 lp mm^{-1}, which is not very far from the real resolution of image tubes used with large aperture optics (10 to 15 lp mm^{-1}).[7,8] This resolution can be increased with the use of a 512×512 electronic matrix.

The linearity is quite good as Fig. 1 shows. The dynamic range extends

---

[†] Manufactured by Thompson CSF, Division Tubes Electronique, Boulogne-Billancourt, France.

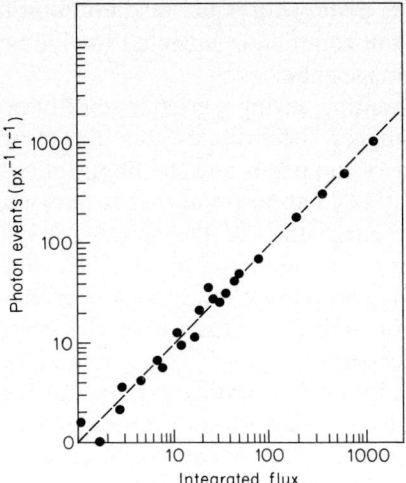

Fig. 1. Linearity curve of the system.

from $2\,\mathrm{ev\,px^{-1}\,h^{-1}}$† to $3000\,\mathrm{ev\,px^{-1}\,h^{-1}}$, corresponding to typical low fluxes. For higher fluxes the response rises and then falls abruptly.

The dark noise is at present about $2\,\mathrm{ev\,px^{-1}\,h^{-1}}$, but can be decreased by real time acquisition on a computer. We hope to reach the thermal emission limit of a trialkali photocathode cooled to $-15°C$ (about $0\cdot5\,\mathrm{ev\,px^{-1}\,h^{-1}}$ for this pixel size). The noise statistics are very close to Poissonian.

The total response to uniform illumination is not very good (factor of 2 over the field) due to the electrostatic focusing and to photocathode inhomogeneities. Corrections are done later, using a computer, the main effect being that some parts of the image are statistically less valuable than others.

The Use of Photon Counting

The photon counting system used here is very well adapted to low flux observations ($<2000\,\mathrm{ev\,px^{-1}\,h^{-1}}$). For low fluxes photon counting provides the best signal to noise ratio, but observations of higher fluxes are well covered by other receptors (IIIa-J plate, electronographic cameras, analogue TV or CCDs).

It appears that the gain available with photon counting allows the use of a 2 m telescope in place of one of 3 m with an image tube, or an 11 m

† Photon events per pixel per hour.

telescope with a 103a-E plate, with equal resolution on the sky. Used on a 3·6 m telescope, photon counting is expected to give better results than a 5 m telescope with image tube.

The use of photon counting, giving a greater sensitivity but also a direct digitalization of the images, necessitates the use of specially adapted optics to match the size of the pixels and the fibre optics window. A good focal ratio seems to be $f/2$ ($f/3$ at maximum) as is shown in Fig. 2 in which one can see a 30 min integration of the spiral galaxy M51 through a narrow band interference filter centered on H_α. The chain of HII regions is as well defined as with ordinary photographs. Correction of the image with respect to the non uniform response of the receptor to uniform illumination is also necessary.

The high sensitivity of photon counting is shown in Fig. 3(a–c). Figure 3(a) is a comparison of the observation of a narrow H_α band (0·4 nm in the centre of the spiral galaxy M33) between photon counting ($f/3$) and an RCA 2-stage image tube ($f/2$). The exposure times are respectively 20 min and 1 h 30 min. All the details of morphology observed with the image tube are visible on the photon counting integration, but also the chain of HII regions towards the bottom right corner is very well defined

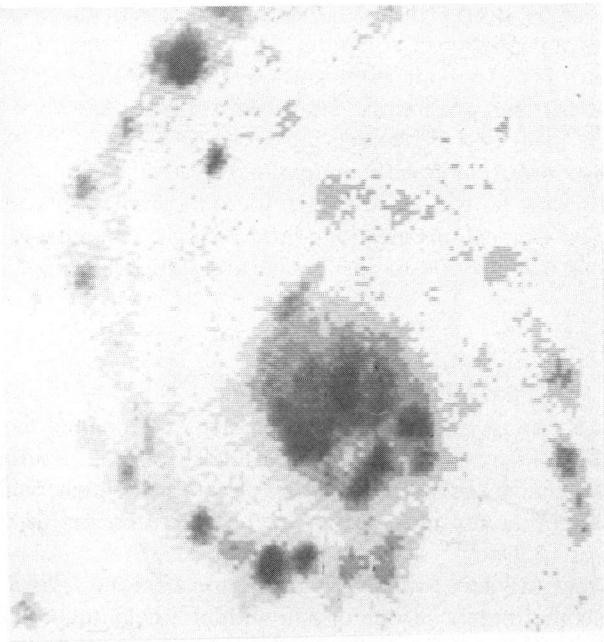

FIG. 2. H_α integration of M51 (exposure time 30 min, $\Delta\lambda = 1$ nm, 193 cm telescope, $f/3$).

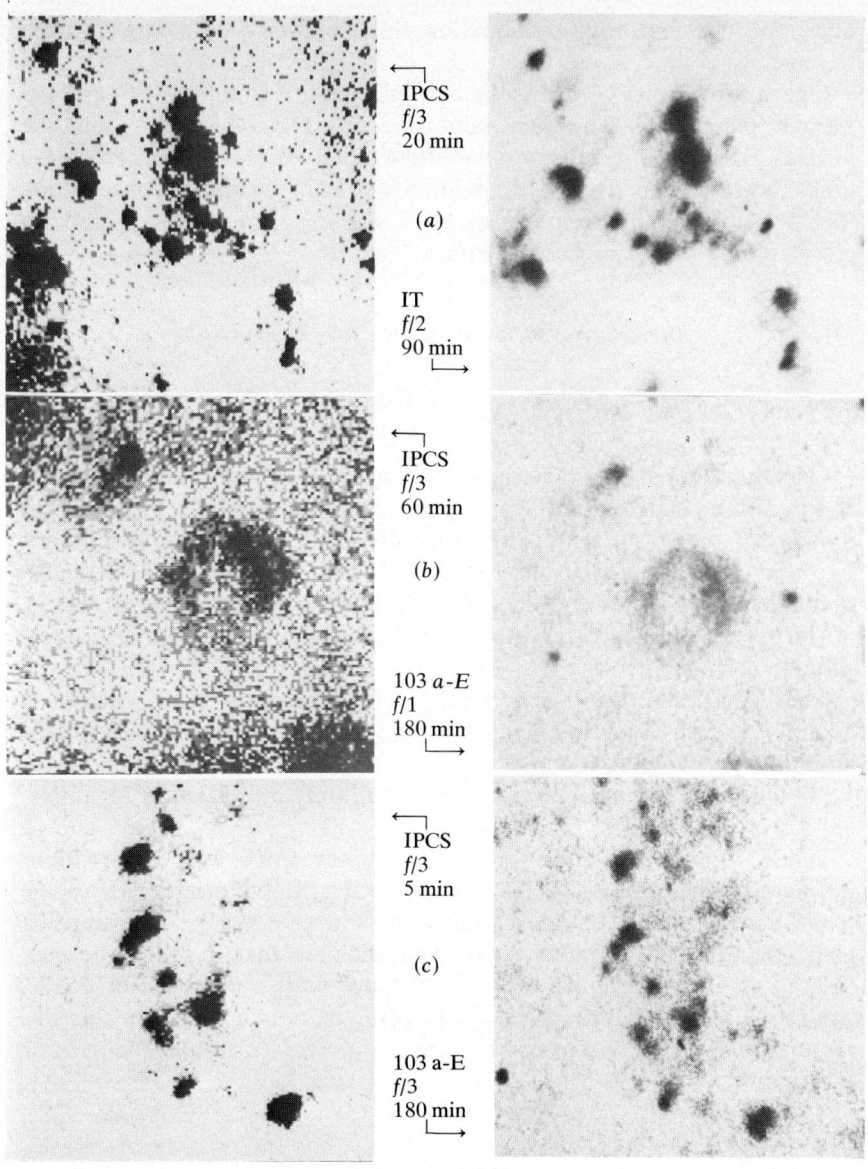

FIG. 3. Comparison of the IPCS (left hand images) with other detectors in observations on M33. (a) RCA cascade image tube; (b) and (c) Eastman Kodak 103a-E plates. Exposure times and telescope apertures as shown.

here, for the first time invalidating the possibility of a stream from the nucleus.

Figure 3(b) shows comparisons between photon counting (f/3) and the 103a-E plate (f/1). The very faint ring-like HII region far from the nucleus (35 arcmin) is also observed here in H_α ($\Delta\lambda = 0.5$ nm). Exposure times were respectively 1 h and 5 h 30 min. Another observation (Fig. 3(c)) of the southern arm of M33 with exposures of 5 min and 3 h gives an idea of the sensitivity gain.

Perspectives Opened by Photon Counting

Morphology and Photometry

On observations like that shown in Fig. 2, a complete morphological study can be carried out (catalogue, sizes, intensity of HII regions, drawing of the spiral arms, central details, etc.). But photon counting is also very well adapted to photometric studies. With its sensitivity gain, photon counting offers also the following advantages over other detectors.

The electronic camera: for every integration, the same photocathode is used.

The image tube: the digitized images are always the same, each pixel being, for each integration, at the same location on the sky (image multiplex).

Photographic plates: image multiplex and only one detector for different wavelengths.

Figure 4 gives an illustration of multiplex work with observations through different filters (H_α, H_β, [OIII], [NII], [SII]) of a large HII region (NGC 604) in M33. Four excitation knots were found, confirming the giant character of this HII region and showing that it cannot be considered as a statistical criterion for distance scale calibration of nearby galaxies. Figure 5 shows typical photometric work in a very faint ring-like region in the northern part of M33 where one can see a double pattern on the right.[2]

Interferometry

Three observations confirm the sensitivity of the system. (i) Measurement of diffuse H_α rings in the disc of M33 for which 7 h, exposures are needed at f/1 with a 103a-E plate. The signal is good in 30 min at f/3.

(ii) [OIII] detection in the disc of M33 in 3 h. Photometry leads to an interpretation of the excitation mechanism of the interarm gas.

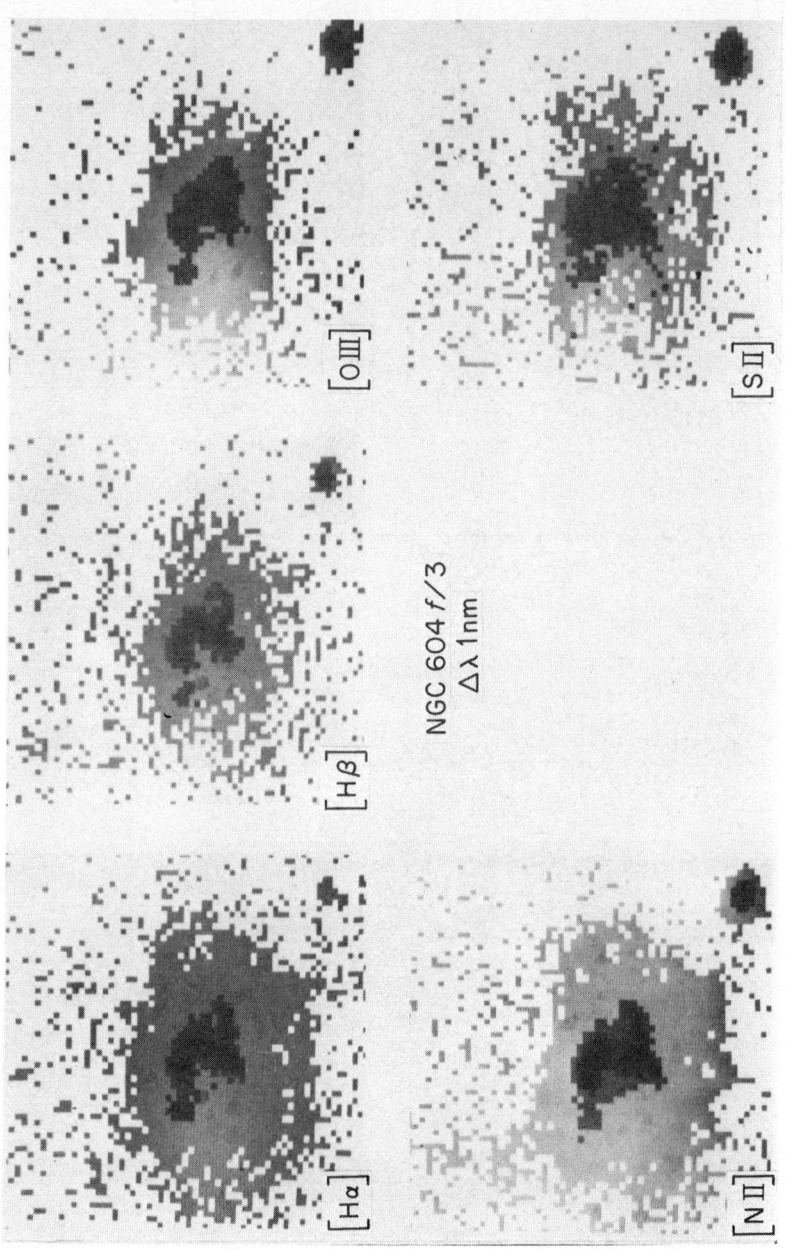

FIG. 4. Example of image multiplex (NGC 604 in several emission lines).

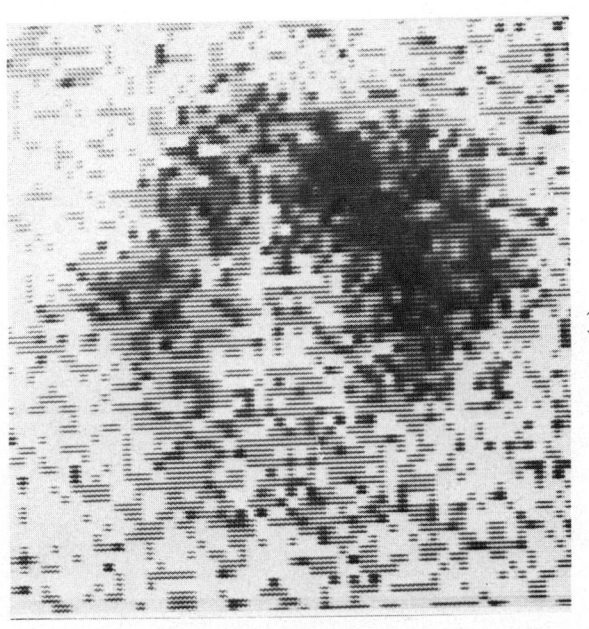

FIG. 5. Example of H_α photometric work on a very faint HII region in M33. (a) IPCS image; (b) contour plot from IPCS data.

(iii) [NII] observation of NGC 604 with a high interference order ($p = 4800$) showing in 1 h 30 min of integration the central kinetic turbulence and a double [NII] ring.

Spectrography

Observations with the nebular spectrograph of the Marseille Observatory were carried out.

In the red (2·3 nm mm^{-1}, $f/1·6$): in Fig. 6 can be seen the central part of the spiral galaxy M51 (from bottom to top: [NII] 6548 nm, H$_\alpha$, [NII] 6584 nm), with a large rotational effect (curvature of the lines). The H$_\alpha$/[NII] ratio can also be computed: in HII regions it is >1, but in the nucleus it is, as expected, <1. A good test of linearity consists in

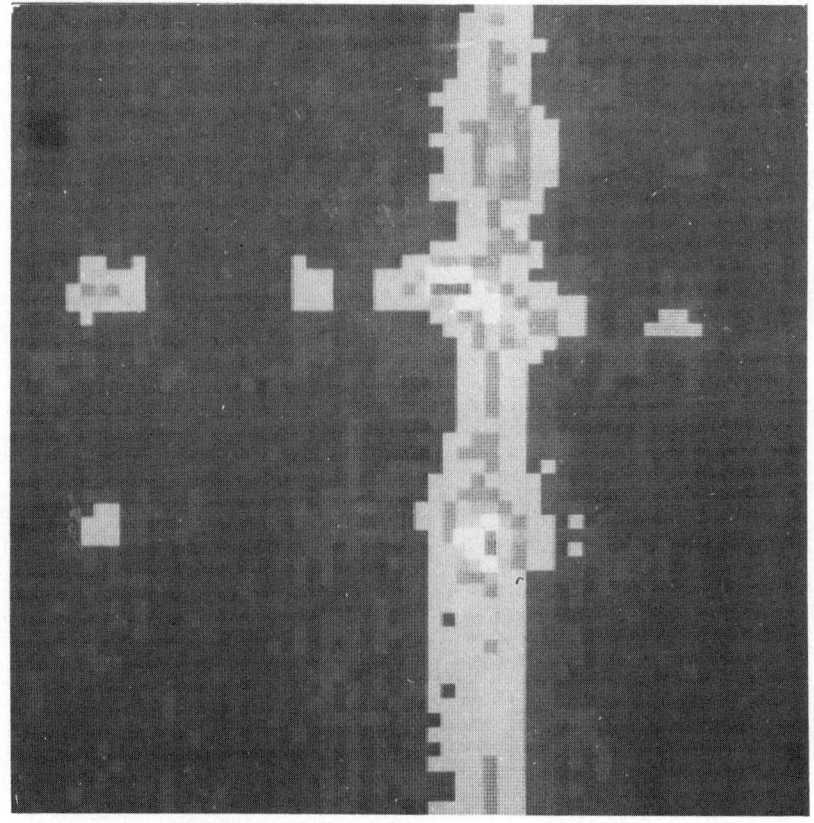

FIG. 6. Spectrographic integration of M51 (3 nm mm^{-1}).

computing the ratio of the [NII] lines, which is theoretically 3; it is found to lie between 2·92 and 3·03. The interarm medium was also detected in 1 h 30 min of integration.

In the blue (1·8 nm mm^{-1}, $f/1\cdot 5$), central parts of galaxies (NGC 4736, NGC 7331) were observed in absorption showing numerous metallic lines.

In the red (8 nm mm^{-1}) quasars of $m_V = 17$ were observed in 15 min.

Scanning Fabry-Perot Interferometer

Such an instrument, with multiplexing of several tens of images, needs a real time detector such as photon counting. This instrument is presently being built at the Marseille Observatory for further observations with the 3·6 m telescope.

ACKNOWLEDGMENTS

These observations are the result of cooperative work between the Observatoire de Marseille and the Laboratoire d'Astronomie Spatiale, which has built the indispensible centroiding electronics.

I am most indebted to MM. Cenalmor, Nguyen and Perrin (LAS) and MM. Di Biaggio and Fort (OM) for their efficient technical support.

REFERENCES

1. Boksenberg, A., *In* "Astronomical Use of Television Type Image Sensors", Princeton Symposium (NASA SP-256) (1970).
2. Boksenberg, A., *In* "Astronomical Applications of Image Detectors with Linear Response" (IAU Colloquium No. 40), ed. by M. Duchesne and G. Lelièvre p. 13–1. Paris-Meudon Observatory (1976).
3. Blazit, A., Bonneau, D., Koechlin, L. and Labeyrie, A., *Astrophys. J. Lett.* **214**, L79 (1977).
4. Cenalmor, V., *In* "Astronomical Applications of Image Detectors with Linear Response" (IAU Colloquium No. 40), ed. by M. Duchesne and G. Lelièvre, p. 16–1. Paris-Meudon Observatory (1976).
5. Lamy, Ph., Nguyen, T. T. and Perrin, J. M., *In* "Astronomical Applications of Image Detectors with Linear Response" (IAU Colloquium No. 40), ed. by M. Duchesne and G. Lelièvre, p. 17–1. Paris-Meudon Observatory (1976).
6. Bijaoui, A. and Boulesteix, J., "Journées d'études du CDCA" (A1-A5-H2), F-Lans-le-Bourg (1977).
7. Deharveng, J. M., Thèse de Doctorat, Université de Provence (1971).
8. Deharveng, J. M. and Pellet, A., *Astron. & Astrophys.* **9**, 181 (1970).

Photon Counting with Intensified Solid State Arrays

T. E. STAPINSKI, A. W. RODGERS and M. J. ELLIS

Mount Stromlo and Siding Spring Observatories, Australian National University, Canberra, Australia

INTRODUCTION

The rapidly increasing sky brightness due to city lighting near Mount Stromlo has made the need for a linear response, photon counting, high resolution spectroscopic detector system, a crucial one to maintain the site as a productive astronomical facility. The Observatories have therefore developed a detector system based on the work of Shectman and Hiltner.[1] The aim of this programme has been, firstly to place in operation the system based on intensified linear arrays (Reticons†) and secondly to develop the system to have a 2-dimensional capability using CCD arrays. This second development is aimed at astronomical problems involving two-dimensional formats, such as echelle spectrophotometry. The current status of this programme is that the intensified Reticon array system is in regular use on both the Cassegrain and coudé spectrographs of the 1·9 m telescope at Mount Stromlo and the testing of the intensified CCD system is underway. This paper describes the systems in use and the proposed development.

THE PHOTON COUNTING ARRAY (PCA) SYSTEM

The PCA system has the following objectives: (i) count every photoelectron; (ii) count these only once; (iii) centre the photon event pulse to obtain maximum resolution; (iv) eliminate all non-photon noise. The Shectman–Hiltner[1] concept appeared to us to meet these specifications and have most promise of further development.

The system consists of the following elements. The spectrum is imaged onto a tandem stack of six electrostatically focused, fibre optically coupled, 40 mm diameter image tubes. This is assembled from individual

---

† Reticon Corp., 450E Middlefield Rd., Mountain View, CA 94040, U.S.A.

modules which are interchangeable to give a choice of first tube and photocathode response. Electrostatic tubes also have the advantage of being relatively ion free, available in large numbers and cheap. The stack is run at 62 kV and provides a light gain of 10^7. The photon pulse half width at the final P·20 phosphor is about 50 μm and the phosphor image is transferred to the diode array through two Nikon $f/1.2$, 55mm lenses run back to back. A light loss of about a factor sixteen occurs in the transfer optics. The detector is the Reticon dual 1024 element array. The six tube stack introduces pincushion distortion; this is corrected by introducing barrel distortion into the optical transfer system. Mechanical adjustments are provided for focus of the transfer system and rotation and translation of the array mounting. Each element of the array is 25×430 μm^2 and the two arrays are separated by 2·5 mm.

Photon Event Detection

The image tubes provide sufficient gain so that the video pulses from the diode array are much larger than the preamplifier noise. A feature of the MOS arrays is that the video signal is superimposed on a coherent clock noise level (peak to peak value of 2×10^5 electrons). This clock noise is very stable with a RMS variation of 2000 electrons. The largest photon event pulse peaks are 8×10^5 electrons and larger events due to ions are rare; ion events are not rejected because these occur at a rate of 1/100 that of the dark noise which is 0·01 Hz per element when the first cathode is cooled to $-40°$C.

The video signal from the diode arrays is amplified, integrated and digitized to eight bits. A one-line buffer stores each frame, and the previous frame is subtracted from the current frame to remove the stable clock noise. The photon pulse spreads over approximately four elements and, because the signal to noise ratio is high, the centre of the photon event can be located with an accuracy of half an element. Thus 2048 resolution bins are obtained from each of the 1024 element arrays. The photon pulse appearing as the difference between two succeeding frames must be higher than a threshold to be counted. Phosphor decay of the last image tube has a time constant of about 10 msec and the frame time is 2 msec which is a compromise between diode integration and decay times. Thus the photon pulse lasts over several frames. The decayed photon pulses cause no problem as they are smaller than the pulse in the preceding frame and are removed in the frame subtraction circuit. The exception occurs when a photon event arrives towards the end of an integration period: the first frame pulse will be small but could be detected and the second frame pulse will be larger than the first, so that a photon could be counted twice. This problem is eliminated by gating the photon increment pulse (at the output of the photon detection circuit) with a one frame delayed version of it. A schematic block diagram of the system is shown in Fig. 1.

The two arrays are treated separately until the photon increment pulses are generated, when they are combined and sent as one stream to be accumulated in the fast incrementing memory. This memory was developed to increment a word in memory in 200 nsec.

The existing memory contains 4 K words and was designed for future expansion for the two-dimensional system. The memory is interfaced to a PDP-11 series computer through the unibus system and is equivalent to normal memory. The operating system (RT11) resides on a dual floppy disc drive; the control programs were written in Fortran with Assembler embedded subroutines for hardware control. A VDU provides an interactive graphics facility so that during the course of the observation the accumulated counts can be displayed and examined in detail. The data is recorded on magnetic tape. An observation generally consists of an exposure of the star to one array and then to the other.

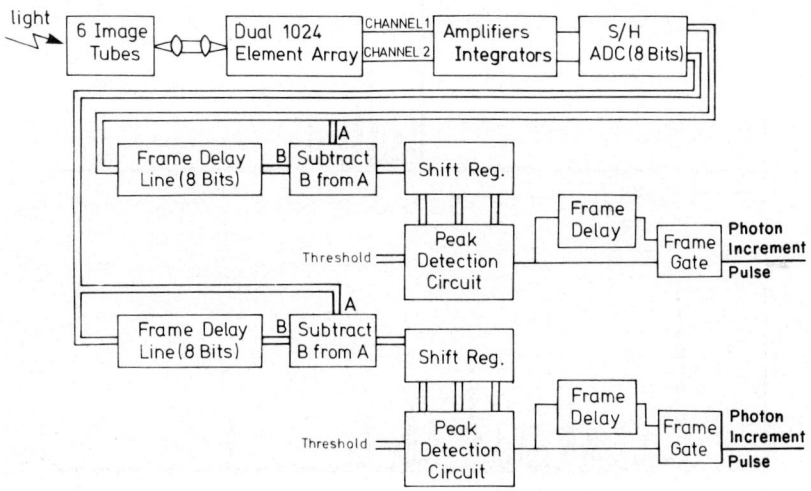

FIG. 1. Schematic block diagram of Photon Counting Array System.

Performance

To ensure the system is photoelectron counting, count rates at a variety of high voltage (on the intensifier stack) and threshold levels were applied. A plateau was found to exist at 60 kV with a threshold set at 6% of the peak video signal. With these values, a star of $B = 14·2$ gave 10 Hz nm^{-1} at 400 nm wavelength on the Cassegrain spectrograph of the 1·9 m telescope, used slitless. This is within 20% of the expected count rate. The dark current count rate is 0·005 Hz per bin. The resolution of the PCA appears to meeet the limit achievable with a memory bin size of 12·5 μm. Optical image size formed by the spectrograph camera is about 10 μm and this is degraded by the fibre optics at the first cathode, the individual fibres having diameters of about 8 μm. Figure 2 shows the spectrum of a Neon lamp obtained with a dispersion of 58 nm mm^{-1}. Doublets as small as 1·8 nm separation are resolved. The resolution at this level depends on the spatial phase relationship between the emission lines, the fibre optics and the resolution elements. The coherent noise of the PCA reflects the uniformity of the first cathode and the high frequency modulation is less than 3%. This noise plus the low spatial frequency shading can be removed using flat field data but for the

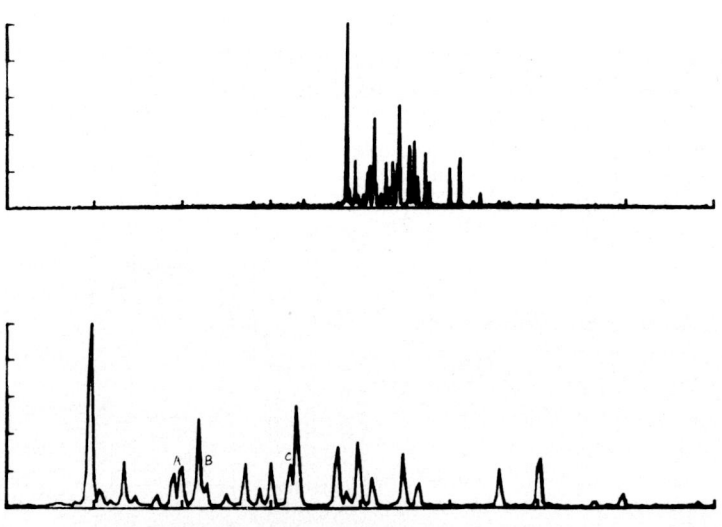

FIG. 2. Neon spectrum between 600 and 660 nm. Doublets A, B and C are 2·2, 2·0, and 1·8 nm apart. The lower spectrum is expanded × 4 over the upper one.

majority of uses this tedious procedure is not necessary. The maximum counting rate is set by two effects. Firstly there is the limit set by the frame time and secondly there is the effect of phosphor intensity pile up which causes saturation of the eight bit digitization of the video signal. The system accepts continuum counting rates of 20 Hz per bin with a 3% correction.

Observational Results

The dual linear PCA has been in regular use on the 1·9 m telescope for six months, working on a variety of programmes. Most of these involved relatively low dispersion spectroscopy of galaxies and took advantage of the sky subtraction facility of the PCA. Others have involved its use on the highest dispersion (echelle) spectrograph where the system resolution of 600 000 at 0·05 nm mm^{-1} has been used in a study of fine structure of interstellar NaID and the LiI doublet lines.

We show in Fig. 3 a hard copy of the VDU terminal display of a spectrum at an original dispersion of 28 nm mm^{-1} of a blue emission line irregular galaxy ISZ 182 covering the wavelength region of 350 to 780 nm. The galaxy emission lines of H_0, [NII], [OIII], H_β, [OII] and the

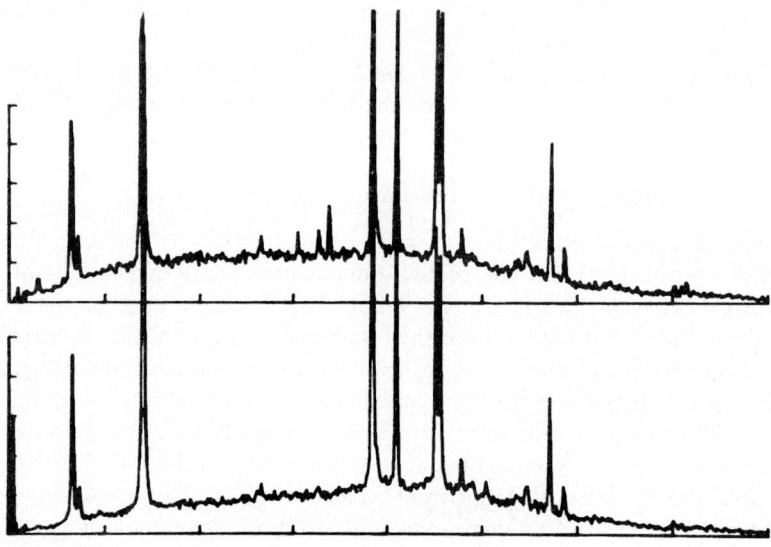

FIG. 3. Spectrum of the Zwicky blue compact galaxy IZw 182 described in the text with sky spectrum in lower panel.

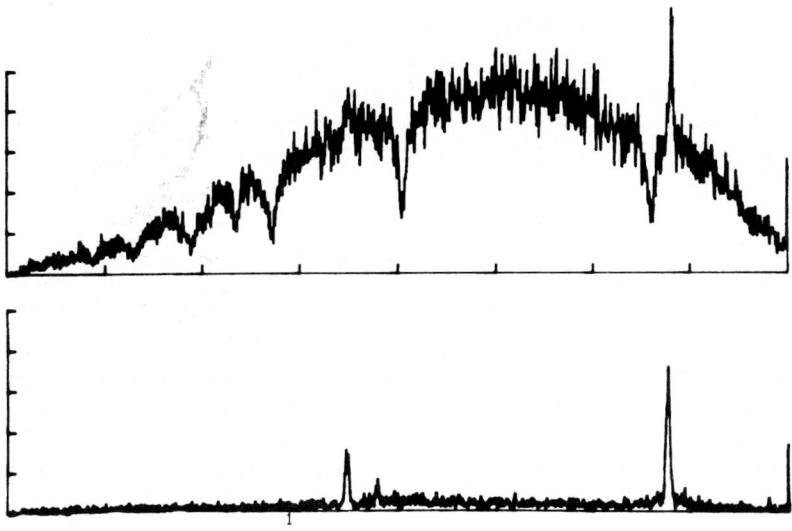

Fig. 4. Spectrum of BU Leo, 7 nm mm^{-1} dispersion. The bright sky lines are HgI 404·6 and HgI 435·8 nm.

blue continuum of this $V = 16·8$ object are easily seen together with the enormously strong city light HgI emission lines.

Figure 4 shows, by courtesy of Dr J. E. Norris, the spectrum of BU Leo, an RR Lyrae star at intermediate phase, at a dispersion of 7 nm mm^{-1}. The spectral range is from 350 to 440 nm.

DEVELOPMENT OF THE TWO-DIMENSIONAL SYSTEM

Our experience with linear arrays encourages us to believe that a similar approach to the use of two-dimensional CCD arrays will prove fruitful. For spectroscopic use we believe that a system such as the PCA where detective quantum efficiency is nearly equal to the theoretical quantum efficiency has advantages over the use of the unintensified CCD with its high theoretical QE (the advantage of which mainly is significant above 800 nm) and attendant problems of charge leakage, diode non-uniformity, high dark current and low noise preamplifier requirements. Further, the PCA has the fundamental advantage of event centring while the resolution of the CCD alone depends on diode spacing. Our system is designed to take 500×1000 resolution elements but currently we are working with a 244×180 Fairchild chip.† The logic is designed to centre

† Fairchild Semiconductors Group, 464 Ellis Street, Mountain View, CA 94042, U.S.A.

events in the direction of dispersion and at right angles to this direction. We double bin the diode elements in dispersion but not in the orthogonal direction.

Acknowledgment

We thank Mr J. Hart for his help in solving the optical problems associated with placing the PCA in operation.

Reference

1. Shectman, S. A. and Hiltner, W. A., *Publ. Astron. Soc. Pac.* **88,** 960 (1976).

Quantum Noise Limited Readout of Spectrographic Data Using Image Intensifiers and a Reticon Photodiode Array

E. K. HEGE, R. H. CROMWELL and N. J. WOOLF

Steward Observatory, University of Arizona, Tucson, Arizona, U.S.A.

Introduction

The design of one-dimensional quantum noise limited detectors for use in astronomical applications may take either a photon counting approach or an analogue detector approach. Several schemes for implementing photon counting techniques have been described including intensified image dissector systems,[1] and intensified Plumbicon systems,[2] as well as intensified Reticon systems[3] and Digicon systems.[4] Analogue detector systems including the self scanned Digicon[5] and direct detection Reticon systems[6-10] have also been reported.

Analogue detectors of the self scanned Digicon type, which provide sufficient gain so that the inevitable effects due to event spreading and readout noise are minimized,[5] have the dual advantages of quantum noise limited performance and very large dynamic range. Reticon silicon $p-n$ junction photodiode arrays have been shown[9] to respond linearly over a dynamic range of 10^4 to 1 or greater but to be limited by readout noise of about 10^3 electron–hole pairs per pixel.

Although analogue systems have a far wider dynamic range than digital systems, their signal to noise ratios may be poorer for three reasons: first, if the pulse height spread of the system is very broad, the unequal weighting of events causes an excess noise; second, at very low light levels, there is a noise contribution from cosmic rays and ion events, both of which produce very large pulses, which give more noise in analogue systems than in digital systems; third, digital systems have a pulse height discriminator which cuts out large numbers of small pulses and prevents them from adding to the noise.

In the analogue systems described here, these effects are all rather unimportant. The large gain per stage of image intensifiers, and the rather uniform photon event pulse height make the first and third items unimportant. The data acquisition algorithm identifies and reduces ion and cosmic ray events to a level similar to that of photon events.

In this paper we describe the results of laboratory tests and actual performance of several analogue detector schemes. Image intensifiers, magnetically focused, electrostatically focused or proximity focused, are the primary detector devices employed in these systems. Reticon photodiode arrays, either lens coupled or fibre optic coupled, are used for analogue conversion of the image intensifier output.

We have extended the versatility of these detectors by developing systems in which the photocathode device used as the primary photon detector and the electronic analogue readout device are separate assemblies. This yields an advantage of flexibility in the design of systems for special purposes, especially in the choice of photocathodes in order to optimize particular spectral responses and to effectively utilize existing, proven components.

We also describe the use of these detector systems in spectrophotometric observations using a Boller and Chivens spectrograph at the Cassegrain focus of the Steward Observatory 2·3 m telescope. Results of actual system performance in astronomical observation are presented. The unique features of analogue systems are emphasized, especially the large dynamic range of the system. Performance near the theoretical limit is achieved for event rates of the order of $0·1\,\text{pixel}^{-1}\,\text{sec}^{-1}$ with a detector which can accommodate event rates more than 10^5 times greater.

The Detector Systems

The detector configurations which have been evaluated are shown schematically in Fig. 1. Performance characteristics of image tubes similar to those used in these systems are reported by Cromwell and Dyvig.[11]

System A consists of four Varo type 8605 electrostatically focused tubes coupled fibre optic output to fibre optic input and potted in a silicon based RTV compound using techniques developed at Steward Observatory to make the multistage tube operate reliably.† The photocathode quantum efficiency of the first stage is selected for the overall spectral region 500 to 800 nm and is better than 10% over 480 nm to 600 nm, dropping gradually to about 5% at 800 nm. The four stage intensifier produces about 3×10^7 photons at the final phosphor output for each primary photoelectron. This gives ample gain for transfer lens coupling to a solid state readout device. The tube is housed in a magnetic shield comprised of low carbon steel and mu-metal.

Systems B and C both utilize the four stage intensifier of system A to which a 1:2 fibre optic magnifier boule has been attached. These systems were constructed specifically to intercompare the performance of a Varo

† See p. 183.

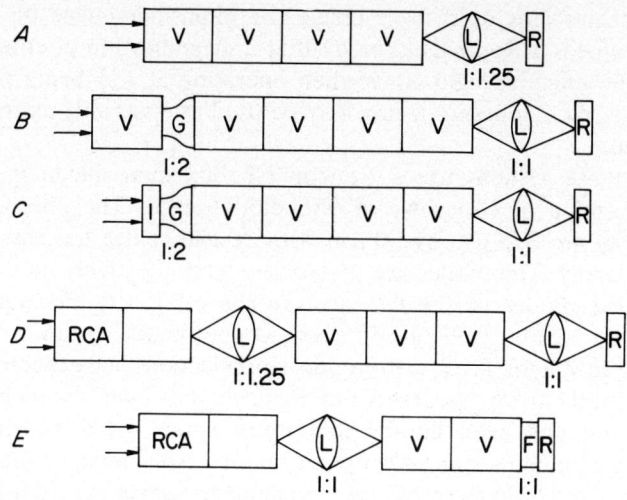

FIG. 1. Schematic diagrams of detector configurations studied. V = Varo type 8605. L = Repro-Nikkor f/1·0. R = Reticon CP1001. I = ITT F-4109. G = Gallileo D-80 fibre optic magnifier. RCA = RCA C33063 two-stage intensifier. F = fibre optic image transfer.

8605 electrostatic intensifier and an ITT F-4109 proximity focused intensifier.

Systems D and E utilize a two stage magnetically focused RCA type C33063 intensifier with 38 mm active diameter S·20 type photocathode deposited on a sapphire window and processed for especially good blue response. The quantum efficiency in the region 300 nm to 400 nm is about 22%. The output phosphor releases approximately 4×10^4 photons for each primary photoelectron which is insufficient, when transfer lens coupled, to drive most currently available solid state readout devices. The magnetic field is provided by a specially designed array of permanent magnets and pole pieces which produce a uniform field with short extension beyond the photocathode to enable use of the system with a semi-solid Cassegrain Schmidt camera, a 120 cm Ritchey–Chretien, or a 40 cm folded off-axis Schmidt.

Two intensifier booster systems have been fabricated to provide amplification of the RCA output image. They are both magnetically shielded systems composed of Varo intensifiers constructed similarly to the four stage system A. System D uses a three stage booster to provide sufficient gain for transfer lens coupling to the solid state detector. System E requires only two subsequent stages to produce sufficient gain when the image is directly coupled to the surface of the solid state detector using fibre optics.

The transfer lens used to re-image the intensifier image on the solid state detector is a Repro-Nikkor $f/1·0$ at a magnification of either $1:1$ or $1:1·25$, the effective light cone when operating at $1:1$ being $f/2·0$. The lens has an iris diaphragm which may be used as a variable attenuator for the system.

All of these systems use a Reticon CP1001 dual silicon $p-n$ diode array consisting of two rows of 936 diodes each. The photodiodes in either array are 375 μm by 30 μm with 30 μm centre to centre spacing along the array. The diodes are juxtaposed with effectively no dead space between the diodes or the two arrays. The electronic and photometric properties of similar Reticon self scanned photodiode arrays used in high precision, low light level astronomical applications have been well described.[9] Both arrays are read out through only two channels; corresponding odd and even numbered outputs share the same channel by being wire summed at the preamplifier inputs. After independent filtering and sample and hold circuitry, the two analogue channels are multiplexed by an analogue gate and digitized by a 14 bit analogue to digital converter. The data are read out of the arrays at 4 kHz each channel, giving an effective data rate of 8 kHz. The digitized output is acquired by a Data General Nova 800 minicomputer system with 32 K words memory which also controls the scanner logic and associated observing instrumentation

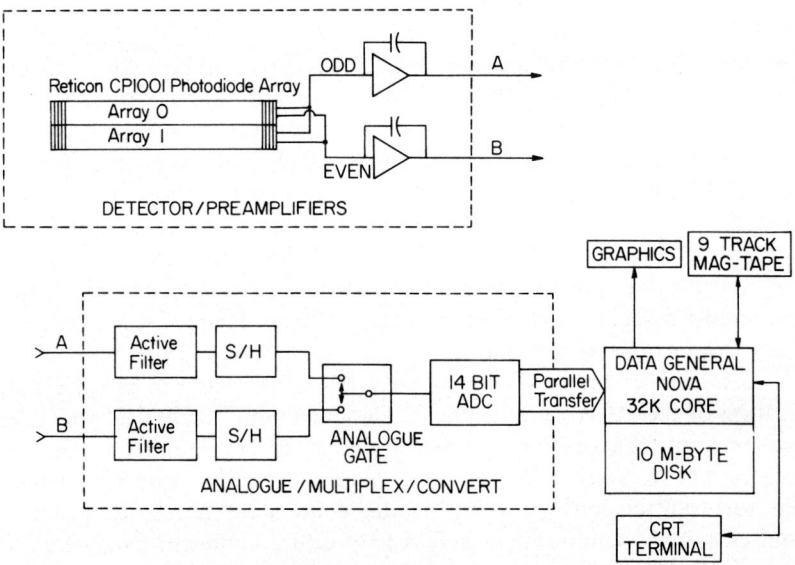

FIG. 2. Schematic of data readout system.

and supervises the observing programme. The system includes a 10 M-byte disc, nine-track magnetic tape for data recording and Tektronix vector graphics for data display. The data acquisition system is shown schematically in Fig. 2.

In all systems the image tubes are cooled by circulating a refrigerated liquid through tubing in thermal contact with the flanges and potting materials of the image tube assemblies. The packaging of the Varo intensifiers includes pairs of magnetic deflection coils (substepping coils). Mechanical motions for precise alignment and focusing of the solid state array are provided.

The Reticon array is cooled to −30°C by thermal contact with a copper heat sink cooled in turn by an R-22 refrigeration system. No temperature control other than the evaporation of the R-22 refrigerant at constant pressure is provided. At this temperature the photodiode dark current[8,9] is well suppressed and 30 sec on-chip integrations are feasible.

The Observing Algorithm

Short period (usually 0·5 to 30 sec) on-chip integrations during which the image tube phosphor output directly discharges the photodiode pixels are obtained by reading the charge necessary to recharge the photodiode pixels. Successive short scans are summed for longer intervals in the computer memory until the desired signal to noise ratio is achieved.

The dual array observing programme provides for observation of image tube and diode array dark currents and sky background simultaneously by observation of the object through two apertures positioned to correspond to the centres of the two diode arrays as the object is placed alternatively in the apertures. The programme enables use of the full dynamic range of the system, provides a substepped oversampling of the instrumental resolution profile by magnetically deflecting the signal past the 30 μm wide photodiodes in 15 μm substeps, provides a deglitching algorithm for suppression of image tube ion events, nullifies the effects of baseline drifts induced by small drifts in detector temperature, and averages the responses of the two analogue channels. The detector readout is synchronized with the 60 Hz power mains so that any residual 60 Hz component in the analogue electronics will be removed by the background subtraction algorithm. The signal is oscillated optically perpendicular to the dispersion, along the long dimension of the diodes, for spatial averaging in order to reduce the high frequency components of the image tube fibre optic patterns as well as to smooth over other spatial non-uniformities.

Short, identical readouts of the image tube output permit cancellation of the image tube ion event noise spikes which typically produce signals about 50 times those of photoelectron events. A given readout is compared element by element to one separated by one readout interval in order to avoid confusion by phosphor afterglow induced by these anomalously large events. Data elements in either readout which deviate by more than, typically, three standard deviations from the mean of the difference are replaced by the corresponding elements in the readout: data always replaces data—no arbitrary replacements are made. This deglitching procedure produces an algorithm in which the basic scan unit is four identical readouts of either signal plus background or background alone. The observing kernel is thus two sets of such groups of four deglitched readouts with the roles of the two arrays complemented.

A complete observing cycle then consists of a set of replications of the kernel observation at each of the substepped positions. In order to eliminate odd–even effects in the data due to mismatch in the two readout channels, four half-diode substeps are necessary to permit each element of the signal to be observed by two adjacent diodes. The program is completely symmetrized in that the substepped positions are repeated in reverse sequence returning to the original position. This cycle is the basic signal integration unit requiring from 1 to 30 min, depending upon the on-chip integration time, for four substep positions. A single observation may consist of one or more of these cycles.

This observing algorithm is presented schematically as a set of operational sequences along a time line in Fig. 3. The four dual array detector scans are contiguous with no dead time between them. The spectrograph apertures are normally placed along the east–west axis so that the telescope wobbling is done in right ascension to avoid coherent wear in the telescope drive train. The 20 arcsec wobble takes 0·5 sec at 0° declination which inserts an unavoidable dead time between the groups of four scans. Duty cycles of 95% or better are usually achieved, however. The observing software programmed in FORTH provides two observing modes, STAR/SKY and CALIBRATE. A mode flag is set to choose the appropriate form of the observing kernel. In STAR/SKY mode the telescope position is complemented after each group of four scans and the shutter remains open. T-PAUSE is set to correspond to the telescope wobble time. After each group of four scans in CALIBRATE mode the shutter position is complemented and the telescope position is stable.

The oscillation of the two optical beams is accomplished by using a stepping motor to rock a quartz optical flat mounted nominally perpendicular to the optical axis immediately behind the apertures. The amplitude of the oscillation is chosen to be comparable to the magnitude of

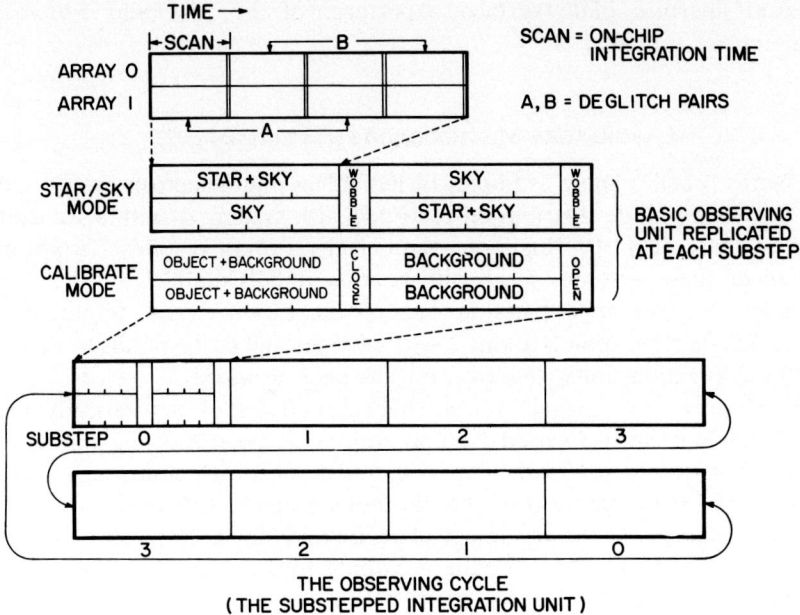

Fig. 3. Schematic diagram of observing algorithm.

the translation produced by the substep coils. The oscillation position is altered after each detector scan.

The image tube fibre optics, variations in individual photodiode responses, and other non-uniformities induce spatial fixed pattern modulations in the detector readout. These are mostly high frequency modulations and may be removed by a calibration with a continuum spectrum observed with good statistics. To establish an absolute instrument response function, spectrophotometric standard stars must also be observed. The wavelength scale must be established by observing a known emission line source. Because of the large pincushion distortion of the electrostatic image tubes, a third degree polynomial is required to fit the wavelength scale with sufficient accuracy. Models for atmospheric extinction[12] are used to make air mass corrections. The data reductions, also programmed in FORTH, which apply these calibrations and corrections in order to produce photometric results are done off-line on a second Data General Nova minicomputer system.

The image is manually guided into the circular spectrograph apertures using an intensified TV system† to view the star field reflected from the

† See p. 339.

polished aperture plate surface. Apertures of 2·5, 3·5 and 5 arcsec diameter are provided.

Laboratory Measurements of Performance

Figures 4 and 5 show examples of individual photoelectron events and the event amplitude distribution recorded with system A with a transfer lens aperture of $f/1\cdot 0$. This large photoelectron event response is typical for all of these systems. Each quantization unit is equivalent to about $2\cdot 5 \times 10^3$ electron–hole pairs in one of the $30 \times 375\ \mu\text{m}^2$ photodiode pixels. Diode saturation is about 2×10^7 charges and corresponds to about 8000 quantization units (the 14th bit has been dropped).

A comparison of signal induced background for electrostatically focused and proximity focused systems was made by projecting onto the array a narrow slit perpendicular to the array of length sufficient to just overfill the long dimension of the photodiode pixels. Integrations of the detector response to this slit image show a relatively large (~16%) signal induced background effect for the proximity focused device of system B compared to about 1% for the electrostatically focused input of system C.

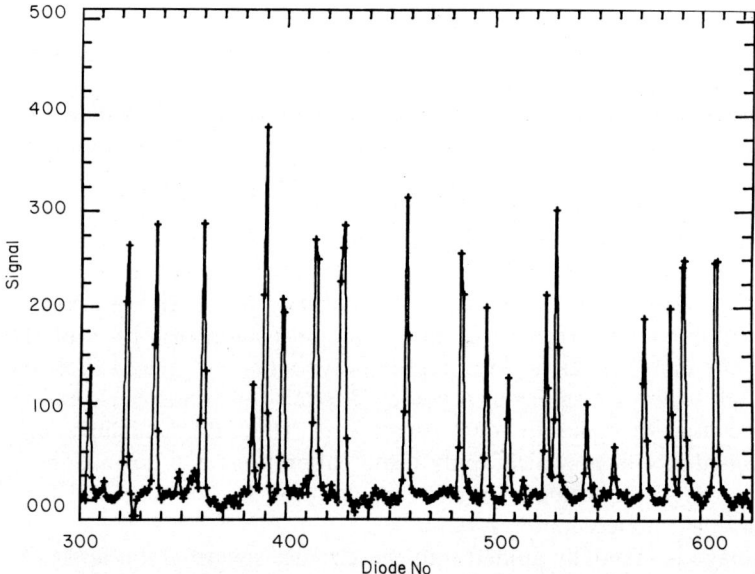

Fig. 4. Individual photoelectron events observed in a single 0·5 sec on-chip integration using system A with transfer lens at $f/1\cdot 0$.

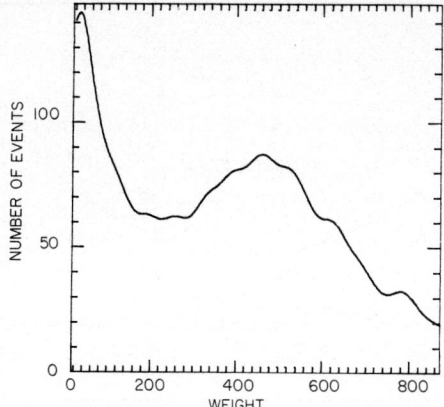

FIG. 5. Pulse-height distribution from an ensemble of observations such as shown in Fig. 4. A smoothed histogram of events with weight equalling sum of signal in five channels centred on event peak for system A.

Systems A, D, and E respond similarly to system C for which the signal-induced background effect is negligible in most applications. For proximity focused systems, significant corrections must be applied to compensate for such background effects.

The resolution of systems A and D was measured by projecting an unresolved spot (25 µm diameter) of light onto the photocathode and measuring the FWHM of the resulting output as a function of the position of the spot on the photocathode. The incremental magnification (local magnification of a small image element) as a function of photocathode position was also measured. The results referred to the primary photocathode for the red sensitive system A and the blue sensitive system D are summarized in Fig. 6. The resolution of these systems is somewhat poorer than the 40 µm resolution of the Cassegrain spectrograph with 2·5 arcsec entrance aperture. The effective mean resolution, referred to the photocathode, of these systems when used in conjunction with the Cassegrain spectrograph is 64 µm for the red sensitive system A and 72 µm for the blue sensitive system D, the difference being due to differences in system magnification.

Absolute photocathode quantum efficiencies have been measured for the primary photocathodes of these systems. A well stabilized incandescent lamp with narrow bank filtered output was used to project a highly uniform, calibrated photon flux onto the photocathode surface. The resulting photocathode current produced in the intensifier diode was measured directly to determine the number of photoelectrons produced.

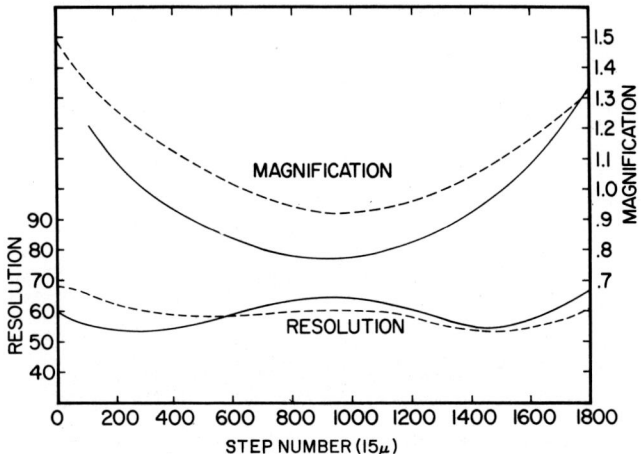

FIG. 6. Resolution and magnification. System A solid lines. System D broken lines. Resolution is read from left hand scale; incremental magnification from right hand scale. Both resolution and magnification are referred to the photocathode.

This gives the absolute photocathode quantum efficiency η for the corresponding filter bandpass. The resulting quantum efficiency curves are shown in Fig. 7.

A sequence of 100 observations of the total system response to the same calibrated photon flux was made for each of systems A, B and C in order to determine their detective quantum efficiencies (DQE). Each observation was well separated in time from the previous one so that image intensifier phosphor output lag would not produce statistical correlations in time. In order to accommodate the spatial spreading of the incident point process events, groups of individual pixel responses were summed into larger bins for statistical analysis. Stochastic sequences of the signal in particular bins were prepared from the set of 100 observations. From the mean value of the signal in a bin, S, and the standard deviation of that signal, σ_S, the number of events in the signal was inferred from the assumption that a Poisson process was sampled: $N = (S/\sigma_S)^2$. From this statistically determined event rate and the measured system magnification, an event rate referred to the photocathode was determined.

From the ratio of the statistically determined event rate to the measured photoelectron flux a Relative Quantum Efficiency (RQE), referred to the absolute quantum efficiency of the photocathode, results. If this ratio approaches unity, we say that the system is "quantum noise

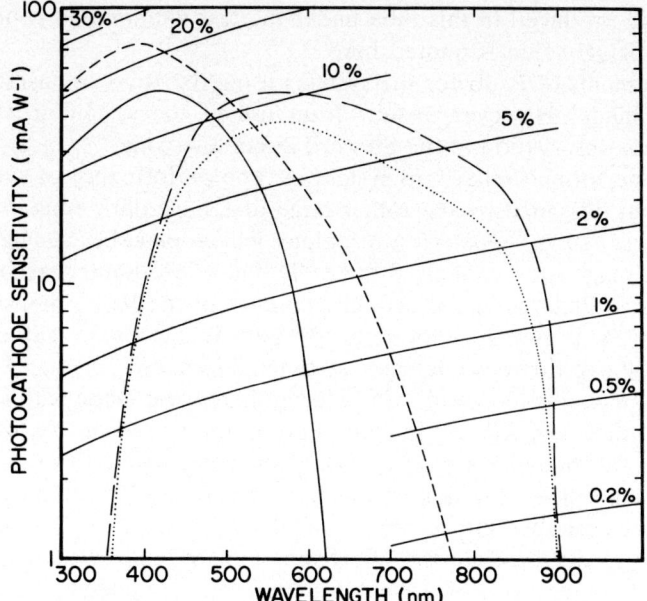

FIG. 7. Photocathode sensitivities. The measured photocathode sensitivities for Steward Observatory intensifiers: Varo type 8605 (long dashes), a second Varo type 8605 (dots), ITT type F-4109 (solid line), and RCA type C33063 (short dashes). Corresponding photocathode responsive quantum efficiencies may be read using curves labelled %.

limited". The results of these measurements are given for systems A and B in Table I. These results show performance impressively close to the shot noise limit. To within the precision of the measurements (about 10%), RQE is unity for both systems A and B. The statistical photoelectron flux measurement for the proximity focused tube is complicated by the presence of a uniquely large (estimated about 25%) signal induced

TABLE I

Relative quantum efficiencies of three systems derived from absolutely determined ($A(N)$) and statistically determined ($S(N)$) photoelectron fluxes

System	$A(N)$	$S(N)$	RQE
A	$6 \cdot 18 \times 10^5$ e cm^{-2} sec^{-1}	$5 \cdot 60 \times 10^5$ e cm^{-2} sec^{-1}	0·91
B	$5 \cdot 67 \times 10^5$	$5 \cdot 55 \times 10^5$	0·98
C	$7 \cdot 09 \times 10^5$	$[6 \cdot 26 \times 10^5]$†	–

† Value not determined due to influence of signal induced background (see text).

background produced in this tube under the test illumination conditions. Hence no RQE value is quoted here.

Measurements of RQE for the systems using the RCA intensifier have not been made. However, results from use of the system in spectrophotometric observation at the Steward 2·3 m telescope[13-15] indicate that system E performs similarly to system A. The performance of system D is somewhat degraded by the rather large first stage dark emission noise of the three stage booster: a problem which may be alleviated by improved image tube cooling. Systems D and E are limited at low light levels by scintillations in the sapphire window of the RCA intensifier.

Table II is a summary of detector performance data gathered as described above for these detector systems. All measurements of system resolution, magnification and gain reported here were made with a 25 nm bandpass filter centred at 425 nm. Photocathode responsive quantum efficiencies at 600 nm are also listed for comparison. The detective quantum efficiencies for this wavelength region can be computed from the RQE as described above.

In order to demonstrate the effectiveness of the ion event deglitching procedure, observations of a weakly illuminated graduated density spot pattern were made using an uncooled Varo system. Figure 8 shows ten successive such observations superimposed with and without the deglitching procedure. Both the necessity for and the effectiveness of the deglitching procedure are well illustrated. The large ion event contributions

TABLE II
Summary of detector performance

System	Photocathode	M	FWHM	$\bar{g}$†	R_D	η_B	η_R
A	Varo 8605 No. 15999	0·857	55 μm	230	0·055 sec^{-1}	0·055	0·10
B	Varo 8605 No. 18900	1·704	50	1600	0·050	0·050	0·10
C	ITT F-4109	1·863	48	~1000	0·030	0·15	0·008
D	RCA C33063	0·862	60	135‡	0·015	0·20	0·04
E	RCA C33063	0·90	48	125	0·015	0·20	0·04

M = Mean magnification for diodes 300–600: $\Delta L_{array}/\Delta L_{photocathode}$; FWHM = mean value of unresolved image referred to photocathode; $\bar{g}$ = mean value of photoelectronic gain at $f/1\cdot4$ ($\bar{g}$ is centroid of pulse height distribution, e.g. Fig. 5); R_D = photocathode dark event rate per pixel with photocathode at 0°C; η_B = responsive photocathode efficiency at 425 nm; η_R = responsive photocathode efficiency at 600 nm.

† Pixel saturation = 8000 units corresponds to 2×10^7 electron–hole pairs. The quantization unit is nearly equal to the electronic readout noise = $2\cdot5 \times 10^3$ electron–hole pairs.
‡ First transfer lens at $f/2\cdot0$.

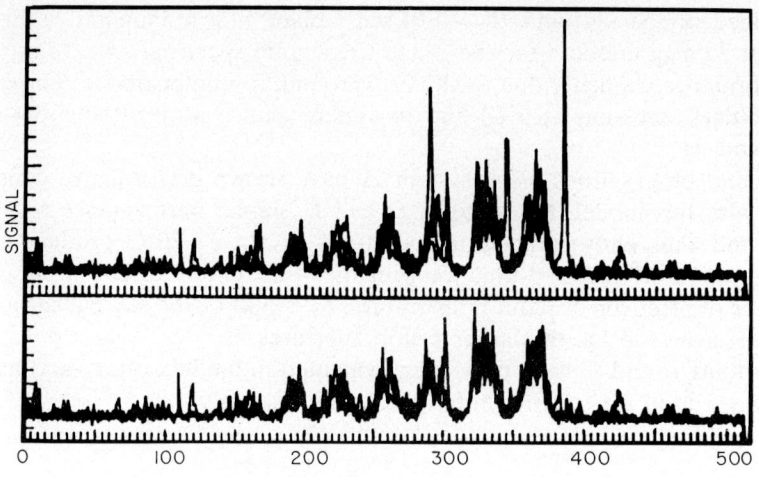

FIG. 8. Test of deglitching algorithm. The upper curves show data before deglitching which include several image tube ion events. The lower curves show the deglitched results.

are typically reduced to a signal comparable to two or three photoelectron events. The central peaks of the glitches are completely removed.

The expected relative system efficiency predicted for a typical application in which system A is the detector is modelled graphically by Fig. 9 for a photocathode dark emission rate of $R_D(0°C) = 0·1 \sec^{-1} \text{pixel}^{-1}$. A very dark sky under conditions of very good seeing was assumed to

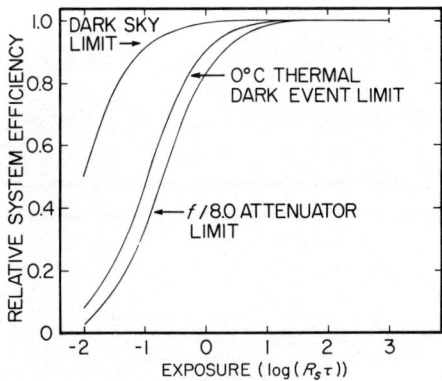

FIG. 9. A performance model for system A. The relative system efficiency is an estimate of the factor by which the signal to noise ratio expected in observations at the telescope will differ from the quantum noise limit of the detector. R_S is the source photon rate and τ the exposure time.

produce sky background $R_B = 0\cdot01\,\text{sec}^{-1}\,\text{pixel}^{-1}$ (at $4\cdot5\,\text{nm nm}^{-1}$, this is about 22 mag arcsec^{-2} for the 2·3 m Cassegrain spectrograph). Limits of performance explicitly due to sky background, to photocathode temperature (dark emission at 0°C) and to system gain (f-stop attenuator) are shown.

Actual observations using system A have shown performance consistent with this model. For systems D and E, similar performance results. We find that photocathode temperatures of $T_D = -20°C$ produce sky background limited performance with the Varo photocathode of system A. For the Reticon readout temperature $T_R = -30°C$, the sky background limit is achieved for the larger f-stop apertures.

Systems B and C have only been evaluated in the laboratory and have not been used for spectrophotometric observations.

Observations at the 2·2 m Telescope

Figure 10 shows a 7·5 min integration of the $m_v = 10\cdot8$ spectrographic standard star BD +33 2642 using system D with the Cassegrain spectrograph. The dispersion is such that the instrumental resolution is 0·7 nm

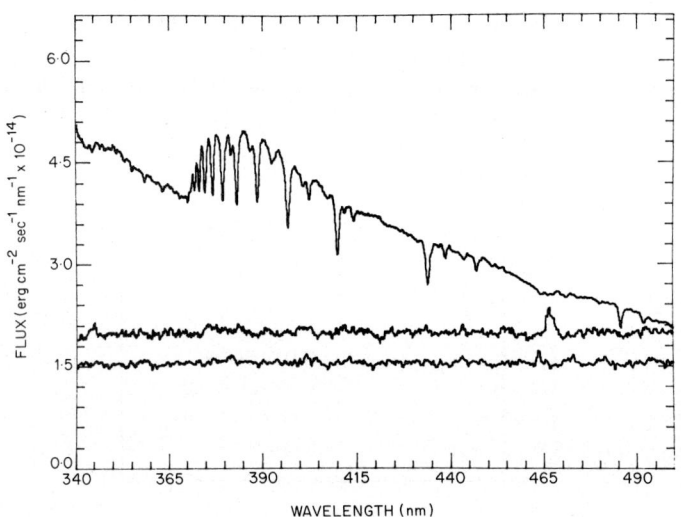

Fig. 10. BD +33 2642 observed with system D. The upper curve shows the reduced result. The lower two curves show the diode response functions, which also include the fibre optics contribution, for the two readout arrays (the curve for one array is arbitrarily shifted upward for display).

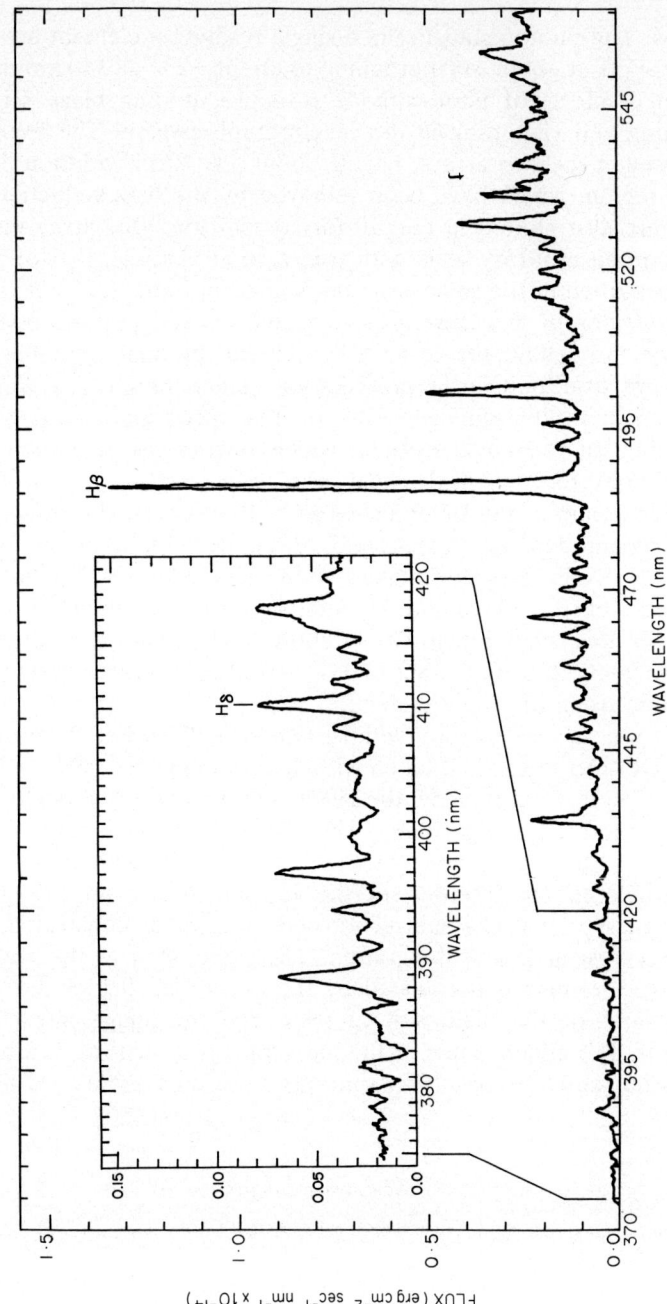

Fig. 11. MWC 349 observed with system E. The region 375 to 422·5 nm is shown on expanded scales in the inset.

FWHM. The photon statistics in a single resolution element are approximately $\pm\frac{1}{2}\%$ at about 400 nm falling to about $\pm 2\%$ at 340 nm due to the combined effects of photocathode response, grating blaze image tube vignetting and vignetting in the spectrograph camera. The fixed pattern responses of the two arrays, having about 5% RMS variation over most of the region, which have been removed by the data reductions procedures, are also plotted in Fig. 10 for comparison. One array response is shown on an arbitrary scale with true zero at abscissa, that of the other array is plotted at the same scale but shifted upwards for comparison. At least a factor of five suppression of these fixed pattern responses is achieved throughout the co-added result for the dual array observation.

The practical limit in the precision of systems D and E is imposed by small magnetically induced shifts of the spectrum, of 10 to 30 μm depending on telescope attitude, which restrict the reduction of fixed pattern noise to $\pm 1\%$ in the final result.

The spectrum of the IR object MWC 349 has been chosen to illustrate the excellent dynamic response of these detector systems. Figure 11 shows an 18 min integration taken on the 25 August 1977 with system E. The ratio of the maximum H_β emission to the minimum significant feature in this single integration is 3000:1. H_α emission, with a line flux approximately 250 times that of H_β, was also observed directly without saturation using the same system.

Faint object spectroscopy with system E is illustrated by observations of the DQ Her nova shell taken with system E in June 1977.[16] Figures 2 and 3 of Reference 16 show the effectiveness of both the deglitching and the sky subtraction algorithms. The intensity of the night sky emission observed with the same system shows [OI] 557·7 nm with intensity roughly comparable to the intensity of the H_α emission from the shell. The effectiveness of the sky subtraction technique is demonstrated; no [OI] emission is seen in the integration. The noise seen in the baselines of these observations is dominated by the event statistics of the zero sum subtractions of the night sky emission and the image intensifier dark emission. The effectiveness of the ion suppression scheme is also evident in that no randomly occurring spurious emission features are seen.

Acknowledgments

This work was supported in part by NSF grant AST-75-00525 and in part by AFGL grant F19628-77-C-0063.

The contributions of G. R. Gilbert to the design of the readout electronics system as well as his encouragement in the initiation of this project are greatly appreciated. Also, the helpful and constructive criticisms and discussions with P. A. Strittmatter and J. R. P. Angel contributed much. The data acquisition programmes reflect the patient counselling of T. A. Sargent, and the data reduction programming is largely the work of R. L. Moore.

Finally E. K. Hege expresses thanks to Hollins College for sabbatical year support during the initiation of this detector development project and to E. A. Periman for helpful discussions of the statistics of point processes.

REFERENCES

1. Miller, J., Robinson, L. and Wampler, E., *In* "Adv. E.E.P." Vol. 40B, p. 693 (1976).
2. Boksenberg, A. and Burgess, D., *In* "Adv. E.E.P." Vol. 33B, p. 835 (1972).
3. Schectman, S. and Hiltner, W., *Publ. Astron. Soc. Pac.* **88,** 968 (1976).
4. Beaver, E., Harms, R. and Schmidt, G., *In* "Adv. E.E.P." Vol. 40B, p. 745 (1976).
5. Tull, R., Choisser, J. and Snow, E., *Appl. Opt.* **14,** 1182 (1975).
6. Weckler, G., *IEEE J. Solid-State Circuits* **SC-2,** 65 (1967).
7. Livingston, W., Harvey, J., Slaughter, C. and Trumbo, D., *Appl. Opt.* **15,** 40 (1976).
8. Geary, J., *In* "Astronomical Applications of Image Detectors with Linear Response" (IAU Colloquium No. 40) ed. by M. Duchesne and G. Lelièvre, p. 28–1. Paris-Meudon Observatory (1976).
9. Vogt, S., Tull, R. and Kelton, P., *Appl. Opt.* **17,** 574 (1978).
10. Horlick, G., *Appl. Spectrosc.* **30,** 113 (1976).
11. Cromwell, R. and Dyvig, R., *In* "Adv. E.E.P." Vol. 33B, p. 677 (1972).
12. Hayes, D., *Astrophys. J.* **159,** 165 (1970).
13. Liebert, J., Stockman, H. S., Angel, J. R. P., Woolf, N. J. and Hege, K., *Astrophys. J.* **225,** 201 (1978).
14. Koester, D., Liebert, J. and Hege, K., *Astron. & Astrophys.* (1979) (in press).
15. Liebert, J., Dahn, C. C., Gresham, M., Hege, K., Moore, R., Romanishan, W. and Strittmatter, P., *Astrophys. J.* **229,** 196 (1979).
16. Williams, R., Woolf, N., Hege, K., Moore, R. and Kopriva, D., *Astrophys. J.* **224,** 171 (1978).

Echelle Spectroscopy with Electronographic and Solid State Detectors

F. H. CHAFFEE, JR.

Smithsonian Institution, Mount Hopkins Observatory, Amado, Arizona, U.S.A.

INTRODUCTION

We have been pursuing a programme of high resolution echelle spectroscopy with the 1·5 m reflector at the Mt. Hopkins Observatory (MHO) since 1973. Our primary detector has been a Kron electronic camera, and a number of papers, dealing primarily with stellar atmospheres and the intersteller medium, containing results from this system have been published or are in press.[1-13]

In the last two years the Center for Astrophysics (CfA) has embarked on a major instrumentation development programme, largely in anticipation of the completion of the multiple mirror telescope (MMT). One of these instruments is a moderate resolution spectrograph designed to measure redshifts of bright ($m_v < 14·5$) galaxies, whose detector, the I-Ret, consists of a three stage image tube/Reticon combination. The I-Ret has been adapted to the MHO echelle, and this paper compares the relative performance of the two detector systems for work on problems requiring high signal to noise (S/N) spectrophotometry.

THE ECHELLE SPECTROGRAPH

The MHO Cassegrain echelle spectrograph was designed by D. J. Schroeder and has been described in detail by Chaffee.[14] It consists of an $f/9$ collimator whose 2·5 in. beam illuminates a 31·6 grooves mm^{-1} echelle grating. Cross dispersion is achieved by 1st order gratings and an $f/13$ camera mirror forms the image. The angles between optical elements are such that 400 μm of residual astigmatism is produced in the direction perpendicular to the echelle dispersion. In practice the stellar image is not trailed along the entrance slit; the widening is provided by the seeing disc and the astigmatism. The dispersion at the focal plane is 0·16 nm mm^{-1} at

$\lambda 400$ nm and the spectrograph is used in the resolution range $10^4 \le R \le 10^5$. The lower resolution is produced by a 3·4 arcsec entrance slit which corresponds to 250 μm at the focal plane of the 1·5 m telescope. Because the spectrograph demagnification along the echelle dispersion is unity, the slit has the same size at the spectrograph focal plane. The high resolution limit is achieved by using an entrance slit a factor of ten smaller.

The Kron Camera

The Kron camera[15] is an electronographic detector which has a resolution of 80 line pairs mm^{-1}, but is used at only half this resolution in the highest resolution application with the echelle spectrograph. The echelle format is such that about 100 nm (30 echelle orders) of the spectrum is covered by the 40 mm field of the camera. If the combined seeing conditions and resolution requirements are such that no light is lost at the entrance slit, the total system efficiency (telescope mirrors, spectrographic optics and S·11 photocathode) is 2% at $\lambda 400$ nm. Under such conditions 10^4 photoelectrons per resolution element can be recorded in 2 hours from an 8th magnitude star. Echellograms on nuclear track emulsions are reduced using the PDS microphotometer at Kitt Peak, the David Mann microphotometer at CfA or the PDS system at Lockheed. Each reduction scheme is different and they are described in Refs. 16, 17, and 3, respectively.

The Photon Counting Intensified Reticon Detector (I-Ret)

The Harvard I-Ret was built during 1977–78 and is similar to the system originally designed by Shectman.[18] It is used primarily on a moderate resolution spectrograph to measure galaxy redshifts and is designed for low photon rate spectroscopy. The detector package consists of three fibre optically coupled image tubes followed by a Reticon self-scanned photodiode array which is expoxied to a fibre optic plug contacted with immersion oil to the output of the final image tube. The photocathode of the first image tube is cooled thermoelectrically to $-15°$C; the other tubes operate at ambient temperature.

The Reticon is a 936×2 element array that is scanned continuously at 500 frames sec^{-1}. The diode dimensions are 375×30 μm^2 with no inter-diode dead space. Individual photoelectrons from the first photocathode produce a light flash of high enough S/N at the final phosphor that the readout noise of the Reticon is negligible and a centroiding circuit locates the position of an arriving photon to an accuracy of half a diode. The position of the detected photon is sent to a Nova 2 computer which acts as a multiscaler, and the observed spectrum is displayed on a Tektronix graphics terminal.

For the moderate resolution spectrograph, which is stigmatic, one diode line acts as a sky brightness channel while the other records the spectrum of the source. For the echelle spectrograph, which is astigmatic, the widened spectrum is 500 μm tall and covers part of both diode lines.

Relative Spectrophotometric Performance

In order to compare the performance of the two detectors, exposures of ζ Oph near λ393·3nm were obtained with each during successive nights on the 1·5 m telescope. ζ Oph was selected because at high resolution its spectrum appears continuous, interrupted with only a few sharp interstellar lines, many of whose strengths have been measured with great precision by other techniques. The purpose of the test was twofold: (1) to compare the measured strengths of moderate-to-weak absorption lines and (2) to compare the noise level of the spectrum to that expected from pure Poisson statistics of the counted photoelectrons. For this comparison the data from the better of the two I-Ret nights have been selected.

The measured instrumental width (FWHM) of the I-Ret system is 40 μm, and the spectrograph slit was set to achieve this resolution for both sets of exposures. The Nova 2 has spectral analysis software which allows fitting of absorption lines with Gaussian profiles, and provides statistical analysis of noise levels. The I-Ret data are recorded directly on floppy discs at the telescope. The Kron camera plates were scanned on the Kitt Peak microphotometer as described by Chaffee and Schroeder,[16] and were transferred to floppy discs for analysis by the same software system. Measured density was converted to counted photoelectrons from an independent measurement which determined that eleven 24 keV photoelectrons μm^{-2} will produce unit density on the Ilford L4 emulsion. Exposures were taken so that $1·3 \times 10^4$ counts per sampling interval (three samples were taken per resolution element in each case) were recorded by each detector.

Each absorption line and its adjacent continuum were fitted by a sloping linear continuum modulated by a Gaussian line, and the final fit is that which minimizes χ^2 in all fitting parameters. Table I contains measurements of equivalent widths of interstellar lines from ζ Oph. W_{KEC} are equivalent widths measured with the Kron camera; W_{IRET} are those measured with the I-Ret detector; W_{STD} are the best values available from other techniques as listed in the column labelled "Ref".

The measured statistical uncertainty of the equivalent widths in the

Table I
Intercomparison of measured equivalent widths

Line (nm)	W_{KEC}	W_{IRET}	W_{STD}	Ref.
393·3664	3·0	2·8	3·5	20
395·7700	1·2	1·0	1·3	21
396·8470	2·1	1·8	2·1	20

columns W_{KEC} and W_{IRET} is ±5%, and the agreement among the measurements is good, though there is a statistically significant tendency for the I-Ret values to be smaller than the others.

Finally, from the values of χ^2 one can estimate how closely the measured noise level approaches that expected from Poisson statistics of the photoelectrons leaving the first photocathode. For both detectors, from the analysis of two complete sets of absorption line measurements, the statistical uncertainties are 1·6 times those expected from pure Poisson statistics.

Conclusions

The performances of the Kron camera and the I-Ret detector for 2% precision spectrophotometry at high resolution are indistinguishable. Both fail to obey Poisson statistics by the same factor, but for different reasons.

The Kron camera performance is limited by the quality of Ilford nuclear track emulsions; on a scale of tens of microns enough plate flaws occur to degrade the camera's performance. A possible, admittedly tedious, solution to this problem would be to add multiple exposures. Since the plate flaws are random, their effect would be minimized by such a process.

The I-Ret performance is limited by an unstable asymmetry in the odd–even video lines which introduces a slightly unstable fixed pattern noise between alternate diodes. This pattern is thus not completely eliminated even when the recorded spectrum is normalized by the "flat" spectrum of an incandescent lamp a normalization which is routinely performed on all data. Furthermore this pattern is not random and multiple exposures will not reduce its effect. The cause of this problem is believed to be understood and the circuitry responsible for it is being appropriately modified.

The smaller equivalent widths measured by the I-Ret result from the shape of the Reticon diodes. They are sufficiently tall, and in this spectral region the echelle orders are sufficiently close together, that light from an adjacent order can be detected by the I-Ret and will fill in absorption lines. The echelle format is being changed in anticipation of its use on the MMT and doing so will separate the orders sufficiently to eliminate this problem.

The dark count from the two detectors is similar: $0\cdot 2$ counts $\sec^{-1}$ per diode for the I-Ret and $0\cdot 15$ counts $\sec^{-1}$ per spectral resolution element for the Kron camera. An improved cooling package for the I-Ret is expected to reduce its dark count rate by a factor of ten. The efficiency of the I-Ret telescope/echelle combination has been measured to be 3% at $\lambda 670$ nm though it is less at $\lambda 400$ nm because of non-UV fibre optics in the first image tube.[19] The equivalent figure for the Kron camera is 2%.

The resolution with the Kron camera is a factor of two higher than that achievable with the I-Ret since no phosphor exists to degrade its resolution.

Thus for a certain class of problem in high resolution spectrophotometry e.g., a problem for which small spectral coverage is satisfactory, no better than 40 μm resolution by the detector is needed, and real time data are required, the I-Ret system, because of its convenience, is superior to the Kron camera system. When higher resolution is required, or when spectral coverage of more than 3 nm is needed, the panoramic advantage of the Kron camera (6×10^4 spectral resolution elements compared to 10^3 for the I-Ret) make it clearly preferable.

Finally for low S/N applications the Kron camera is not useful because even the fastest nuclear track emulsion would not produce a high enough density to permit such exposures to be measured with conventional microphotometers. The I-Ret, on the other hand, is ideally suited for the low S/N regime.

Acknowledgments

The I-Ret detector system was developed under NSF grant Ast 76–22675 to Harvard University for a programme to measure redshifts of galaxies. The results presented here would not have been possible without many man months of effort on the part of those who developed the detector package, primarily Drs Marc Davis and David Latham. I am greatly indebted to them for their assistance and their encouragement in adapting the detector to the echelle spectrograph.

I am also indebted to Arthur Goldberg for his enthusiastic assistance in the operation of the device at the telescope and for developing the χ^2 analysis software which I have used extensively.

References

1. Brown, R. A. and Chaffee, F. H. Jr., *Astrophys. J.* **187**, L125 (1974).
2. Chaffe, F. H., Jr., *Astrophys. J.* **199**, 397 (1975).

3. Peterson, R. C. and Title, A. M., *Appl. Opt.* **14,** 2527 (1975).
4. Hearnshaw, J. B. *Astron. & Astrophys.* **51,** 71 (1976).
5. Hearnshaw, J. G., *Astron. & Astrophys.* **51,** 85 (1976).
6. Dupree, A. K., Baliunas, S. L. and Lester, J. B., *Astrophys. J.* **218,** L71 (1977).
7. Peterson, R. C., *Astrophys. J. Suppl.* **30,** 61 (1976).
8. Peterson, R. C., *Astrophys. J.* **206,** 800 (1976).
9. Peterson, R. C., *Astrophys. J.* **222,** 181 (1978).
10. Peterson, R. C., *Astrophys. J.* **222,** 595 (1978).
11. Peterson, R. C., *Astrophys. J.* **224,** 595 (1978).
12. Peterson, R. C. and Sneden, C., *Astrophys. J.* (in press).
13. Chaffee, F. H., Jr. and Dunham, T. Jr., *Astrophys. J.* (in press).
14. Chaffee, F. H. Jr., *Astrophys. J.* **189,** 427 (1974).
15. Kron, G. E., Ables, H. D. and Hewitt, A. V., *In* "Adv. E. E. P." Vol. 28A, p. 1 (1969).
16. Chaffee, F. H. Jr. and Schroeder, D. J., *Annu. Rev. Astron. & Astrophys.* **14,** 23 (1976).
17. Chaffee, F. H. Jr. and Lutz, B. L., *Astrophys. J.* **213,** 394 (1977).
18. Shectman, S. A. and Hiltner, W. A., *Publ. Astron. Soc. Pac.* **88,** 960 (1976).
19. Brown, R. A., *Astrophys. J.,* **224,** L97 (1978).
20. Shulman, S., Bortolot, V. J. and Thaddeus, P., *Astrophys. J.* **193,** 97 (1974).
21. Herbig, G. H., *Z. Astrophys.* **68,** 243 (1968).

A Review of Solid State Image Sensors

J. L. LOWRANCE

Princeton University Observatory, Peyton Hall, Princeton, New Jersey, U.S.A.

INTRODUCTION

Solid state image sensors are distinguished by their exceedingly wide dynamic range, linear response and high quantum efficiency in the green to near infrared spectrum. With the exception of Japan,† most current government and industrial sponsored research in photoelectronic devices appears to be in this area. Solid state image sensors encompass charge coupled devices (CCDs), charge injection devices (CIDs), and addressable photodiode arrays such as Reticon devices. While many of the publications in this area have been proceedings of special symposia,[1-6] there are now several books describing the subject in detail.[7-8] This review attempts to give a brief summary of the various competing devices and a qualitative discussion of characteristics unique to solid state arrays. The references should be consulted for detailed quantitative analysis of the various points mentioned in this review.

DESCRIPTION OF TYPES OF SOLID STATE IMAGE SENSORS

In all these devices photons absorbed in the silicon material generate an electron hole pair. This charge is collected in spatially discrete potential "wells" in the silicon. The distinction between the types of sensors is, primarily, in the readout process.

In the CCD the collected charge is scanned out by sequencing the phased electrode voltages that establish the electric field potential wells within the silicon. In this fashion the charge in a given well can be moved laterally from well to well and out to a charge sensitive amplifier built into the same silicon wafer. In surface channel CCDs the potential well is at the interface of the silicon and silicon dioxide insulating layer. Imperfections at this interface results in charge trapping which seriously degrades

† Japanese scientists continue to make significant improvements in electron beam readout image sensors.

the performance at low signal levels. An electrical or optical bias charge termed "fat zero" is required with surface channel devices to keep these traps filled. By ion implantation the potential well zone can be moved away from this oxide interface into the silicon to form a "buried" channel with negligible charge trapping. Buried channel CCDs exhibit better charge transfer efficiency, lower noise, and do not need a fat zero bias.

In the CID the charge is accumulated at each pixel in two closely spaced MOS capacitors. For readout the displacement current is sensed as the charge is dumped into the substrate or moved between the two capacitors in a non-destructive readout mode.

In the photodiode array, multiplex switches are turned on and off in sequence by shift register scanning circuits to switch each photodiode into the video amplifier input. The video signal constitutes the charge removed from each elemental capacitor in the photo-induced reverse current flowing in the associated photodiode. A photodiode array incorporating non-destructive readout is discussed in a following paper.†

Which to Choose

The merits and demerits of each generic type are too complex to yield to concise summary. However, a few generally accepted comments are offered.

The noise threshold in each device is closely tied to the shunt capacitance at the input of the on-chip preamplifier. The CCD has a unique advantage in that, by moving each elemental packet of charge over to the preamplifier input prior to sensing this charge, the shunt capacitance can be made in the order of 0·1 pF. Readout noise levels substantially less than 10^2 electrons RMS per picture element have been reported for cooled, buried channel CCDs. Surface channel CCDs are noisier because of charge trapping at the silicon–silicon dioxide interface. The shunt capacitance of both the CID and photodiode array is two orders of magnitude greater than that of the CCD such that the readout noise is typically one to two orders of magnitude greater. The non-destructive readout feature of the CID and some photodiode arrays makes it possible to reduce the noise by averaging successive readouts in those applications where this scheme is practical.

One attraction of the CID and photodiode array, especially in broadcast television applications, is the ability to read out the image without resorting to a separate storage register to prevent smearing the image. The integrated charge image remains in position and can be addressed in sequence much as one reads out an image in a conventional television

† See p. 431.

camera tube by electron beam raster scanning. The lateral transfer of the charge image in the CCD readout process requires that charge image be shifted from the image area to an unilluminated area of comparable size in a time short compared to the frame rate so that the exposure can continue. In a Fairchild CCD this is accomplished by alternating light sensitive and opaque strips across the CCD at the pixel pitch. In the R.C.A. and Texas Instrument CCDs, designed for conventional television applications, there is a frame storage area contiguous to the image area. This area can also be used for imaging in sequential expose–readout operation where continuous exposure is not required.

Since the silicon chip area required for integrated circuits strongly affects the manufacturing yield and the corresponding cost, the area required for a given concept is an important tradeoff in its development and application. A number of companies are working on CCD devices. At this time General Electric (U.S.A.) is the only company offering CID image sensors sensitive in the visible. The CIDs may have more application in the IR where the readout noise is usually small compared to the background fixed pattern and random noise. Photodiode arrays are available from several companies, especially Reticon. Hitachi has recently reported on a photodiode array with 484×384 pixels for a single chip colour television camera.[9] The active area is 60% of the total area with row and column pitch of 20 and 34 μm respectively. The design provides good anti-blooming with negligible loss in active area and as mentioned earlier, the total silicon area is limited to the image area, i.e., no storage register is required.

Spectral Sensitivity

Nearly all solid state image sensors have been made of silicon. The spectral response is limited by the transparency of silicon to about 1·1 μm. The response at short wavelengths is limited at about 400 nm by the absorption of the photoelectrically inactive surface layers of SiO_2 and polysilicon on the silicon that make up the electrodes and insulating layers. By thinning and back illuminating the device the short wavelength response can be extended somewhat. And in some devices such as Reticon photodiode arrays the area over the photodiodes can be made to have a very thin surface dead layer. In the Space Telescope Wide Field Camera the CCD short wavelength response is being extended down to $\sim$100 nm by coating the CCD with a fluorescent wavelength converter (phosphor). This technique can probably be employed at even shorter wavelengths. In the soft X-ray region the silicon becomes sufficiently transparent that the silicon solid state detectors can be used from about 1 to 10 keV. At higher energies the silicon again becomes too transparent.

Returning to long wavelengths, infrared solid state image sensors are being intensively developed for military applications. Using CCD, CID and photodiode array techniques for multiplexing and signal processing it is becoming practical to make IR focal planes with thousands of IR detectors and focal plane filling efficiencies of 10%.[10] There are a number of approaches being pursued for charge transfer devices in infrared focal planes. The Shottky barrier IRCCD described in a following paper describes one such approach.† In monolithic extrinsic silicon arrays, an extrinsic substrate provides the infrared sensitivity. The photo-induced charge is injected into a lightly doped expitaxial layer for charge transfer multiplexing and readout. In monolithic intrinsic arrays the infrared photons are absorbed in a narrow bandgap semiconductor such as InSb. Charge transfer and readout occur either in the same material or signal charge is transferred to a wider bandgap layer for CCD readout. Hybrid arrays employ separate sensing and charge transfer media. The detector array, fabricated in the photosensitive medium, is electrically and mechanically coupled to a silicon CCD for signal multiplexing (readout).

Spatial Frequency Response

All of these solid state image detectors differ from most other electronic image devices in that the electrical signal is spatially sampled in the image integration process by the discrete potential wells that constitute each pixel of the image. While this is not necessarily bad, it does set definite limits on the number of pixels and the spatial frequency response.

Spatial sampling, as in these solid state detectors, constitutes spatial filtering of the image: the attenuation A is given by

$$A = \frac{\sin(\pi f/f_c)}{\pi f/f_c} \cos \theta,$$

where f is the spatial frequency of interest, f_c is the number of sample pairs per unit length and θ is the phase of the image pattern relative to the pixel position.

The same equation generally applies to vertical sampling by electron beam scanning in a conventional television type image sensor. Horizontal sampling is also present in digitized video signals. The sampling spacing is at the discretion of the system designer and usually constitutes a tradeoff of spatial filtering versus digital bit rate. The $(\sin x)/x$ attenuation can be circumvented in the horizontal direction by making the samples of the bandwidth limited analogue short compared to the spatial frequency period.

† See p. 495.

The spatial frequency response of the solid state arrays is also reduced by scattered light within the silicon substrate and by lateral charge diffusion prior to being collected in the wells. This is wavelength dependent since the depth of absorption varies with wavelength. During the readout process any charge transfer inefficiency reduces the spatial frequency response as well. In spite of these effects the present state of the art is such that the spatial frequency response often approaches the limit set by the discrete sampling process as discussed above.

CCD Preamplifiers

As discussed earlier, the ultra-low input shunt capacitance afforded by charge coupled readout combined with on-chip amplification make the readout noise much lower than has been previously possible without resorting to image intensification techniques. Two types of CCD on-chip preamplifiers have been developed, the precharge and the floating gate.

In the precharge amplifier, the input gate of a field effect transistor is biased to a reference voltage by charging the input shunt capacitance through a field effect transistor switch. The charge from a CCD well is subtracted from this reference level and the change in output is detected as the video signal. The reference is then reset for the next pixel, etc. Thermal switching or kTC noise in resetting the input reference is typically 300 electrons RMS, and for low noise operation is removed by correlated double sampling of the amplifier output before and after the signal charge is added to the preamplifier input.

In the floating gate amplifier, the gate of the preamplifier senses the charge non-destructively. The floating gate approach allows the same signal to be read out by successive preamplifiers for special signal processing, or averaged to reduce the amplifier noise.[11]

Below about 1 MHz pixel rate the noise performance of the precharge and floating gate amplifiers are comparable. Above 1 MHz the correlated double sampling becomes increasingly difficult to implement effectively because of transients, and there appears to be an advantage in employing the floating gate amplifier where reset can be done on a line by line basis with minimal degradation in signal fidelity.

CCD On-chip Signal Processing

Adding the signal in adjacent pixels of the CCD can be implemented prior to readout by loading more than one line into the horizontal shift register prior to shifting the signal out to the on-chip preamplifier. Pixels in the horizontal register can be summed by not resetting the precharge type of preamplifier input after each pixel transfer. These modes of operation appear to be essentially noise free and can be used to minimize the readout noise component and the quantity of data in those cases where the shape or size of the picture element of interest is larger than the CCD pixel size. Slit spectrograph readout is one example where the information of interest is one-dimensional and adding pixels along the slit prior to readout can make significant reduction in the readout noise per spectral element. The technique is, of course, limited at high signal levels by saturation effects, but at high signal the readout noise is less important.

Time delay integration (TDI) is a similar technique in which the charge image accumulating during an exposure is shifted laterally to track a moving optical image and thereby extend the exposure without smear. This technique is attractive for aerial photography, panoramic scanning, and facsimile type application.[12]

In some applications, such as star trackers or photon counting, centroiding is used to locate the centre of the point image. The random location of the image on the reticle-like CCD focal plane makes it necessary to compare the signal in adjacent pixels. While this function can be performed after readout of the signal it has been implemented on the CCD chip by a set of nine floating gate amplifiers separated by CCD delay lines such that the amplifiers simultaneously see a 3×3 pixel matrix as the CCD is read out.[13]

THERMAL DARK CURRENT

The silicon thermal dark current in CCDs is typically 10^{-9} A cm^{-2} at 20°C. This corresponds to about 10^5 electrons sec^{-1} pixel^{-1}, or approximately 10% of the dynamic range for a one second exposure time. At standard television rates of 50 to 60 fields per second this dark current is tolerable. However, for most scientific applications the detector must be cooled. Cooling the CCD to -100°C yields typical dark currents of 1 electron sec^{-1} pixel^{-1} and cooling to -120°C reduces the dark current to 0·1 electron sec^{-1} pixel^{-1}.[14]

Cooling the CCD does have the liability of reducing the quantum efficiency for wavelengths longer than one micron.[15]

INTENSIFIED SOLID STATE SENSORS

Solid state linear and area arrays have been successfully coupled optically to image intensifiers. Several examples are described in the preceding section and in other papers presented at the Symposium.† The advantage of these systems over earlier versions employing television camera tubes is usually in the lower readout noise exhibited by the solid state detector. Also the solid state detector is physically smaller and, more important, neither requires, nor is sensitive to, magnetic fields. They are, however, more susceptible to damage from arcs and other electrical transients.

For quantum noise limited operation or photon counting applications the solid state sensor image intensifier systems must employ several image intensifier stages with the attendant problems of high voltage, size, etc. For these reasons there is substantial interest in intensified solid state sensors where the photoelectrons from the first photocathode strike the silicon detector directly, instead of a phosphor. The Digicon tube incorporates an array of individual silicon diodes each connected to separate preamplifiers outside the image tube;[16-18] this was the first intensified solid state detector employing electron bombardment to detect single photon events. It was followed by versions of the Digicon in which a Reticon self-scanned linear photodiode array replaced the individual silicon diodes.[19] The readout noise of the Reticon prevents single event discrimination; however the electron bombardment gain is sufficient to make the sensor quantum noise limited over its entire dynamic range.[20]

Early attempts to make similar tubes incorporating CCDs met with problems of compatibility between the CCD and the photocathode processing. Recent progress reported at this Symposium indicates that intensified CCDs are feasible and the expected single photoelectron pulse height distribution is sufficiently peaked to detect single events unambiguously.‡

In these photon counting applications the intensified CCD must compete with other schemes such as crossed-wire grid readout and resistive anode readout of microchannel plate image intensifiers.[21] All of these devices including the intensified CCD can be considered experimental at the time of this Symposium and it will be interesting to see their status at the next Symposium.

The intensified CCD is, of course, of interest for other applications as well. A streak camera tube employing a CCD in place of the phosphor has been built by R.C.A.[22] And an intensified CCD tube built by Varian using a Texas Instruments CCD is being employed in fusion diagnostic instrumentation.§

---

† See pp. 389, 397 and 481.
‡ See pp. 441 and 463.
§ See p. 441.

Summary

Serious work on solid state image detectors of the sort discussed in this paper began about 20 years ago, and there was a significant advance ten years ago with the invention of the CCD. One can expect continued invention and innovation in this field during the next decade, with particular emphasis on conventional television and colour television applications. Infrared solid state image sensors are especially active areas of research and development with much of the present work classified. This can be expected to yield useful devices in a few years for astronomical, medical and other scientific applications.

References

1. Proceedings of the International Conference on Technology and Applications of Charge Coupled Devices, Edinburgh, Scotland, September, 1974.
2. Proceedings of Symposium on Charge Coupled Device Technology for Scientific Imaging Applications, JPL, California Institute of Technology, Pasadena, March, 1975.
3. Proceedings of International Conference on Applications of Charge Coupled Devices, Naval Electronics Laboratory Center, San Diego, October, 1975.
4. Proceedings of Conference on Charge Coupled Device Technology and Applications, Washington D.C., November, 1976. (NASA Office of Aeronautics and Space Technology and Jet Propulsion Laboratory, California Institute of Technology.)
5. International Electron Devices Meeting, Washington D.C., December 1978. (Electron Devices Society of IEEE.)
6. Proceedings of International Conference on Applications of Charge Coupled Devices, Naval Electronics Laboratory Center, San Diego, October 1978.
7. "Solid State Imaging", Proceedings of the NATO Advanced Study Institute, Louvain, ed. by O.P.G. Jespers, F. van de Wiele and M. H. White, Noordhoff Publishing, Leyden, Netherlands (1976).
8. "Optical and Infrared Detectors". Topics in Applied Physics, Vol. 19, ed, by R. J. Keyes, Springer-Verlag, New York (1977).
9. Kuike, N., *In* Technical Digest Solid State Circuits Conference, p. 192, February 1979.
10. "Optical and Infrared Detectors". Topics in Applied Physics, Vol. 19, p. 197, ed. by R. J. Keyes, Springer-Verlag, New York (1977).
11. Amelio, G. F. and Dyck, R. H., Fairchild Camera and Instrument Corporation, Palo Alto, p. 605–614.
12. Ibrahim, A. A., *In* Proceedings of the International Conference on Applications of Charge Coupled Devices, Naval Electronics Laboratory Center, San Diego (1978).
13. Hall, J. E., Breitzmann, J. F., Blouke, M. M. and Carlo, J. T., *In* Technical Digest IEEE International Electron Devices Meeting, Washington D. C., December 1978.
14. Chodil, G. J., Hoshiko, H. H., Johnson, J. J. and McConaughy, R. M., *In* "Seventh Symp. P.E.I.D. Preprints", p. 369 (1978).
15. Dreux, M., Fauconnier, T. and Fort, B., *In* "Seventh Symp. P.E.I.D. Preprints", p. 367 (1978).
16. Beaver, E. A. and McIllwain, C. E., *Rev. Sci. Instrum.* **42**, 1321 (1971).
17. Beaver, E. A., McIllwain, C. E., Choisser, J. P. and Wysoczanski, W., *In* "Adv. E.E.P." Vol. 33B, p. 836, (1972).

18. Beaver, E. A., Harms, R. J. and Schmidt, G. W., *In* "Adv. E.E.P." Vol. 40B. p. 745 (1976).
19. Choisser, J. P., *In* "Adv. E.E.P." Vol. 40B, p. 735 (1976).
20. Tull, R. G., Choisser, J. P. and Snow, E. H., *Appl. Opt.* **14,** 1182 (1975).
21. Lampton, M., *In* "Astronomical Applications of Image Detectors with Linear Response" (IAU Colloquium No. 40) ed. by M. Duchesne and G. Lelièvre, p. 32-1. Paris-Meudon Observatory (1976).
22. Cheng, I., Tripp, G. and Coleman, L. *In* "Proceedings of International Conference on Charge Coupled Devices", Naval Electronics Laboratory Center, San Diego (1978).

A Novel Photodiode Array System for Direct Astronomical Spectroscopy

A. W. CAMPBELL, A. R. HEDGE, G. R. HOPKINSON, A. HUMRICH
and J. M. BREARE

Department of Physics, University of Durham, Durham, England

Introduction

The use of linear photodiode arrays for direct astronomical spectroscopy has been extensively reported.[1-8] The advantages of such arrays are, chiefly: a very high quantum efficiency, wide spectral response and geometrical stability of pixels. Most workers have used Reticon† diode arrays. We describe an array developed by the Plessey Co. Ltd.‡ in conjunction with the Royal Greenwhich Observatory which has a potentially superior noise performance to the Reticon devices and also offers non-destructive readout for monitoring during an exposure.

The Plessey Array

Reticon arrays operate in the recharge sampling mode in which the amount of charge required to restore the voltage on the photodiode is measured. This form of readout is necessarily destructive. Arrays can be constructed which operate in a voltage sampling mode, where the voltage on the diode is measured as it is discharged. These arrays offer the possibility of non-destructive readout (NDRO) but suffer from inherent non-linearity of photoresponse. The Plessey array incorporates a charge amplifier transistor to improve the linearity. The circuit of a single element of the array is shown in Fig. 1. Preliminary tests on a single element were carried out using electron bombardment by McMullan *et al.*[9]

Transistor T_1 is the charge amplifier. A constant negative voltage, V_{ref}, in excess of V_T, the threshold voltage, is applied to the gate and the

---

†Reticon Corp., 910 Benicia Avenue, Sunnyvale, CA 94040, U.S.A.
‡Allen Clark Research Centre, The Plessey Co. Ltd., Casswell, Northants, England.

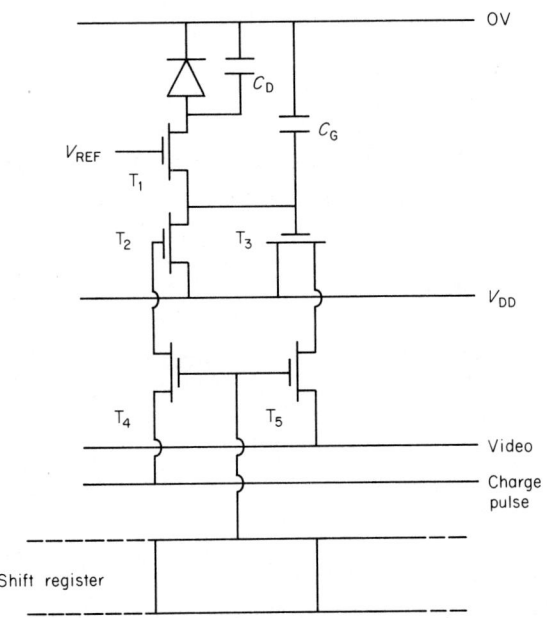

Fig. 1. Circuit diagram of a single array element.

transistor conducts so as to maintain a constant voltage across the photodiode. Hence the diode is always fully charged and thermally and optically generated currents discharge the gate capacitance C_G of the source follower transistor T_3. C_G is typically 0·1 pF and with a diode capacitance C_D of ~1 pF there is an effective voltage gain of ~10 at the gate of T_3. Since T_1 maintains a constant bias voltage across the diode, the depletion layer width does not vary as the array is exposed. In other voltage sampling arrays the change in depletion layer width as the diode discharges cause variations in diode capacitance and in thermal leakage current, leading to non-linearities in photoresponse of as much as 20%. It has also been suggested[1] that the blue response is affected by changes in depletion layer width.

After exposure the gate capacitance C_G is recharged to V_{DD} through T_2. T_4 and T_5 are multiplexing transistors which are addressed by a two-phase low power shift register.

In "double sampling" readout a charge pulse is applied whilst T5 is on. The video output voltage before the charge pulse corresponds to the "signal" level and the voltage after the charge pulse is the "zero" level. Subtraction of one from the other gives the true signal and eliminates

Table I
Characteristics of the Plessey array

No. of video lines	4
No. of elements per line	256
Diode area	$200 \times 40\ \mu m^2$
Diode pitch	$50\ \mu m$
Package	24 pin DIL

offset (or fixed pattern) noise. Additionally, by inhibiting the recharge pulse, the signal can be read out without recharging at any time, thus providing non-destructive readout.

In fact the above description is somewhat simplified in that the diodes are addressed in pairs (this reduces the length of the shift register and so saves space on the chip). Thus what has been termed video "signal" is always the sum of the outputs from two adjacent diodes. The individual diode outputs can be obtained in the double sampling (but not NDRO) mode since the diodes can be recharged separately. We measure three levels: the sum of two diode signals, the sum with one diode recharged and the sum with both recharged (i.e., the zero level). The disadvantage is that the resolution is halved with NDRO and also the dynamic range of the device is reduced.

The physical characteristics of this device are given in Table I.

Thermal Leakage Current

A thermal leakage of $\sim 3\ \mu V\ sec^{-1}$ has been measured at a temperature T of $-120°C$; this is small enough to permit exposures of several hours. The leakage current (I_L) seems to follow a law of the form

$$I_L \propto 10^{-2000/T}$$

so that at $0°C$ a factor of 2 change in I_L is caused by a temperature change of $\sim 8°C$. Similar results have been found for Reticon arrays by other workers.

Responsivity

The photoresponse of the array has been measured using a narrow band filter and a calibrated phototransistor. The responsivity was measured to be $\sim 1 \cdot 0\ \mu V$ per detected photon at $\lambda = 543 \cdot 4$ nm when operating with a head amplifier with a 47 kΩ feedback resistor. Using typical values

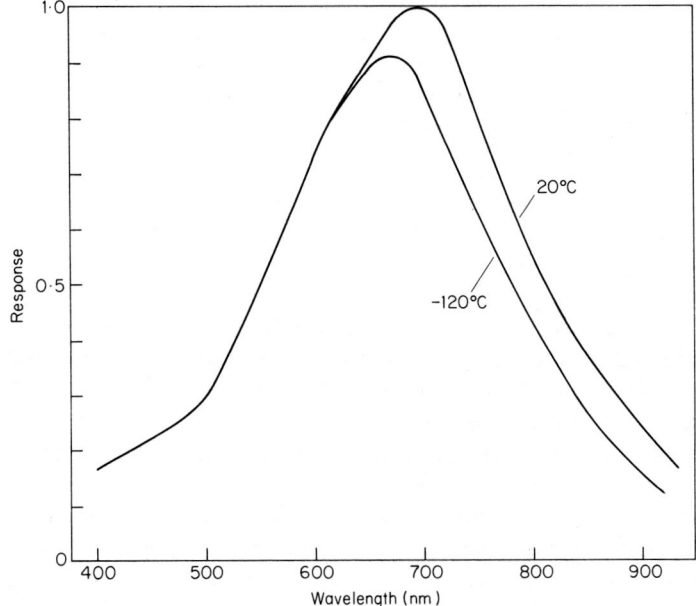

Fig. 2. Spectral response curves for the Plessey array (normalized to 700 nm at 20°C).

of $1 \cdot 0$ pF for C_G and $20\ \mu\mathrm{A\ V}^{-1}$ for the transconductance of T_3, the calculated value is $\sim 1 \cdot 5\ \mu\mathrm{V}$ per detected photon.

Spectral response curves are shown in Fig. 2; as has been noted previously by several authors, the near infrared response is reduced by cooling.

Noise Sources and Measurements

Apart from the photon shot noise the principal noise sources in a diode array are shot noise on the thermal leakage current and readout noise. The main sources of readout noise in this array are:

(a) low frequency $(1/f)$ noise in T_1 and T_3;
(b) thermal reset noise on C_G;
(c) charge pumping shot noise in T_2.

Figure 3 shows predicted noise values compared with the photoresponse at an operating temperature of $-120°\mathrm{C}$ (assuming $C_G = 0 \cdot 1$ pF, area of gate capacitance $= 10^{-6}\ \mathrm{cm}^2$ and that the surface trap density at the Si–SiO$_2$ interface of the MOS transistors is $\sim 10^{11}\ \mathrm{cm}^{-2}$). At room

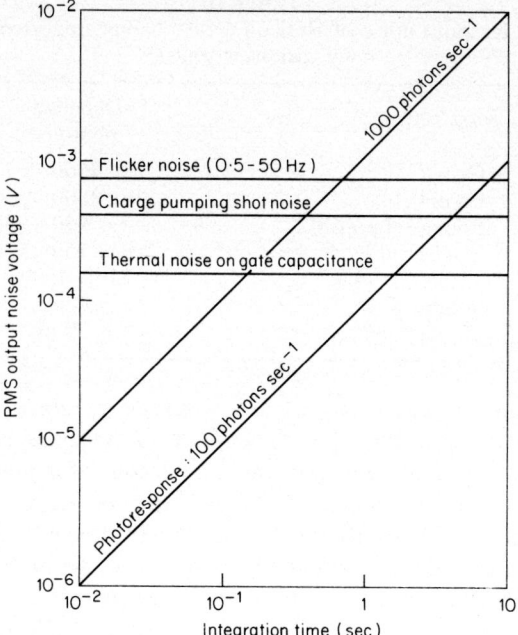

FIG. 3. Predicted noise voltages at $T = -120°C$.

temperatures the shot noise on the thermal leakage current is no longer negligible but this seems to be the only additional noise source.

We have found experimentally that using the double sampling technique the low frequency noise on the signal and zero levels is well correlated and so can be removed. The characteristic frequency above which the noise is no longer correlated lies in the range 100–500 Hz. With double sampling we have measured an RMS noise level (at $T = -120°C$) of 0·35 mV or 350 detected photons. For non-destructive readout this rises to 700 photons. The noise values published by workers using Reticon arrays are compared in Table II.

Response Time

It has been observed that, at low temperatures and low signal levels, the video output has a slow response to changes in illumination and suffers from image retention. This effect has not been observed with Reticon arrays.[10]

It can be shown that transistor T_1 has a slow charge transfer response when the leakage current in the diode is small. The time constants

TABLE II
Readout noise of Reticon arrays: values reported by various workers

	Detected photons
Geary[1]	4700
Campbell[2]	3800
Livingstone et al.[4]	950
Vogt et al.[5]	750
Walker et al.[6]	1100–3600
Smithson[7]	<7000
Dravins[8]	~2000

however are in the region of 10^{-2} sec, whereas the observed lag is of the order of seconds. That the lag is dependent on the signal level is shown in Fig. 4. A slower process must be found to account for this effect and it is suggested that the existence of "slow" surface traps at the Si–SiO$_2$ interface in the MOS transistor T_1 may be responsible. Several models have been suggested[11] in connection with $1/f$ noise in MOS transistors, some of which produce time constants of the correct order.

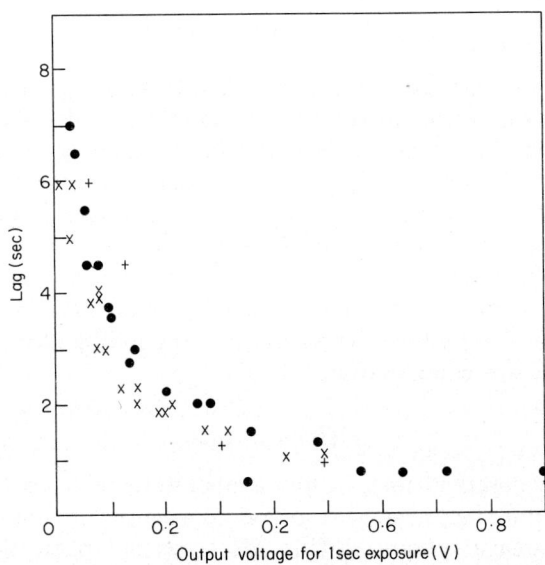

FIG. 4. Signal lag as a function of output voltage. $+$, $T = -120°C$ (preliminary run); ●, $T = -120°C$; ×, $T = -43°C$.

Signal Processing and Data Acquisition

The control and acquisition logic is shown as a block diagram in Fig. 5. Signal "tailoring" for a voltage sampling array is considerably simpler than that required for recharge sampling readout. Synchronous noise due to clock breakthrough is easily avoided by using a sample and hold circuit. This is followed by a 12-bit analogue to digital convertor (ADC). The digitized output is transferred to a buffer memory in CAMAC. Control of the readout sequence and interfacing of all peripherals is via CAMAC; at present we have a visual display unit, storage display and fast paper tape reader and punch. The control computer is a PDP11/03 programmed in ACSL's CATY 2 language. Further details of the system have been presented elsewhere.[12] Fixed pattern offset noise can be subtracted and the spectrum displayed at run time; however there is further fixed pattern noise present in the form of diode-to-diode responsivity variations. This can be removed by division by a flat field spectrum but, at present, this has to be done off-line because of a shortage of computer memory space.

Observations

During August 1978 the diode array system was installed at the coudé spectrograph of the 30 in. telescope of the Royal Greenwhich Observatory. The array was cooled to $\sim -120°C$ by a cold finger dipped into a dewar of liquid nitrogen. Alignment of the array was not found to be difficult. However a splitting between the video outputs from odd and even channels was produced whenever light was allowed to "spill over" the edge of a line of diodes onto the adjacent MOS transistors. This is because the transistors associated with odd or even diodes lie on opposite sides of a line and a "signal" is produced in the MOS junctions. Odd–even splitting was produced on changing from calibration lamp to star and on varying the grating setting thus necessitating time-wasting realignments of the array. We are in the process of fitting a cylindrical Fabry lens to image the telescope primary onto the array, and it is hoped that this will eliminate the problem. A similar solution was found to be necessary with the Reticon system of Vogt et al.[5]

Figure 6(a) shows a spectrum obtained in a two minute exposure of Vega. The lower trace is the raw data and the upper trace has had the fixed pattern offset noise subtracted. The prominent absorption feature is the H_α line and the horizontal scale represents approximately 10 nm (the resolution is $\sim 0\cdot 05$ nm per diode). Figures 6(b) and (c) show an expanded display of the reduced spectrum before and after division by a flat field

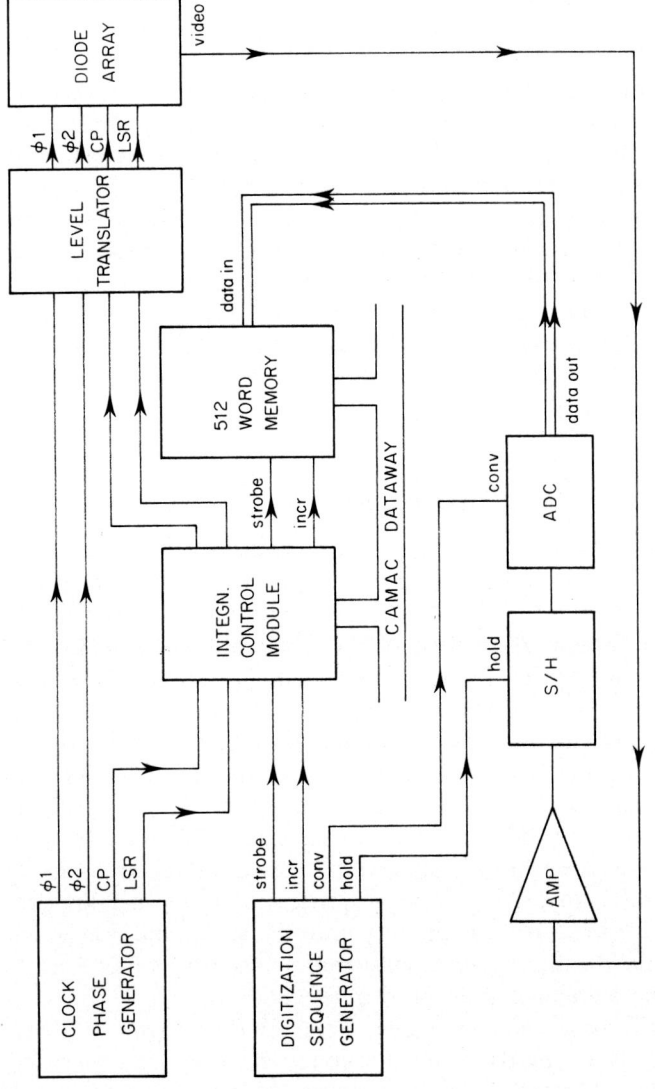

FIG. 5. Block diagram of the control logic: φ1 and φ2 are clock pulses, CP is the recharge pulse and LSR is the word shift register pulse issued at the beginning of each frame to initiate a scan.

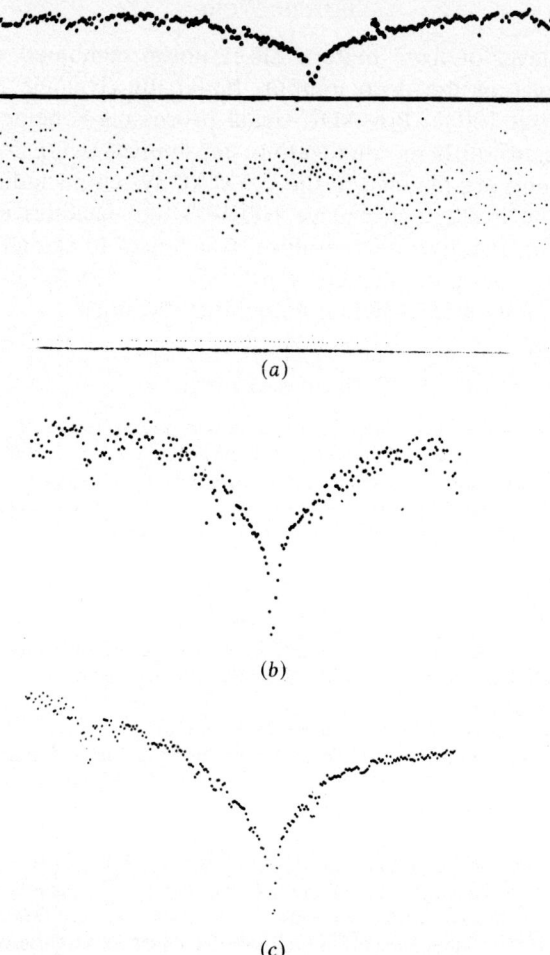

FIG. 6. Raw and reduced spectra of the H_α line in Vega. (a) Raw data (lower trace) after subtraction of fixed pattern noise (upper trace). (b) and (c) Expanded display of reduced spectrum before and after subtraction of flat field.

spectrum. The high noise level at the left hand (blue) end of Fig. 6(c) is due to the odd–even splitting mentioned above. The height of the continuum corresponds to $\sim 10^6$ photons per diode indicating an overall light detecting efficiency of 0·5% (this will be a product of atmosphere, telescope, spectrograph transmissions and the quantum efficiency of the array).

Future Work

The high level of fixed pattern offset noise, combined with the low saturation level of the array output, limits the dynamic range of the system to about 100:1. Pre-ADC signal processing is being investigated to extend this towards the digitization step-limited figure of 4000:1.

Improvements are planned to the CAMAC system including a circulating memory to average successive NDRO's and facilities for automatic logging of time and array temperature. It is hoped to extend the computing system to include a magnetic type drive, for recording data, and an increased memory size including floppy disk facilities.

Acknowledgments

The authors would like to acknowledge the support and assistance of the Director and staff of the Royal Greenwich Observatory, in particular Dr D. McMullan for his close association with the project. Professor A. W. Wolfendale is thanked for his interest and support and a Research Grant from the Science Research Council is gratefully acknowledged.

References

1. Geary, J. C., In "Astronomical Applications of Image Detectors with Linear Response" (IAU Colloquium No. 40), ed. by M. Duchesne and G. Lelièvre, p. 28–1, Paris-Meudon Observatory (1976).
2. Campbell, B., *Publ. Astron. Soc. Pacific* **89,** 728 (1977).
3. Livingstone, W. C., In "Astronomical Applications of Image Detectors with Linear Response" (IAU Colloquium No. 40), ed. M. Duchesne and G. Lelièvre, p. 22–1, Paris-Meudon Observatory (1976).
4. Livingstone, W. C., Harvey, J., Slaughter, C. and Trumbo, D., *Appl. Opt.* **15,** 50 (1976).
5. Vogt, S. S., Tull, R. G. and Kelton, P., *Appl. Opt.* **17,** 574 (1978).
6. Walker, G.A.H., Buchholz, V., Fahlman, G. G., Glaspey, J., Lane-Wright, D., Mochnaki, S. and Condal, A., In "Astronomical Applications of Image Detectors with Linear Response" (IAU Colloquium No. 40), p. 24–1, Paris-Meudon Observatory (1976).
7. Smithson, R. C., *Sol. Phys.* **40,** 241 (1975).
8. Dravins, D., In "Image Processing Techniques in Astronomy," ed. by C. De Jager and H. Nieuwenhuijzen, p.97. Riedel, Amsterdam (1975).
9. McMullan, D., Wellgate, G. B., Ormerod, J. and Dickson, J., In "Adv, E.E.P." Vol. 33B, p. 873 (1972).
10. Horlick, G., *Appl. Spectrosc.* **30,** 113 (1976).
11. Soares, R. A., *Design Electronics* **8,** 24 (1971).
12. Hedge, A. R., Breare, J. M., Campbell, A. W., Hopkinson, G. R. and Humrich, A., In Proc. ESO/SRC Conf. "Applications of CAMAC to Astronomy", CERN, Geneva (1978).

ICCD Development at Princeton

J. L. LOWRANCE, P. ZUCCHINO, G. RENDA and D. C. LONG

Department of Astrophysical Sciences, Princeton University Observatory, Princeton, New Jersey, 08540, U.S.A.

INTRODUCTION

Intensified Charge Coupled Devices (ICCDs) are promising detectors for both laboratory measurements of high temperature plasmas in Tokamak nuclear fusion machines and in astronomy. This paper reports in three sections on recent developments in these areas: (1) an image tube detector for plasma instrumentation; (2) a windowless ICCD detector configuration with potential application to both plasma instrumentation and astronomy; and (3) laboratory results on the single electron response of CCD arrays.

ICCD IMAGING DETECTOR FOR PLASMA INSTRUMENTATION

Detector Description

Princeton University has developed an Intensified Charge Coupled Device image tube with a 130 mm diagonal S·20 photocathode and a 160 pixel by 100 line CCD. The image section operates at 20 kV and has an adjustable demagnification, with 30:1 maximum, from photocathode to CCD. With an EBS gain of 2500, near quantum limited photometric performance is obtained.

Figure 1 is a schematic of the ICCD image tube, and includes the major dimensions and electrode potentials. The large physical size of the extended red S·20 photocathode is dictated by the optical requirements of the Thompson scattering imaging spectrograph in which this detector is used. The high demagnification is needed to match the 130 mm diameter photocathode to the 4·3 mm diagonal of the available CCD.

The 160×100 CCD was produced by the Central Research Laboratories of Texas Instruments, Inc. It was the largest CCD known to

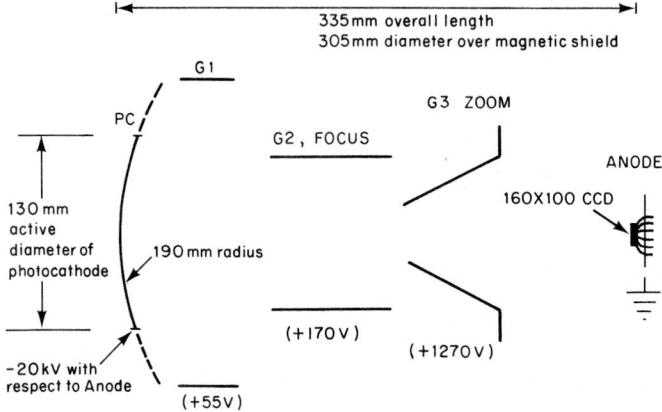

FIG. 1. Diagram of the 30:1 demagnifying ICCD system. Positive electrode potentials are relative to the photocathode.

us to be suitable for electron bombardment operation at the time the ICCD was developed. Although the small format ($3\cdot7 \times 2\cdot3$ mm^2) required the 30:1 demagnification in the electron optics, this number of pixels was adequate for the relatively low spectral and spatial resolution requirement of the plasma diagnostics.

Except for the CCD, the detector was designed and produced by Varian LSE. The electron optics design was based on their earlier development[1] of continuous zoom X-ray image intensifiers which zoomed from 12:1 to 6:1 demagnification with a maximum input diameter of 230 mm. Varian employed their imaging tube computer program to revise the electron optics of the zoom X-ray tubes for this application. In the 30:1 demagnifying design all aberrations are effectively below the resolution limit set by the 23 μm size of the CCD pixels.

A total of seven completed ICCD image tubes have been built for this project. Three tubes have both the high red quantum efficiency needed by the experiment and good, nearly pixel defect free, CCDs. Most of the remaining tubes function well enough for test purposes, but either their photocathode or CCD quality is not up to the system requirements.

Gating Considerations

Since the Thomson scattering experiment requires gathering the few photons of interest during a 30 nsec pulse of laser light, while rejecting the substantial continuous background light from the plasma, means for fast gating or shuttering of the detector are needed.

The previously established method of gating the X-ray intensifier tubes was to switch the focus electrode, G2, potential from its normal value to about 150 V more negative than the photocathode. In our tests we found that gating with G2 succeeded in preventing any photoelectrons from the photocathode from reaching the CCD. However a minute electron input to the CCD still remained which we attributed to low level photoemission from the focus electrode itself. This background electron flux was measured to be less than one part in 10^4 of the ungated photocurrent. Since a gating ratio of better than 10^6 is required, G2 gating is not sufficient for this application. Another difficulty with the G2 gating scheme, is that the 100 n sec wide pulse of 400 V amplitude that is applied to G2 to switch the tube from the shuttered state to the imaging state, must be flat-topped and accurate to within 0·5% (2 V) to maintain the image in focus.

It appeared from the shape and arrangement of the electrodes, that gating on the zoom electrode, G3, would avoid any problems associated with photoemission from internal electrodes. Because the normal ungated G3 potential (with respect to the photocathode) is +1270 V, the pulse amplitude required to swing G3 from below cutoff (−200 V) to the ON state, is 1500 V. The pulse flatness and accuracy requirements in this case are set by the maximum allowable size (zoom) error, as opposed to the focus error considerations associated with G2 gating. A pulse tolerance of one percent (15 V) corresponds to a 0·6 pixel size error at the sides of the format. Although G3 gating requires nearly four times the pulse amplitude of G2 gating, the required pulse accuracy is relaxed by a factor of two. We found that G3 gating totally blocked all electron input to the CCD, and it was the method chosen for this application.

Photoresponse

The Thomson scattering experiment requires that the photocathode quantum efficiency be high, especially in the red. The measured quantum efficiency versus wavelength for the best tube, and the experimenter's specifications are tabulated in Table I.

TABLE I
Photoresponse of Tube 008

% Q.E.	Wavelength (nm)				
	400	500	600	700	800
Measured	34·1	19·8	10·8	4·8	0·74
Specified	14·0	12·4	9·3	4·4	—

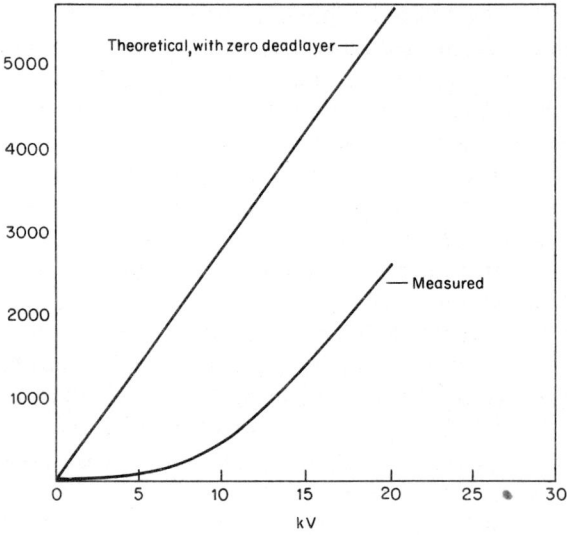

FIG. 2. Electron gain versus photocathode voltage for Varian tube 001 and Texas Instruments 160×100 CCD 220-6-8.

Electron Gain

The photoelectrons are accelerated to 20 keV before impinging upon the thinned back side of the CCD. Theoretically each 20 keV photoelectron should produce nearly 5500 electron–hole pairs in the silicon of the CCD. We measure an actual electron gain of 2500 at 20 keV. This result suggests a dead layer of 11 keV equivalent thickness on the surface of the CCD. Indeed our measured curve (Fig. 2) of gain versus photocathode voltage does not begin to rise significantly until the photocathode voltage is above 11 kV.

Dynamic Range

If the ICCD detectors are operated without benefit of significant cooling of the CCD, the lower exposure limit is set by the shot noise in the CCD dark current. At 20°C the dark current per pixel is approximately 5×10^5 electrons sec^{-1}. We read the CCD at a pixel rate of 10^5 pixels sec^{-1}. With allowances for various system overhead times, the frame rate for the 160×100 pixel CCD is 5 frames sec^{-1}. Therefore the dark count per pixel caused by dark current is 10^5 electrons with a corresponding shot noise of 300 electrons RMS. This is the dominant readout noise. Since each photoelectron produces an average signal of

2500 electrons (which may be shared between adjacent pixels in some cases), the system is limited by only the shot noise in the photoelectron input for even the weakest exposures.

The maximum exposure is set by the CCD pixel capacity of 10^6 electrons per pixel, and by the average EBS gain. Since we are primarily interested in fairly weak exposures, we operate at a gain of 2500. The maximum exposure per pixel is then 360 photoelectrons. At all exposures the signal to noise ratio is determined by the quantum noise in the photoelectron flux.

Windowless Intensified CCD Development

There has been an interest for some time in image intensifier configurations that allow the use of opaque substrate photocathodes which exhibit higher quantum efficiency than semi-transparent photocathodes, especially in the ultraviolet and near infrared.[2,3] As shown in Fig. 3, the oblique, or reflection mode image tube, deflects the photoelectrons along a path at an angle to the optical axis by tilting the magnetic focus field, B. This results in an additional displacement perpendicular to the plane formed by the B and E field such that the image at the photocathode is

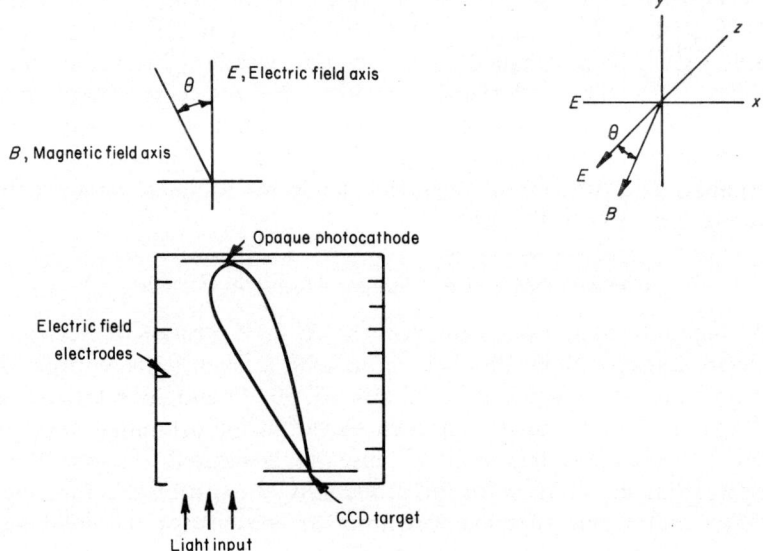

Fig. 3. Schematic diagram of windowless image intensifier configuration showing oblique magnetic and electric field vectors.

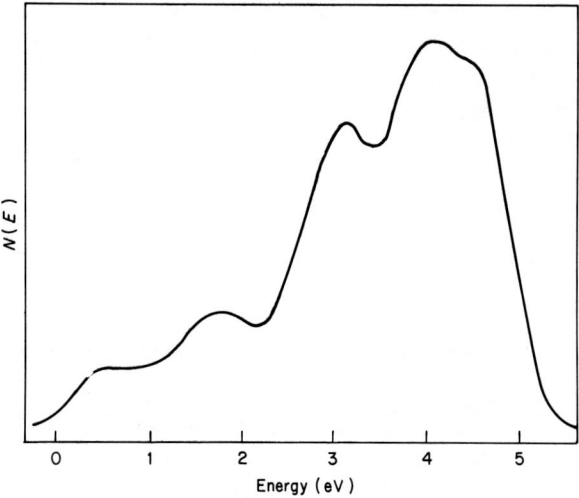

Fig. 4. Photoelectron energy distribution for 101·7 nm excitation of opaque CsI photocathode.[3]

shifted in X and Y as it is translated along the Z axis (E field direction). While this translation occurs without distortion, some aberration has been predicted.[2] One of the important applications for such image tubes is in the XUV below the LiF window cutoff at 105 nm. The emission energy dispersion of XUV generated photoelectrons from UV photocathodes such as CsI, is higher than that from typical visible light excited photocathodes. This energy dispersion reduces the resolving power of the magnetic focusing lens. Using computer generated electron trajectories, the point spread function of an oblique magnetic imaging system has been determined as a function of deflection angle for a typical photoelectron energy distribution in the XUV.

Computer Model of Oblique Magnetic Focusing

The digital computer program plots the X, Y, Z position of electrons in an electromagnetic field. Photoelectrons with a given initial energy were traced for elevation angles, ϕ, of 15, 30, 45, 60, 75 and 90° relative to the X, Y (photocathode) plane. At each elevation the azimuthal angle was varied from 0 to 360 degrees at 15° intervals measured from the X axis. Using a cosine distribution for the probability of emission as a function of elevation and a sine function weighting to account for the solid angle represented by each electron, each trajectory is given a $\cos 2\phi$ weighting. The mean position of the electrons was calculated as well as the RMS deviation (σ) from the mean over a range of Z to determine focus.

Figure 4 shows the photoelectron energy distribution for 100 nm excitation of a caesium iodide photocathode.[4] Based on these data, trajectories for photoelectrons of 2, 3, 4 and 5 eV energy were calculated. The RMS deviation at each photoelectron energy was given a weight indicated by the integral of the distribution about that energy. The net RMS deviation (σ_T) is calculated by the equation

$$\sigma_T = \tfrac{1}{3}[0\cdot 6\sigma_2^2 + 0\cdot 8\sigma_3^2 + 1\cdot 0\sigma_4^2 + 0\cdot 6\sigma_5^2]^{1/2}. \qquad (1)$$

Results of Computer Simulation

Figure 5 shows a typical graphic output at three focal plane positions for a photoelectron energy of 4 eV and a tilt angle, θ, of 30°. In this set of trajectories the electrons are emitted at an elevation of 30° at 15° intervals in azimuth measured from the X axis. The point spread function is symmetrical about a line parallel to the X axis. The electrons form two sets of points dependent on the sign of the X component of the electrons' emission energy. Each set has an optimum focus; in this case about 1 mm apart. The mean $\bar{X}$ position falls on the magnetic field vector and is given by

$$\bar{X} = Z \tan \theta. \qquad (2)$$

The nominal focus is at

$$Z = (2m/e)^{1/2} (\pi N V^{1/2} \cos \theta)/B,$$

where N is the number of focus loops. The mean $\bar{Y}$ position is given by

$$\bar{Y} = \frac{Z \tan \theta}{\pi N \cos \theta}, \qquad (3)$$

which is consistent with earlier analysis of oblique magnetic focusing.[2]

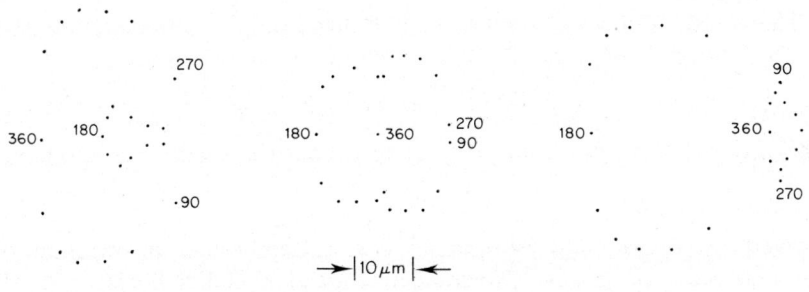

FIG. 5. Electron locations at three focal plane positions: $V = 4$ eV, $\theta = 30°$, $B = 18\cdot 75$ mT, $E = 3000$ V cm^{-1}. Elevation angle 30°, azimuth angles from 15° to 360°.

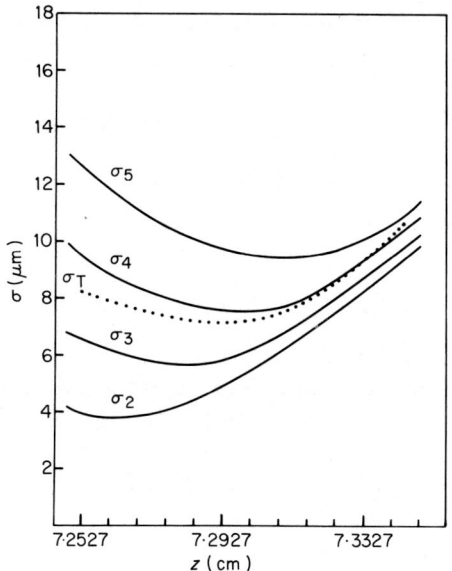

FIG. 6. RMS deviation from mean position of photoelectrons emitted with a Lambertian angular distribution as a function of focal distance. Subscripts denote photoelectron energy in eV.

Figure 6 shows the RMS deviation from the mean as a function of focal plane position for several values of photoelectron emission energy. In this case the magnetic field was 18·75 mT at an angle of 30° to the electric field, E, of 3000 V cm^{-1}. The weighted RMS deviation, σ_T, is also shown for the photoelectron emission energy distribution shown in Fig. 4. One notes a fairly broad focus region created by the range of axial energy of the photoelectrons.

The variation of σ_T with tilt angle is shown in Fig. 7 with data points at 0, 15, 30 and 40 degrees. These data closely fit the equation

$$\sigma_T = 3\cdot 2\,[1 + 3\cdot 6^2 \tan^2 \theta]^{1/2}\ \mu\text{m}. \tag{4}$$

The zero tilt RMS deviation of 3·2 μm is consistent with the equation,

$$\sigma = 0\cdot 31\ \nu/E \tag{5}$$

derived by Beurle and Wreathall,[5] for a Lambertian distribution of photoelectrons of energy ν, focused by coaxial E and B fields.

Doubling the magnetic field, doubles the number of focus loops per unit distance and as indicated by Eq. (3) decreases the $\bar{Y}$ displacement accordingly. The RMS deviation σ at the second focus is the same as that

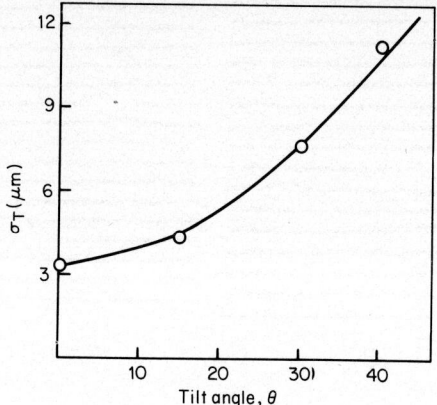

FIG. 7. Net RMS deviation σ_T as a function of tilt angle θ; $B = 18\cdot75$ mT, $E = 3000$ V cm^{-1}, $L = 7\cdot3$ cm.

found for single loop focus provided the focal plane to photocathode spacing and E field remain constant. Reducing the electric field E by a factor of 2 doubles σ for the same focal distance, (B is reduced by $2^{1/2}$). This is consistent with zero tilt magnetic focus systems.

Since the purpose of oblique magnetic focusing is to obtain a given lateral displacement of the focal plane from the optical axis, one must consider the trade-off of increased focal distance versus larger tilt angles. For a given accelerating voltage the aberrations increase with focal distance. From Eq. (4) one notes that the aberration is proportional to the tangent of the tilt angle θ except for small angles when the zero tilt abberations are a significant fraction of the total. Thus, the RMS deviation σ is inversely proportional to E and proportional to the lateral displacement, $\bar{X}$. It would appear desirable to employ higher tilt angles in order to minimize the magnetic field volume required. However, this is not a clear advantage because larger tilt angles require larger inside diameter magnetic focus assemblies to clear the electrode structure and the optical input beam.

ICCD RESPONSE TO SINGLE 20 KEV ELECTRONS

Figure 8 shows the video signals generated by single electron events in three 16×16 pixel patches in a 160×100 CCD.[6] The numerical display shows the numerical values of the digitized video signal for the single electron event at the lower left of patch 84.

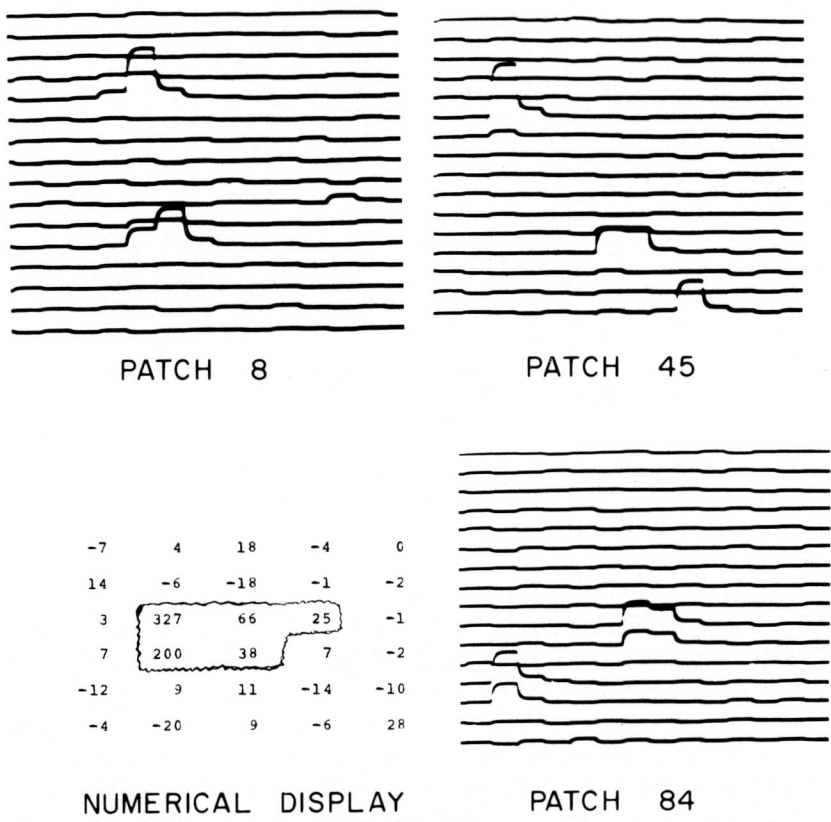

FIG. 8. Response of T1 160×100 ICCD to single 20 keV electrons. Numerical display shows signal generated by the single electron event at the lower left of patch 84; scale is 6·4 electrons per count.

The topmost event in patch 45 is representative of the cases where most of the charge generated by a single 20 keV electron event is collected by one pixel of the CCD. In this event, 3400 electrons were collected in the peak pixel, while the balance of the 4300 total charge collected is found in the pixel to the right and in the pixel directly below which collected respectively 560 and 340 electrons each.

The event in the centre of patch 84 is representative of the cases where the charge generated by a single electron event is nearly equally split among four pixels in a square block, with the electron event nearly hitting the four corner intersection of four pixels. In this case the peak pixel received a charge of 1500 electrons while the adjacent three received approximately 900 each, in addition to a small tail of 200 electrons appearing in a fifth pixel.

Computer analysis of several thousand single electron event signals, showed the total integrated charge collected in the CCD from the impact of a single 20 keV photoelectron to be 4300 electrons. Depending on the relationship between an event's location and the CCD pixel grid structure, each event occupies from three to six pixels.

Figure 9 is a histogram of the frequencies of occurrence of single

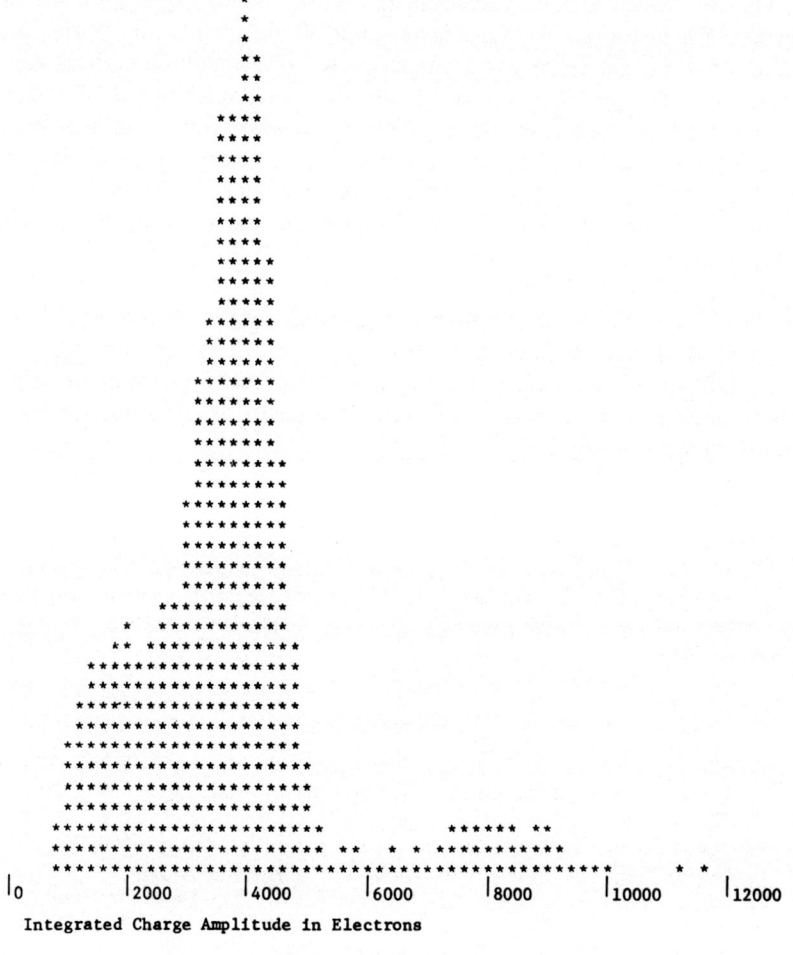

FIG. 9. Histogram of single electron events.

electron events as a function of the integrated charge found in each event. The peak at 4300 electrons is clearly shown as well as a second peak at 8600 electrons. This second peak represents double hits, i.e., events which include the charge of two photoelectrons. It is not necessary for the two photoelectrons to have landed in exactly the same pixel in a given data frame for a double event to be produced. It is sufficient merely that the groups of pixels containing any detectable charge of two nearby events touch or overlap slightly. In those cases the computer program that looks for events and gathers up all the charge in adjacent pixels treats the two events as one.

The full width at half maximum (FWHM) of the single photoelectron peak in the histogram of Fig. 9 is about 1400 electrons wide. With a mean value of 4300 electrons for a single event, single photoelectrons can be counted with high confidence. Note that the splitting of the charge among adjacent pixels causes a much larger variation in the single pixel peak signal generated by a single photoelectron event. This suggests that photon counting detector systems must process adjacent pixels and be sensitive to the total integration charge from each event for maximum effectiveness.

An aspect of the data in the histogram of Fig. 9 that deserves further study is the small but certainly not negligible number of events that have integrated charges of less than about 2500 electrons. Whether they are the result of secondaries, related to gas in our test apparatus, or primary photoelectrons that produce unusually low yields, is an important but not yet resolved question.

ACKNOWLEDGMENTS

This work was supported by U.S. Department of Energy contract EY–76–C–02–3073 and NASA grant NSG 5277. The authors gratefully acknowledge the assistance of Mark Greeley in carrying out the computer simulation and John Opperman in making the laboratory measurements.

REFERENCES

1. Robbins, C. D., Enck, R. S., Jr. and Sackinger, J. P., *In Proc. SPIE* Vol. 35 (1972).
2. Picot, J. P., Combes, M., Felenbok, P. and Fort, B., *In* "Adv. E.E.P." Vol. 33A, p. 557 (1972).
3. Johnson, C. B. and Hallam, K. L., *In* "Adv. E.E.P." Vol. 40A, p. 69 (1976).
4. Di Stefano, T. H. and Spicer, W. E., *Phys. Rev. B* **7**, No. 4 (1972).
5. Beurle, R. L. and Wreathall, W. M., *In* "Adv. E.E.P." Vol. 16, p. 333 (1962).
6. Antcliffe, G. A., Hornbeck, L. J., Chan, W. W., Walker, J. W., Rhines, W. C. and Collins, D. R., *IEEE Trans. Electron Devices* **ED-23**, 1225 (1976).

Astronomical Imagery with Solid-State Arrays

G. G. FAHLMAN, S. W. MOCHNACKI, C. PRITCHET, A. CONDAL
and G. A. H. WALKER

Department of Geophysics and Astronomy, University of British Columbia, Vancouver, Canada

Introduction

In this paper we will discuss optoelectronic characteristics and astronomical applications of two panoramic solid state imaging arrays: (i) a 50×50 photodiode array manufactured by the Reticon Corporation; and (ii) a 400×400 CCD (charge coupled device) manufactured by Texas Instruments.

Photodiode Array and Observations

The 50×50 photodiode array plus associated electronics and software were assembled by one of us (S.W.M.) as part of the requirements for a Ph.D. dissertation. This system has been extensively discussed elsewhere.[1,2] Some operating characteristics are summarized in Table I. Of particular interest are the dead space (from surface electrodes on the array), and the large readout noise and fixed pattern amplitude. The fixed pattern arises from clock signals which are coupled onto the video line, and is quite stable from readout to readout. It may therefore be successfully removed by subtracting a "dark" readout taken at about the same time as the "data" readout (at the expense of $\sqrt{2}$ degradation in noise). Recently the amplitude of fixed pattern noise has been reduced by a factor of 10 simply by laying out the video circuit board more carefully.

This system has been used quite successfully in observations of the nuclear regions of nearby galaxies, an observational area for which high dynamic range and signal to noise ratio are required, and for which there is no shortage of light. To illustrate this, we will consider some observations of the bright Type II Seyfert galaxy NGC 1068 which were made with the 1·8 m telescope of the Dominion Astrophysical Observatory, Victoria, Canada in September 1977. The observations consist of two

TABLE I

Device properties

Type of device	Diode array	CCD (rear illuminated)
Size	5×5 mm^2	9×9 mm^2
Format	50×50 pixels	400×400 pixels
Centre to centre spacing	102 μm	23 μm
Dead space	50%	none
Saturation charge	5×10^6 electrons	3×10^5 electrons
Readout noise	4000 electrons	27 electrons
Fixed pattern amplitude	~1/2 of saturation	very small
Operating temperature	$-76°$C	$-100°$C
Dark current at T_{op}	50 electrons sec^{-1} pixel^{-1}	<0.5 electrons sec^{-1} pixel^{-1}
Peak Q.E.	70% at 0.7 μm	70% at 0.7 μm
Thickness	300 μm	10 μm

unfiltered 5 min exposures of an area 30×30 arcsec^2; the seeing during these exposures was 2 arcsec.

A convenient way to analyse the observations is to measure the "effective radius" $r^* = \sqrt{A/\pi}$ of a contour of constant image brightness. In Fig. 1 we plot the logarithm of contour intensity against r^*. This shows that the inner 7 arcsec of NGC 1068 may be separated into at least two components: (i) a region displaying an exponential fall-off of intensity with radius (which Freeman[3] would call a "disk"); and (ii) an inner excess.

The disk component of NGC 1068 has a scale length of about 5 arcsec, or 230 pc at an assumed distance of 10 Mpc (H = 100 km s^{-1} Mpc^{-1}). This is similar to the inner disk scale lengths found for two other Seyferts: NGC 1566 and NGC 4151.[4,5] The central surface brightness of the inner disk is $\mu_R \simeq 15.0$ mag arcsec^{-2}.

Subtracting this disk from the observations yields a nearly Gaussian component which follows the seeing profile quite closely (Fig. 2). The total brightness of this inner point source is $R = 12.2$, and its *intrinsic* size is less than 1 arcsec.

Is there any structure in the inner regions of NGC 1068 that is masked by the extremely bright nucleus and disk? To answer this question we have subtracted the stellar component from the observations. This procedure reveals a bar-like feature about 7 arcsec long at a position angle of 30°. It is interesting to note that this feature is similar in physical size to a bar[1,6] found in the nuclear regions of NGC 4736—a galaxy which possesses some morphological similarities to NGC 1068.

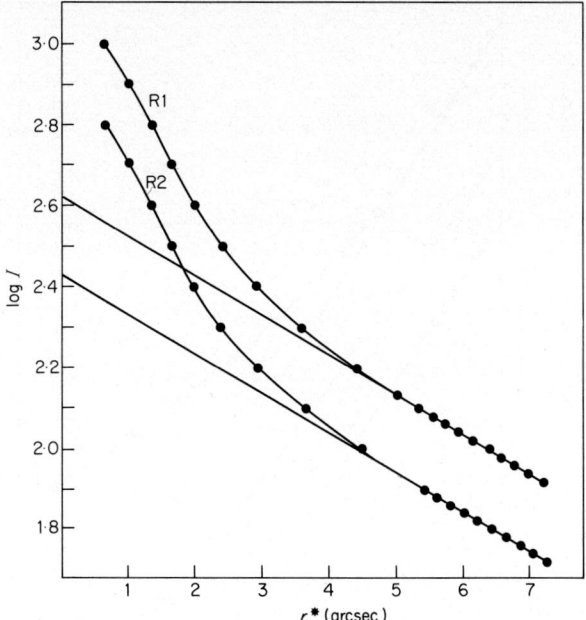

FIG. 1. Luminosity profile of NGC 1068. Two separate exposures are indicated as R1 and R2.

CCD ARRAY

The Texas Instruments CCD array described in this section is part of a touring system kindly made available to several observatories by the Jet Propulsion Laboratory, Pasadena, California. Two of us (A.C. and C.P.) were the first users of this system at Cerro Tololo Inter-American Observatory in January 1978, and were, as far as is known, the first users of a CCD in the Southern hemisphere.

The characteristics of the system are shown in Table I. Note especially the very low readout noise, and also the thinness of the array (10 μm, as compared to 300 μm for the diode array discussed above). Since the array is rear-illuminated, it must be thinned to provide acceptable blue response. Probably as a result of this thinning, the surface of the array is not flat; even to the eye, the array has a pronounced "buckled" appearance.

Fringing

When illuminated by a uniform monochromatic ($\lambda \gtrsim 900$nm) field, most silicon-based imaging arrays show fringe-like pixel to pixel variations in

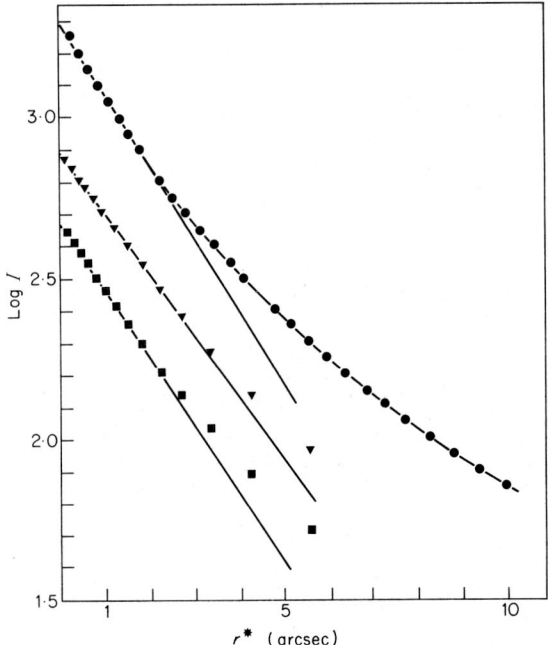

FIG. 2. Residual brightness in NGC 1068 after subtracting exponential disk. Results from two exposures are compared with a stellar image. ●, HD 196882A; ▼, NGC 1068: R1; ■, NGC 1068: R2.

their response. These interference fringes are due to thickness fluctuations in the array, coupled with the near transparency of silicon at long wavelengths. This problem is exacerbated in CCDs due to the extreme thinness of the arrays.

In the present array, fringing was present for monochromatic observations with $\lambda > 600$nm. At 656nm, fringe amplitude was 10 to 20%; at 953 nm, fringe amplitude was greater than 50%. This extreme fringing is demonstrated in Fig. 3, which shows an image of the Orion Nebula taken in the light of [S III], 953 nm. Fringing was *not* a problem for observations made through broad-band filters, since the bandwidth of the filter substantially reduced the coherence of interference.

In principle, it should have been possible to remove fringing in our monochromatic CCD images simply by dividing by a flat field taken through the same filter. In practice, we were always left with small (but distressing!) residual fringes with an amplitude 0·1 to 0·2 of the original fringes. This is at least partly due to the fact that flat field sources (lamp, sky, etc.) generally have continuous spectra, and hence fill the (finite)

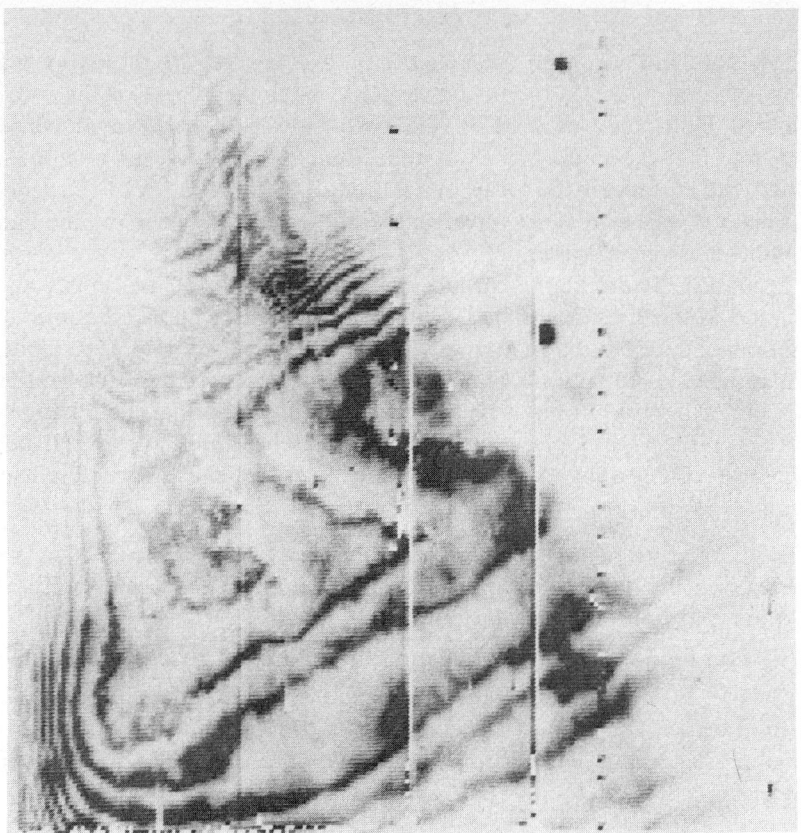

FIG. 3. Image of Orion Nebula taken in the light of [S III] at 953 nm. Detector is the CCD array.

bandwidth of a filter in a different manner than does a (monochromatic) astronomical source.

(It is interesting to note that fringe amplitude appeared lower at wavelengths longer than 1 μm. The transparency of silicon at long wavelengths is large enough for multiple reflections to occur. Both resolution and fringe amplitude are therefore degraded as a result of the finite solid angle of the telescope beam.)

One possible solution to the problem of fringing is to build thicker arrays. Such arrays have lower fringe amplitude (again, a solid angle effect) at the expense of response to blue light. To improve the blue response, one might then coat the array with an appropriate red emitting fluorescent compound.[7,8]

Flat Fields

The flat field response (blue light) in two corners of the array was approximately 0·2 relative to the response near the centre of the array. For red light, the response in the corners was 0·5. Lowered corner response may have been due to reflection losses, since, as mentioned above, the surface of the array was not flat. Since the index of refraction of silicon is higher at short wavelengths, the corner response in blue light should be lower, as observed.

In another paper at the Symposium, Baum[9] discussed the importance of, and difficulties associated with, calibrating flat field response of panoramic detectors to an accuracy better than 1 part in 10^3. The stability of flat fields taken with the present CCD array was very much worse than this target figure. During a typical night, variations in flat field response were as large as 2–3 per cent: that is, the global *shape* of the flat field response changed by several per cent. Possible causes are discussed below.

(i) If pixel response were slightly polarization dependent, then flat fields using the dusk or dawn sky as a source might vary with telescope position and hour angle of the sun.

(ii) Flat fields taken by illuminating the inside of the dome might not have been perfectly uniform.

(iii) In view of problems encountered with the cooling system, it is possible that a thin layer of frost may have built up on the array during the night.

Wavelength Response

We obtained useful observations over a wavelength range 440 nm (broadband B) to 1·083 μm (He I). Response at B was about 0·30 of the peak response (at 700 nm). Response at 1·083 μm was less than 0·01 of peak, presumably due to cooling the array.

"Cosmetic" Problems

The array used in the present work contained a number of defects, including dead columns and pixels, "hot" pixels, that is pixels with higher than average dark current rate, and columns prone to streaking. Some of these defects were permanent; others were transient, for reasons unknown. The severe impact of these defects on subsequent picture processing may be gauged by examination of Fig. 4, which shows a contour map of a slightly worse than average frame of the S0 galaxy NGC 5102.

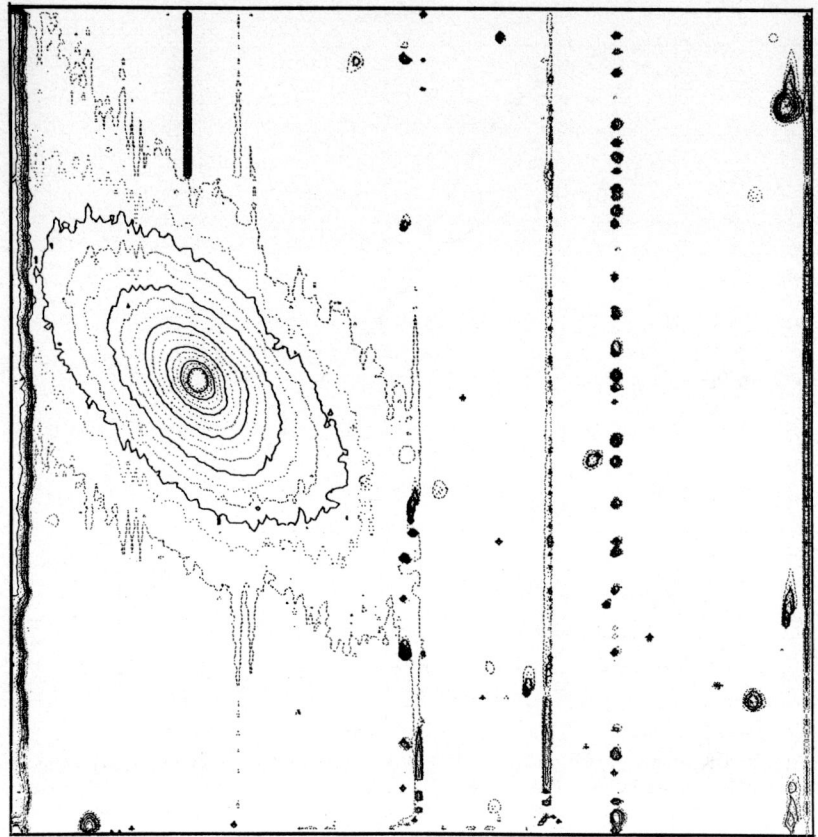

FIG. 4. Contour map of NGC 5102 from 4-min exposure with CCD.

Astronomical Results

Because of the above mentioned array defects, and also because of problems with the cooling system, observations were generally short exposures of bright objects. We were still able to take advantage of the large dynamic range of the array.

In this section we will briefly discuss observations of a bright southern galaxy, NGC 5102. Further details will be published elsewhere.

NGC 5102 is an S0 galaxy with at least two peculiarities: (i) strong 21 cm emission, with $M(\text{HeI})/L$ typical of an Sb galaxy;[10] and (ii) a strong nuclear colour gradient, in the sense that the nucleus is *bluer* than the surrounding bulge.[11,12] Observations to delineate the nuclear structure of

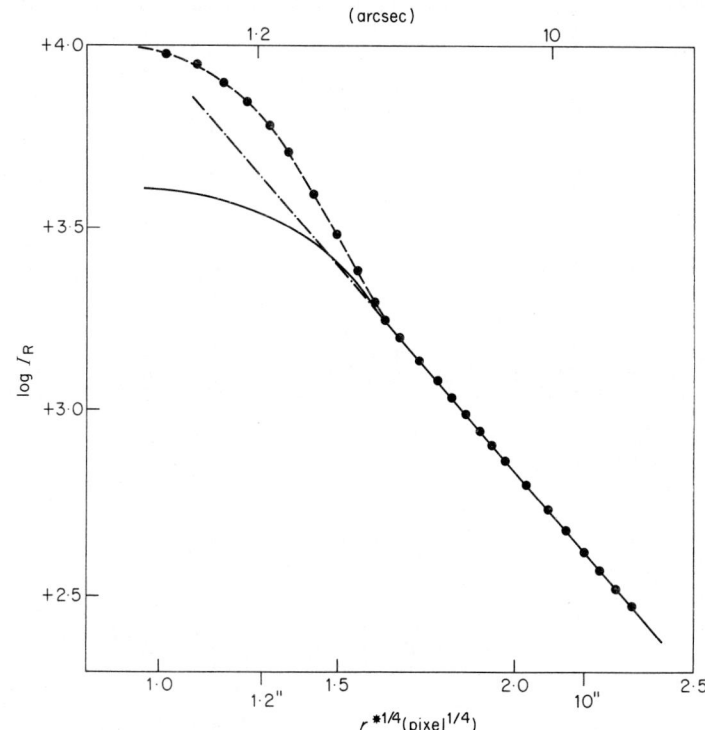

FIG. 5. Brightness profile of NGC 5102 (log I versus $r^{1/4}$). Observations are compared with a de Vaucouleurs law convolved with the seeing profile. $-\cdot-$, dV law; —, dV convolved with seeing; ---, dV convolved with seeing plus central object; • • •, observations.

this object were ideally suited to a CCD array because of this detector's wide dynamic range and (potentially) high photometric accuracy.

Observational data consisted of four 4-min exposures through B and R filters using the CTIO 1·5 m telescope. The image scale was 0.42 arcsec per pixel; seeing was approximately 2 arcsec.

Data reduction was performed on the U.B.C. Amdahl computer using a software package developed by one of us (C. P.). Preliminary data analysis consisted of : (i) dark current removal; (ii) division by flat field; (iii) removal of sky background; (iv) "patching" of defective pixels; (v) contour plotting with an estimate of equivalent radius, ellipticity, and orientation for each contour.

The disk of NGC 5102 does not contribute significantly to the light from the area of the galaxy that we observed. Hence we have compared our observations directly with a de Vaucouleurs law:

$$\log I = a\, r^{1/4} + b,$$

which has been shown[13,14] to provide a good fit to the luminosity profile of spheroidal systems (including "normal" S0 bulges) even within the seeing disk. Figure 5 shows that for $r \lesssim 5$ arcsec, NGC 5102 indeed follows a de Vaucouleurs law. At smaller radii, the observations must be compared with a *seeing-convolved* de Vaucouleurs law. Such a comparison (Fig. 5) shows that NGC 5102 possesses a significant excess of light in its central regions relative to a de Vaucouleurs law. This excess has a luminosity profile indistinguishable from the seeing profile (i.e., its size is certainly less than 1 arcsec).

An analysis for blue light images yields similar results. However, while the spheroid appears to have a negligible colour gradient, the "nucleus" (the region of excess over a de Vaucouleurs law) is quite blue ($B - R \simeq 0.7$). The nuclear regions of NGC 5102 therefore consist of a normal bulge, superimposed on a small, bright, blue object. Seeing effects produce the colour gradient observed with aperture photometry.

Acknowledgments

The authors wish to thank the directors and staff of Cerro Tololo InterAmerican Observatory and Dominion Astrophysical Observatory for granting observing time and assisting with the observations reported in this work. Jet Propulsion Laboratory, and, in particular, Dr James Janesick, made the CCD array available to us. Vern Bucholz, Mike Creswell, and Greg Cozza provided invaluable technical assistance with the arrays. Peter Dewdney and Carmen Costain provided the use of the plotting facilities at the Dominion Radio Astrophysical Observatory, where Figs. 1 and 4 were produced. Finally we wish to acknowledge fruitful and stimulating discussions with Bruce Campbell, Paul Hickson and Pat Monger. This work was supported by the National Research Council of Canada through operating grants to G. A. H. Walker, G. G. Fahlman, and J. R. Auman.

References

1. Mochnacki, S. W., Ph. D. Dissertation, University of British Columbia (1977).
2. Walker, G. A. H., Buchholz, V., Fahlman, G. G., Glaspey, J., Lane-Wright, D., Mochnacki, S. W. and Condal, A. *In* "Astronomical Applications of Image Detectors with Linear Response" (IAU Colloquium No. 40) ed. by M. Duchesne and G. Lelièvre, p. 24–1. Paris-Meudon Observatory (1976).
3. Freeman, K. C., *Astrophys. J.* **160**, 811 (1970).
4. de Vaucouleurs, G., *Astrophys. J.* **181**, 31 (1973).
5. Simkin, S. M., *Astrophys. J.* **200**, 567 (1975).
6. Mochnacki, S. W. (in preparation).
7. Westphal, J. A. *In* "Wide-Field/Planetary Camera for Space Telescope", Technical proposal submitted to NASA.
8. Campbell, B. and Walker, G. A. H., *In* "UV-Sensitization of Solid-State Imaging Arrays", Technical proposal to Space Science Coordination Office, NRC (1978).
9. Baum, W. A., *In* "Seventh Symp. P.E.I.D. Preprints" p. 265 (1978).

10. Gallagher, J. S., Faber, S. M. and Balick, B., *Astrophys. J.* **202,** 7 (1975).
11. van den Bergh, S., *Astron. J.* **81,** 795 (1976).
12. Alcaino, G., *Astron. & Astrophys. Suppl.* **26,** 261 (1976).
13. de Vaucouleurs, G., *In* "The Formation and Dynamics of Galaxies" (IAU Symp. No. 58) ed. by J. R. Shakeshaft, p. 335. Reidel, Dordrecht (1974).
14. de Vaucouleurs, G., *Astrophys. J. Suppl.* **29,** 193 (1975).

Photon Detection Experiments with Thinned CCDs

R. G. HIER, E. A. BEAVER and G. W. SCHMIDT

Physics Department, University of California, San Diego, La Jolla, California, U.S.A.

and

G. D. SCHMIDT

Steward Observatory, University of Arizona, Tucson, Arizona, U.S.A.

Introduction

With the trend in present day astronomy toward the capabilities offered by digital data acquisition and processing techniques, it has been realized that CCDs, with their excellent linearity, dynamic range, and quantum efficiency, have great potential as the sensing element at the heart of such a system. This paper describes results of a programme to develop high speed photon noise limited astronomical imaging techniques using CCDs.

CCD Operating System

The details of the computer controlled and operated CCD system as shown in Fig. 1 will be discussed more fully elsewhere,[1] but briefly, the philosophy in its design and construction has been to make it as general and expandable as possible. The present implementation was enabled through the use of the already existing NOVA-CAMAC-FORTH framework developed for the Digicon programme.[2]

The CCDs presently being used are Texas Instruments 160 × 100 backside-illuminated arrays thinned to approximately 10 μm, but the CAMAC based drive electronics (Fig. 2) can handle many other chips by simple software input changes, and most of the rest by insertion of new clocking pattern ROMs. This circuitry also contains a provision for "windowing" sections of the array data, so that rapid time slice procedures, requiring multiple images in memory, or larger arrays may be accommodated by our present rather limited amount of core space (16 k words for data). Readout rate is variable, but presently limited to about 25–30 microseconds per pixel by NOVA memory access time. In addition, if a photon counting mode is desired (for intensified CCD systems),

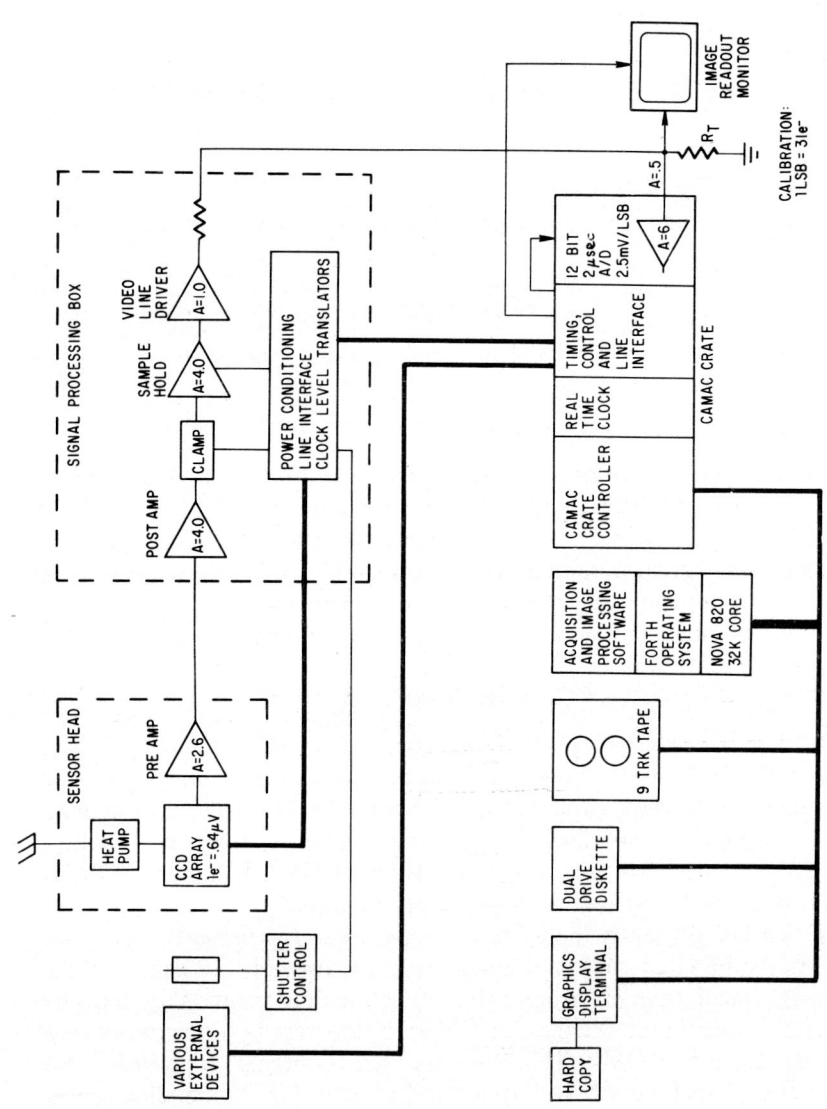

Fig. 1. CCD control and data handling system.

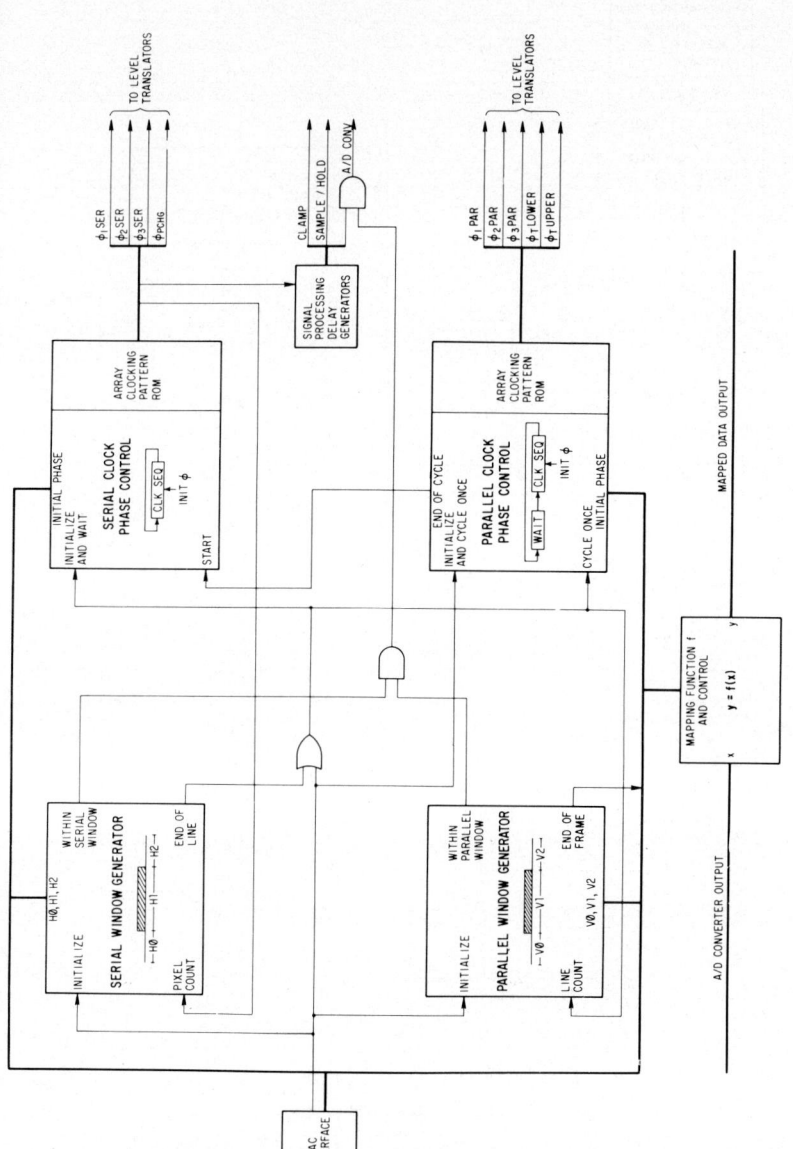

FIG. 2. CCD timing and control block diagram.

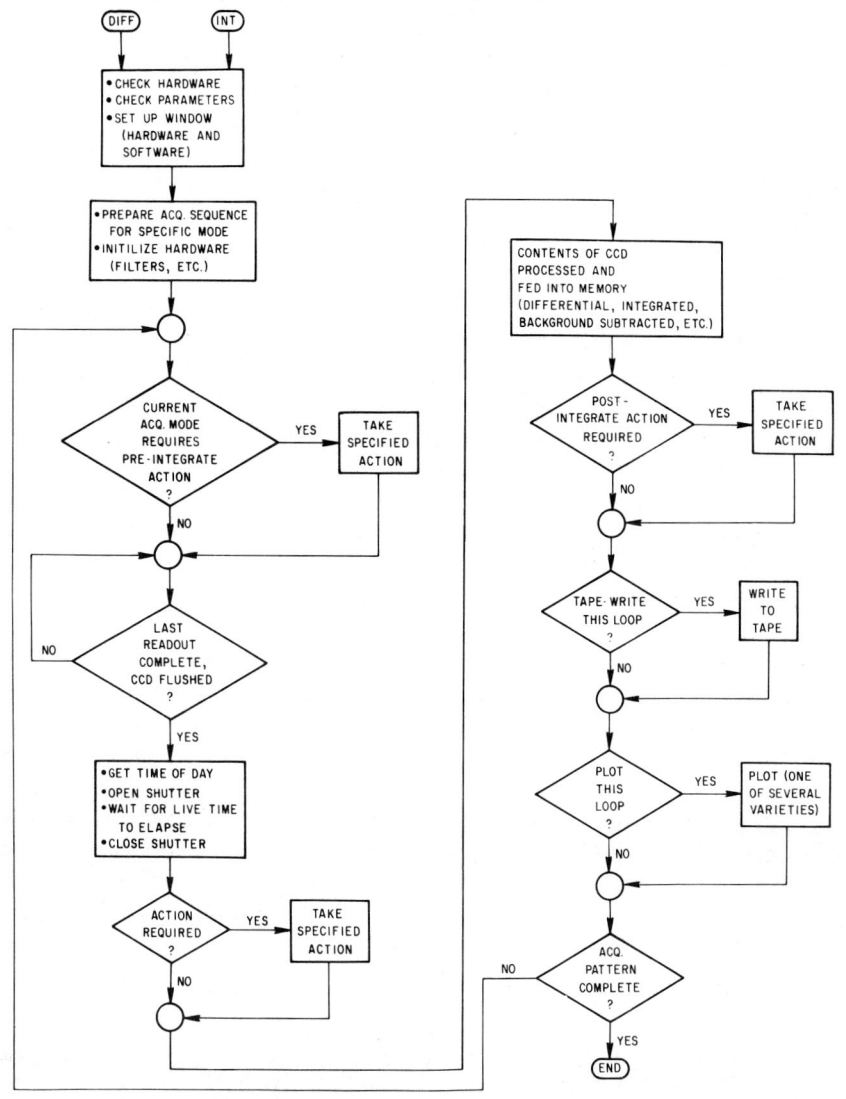

FIG. 3. CCD acquisition flowchart.

the hardware mapping controller seen here may be used effectively as a whole set of software settable discriminator levels to map raw data into photon counts. Alternatively it can be set up to simultaneously count photons in low light level pixels and work in an analogue mode for high levels (the desired breakpoint depending on the particular pulse height distribution), thereby preserving dynamic range.

The off-chip pre-amplifier is bandwidth tailored to match the array readout rate, and correlated double sampling is employed. By clocking the CCDs serial register in reverse, the readout electronics noise has been found to be equivalent to about 30 electrons RMS on the CCD, and this agrees well with theoretical predictions. Most of this is device noise in the pre-amplifier. The array is cooled to reduce thermal generation of charge; at our typical operating temperature of $-30°C$, we see a rate of 400–500 electrons sec^{-1}, so the RMS noise due to this effect is less than the readout electronics noise for integration times less than a couple of seconds (which is acceptable for our relatively high speed applications). An electronically controlled shutter blanks the array during readout, as well as providing a means to sample only the thermal background for subtraction.

The software which controls the overall operation of the system, including data acquisition, storage, and on-line reduction and display, along with the FORTH operating system, resides in the remaining 16 k of the NOVA minicomputer. The acquisition procedure flowchart is shown in Fig. 3. The "specified actions" are quite general and may include virtually anything required by the particular acquisition mode; also a different action may be specified for each pass through the loop. Note in particular that these, through the CAMAC interface, control the "various external devices" of Fig. 1, typically items such as polarizers, filters, and potentially even the telescope (chopping secondary, etc.), as well as setting up background subtraction and other internal procedures.

GAIN FOR SINGLE PHOTON DETECTION

CCD imaging in the direct optical input mode is quite useful, and has important applications in astronomy. However, in many cases, notably high speed applications, single photon detection is desired, and for this, considerably more gain is required. Also, a problem arises when thinned CCDs are used directly to detect photons concentrated in narrow wavelength regions (imaging objects with emission lines, spectrographic work, etc.). In spectral regions where the thinned CCD is not optically thick, interference patterns resulting from internal reflections may result. This difficulty can also be alleviated by the insertion of a suitable gain stage.

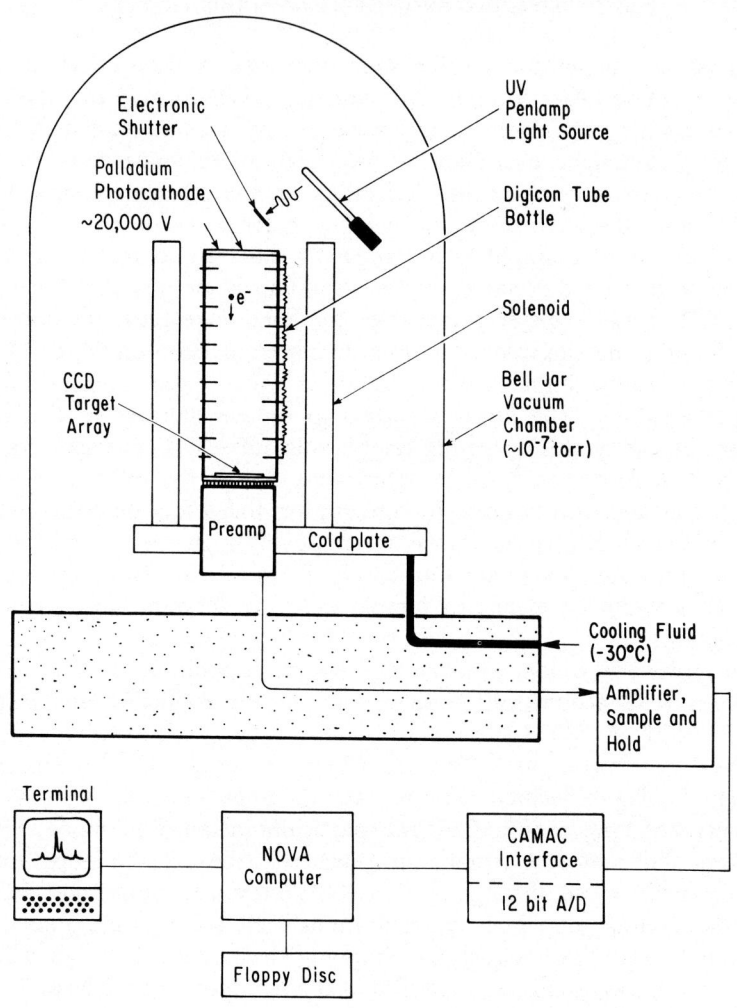

FIG. 4. CCD demountable Digicon configuration.

A clean way of obtaining sufficient gain is by inserting a CCD at the target end of a Digicon tube. This technique offers very low noise gain, low dark rate, and good imaging resolution, along with the added capability of magnetically scanning the image both for substepping purposes and for larger area coverage.[2,3] Progress has been made towards this end over the past couple of years,[4-8] but in the interim before reliable tubes are readily available, programmes have been undertaken here to show that the CCD-Digicon concept is viable and to immediately apply high gain CCD imaging to astronomy.

DEMOUNTABLE DIGICON IMAGING TESTS

In order to investigate CCD-Digicon photoelectron imaging characteristics without being subject to complications and expense due to phototube processing, a demountable Digicon system (Fig. 4) was utilized during the fall of 1977. The demountable Digicon consists essentially of an open-ended Digicon tube which can be sandwiched between a photocathode and the target to be tested. The entire assembly is then placed in a laboratory vacuum chamber, along with the necessary electromagnets, light source, shutter, cooling mechanisms, etc.

No significant system noise was introduced by the demountable system, although it should be noted that at the time of these tests an A/D converter with a factor of 4 less resolution than our present device was being used, so that a large component of system noise was the quantization error[9] inherent in the digitization process. In these tests, the image was provided by a 20 μm wide photoetched slit overcoated with a semitransparent palladium photocathode, illuminated by the 253·6 nm Hg line. The image was magnetically scanned across the CCD in various orientations, and good imaging characteristics of the system were verified, with resolution of approximately 15 μm.

As expected, photoelectrons from the demountable Digicon produce approximately one electron in the CCD silicon for each 3·5 eV remaining after passing through a roughly 4 keV dead layer, and hence single photoelectron events are readily discernible at energies above about 10 keV. However, the resulting pulse height distribution (Fig. 5) is seen to be quite flattened. Note, by the way, that one abscissa scale results from an independent gain calibration of the CCD system using a value of 0·25 pF for the CCD output capacitance,[10] and the other is a linear fit to the maximum response i.e. the sharp cutoff seen in the pulse height distribution, to various input electron energies from 14 to 23 keV. The two scales agree to within a few percent. The "tail" seen to the right of the single peak is due to double events, and these are presumably also

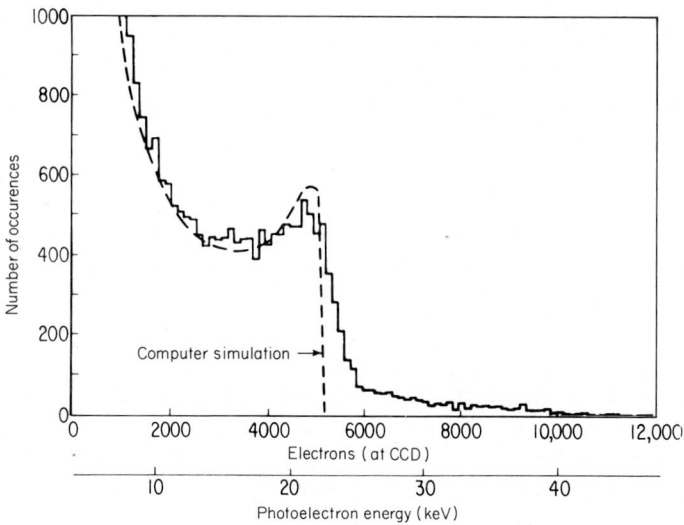

FIG. 5. 23 keV photoelectron pulse height distribution. The computer simulation curve is the same as the 2/9 (5 μm) curve in Fig. 8.

polluting the rest of the curve slightly, but we find that most of the shape of the curve is due to the spreading of the charge packet formed by the impinging photoelectron as it diffuses from the back of the chip through a few microns of undepleted silicon before reaching the potential wells formed by the electrode structure on the front side. A pulse height distribution for a frontside electron bombarded CCD has been obtained by Choisser,[11] and since no diffusion is involved, this case shows a more peaked structure. However, frontside bombardment gives rise to degraded performance very rapidly.[11]

The charge diffusion explanation is intuitively consistent with the geometry of the situation (pixels 23 μm square, undepleted silicon thickness 6–8 μm).[6] It should be pointed out that this is not as incompatible as it might seem with the good point spread and modulation transfer functions exhibited by these devices, but rather the discrepancy with the previous conclusions that diffusion effects would be negligible[6,7] is understood by taking into account the fact that photoelectrons are much more likely to fall close to a pixel edge (an area effect) than is evident from a one-dimensional treatment.

That charge spreading is actually taking place can be verified by looking more closely at the data. Figure 6(a) suggests that in a significant number of cases, most of the charge packet formed does not fall in a

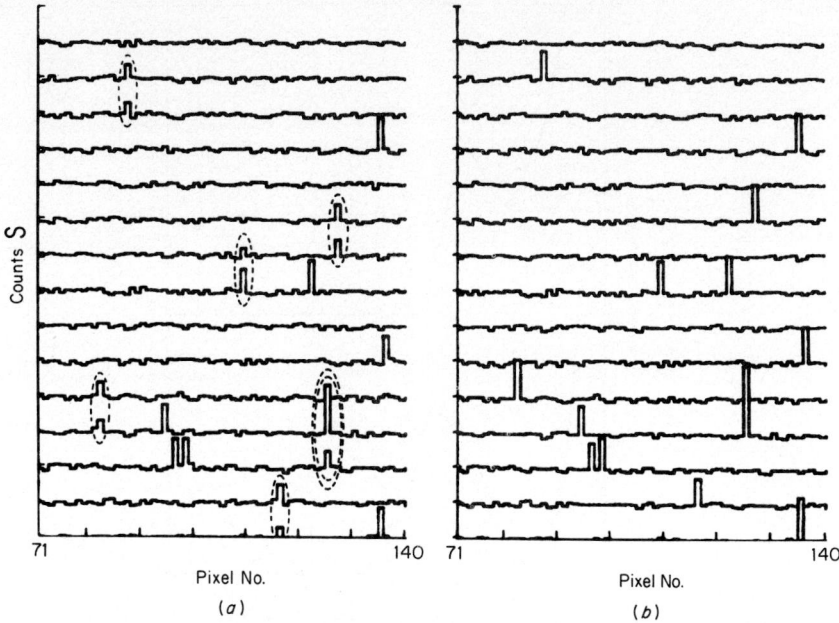

FIG. 6. Single 20 keV photoelectron events: (a) raw, and (b) reconstructed. Lines 44–58 are shown stacked so that line spacing is about 4500 electrons.

single CCD cell, but is distributed between 2, 3, or 4 neighbouring cells. The pulse height distribution resulting from the frame of data of which Fig. 6(a) is a part is shown in Fig. 7(a). The single photoelectron events can be "reconstructed" by a procedure which simply finds each peak greater than some significance level and then stacks all of the counts from the neighbouring (in this case, 8) cells onto that peak. The result of this procedure is seen in Fig. 6(b) and the corresponding pulse height distribution in Fig. 7(b). The heights are much more uniform, and even with the poor statistics, this pulse height distribution is a considerable improvement. This procedure also seems to have found a few double events (one of these is shown in Fig. 6) and perhaps a couple of triples. However, for certain spacings, multiple events may be confused in the reconstruction process; also, since in this case only nearest neighbours were considered, a little, probably less than a few percent, of the charge from single events may sometimes be missed. So, if it were possible, a "correctly" reconstructed distribution would probably look rather more sharply peaked.

As a check for possible artifacts of the reconstruction procedure, the same data was used but the pixel positions were first randomized to

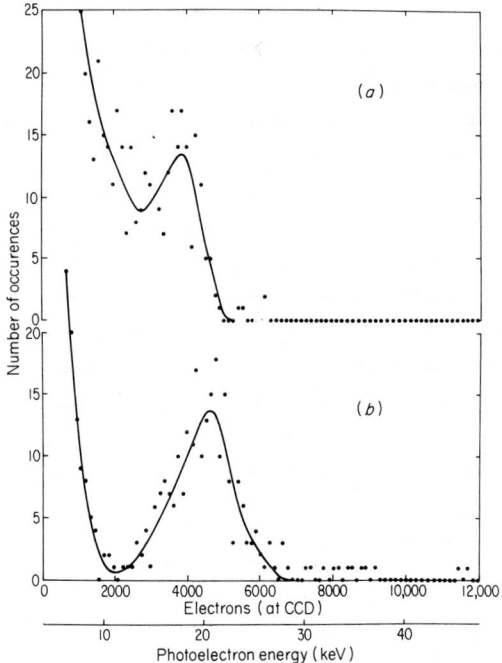

FIG. 7. 20 keV photoelectron pulse height distribution: (a) raw, and (b) reconstructed. The solid lines result from filtering the data points shown.

destroy any association with their original neighbours. The reconstruction then resulted in a distribution essentially the same as the unreconstructed curve shown.

Computer Simulation Results

In order to make some predictions for the behaviour of various possible configurations and to further verify the charge diffusion interpretation of the charge spreading phenomenon, a computer simulation was performed. The simulation calculated the response of a single square pixel to each of a uniform two-dimensional distribution of photoelectrons, using an approximation to the theoretical diffusion point spread function[12,13] parametrized by the ratio of the depth of undepleted silicon the charge must diffuse across to the width of the pixels.

Note that a single cut through the centre of the pixel in this simulation gives results essentially equivalent to, and in good agreement with,

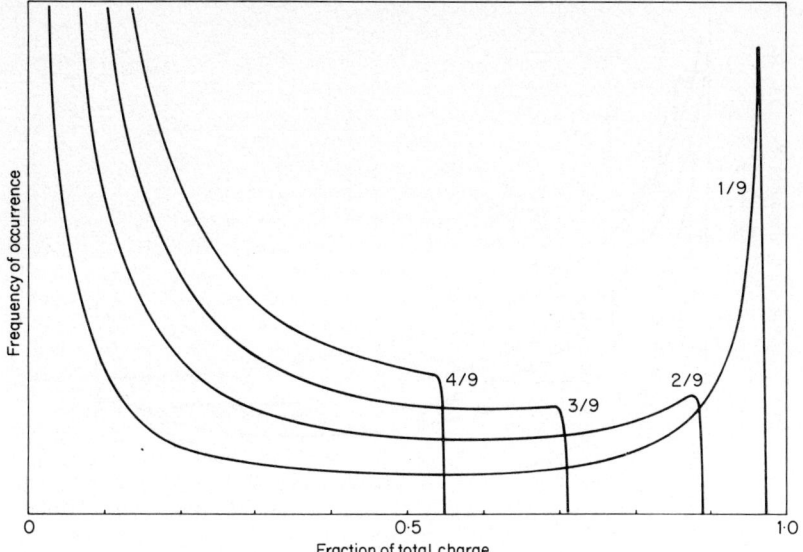

FIG. 8. Pulse height distributions from computer simulation; curves are labelled with ratio of effective diffusion depth/pixel width.

previous treatments, both theoretical[13] and experimental.[14] In fact, we find that pulse height distributions of such cuts are quite sharply peaked.

However, with the full two-dimensional treatment, the importance of the area effect becomes clear. Figure 8 shows that, as expected, for smaller depth to width ratios the pulse height distribution approaches one sharp peak, but as the ratio grows, the distribution is flattened quite quickly. We also note here that the response to even a "dead centre" hit shown by the cutoff in the distribution, varies significantly with the ratio. In order to fit the experimental data to them, these curves were scaled horizontally so that their cutoffs matched the single photoelectron maximum; in addition, semilog versions were used so that vertical scaling could be accomplished by simple sliding. The results are seen in Fig. 9. It is clear that the shape of the experimental data is intermediate to that of the theoretical 5 μm and 7·5 μm curves. Even though the computer simulation includes no noise, virtually all of the "flattened" shape of the data, with the exceptions of the deviation at very low energy and the width of the photoelectron peak, can be accounted for by this simple diffusion model. The "zero keV" (no photoelectrons) curve that was taken in conjunction with the data shown seems to account for the low energy deviation quite well. It should be mentioned that this is wider than

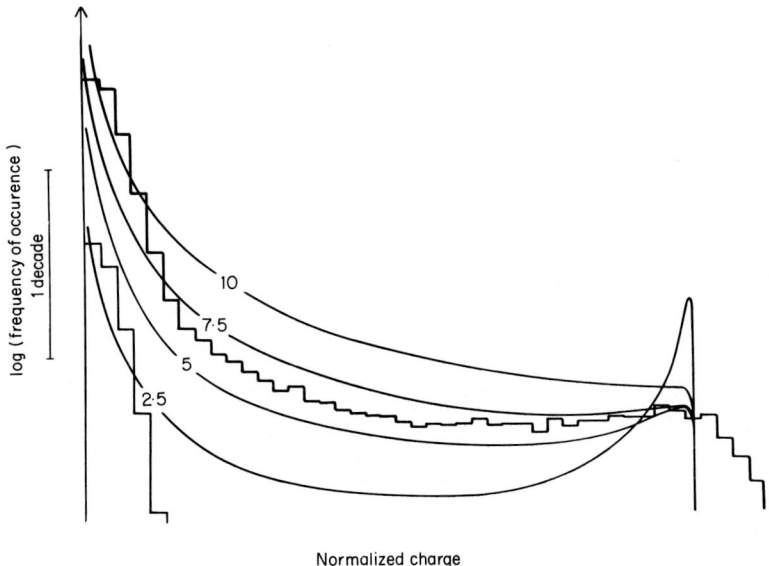

FIG. 9. Fit of experimental pulse height distribution to computer simulation curves of Fig. 8. Computer curves are labelled with effective diffusion depth in microns for the Texas Instruments array used.

the system noise as it contains spatial variations in CCD background which, at the time of these tests, were not subtracted out properly. There are a number of other effects which, even though the photoelectron gain dispersion may be small, cause a spread in the distance over which the charge must diffuse, thus washing out the pulse height distribution. These include non-uniform depletion depths and thinning in the chip, but the largest cause is that, while the peak of the secondary electron generation distribution occurs about a micron or so into the silicon, significant charge is actually spread through the first few microns.[12] Remembering all this, the good fit of the data with these curves suggests an effective diffusion depth of about 6 μm, which is in excellent accord with the 6–8 μm total undepleted region quoted earlier.

The simulation data may also be presented in another way. Figure 10 is a contour plot of the fraction of the charge resulting from photoelectrons landing at a particular position that will be collected by the pixel shown. This data applies to the depth/width ratio of 2/9. For smaller ratios the contours are more square and approach the pixel edges and for larger ratios they are more circular and spread out. In this representation the area effect may be seen more directly.

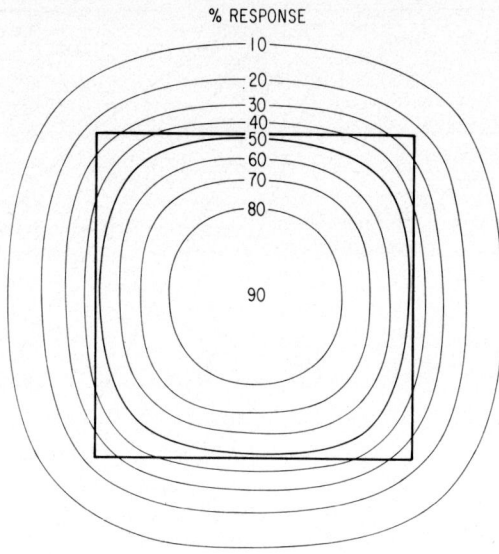

FIG. 10. Contours of equal pixel response to photoelectron events for a depth/width ratio of 2/9.

Another question which may be addressed by this treatment is that of discriminator placement. Even with a somewhat diffusion flattened pulse height distribution, correct photoelectron counting may be done, but with a loss of efficiency. As an example, for the 2/9 depth/width case shown, the contour for 50% of the total charge deposited encloses about 80% of the pixel area. Thus such a discriminator setting would miss about 20% of the photoelectron events (those which land in pixel corners). Ideally, in order to avoid multiple counting of single events while retaining as much efficiency as possible, the discriminator should be placed at the level of the lowest contour which remains entirely within the pixel. As depth/width increases, the maximum efficiency attainable in this way approaches a limiting (circular contour) value around 78·5%. However, even though dark counts due to noise should not be a problem because a noise level of 300 electrons RMS and a discriminator around 30% of the total charge deposited by a 20 keV electron would result in only about one dark count per pixel every 10^6 readouts, the discriminator should be raised a few per cent from this level in order to gain immunity from noise causing incorrect counting of single events. This will also result in some inefficiency. Plots of the computed efficiencies are shown in Fig. 11, in which the curves are cut off at the discriminator level where the

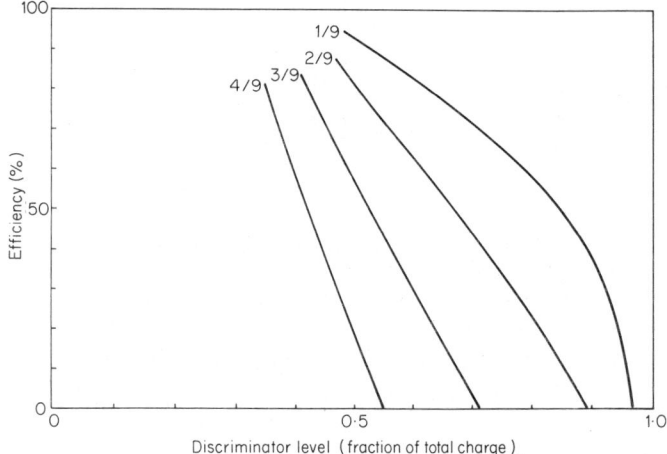

Fig. 11. Photon counting efficiency as a function of discriminator level for the four depth/width ratios shown.

corresponding contour touches the edge of the pixel. We see that as depth/width increases, this mode rapidly becomes impractical, since not only are the attainable efficiencies somewhat lower, but efficiency also varies more rapidly with change in discriminator level.

All in all, we see that there are significant performance gains to be had by further thinning the arrays. Texas Instruments can presently thin their arrays to an overall thickness of 6–7 μm,[15] and this should correspond to an effective diffusion depth of around 1–3 μm. As seen here, this should lead to a much cleaner pulse height distribution, and unambiguous photon counting with reasonably high efficiency should be possible.

On the other hand, if charge centroiding were to be done, a slightly larger depth/width ratio would be beneficial in spreading out the charge for more accurate centre determination. This would buy somewhat better resolution than the one pixel attained with simple counting, but at the expense of limiting the dynamic range to less than one photon per several pixels per frame, a somewhat lower signal to noise ratio and some added complexity.

CCD Astronomy with an Image Intensifier Tube

For astronomical applications, during the last half of 1977 the CCD system was lens coupled with a 3·4 reduction onto a 3-stage Varo image intensifier tube[16] and operated at the Cassegrain focus of the Steward

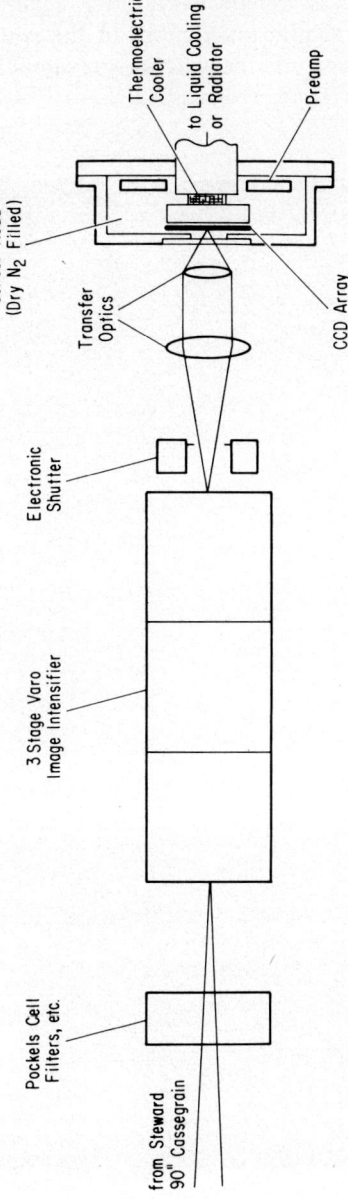

FIG. 12. CCD/image tube configuration.

Observatory 90 in. telescope at Kitt Peak giving a scale of about 0·85 arcsec per pixel (Fig. 12). Single events are again readily observable, although in this case a significant number of the roughly 4000 electrons resulting from each event may be spread, presumably due to the image

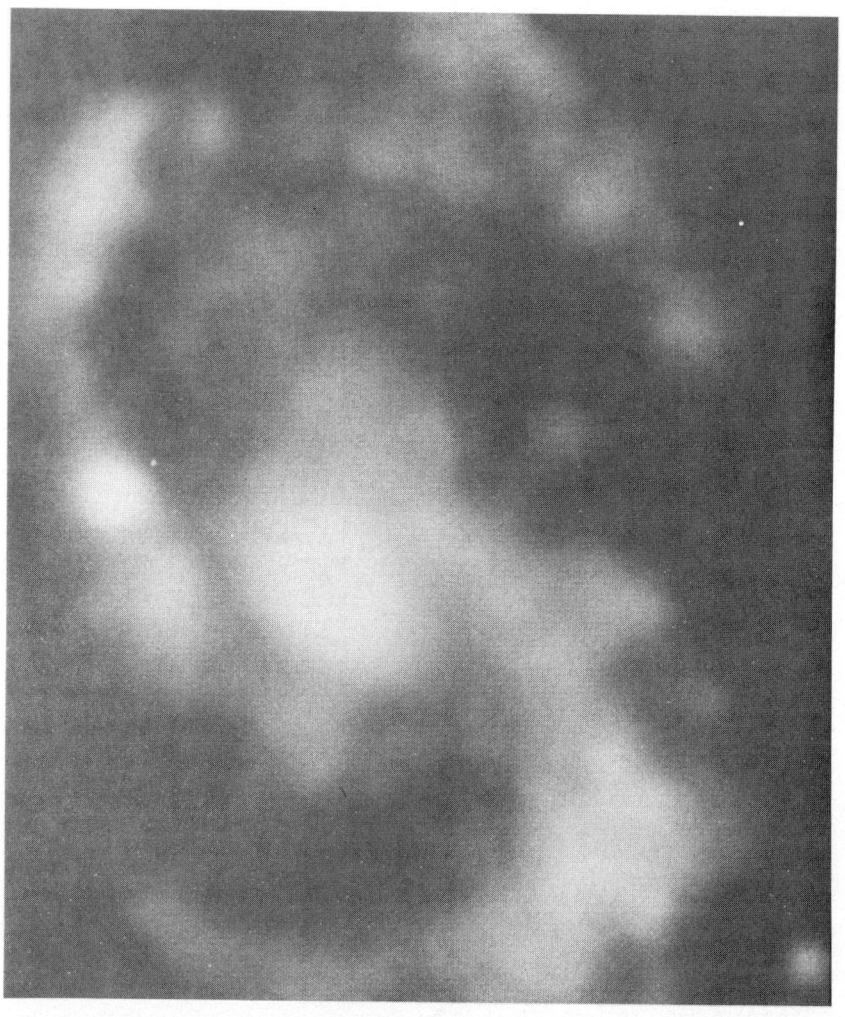

FIG. 13. NGC 157, 20 min exposure through C500 filter. Area is about 82×70 arcsec^2. Top is north, right is west; seeing is 2.5 arcsec.

tube spot size over an area 2–3 pixels square. Pulse height distributions were obtained, but were generally less impressive than the laboratory results, and are not included here for lack of space. The reconstruction process is more difficult here, partly because of the increased spread and partly because the tail of the dark phosphor noise may extend far enough to confuse the processing. It is thought that further work may be able to bring out the peak.

The system does perform well in an analogue mode, and in high speed faint object observations was found to produce photon noise limited statistics. Ion spots are a problem, but their removal may be possible by intercomparing separately recorded frames before adding them together to produce the final image.

So far this configuration has been used for polarization mapping of various galaxies, notably NGC 157 and DA 240, using a Pockels cell[17] to do the rapid polarization toggling, and for very narrow band interference filter comparative photometry of NGC 1275. Provisions were made for doing long slit spectroscopy using the Steward Observatory's Cassegrain spectrograph and for speckle interferometry, but these applications had to be postponed due to lack of time; seeing and weather conditions have not been cooperative.

Figure 13 shows a preliminary picture of NGC 157 as displayed on the Kitt Peak Interactive Picture Processing System[18] taken from the polarization data (hence necessarily using only half of the array area). All polarization states were stacked on top of each other, but no other data processing has been done.

Analysis of the astronomical data obtained is in progress, and detailed results should be forthcoming in the literature. Results have been quite encouraging, and plans are being made for continued use of the system on telescopes.

Conclusions

The general CCD system which has been developed shows promise of being a powerful tool, particularly with the potential addition of fast external memory. The demountable Digicon tests have shown the viability of single photoelectron imaging with CCDs, and, along with the computer simulation studies mentioned here, provide valuable insights into the construction of future systems. The CCD/image tube system, while perhaps not the ultimate detector configuration, has also demonstrated single photon sensitivity and has shown that with care it can be a respectable high speed photon noise limited astronomical instrument.

Acknowledgments

The authors wish to thank Carl McIllwain for suggesting the project and providing facilities and general support, Ken Ando for collaboration in the initial phase, Fred Landauer's group for supplying the arrays and supporting the CCD work, and Roger Angel for encouraging the CCD/image tube observations.

This work was supported by the Caltech President's Fund, grant PF–099; and by NSF grant AST 75-21551.

References

1. Hier, R. G. and Schmidt, G. W. (in preparation).
2. Beaver, E. A., Harms, R. J. and Schmidt, G. W., *In* "Adv. E.E.P." Vol. 40B, p. 745 (1976).
3. Delamere, W. A. and Beaver, E. A., *In* "Seventh Symp. P.E.I.D. Preprints" p. 195 (1978).
4. Lowrance, J. L., Zucchino, P. and Renda, G., "Seventh Symp. P.E.I.D. Preprints" p. 413 (1978).
5. Zucchino, P. and Lowrance, J. L., "Seventh Symp. P.I.E.D. Preprints" p. 415 (1978).
6. Sobieski, S., *Proc. SPIE* **78,** 73 (1976).
7. Williams, J. T., *Proc. SPIE* **78,** 78 (1976).
8. Currie, D. G. and Choisser, J. P., *Proc. SPIE* **78,** 83 (1976).
9. Gersho, S., *IEEE Commun. Soc. Mag.* **15,** 16 (1977).
10. Antcliffe, G. A., Hornbeck, L. J., Chan, W. W., Walker, J. W., Rhines, W. C. and Collins, D. R., *IEEE Trans. Electron Devices* **ED–23,** 1225 (1976).
11. Choisser, J. P., *Opt. Eng.* **16,** 262 (1977).
12. Barton, J. B., *In* Final Technical Report on NVL Contract No. DAAKO2–74–C–0359, p. 54 (1975).
13. Ando, K. J., *In* "Proc. Symp. on CCD Techniques for Scientific Imaging Applications" p. 192 JPL (1975).
14. Collins, D. R., Roberts, C. G., Chan, W. W., Rhines, W. C., Barton, J. B. and Sobieski, S., *In* "Proc. Symp. on CCD Techniques for Scientific Imaging Applications" p. 163 JPL (1975).
15. Walker, J. W., *In* "Int. Conf. on App. of CCDs" NOSC October 25–27, 1978.
16. Gilbert, G. R., Angel, J. R. P., Grandi, S. A., Coleman, G. D., Strittmatter, P. A., Cromwell, R. H. and Jensen, E. B., *Astrophys. J.* **206,** L129 (1976).
17. Schmidt, G. D., Angel, J. R. P. and Beaver, E. A., *Astrophys. J.* **219,** 477 (1978).
18. Wells, D. C., *Computer* **10,** 30 (1977).

Astronomical Applications of a CCD

C. L. DAVIES, B. L. MORGAN, R. J. SCADDAN and R. W. AIREY

Astronomy Group, The Blackett Laboratory, Imperial College of Science and Technology, London, England

and

J. C. DAINTY†

Department of Physics, Queen Elizabeth College, London, England

Introduction

This paper describes the interfacing of a CCD to an Interdata 70 minicomputer and its subsequent performance. The system is currently used in the computation of two-dimensional autocorrelation functions for digital reduction of speckle interferometry data. This method has several advantages over the analogue analysis techniques. The interfaced CCD is also intended for use in a photon counting system using a specially designed microchannel plate image intensifier, which is described below.

Charge Coupled Devices (CCDs) show great promise as area image detectors. Although not yet matching television camera tubes in respect of uniformity of target response, they have many significant advantages: they are small and robust; they require only low voltages and small currents; they exhibit good linearity characteristics and high sensitivity over a wide range of wavelengths.

The CCD which is in use at Imperial College is the Fairchild type 202 device, which has 10^4 image sensing elements, each 18×30 μm^2, arranged in a 100×100 matrix. It is driven by circuitry derived from that suggested by Fairchild. The analogue output is digitized into a two level signal.

Eventually the interfaced CCD will be used in a two-dimensional photon counting system. One of its many applications will be to the photometry of faint extended sources such as galaxies. In the outer

† Now at The Institute of Optics, The University of Rochester, Rochester. N.Y. 14627, USA.

regions of galaxies, the required signal can be as little as one-thousandth of the ambient sky background level and the extreme linearity of the photon counting approach enables the necessary precise subtraction of the sky background to be achieved. Preliminary photon counting experiments are described later in this paper.

APPLICATION OF THE CCD TO SPECKLE INTERFEROMETRY

The limit to angular resolution of a telescope is given by

$$\theta = 1\cdot 22 \ (\lambda/D),$$

where θ is the angular resolution (radians), λ is the wavelength of the light and D is the diameter of the telescope aperture. Thus for a telescope of aperture 2·5 m the angular resolution at a wavelength of 500 nm is about 0·05 arcsec. In practice however, atmospheric turbulence limits angular resolution to about one or two arcsec; consequently for the largest astronomical telescopes, the angular resolution limit may be a factor of fifty better than is actually observed.

Speckle interferometry is a method of deriving diffraction limited resolution from a large telescope despite the presence of atmospheric turbulence. Reviews of the technique have been given by Dainty[1] and Worden.[2]

Labeyrie[3] has shown that short exposure (5–25 msec) images of astronomical objects taken with large telescopes retain diffraction limited information. There are two approaches by which this information can be derived from such images.

Assuming that the quasi-monochromatic imaging equation applies, then

$$I(x, y) = O(x, y) * P(x, y), \tag{1}$$

where $I(x, y)$ is the image intensity distribution, $O(x, y)$ the object intensity distribution, $P(x, y)$ is the short exposure point spread function of the imaging system, and $*$ denotes convolution.

In the first approach, the average squared modulus of the Fourier transform of the image intensity, i.e., the power spectrum, is found:

$$\begin{aligned} W(u, v) &= \langle |i(u, v)|^2 \rangle \\ &= \langle |T(u, v)|^2 \rangle \, |O(u, v)|^2. \end{aligned} \tag{2}$$

In Eq. (2) $W(u, v)$ is the image power spectrum, $|O(u, v)|^2$ is the object power spectrum, $\langle |T(u, v)|^2 \rangle$ is the transfer function and $\langle \ \rangle$ denotes ensemble averaging. It can be shown[4] that the transfer function contains a component which extends to the diffraction limit of the telescope.

In the case of a binary star system, the object power spectrum is a set of cosine-squared fringes whose spacing, orientation and contrast depend on the angular separation, orientation and magnitude difference of the components of the system. These parameters may therefore be determined if the Fourier transforms of short exposure images of the binary system can be obtained. The Imperial College speckle interferometer is designed to record a large number of short exposure images on cine film via an E.M.I. 4-stage cascade image intensifier (Type no. 9912). The system has been described by Beddoes et al.[5] and is shown schematically in Fig. 1.

The two-dimensional Fourier Transform of each image is obtained by an analogue technique, the analysis being effected by a coherent optical system, in which the Fourier Transforms of successive frames of the cine film are co-added onto a single photographic plate. A typical result is shown in Fig. 2.

Good results using this method have been obtained with binaries as faint as $M_v = 9 \cdot 5$ and separations have been determined with errors of only a few per cent.[6,7] The analogue analysis, however, introduces additional sources of noise due to the granularity and surface scattering effects arising in the cine film and to non-linearity in the photographic plate. In practice it has not been found possible to add more than about 10^3 images, limiting observations to fairly bright stars.

In the second approach, the ensemble average spatial autocorrelation of the image intensity is derived from the speckle interferometer cine

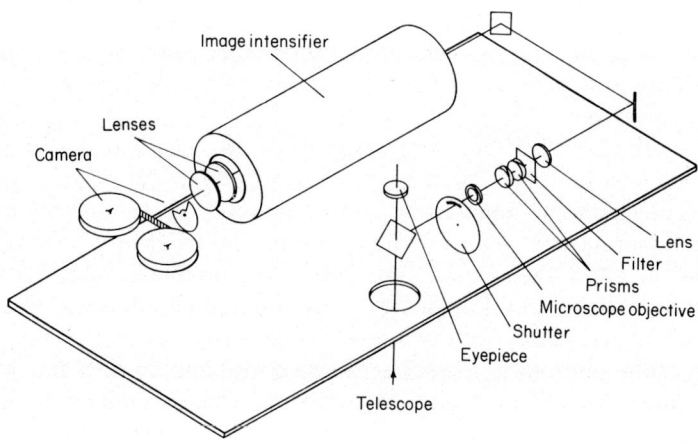

FIG. 1. The speckle interferometer.

Fig. 2. Fringes obtained by adding the Fourier transforms of 700 frames of the binary star system γ Lupi. The fringe spacing corresponds to an angular separation of 0·563 arcsec.

film. In the case of a binary star system, the two-dimensional autocorrelation function contains a central peak, with secondary peaks at locations corresponding to the separation and position angle of the binary system. The amplitude ratio of the central and secondary peaks is determined by the magnitude difference of the stars, although this information may be difficult to recover.[8]

This method lends itself more easily to digital techniques, and off-line digital analysis has been carried out using a CCD in the system shown in Fig. 3. The cine film, mounted on a stop motion projector and illuminated by an incoherent light source, is imaged onto the CCD by a lens, which is movable to allow the scale of the film to be optimally matched to the resolution of the CCD. For photon noise limited cine frames the output of the CCD is digitized into two levels. The images of bright stars which contain many photons per speckle can be correlated by the same process using clipping as a form of data compression. This is possible because the position of speckles in the x, y plane yields valuable results although photometric information is lost and some distortion of the autocorrelation function may be introduced.

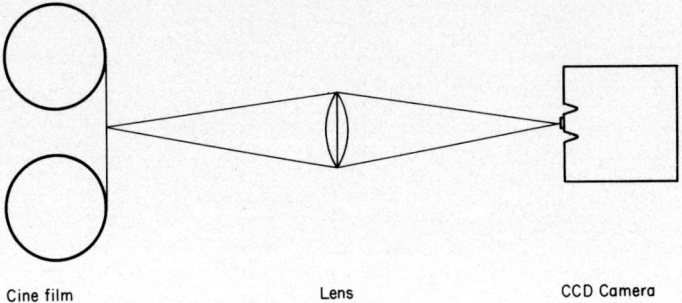

FIG. 3. The digital data analysis system.

It can be shown[9] that in these circumstances the autocorrelation function reduces to

$$C(h, l) = \sum_i \sum_j (x_i - x_{j-l}, y_i - y_{i-h}). \qquad (3)$$

Thus, if the vector separation of every possible pair of speckles is found, a two-dimensional histogram of these separations yields the autocorrelation function directly. Aspects of the performance of the CCD were studied to determine its suitability for this purpose.

CCD Performance

Properties of the CCD relevant to its use as a signal generating device have been widely reported.[10-13] The following characteristics are particularly important in the reduction of speckle data.

Noise Characteristics

The CCD is not cooled and reaches a maximum temperature of 37°C after one hour's use. Consequently the noise performance is dominated by thermal effects; it was felt that cooling was unnecessary in this application. The thermal noise comprises two components: the familiar random thermal background and a fixed pattern signal due to some photosites having a dark current many times greater than the average. This is a serious problem in the computing of autocorrelations. Figure 4 shows a digitized picture of the fixed pattern defects at 37°C, with the digitizing discriminator set at 20% of the diode saturation level. Figure 5 shows the same CCD output with the discriminator level set at 50% of

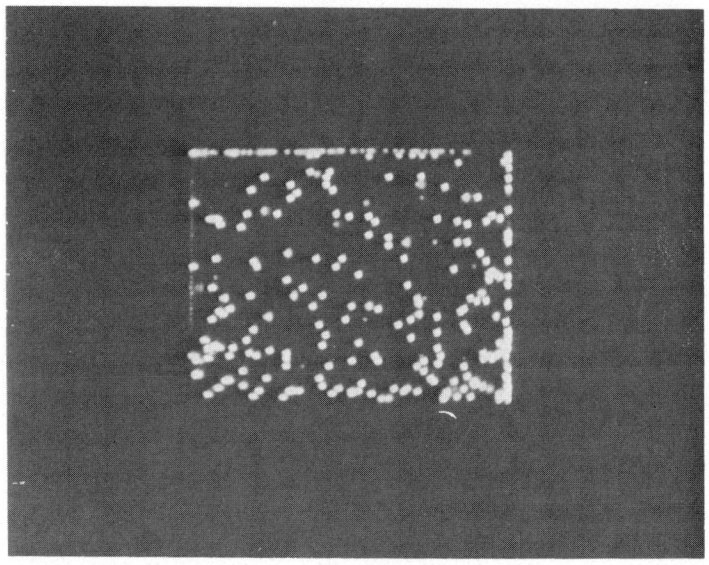

FIG. 4. Dark current defects at 37°C with the discriminator set at 20% of diode saturation level.

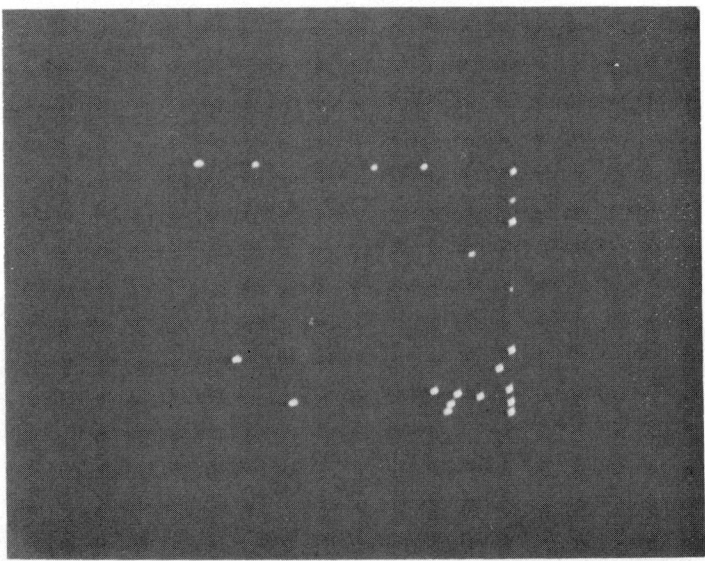

FIG. 5. Dark current defects at 37°C with the discriminator level set at 50% of diode saturation level.

the diode saturation level. Since this noise is "fixed pattern", a mapping of the defects can be stored in the computer and subtracted from each video frame.

Linearity and Dynamic Range

The random thermal background limits the dynamic range which can be achieved using the CCD. Figure 6 shows the relationship between input illumination and output signal level averaged over one line of the array. Photosites displaying the fixed pattern noise mentioned above were excluded from the measurement. It can be seen that the response is linear over a range of at least 100 to 1 at this temperature.

THE DATA PROCESSING SYSTEM

The processing of data from a single cine frame is carried out in four stages: the output of the CCD is firstly written into a buffer store, this store is then read into an Interdata 70 minicomputer where the two-dimensional autocorrelation function is computed and lastly the result is integrated into a disc store. The speed of this process is limited by the time taken for the calculation of the autocorrelation function to about $\frac{1}{2}$ to

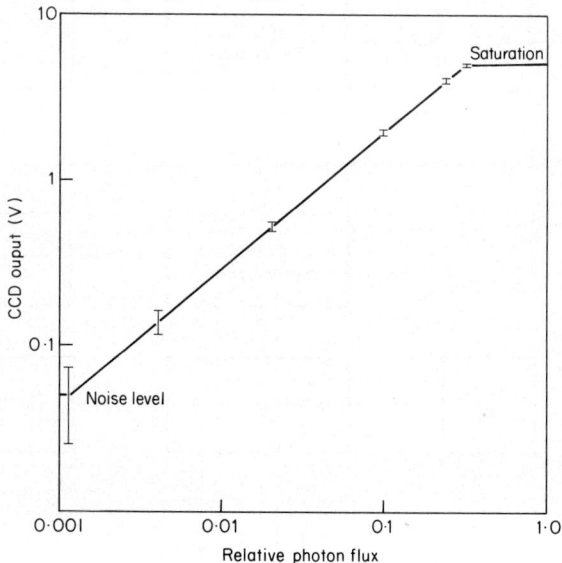

FIG. 6. Graph of CCD output (averaged over one line of video) versus photon flux.

1 min. At the end of this time, the stop motion projector is automatically incremented by one frame. In order to improve the processing time, a hard wired correlator is being constructed which will compute autocorrelations and cross correlations at a rate of 25 frames per second with up to 500 photoelectron events in each frame. This system will operate in real time on the telescope.

The Computing System

The laboratory computing configuration is shown in Fig. 7. The minicomputer is an Interdata 70 which has been programmed using FORTH language.[14] FORTH is an interactive, high level language that makes extremely efficient use of core, thus allowing the manipulation of the large arrays needed in image processing. The CCD, buffer store and CAMAC crate are remotely located and operations are controlled from VDU-1. A 24 bit I/O register CAMAC module is used as a one word store. The address of a photoelectron event is loaded on command from the control logic associated with the buffer store and this establishes a LAM status within the module which is continually sampled in the programme. The module has been modified so that the READ command also clears the LAM bistable and causes a restart signal to be communicated to the control logic of the buffer store.

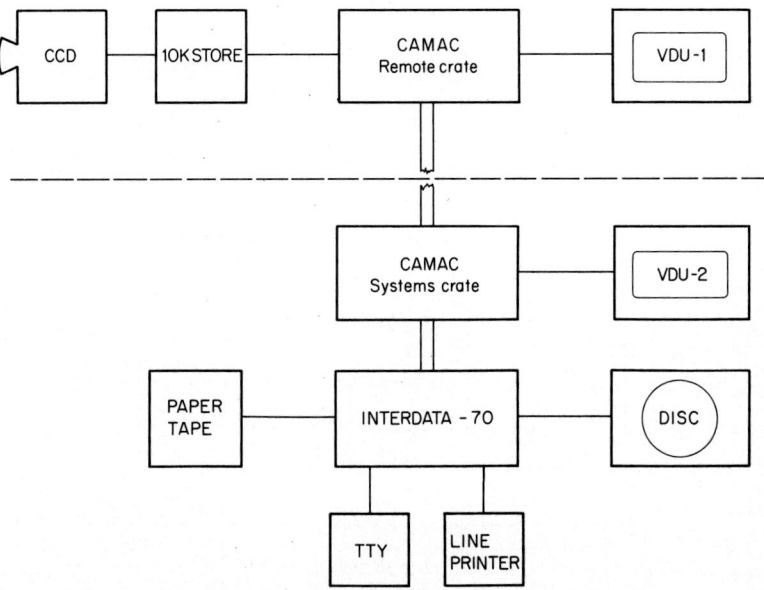

FIG. 7. Block diagram of the laboratory computing configuration.

The Buffer Store

After digitizing the output of the CCD, a complete frame of video is stored in a 10k×1 bit store so that the output bit rate of the CCD ($\sim 3 \times 10^5$ bits sec^{-1}) is matched to the input bit rate of the CAMAC/INTERDATA combination ($\sim 7 \times 10^3$ bits sec^{-1}). The CCD scanning logic continues to operate while the control logic associated with the buffer store searches sequentially through the array for a location with a "1" stored (corresponding to a photoelectron event). It then raises a LAM status in the I/O module and the address is read into the appropriate memory location. A restart command from the module triggers the control logic and the search continues. At the end of readout, the buffer store is rewritten by the CCD. A single frame containing about 1000 events takes approximately two seconds to read into the computer.

AUTOCORRELATION RESULTS

The system is calibrated using a simulated binary star created by filming a single unresolvable star through a calcite crystal to create a double image.[7] Figure 8 shows a typical contour map of the autocorrelation plane for the simulated binary, and Fig. 9 shows a cross section through it. Tables I and II show a comparison of preliminary results using the autocorrelation technique and those obtained using the analogue apparatus.[7]

TABLE I
Comparison of the separations obtained by analogue and digital methods

Object	Analogue	Digital
Simulated binary	0·278″±0·006″	0·277″±0·005″
(η Oph) ADS 10374	0·278″±0·006″	0·277″±0·005″
(β Del) ADS 14073	0·290″±0·006″	0·285″±0·005″

TABLE II
Comparison of the position angles obtained by analogue and digital methods

Object	Analogue	Digital
Simulated binary	7°±2°	8·9°±3°
(η Oph) ADS 10374	276±2°	277°±3°
(β Del) ADS 14073	0°±2°	0°±3°

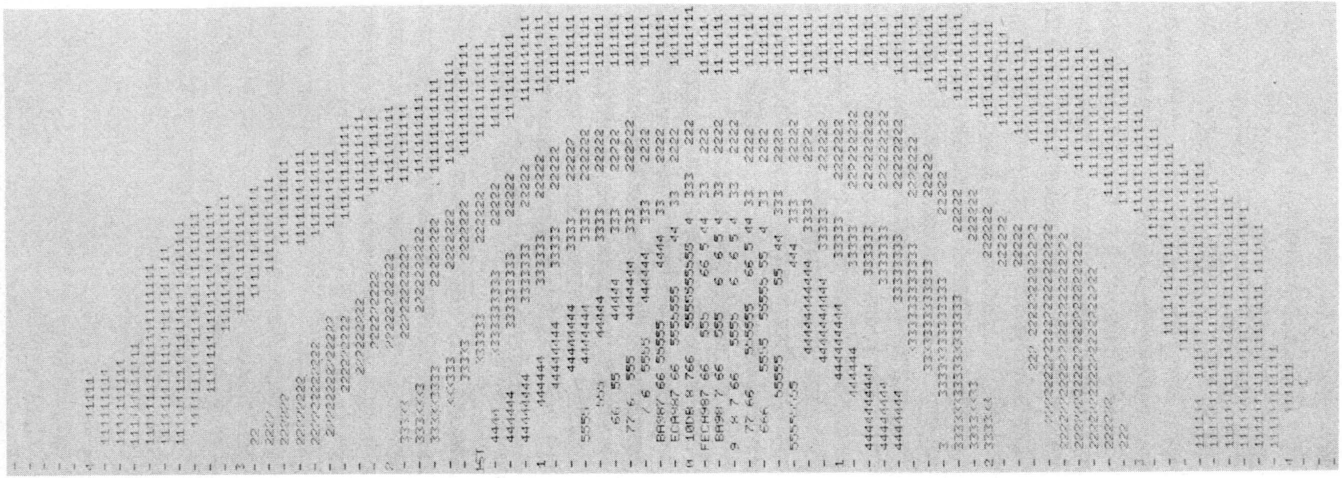

FIG. 8. Typical contour map of autocorrelation plane showing secondary peak.

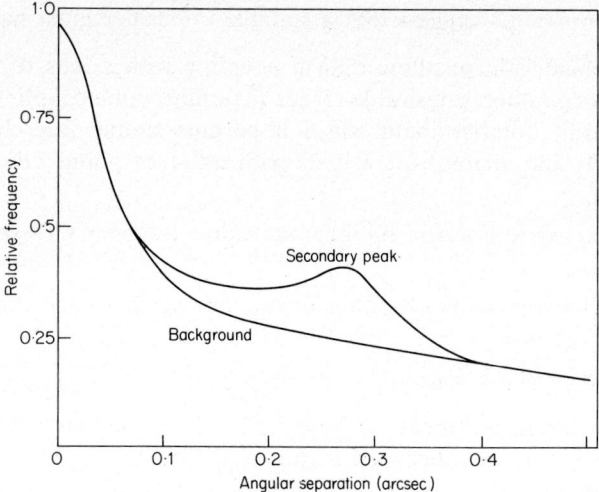

Fig. 9. Cross section through the contour map of Fig. 8 along the line of the secondary peak.

In each case, the digital result is obtained by co-adding less than thirty frames while the analogue method uses at least 250 frames. The digital reduction of speckle data is more efficient than the analogue technique, and a computer simulation by Dainty[8] suggests that objects as faint as 18th magnitude could be resolved in a processing time of about 35 min.

The Use of a CCD in a Photon Counting System

It is eventually intended to use a CCD as part of a photon counting detector on telescopes. For the detection of single photoelectron events it is of course necessary to add a prior stage of image intensification.[15]

In a preliminary experiment, the output of a commercially available image intensifier tube was imaged onto the CCD using a fast lens coupling. Two types of image intensifier were tried.

(i) A four stage, magnetically focused, E.M.I. cascade intensifier with green P·20 output phosphor.

(ii) A three stage, electrostatically focused VARO cascade intensifier with green P·20 output phosphor.

In both cases, images of a granular nature were observed, but it is not certain that single photoelectron events could be distinguished. This and other measurements suggest that a suitable intensifier must have:

(i) Very high gain to allow optical coupling with a lens to the CCD. Lens coupling is desirable rather than fibre optic coupling or direct electron bombardment, since it permits simple interchanging of CCDs and enables intensifiers with different photocathodes to be used.

(ii) A red phosphor for optimal matching to the CCD spectral response.

(iii) A relatively short phosphor decay time so that each event is only counted once.

(iv) Small physical size.

No image intensifier meeting these criteria was commercially available and one was therefore designed at Imperial College,[16] and its construction is now nearing completion. If it is successful, a manufacturing company has expressed interest in producing a commercial version.

The final design (Fig. 10) consists of a short magnetically focused section which images electrons from the photocathode onto the channel plate input face. Two 12·5 μm pore microchannel plates are arranged in chevron configuration and are run in the saturated gain mode. Their output pulses (approximately 10^7 electrons) are proximity focused with a field gradient of 3 kV mm^{-1} onto a rare earth phosphor screen which has spectral emission peaked in the red. This phosphor was chosen for its excellent match to the spectral response of a silicon CCD.

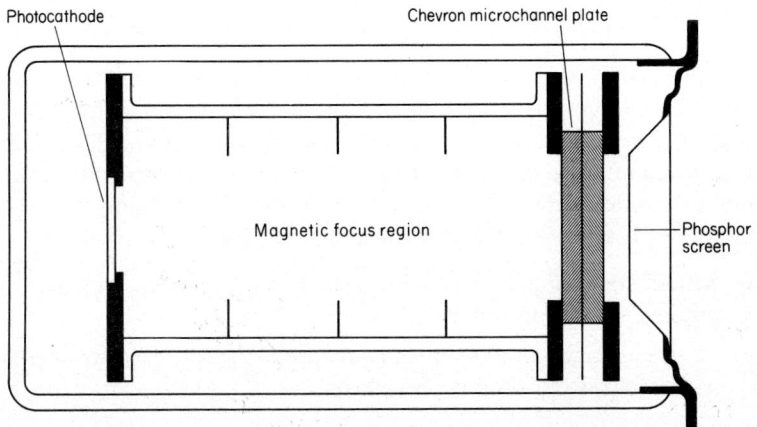

FIG. 10. A schematic cross section of the image intensifier tube.

The photocathode holder, the accelerating electrodes for the magnetically focused region, together with the assembly which holds the microchannel plates and the phosphor screen, are all mounted to form a single rigid structure that can be inserted into the glass vacuum envelope of the tube which is then sealed by means of an argon arc weld at the output end.

The photocathode is prepared separately in a photocathode processing cell, fitted with a special thin glass breakbulb which can be ruptured in vacuum to permit transfer of the photocathode into the tube through a flattened pumping stem, after the tube has been baked out. This photocathode transfer procedure is similar to that which has been well established for the manufacture of the Spectracon electronographic image tube.[17] The advantage of this technique is that alkali vapours used in the photocathode formation process are not liberated within the image tube envelope where they could contribute to spurious electron emissions by lowering the work function of surfaces. Moreover, damage to the microchannel plates by alkali contamination is avoided. The prototype microchannel intensifier has a posted-in S·11 photocathode; later tubes will have S·20 photocathodes. Barium getters are used in the tube to absorb any residual gases liberated by the microchannel plates when the tube is operated. The intensifier is designed to focus its magnetic section in a field of approximately 0·01 T produced by a small solenoid, wound to give a uniform magnetic field over the working region of the tube.

Conclusions

A charge coupled device has been successfully interfaced, via CAMAC, to an Interdata 70 mini-computer for the computation of two-dimensional autocorrelation functions. Although the video picture is digitized with only one bit accuracy, useful information is obtained. The current system allows the off-line reduction of speckle interferometry data to be achieved in a more efficient manner than with analogue techniques. This system is not designed to detect single photoelectrons. However, a magnetically focused, microchannel plate image tube with a red phosphor is being built specifically to intensify astronomical images to a level that will allow detection of single photoelectron events by the CCD.

Acknowledgments

C.L.D. was in grateful receipt of an S.R.C. CASE studentship during the course of this work.

References

1. Dainty, J. C. *In* "Laser speckle and related phenomena (Topics in Applied Physics, Vol. 9)" ed. by J. C., Dainty, p. 255. Springer-Verlag, Berlin (1975).
2. Worden, S. P., *Vistas Astron.* **20,** 301 (1977).
3. Labeyrie, A., *Astron. & Astrophys.* **6,** 85 (1970).
4. Korff, D., *J. Opt. Soc. Am.* **63,** 971 (1973).
5. Beddoes, D. R., Dainty, J. C., Morgan, B. L. and Scaddan, R. J., *J. Opt. Soc. Am.* **66,** 1247 (1976).
6. McAlister, H. A., *Astrophys. J.* **215,** 159 (1977).
7. Morgan, B. L., Beddoes, D. R., Scaddan, R. J. and Dainty, J. C., *Mon. Not. R. Astron. Soc.* **183,** 701 (1978).
8. Dainty, J. C., *Mon. Not. R. Astron. Soc.* **183,** 223 (1978).
9. Blazit, A., Koechlin, L. and Oneto, J. L., *In* "Image Processing Techniques in Astronomy" ed. by C. de Jager and H. Nieuwenhuijzen, p. 79, Reidel, Dordrecht, (1975).
10. Fairchild CCD 202 Data sheet (1976).
11. Loh, E. D. and Wilkinson, D. T., *In* Proc. Conference on "Imaging in Astronomy" Cambridge, Mass., June 18–21 (1975).
12. Livingston, W. C., *In* "Astronomical Applications of Image Detectors with Linear Response" (IAU Colloquium No. 40) ed. by M. Duchesne and G. Lelièvre, p. 22–1. Paris-Meudon Observatory (1976).
13. Dyck, R. H. and Jack, M. D., "Low Light Level Performance of the CCD 201", Fairchild Semiconductor Components Group Information Sheet (1974).
14. Moore, C. H., *Astron. & Astrophys. Suppl. Ser* **15,** 497 (1974).
15. Currie, D. G., *In* "Astronomical Applications of Image Detectors with Linear Response" (IAU Colloquium No. 40) ed. by M. Duchesne and G. Lelièvre, p. 30–1. Paris-Meudon Observatory (1976).
16. Airey, R. W., Morgan, B. L. and Ring, J. *In* "Seventh Symp. P.E.I.D. Preprints," p. 329 (1978).
17. McGee, J. D., Khogali, A., Ganson, A. and Baum, W. A., *In* "Adv. E.E.P." Vol. 22A, p. 11 (1966).

Schottky IRCCD Thermal Imaging

F. D. SHEPHERD, R. W. TAYLOR, L. H. SKOLNIK, B. R. CAPONE
and S. A. ROOSILD

Rome Air Development Center, Deputy for Electronic Technology, Hanscom AFB, Massachussets, U.S.A.

and

W. F. KOSONOCKY and E. S. KOHN

RCA Laboratories, David Sarnoff Research Center, Princeton, New Jersey, U.S.A.

Introduction

Schottky barrier infrared detectors offer the possibility for production of highly uniform, monolithic silicon IRCCD focal plane arrays. Platinum silicide (Pt_xSi) Schottky barriers on p-type silicon have a barrier height of about 0·27 eV which corresponds to an infrared response cutoff wavelength of 4·6 μm.[1] Thermal imaging with platinum silicide IRCCD arrays was described by Shepherd,[1-3] Kohn et al.,[4] Taylor et al.,[5] Skolnik et al.,[6] and Capone et al.[7] In this paper we will describe the construction, operation, and infrared characteristics of a 256 element line sensor[5,6,8] and a 25 × 50 element area sensor.[7,9,10] These devices were designed and fabricated at RCA Laboratories. The reported infrared measurements were made at RADC/ESE.

These IRCCDs can be operated from 50 to 90 K, and are sensitive in the 1·2 to 4·6 μm spectral range. A typical value of the quantum efficiency coefficient (C_1) of the platinum silicide detectors is 5% per eV. A photoresponse uniformity of 0·5% RMS was demonstrated with the 256 element line sensor. Thermal imaging data are reported for both calibrated infrared sources and human subjects. The line and area sensors were demonstrated to have noise equivalent temperatures (NET) of 0·4°C and 0·5°C in a 26°C ambient environment noise equivalent powers (NEP) of 8×10^{-12} W and 4×10^{-12} W, and linear dynamic ranges of 5000 and 13 000 respectively. They were operated with an integration ("staring") time of 30 msec.

Platinum Silicide Schottky Barrier Detectors

The use of Schottky barrier detectors for infrared imaging has been described extensively in the literature.[1-14] The platinum silicide (Pt_xSi) detector is formed by depositing a layer of platinum (typically about 60 nm on a p-type silicon substrate (having resistivity of 10 to 50 Ω cm) and then sintering it at a temperature in the range 200 to 650°C. The composition of the Pt_xSi alloy varies with the sintering temperature and time as well as with the initial thickness of the platinum layer.[14] The reaction between the metal and the silicon places the $Pt_xSi:Si$ junction beneath the original surface of silicon, and thus produces diodes free of surface effects and with highly uniform photorespone. In fact, 1 cm diameter Pt_xSi diodes have been fabricated having RMS photoresponse non-uniformities of less than 0·1%.[3] The photoresponse uniformity of the platinum silicide detectors fabricated thus far have been basically limited, therefore, by the uniformity of the geometric definition of the detectors. The photoyield, Y, for Schottky emission is given by[2]

$$Y = C_1 (h\nu - \psi_{ms})^2 / h\nu \text{ electrons per photon,} \qquad (1)$$

where ψ_{ms} is the barrier height, h Planck's constant, ν is the photon frequency and C_1 a factor determined by the geometrical, optical and transport properties of the silicide contact. With $h\nu$ and ψ_{ms} expressed in eV, C_1 is in reciprocal eV. An example of a photoyield curve for a 60 nm Pt/p-Si diode reacted at 320°C is shown in Fig. 1, where we plot $(EY)^{1/2}$ versus $E(=h\nu)$. From the graph we find a barrier height of 0·268 eV and a C_1 of 3·6% per eV. A typical value of C_1 for Pt_xSi detectors is 5% per eV. It should be noted, however, that the form of the yield curve (Fig. 1) and the values of C_1 ($3 \leq C_1 \leq 11$% per eV) are highly dependent on reaction conditions and initial Pt layer thickness. Effects such as electron–phonon mean free paths and enhanced thin film reflections can influence the energy dependence of the photoresponse.[12,13] The effective quantum efficiency for thermal imaging of platinum silicide Schottky barrier detectors is rather small, i.e. of the order of 0·1%. However, the demonstrated high uniformity of photoresponse (0·5% RMS and better)[3-7] combined with low noise readout by buried channel CCDs made the Schottky barrier IRCCDs attractive alternatives for many thermal imaging applications. The rather low quantum yield in Schottky detectors can be compensated by operating the arrays either in the "staring" or time delay integration mode. Calculations[3] based on present estimates indicate that these devices with detectors of area $6·25 \times 10^{-5}$ cm^2 should be capable of resolving 0·1°C sources against a 27°C background at standard video frame rates.

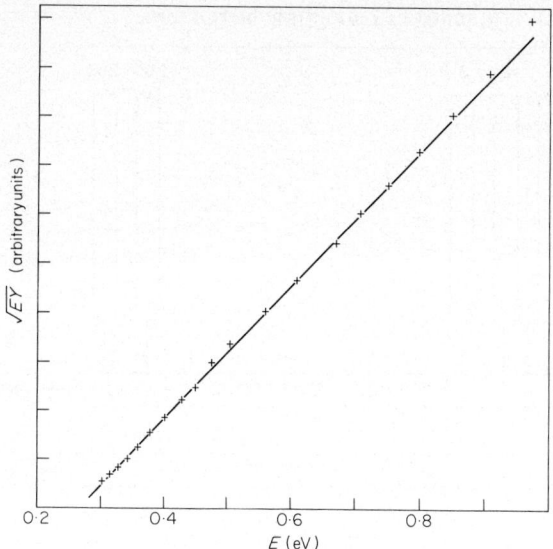

FIG. 1. Photoyield curve for 60 nm Pt_xSi diode.

DEVICE CONSTRUCTION AND OPERATION

Technology

The devices were processed using two level polysilicon buried channel CCD technology with p^+ channel stops and n^+ diffusions not self aligned to the polysilicon gates. The substrates used were 20 to 50 Ω cm, boron doped, p-type [100] silicon wafers polished on both sides to a final thickness of 250 μm. The platinum silicide detectors were formed by depositing a 50 to 60 nm film of platinum on the Schottky contact holes opened in the oxide (SiO_2) and then sintering at temperatures in the range 320 to 650°C. After the platinum silicide formation, the remaining platinum was etched off and a 1·4 μm aluminium metalization was deposited and defined to complete the device processing. The platinum silicide detectors were formed surrounded by implanted n-type guard rings to reduce excess dark (leakage) current. This was accomplished using a buried channel implant, typically in the form of phosphorus with a dose of 1 to 2×10^{12} cm^{-2} at 180 keV.

256 Element IRCCD Line Sensor

A block diagram of the 256 element IRCCD line sensor is shown in Fig. 2. A cross sectional view of the charge coupling structure between

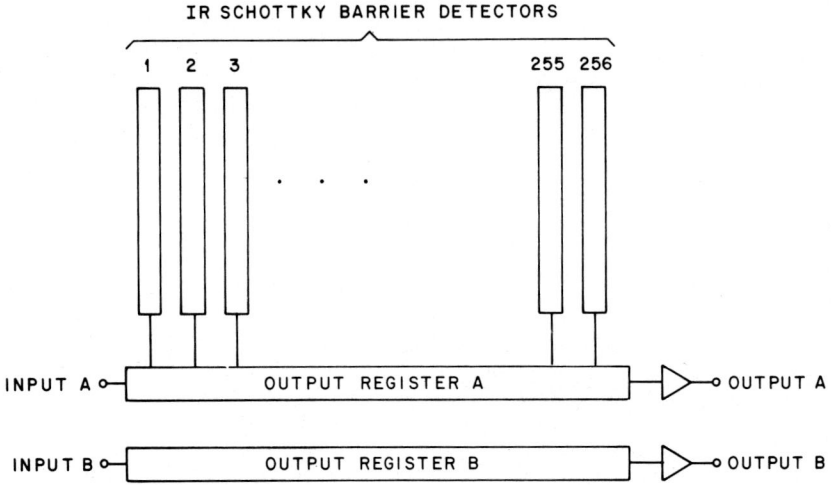

FIG. 2. Block diagram of 256 element IRCCD line sensor.

the platinum silicide detectors and the CCD output register A is shown in Fig. 3. The platinum silicide detectors are surrounded by implanted n-type guard rings and are isolated from each other by 5 μm wide, p$^+$ channel stops. The effective size of the detectors is 10×200 μm^2, and the detectors are spaced on 40 μm centres. The line sensor chip is 11·3 by 1·7 mm^2.

In addition to the buried channel CCD output register A for scanning the infrared detectors, the chip also contains an identical register, output register B, driven by the same clock lines. In the tests reported here, the two registers were normally used together to cancel the common mode clock pickups. However, it is also possible with this sensor to perform on-chip frame comparison schemes such as moving target indication. The output registers are two level polysilicon, two phase BCCDs with 40 μm stages, and have been designed rather conservatively with 15 μm gates, 5 μm spaces, and 5 μm gate overlaps. The readout of charge from the Schottky barrier detectors in the line sensor is illustrated in Fig. 3 using the "continuous charge skimming" mode. The detector array is illuminated from the back. The function of the surface channel gate G_{1T} is to maintain a fixed (DC biased) "charge skimming" barrier that determines the reverse bias voltage for the Schottky diodes. During the integration time of the optical signal, the detected charge is collected in the charge gate G_{3T} (see Fig. 3 (b)). The function of the buried channel gate G_{4T} is

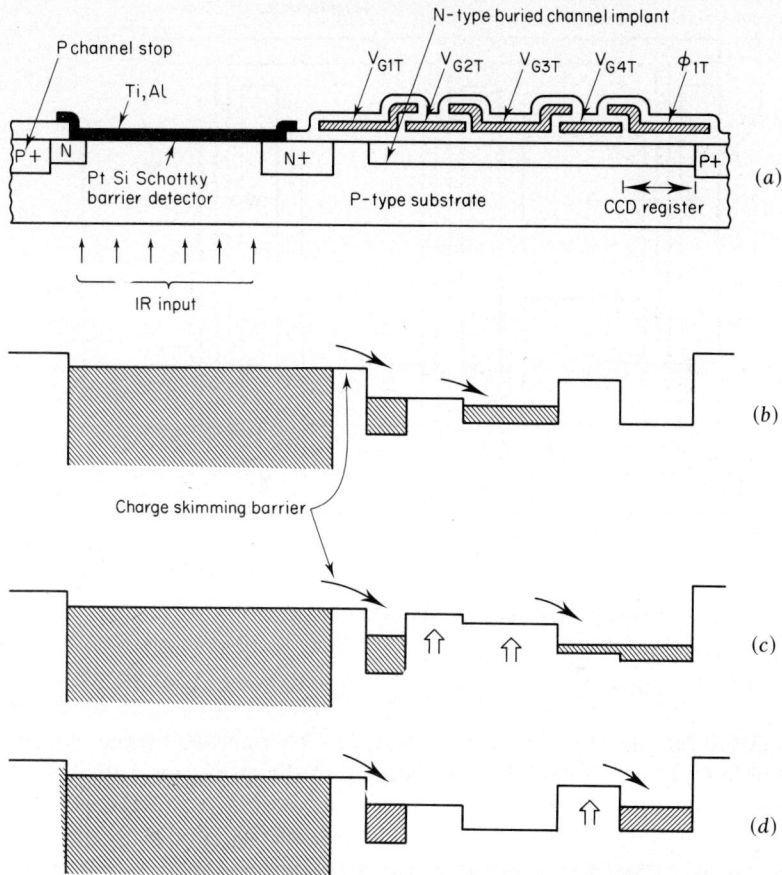

FIG. 3. Schottky barrier detector used in the continuous charge skimming mode. (a) Construction. (b) Collection of charge in charge gate G_{3T} during integration. (c) and (d) Transfer of charge to CCD output register.

to isolate the charge integration well from the CCD output register. At the end of the integration time (variable from 10 to 100 msec) the collected charge signal is transferred from the integration well to the serial CCD output register by a form of push clocking (i.e. one with slow fall time) as illustrated in Figs. 3(c) and (d). The output register clock rate can be varied from 1 kHz to 5 MHz. For most of the measurements reported here, integration and line times were maintained at 30 msec and 23 msec, respectively.

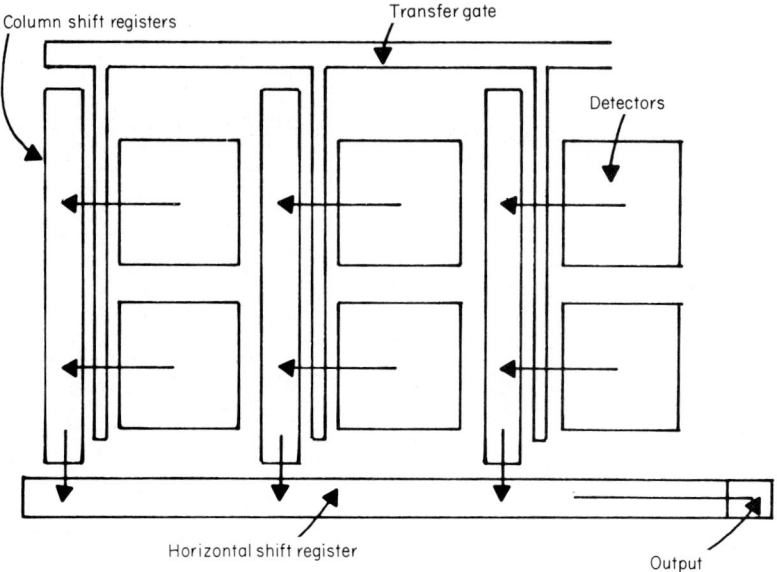

FIG. 4. Block diagram of the 25 × 50 element IRCCD area sensor.

Although the line sensor is designed to operate in the continuous skimming mode, it can also be operated in the more conventional voltage reset mode illustrated later (in Fig. 8) for the area sensor. The continuous charge skimming mode minimizes the noise associated with the resetting of the Schottky diodes. An interesting property of the continuous charge skimming mode is that the Schottky diodes are maintained at a constant potential which can be adjusted to any desired value by controlling the DC voltage V_{GIT}. By maintaining the reverse bias voltage of the Schottky diodes at a low value, the leakage current and the leakage current spikes (which tend to increase exponentially with voltage) can be maintained at minimum levels. However, if desired, the Schottky diode reverse bias voltage can be increased to a value at which the leakage current spikes become appreciable. Our tests show that operation with the larger reverse bias voltage tends to slightly increase the sensitivity of these devices as infrared sensors. Finally, it should be added that construction of the line sensor with separate CCD type, charge integration wells, results in the charge handling capacity of the sensor being limited by the design of the CCD readout structure rather than by the normally smaller depletion layer capacitance of the Schottky diode.

25×50 Element IRCCD Area Sensor

A block diagram of the area sensor chip is shown in Fig. 4. The chip is 5·84 mm square. The area sensor contains 25 columns of detectors with 50 detectors in each. The 25 column charge coupled shift registers are interspersed with the detector columns, and connect to a horizontal register. The vertical detector spacing is 80 μm, while the horizontal detector spacing is 160 μm. The active area of each detector is 2100 μm^2 or 16·4% of the unit cell area. The chip is mounted in a 28 pin dual-in-line ceramic package with a 4·75 mm square opening under the chip for near IR illumination. The chip also contains a 62 element line array and many test devices, which are not discussed here.

This 25×50 element has an interline transfer organization. The infrared detectors are separated from the buried channel CCD (BCCD) column shift register by vertical, surface channel transfer gates. In the "staring" mode of operation, the detected image is transferred from the infrared detectors to the BCCD column registers once every frame time. Then, while the detectors integrate a new frame, the original frame is transferred a line at a time, to the BCCD horizontal output register, from which each horizontal line is read out in series by the floating diffusion output amplifier. The floating diffusion sensing node capacitance of this device is about 0·7 pF, and the voltage gain of the source follower output circuit is about 0·7. The column registers and output register are two level polysilicon, four phase CCDs with 80 μm long stages. The vertical column registers are coupled to every second stage of the output register. A detailed schematic of the platinum silicide detectors of the area sensor is shown in Fig. 5. This figure illustrates how the effective area of the detector of 2100 μm^2 or 3·36 mil^2 is determined.

The area sensor is operated in the voltage reset mode illustrated in Fig. 6. This mode was previously referred to as the "vidicon" mode.[4,9] In this mode the (surface channel) transfer gate is pulsed positively once each frame. The detectors are set to a reverse bias determined by the surface potential under the transfer gate during the transfer pulse. Between transfer pulses, the detectors are discharged by infrared response and by dark current. At each transfer pulse, each detector is set to the same reverse bias as the previous time, so that the charge removed to the CCD well, as shown in Fig. 6(b) and (c), is the true signal from the detector. As in the case of the continuous charge skimming mode, small non-uniformities in transfer threshold and detector capacitance do not cause non-uniformities in the video signal, making possible direct imaging without computer processing of the video signal. By using a surface channel transfer gate between the Schottky diode and the buried channel

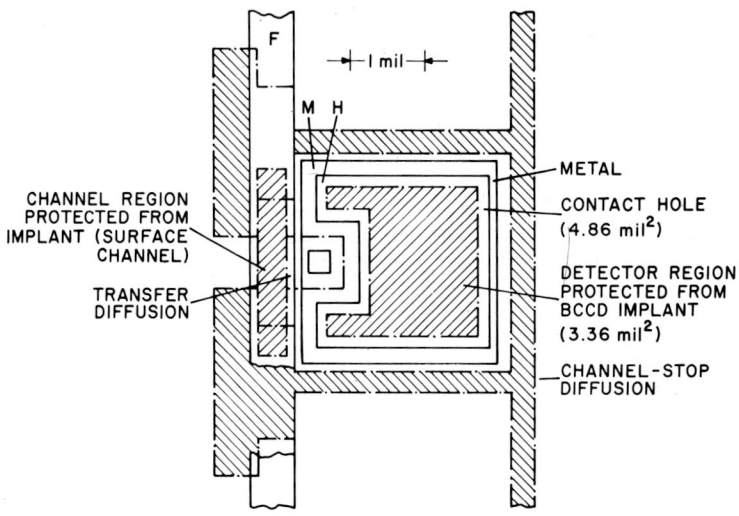

FIG. 5. Scale drawing of the effective detector areas of the 25 × 50 element IRCCD area sensor as defined by the Schottky contact mask and the buried channel or the guard ring mask.

CCD structure, the Schottky diode can be isolated from the BCCD output register and thus the CCD register can be tested at room temperature.

Experimental Results

CCD Characteristics at 77 K

At room temperature the charge transfer loss of the buried channel CCD registers was typically 10^{-5} per transfer for operation with about 10% bias charge and about 2×10^{-5} without the bias charge. At 77 K the transfer inefficiency for the line array was found to be larger, typically 5×10^{-5} with bias charge and 4×10^{-4} without bias charge.

For the infrared measurements reported here the line sensor and the output registers of the area sensor were always operated with a bias charge. However, the column registers of the area sensor have no provision for electrical input and exhibited significant transfer inefficiency.

Dark (Leakage) Current of Platinum Silicide Detectors

The theoretical dark current of infrared sensitive Schottky barrier detectors is due almost entirely to the internal thermionic emission of

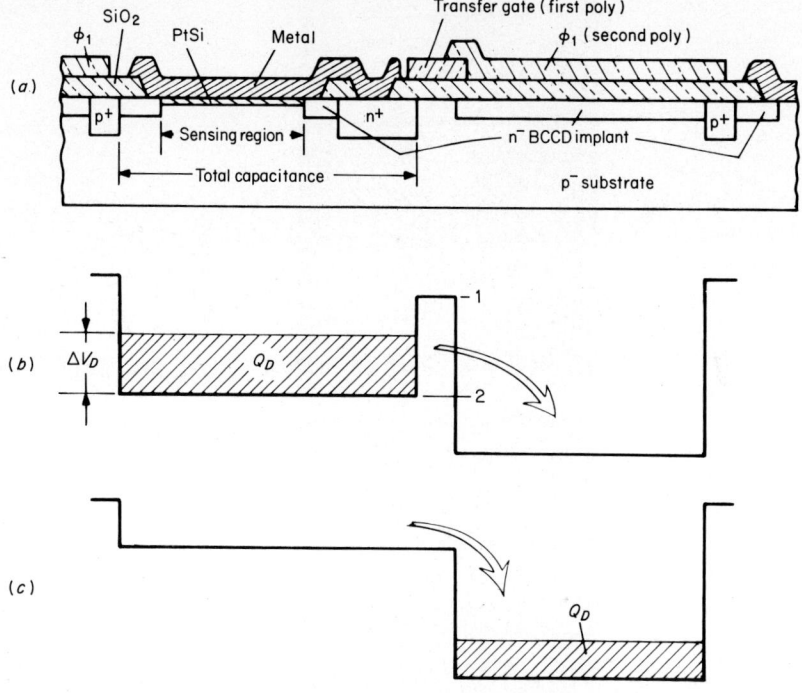

FIG. 6. Construction and operation of area array detectors. (a) Cross section of area sensor showing detector region, transfer structure and shift register gate. (b) and (c) Electron energy profile during voltage reset operation of detector.

carriers from the metal passing over the barrier into the semiconductor. The current density for this process is given[9] by

$$J = A^* T^2 \exp(e\psi_{ms}/kT) \qquad (2)$$

where A^* is ~32 A cm^{-2}K^{-2} for holes in Si, ψ_{ms} is the barrier height, e is the electronic charge, k is the Boltzmann constant, and T is the detector temperature. For platinum silicide Schottky barrier detectors, the above dark current density is 4×10^{-13} and 2×10^{-10} A cm^{-2} for operation at temperatures of 77 and 90 K, respectively. Our experience shows that with the best devices operated at 80K, no dark current is observed for thermal imaging with integration times in the range from 30 to 100 msec.

Infrared Test Set Up

The experimental set up used for infrared measurements consisted of an Electro-Optical Industries source and differential temperature target,

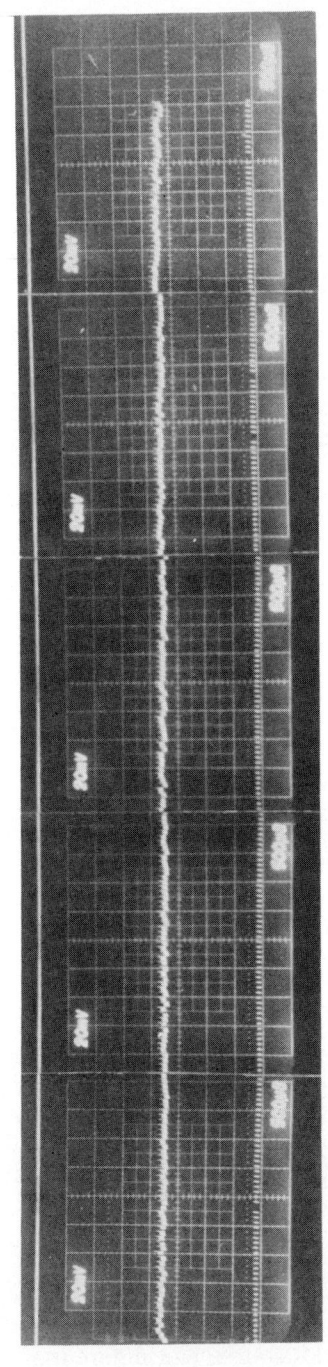

FIG. 7. Background response of 256 element IRCCD line sensor with 3·4 to 4·2 μm band pass.

Space Optics Research Lab IR optics, and an Air Products Displex refrigerator for cooling the arrays. The optics used included a four element $f/1\cdot2$ Si/Ge lens which is antireflection coated for the 2 to 5 μm region.

For the infrared measurement, the sensors were cold shielded and cold filtered (3·4 to 4·2 μm) to minimize the background signal. The optical signals were imaged on the platinum silicide sensor arrays through the back side of the optically polished silicon substrate.

Uniformity

Because of the low contrast in thermal scenes, spatial photoresponse nonuniformity can rapidly degrade the temperature sensitivity (NET) of staring arrays.[17] For example, a Schottky sensor with a 4·5 μm cutoff has a signal to background contrast of 5% $°C^{-1}$ when responding to a 27°C scene. Therefore, a spatial uniformity of 0·5% is required to resolve 0·1°C.[3] As previously discussed, platinum silicide Schottky barrier IRCCDs have inherently uniform response because photoyield is nearly independent of substrate doping concentration and lifetime. Thus it is possible to produce arrays which require no off-chip processing for individual pixel gain and no offset correction. The uniformity achieved with the 256 element IRCCD line sensor is illustrated by the background response shown in Fig. 7. In Fig. 8 we show a typical portion of the line array illuminated by a 230 °C black body source (through a 50% neutral density filter). The measured RMS spatial nonuniformity of 0·3% across the line array is limited only by detector area variations due to the geometric definition of the diodes (i.e., reticulation).

Infrared Transfer Characteristics

The transfer characteristics of the 256 element IRCCD sensor are shown in Fig. 9. The measured signal to noise ratio (S/N) is plotted versus the power density reaching the silicon side of the array. (Since this array has no antireflective coatings, 30% of the incident power is reflected from the surface of the silicon substrate). The noise signal, determined by a superposition of the scope traces, corresponded to 2·0 mV peak to peak of which approximately 1·5 mV corresponds to a random noise and 1·5 mV to a coherent type of pickup noise. From the curve in Fig. 9 we determine the noise equivalent irradiance (NEH) in the 3·4 to 4·2 μm band as $4\cdot5 \times 10^{-7}$ W cm^{-2} which corresponds to the noise equivalent power (NEP) per pixel of 8×10^{-12} W. The measured dynamic range of the line sensor is 5000. A similar measured transfer curve for the area

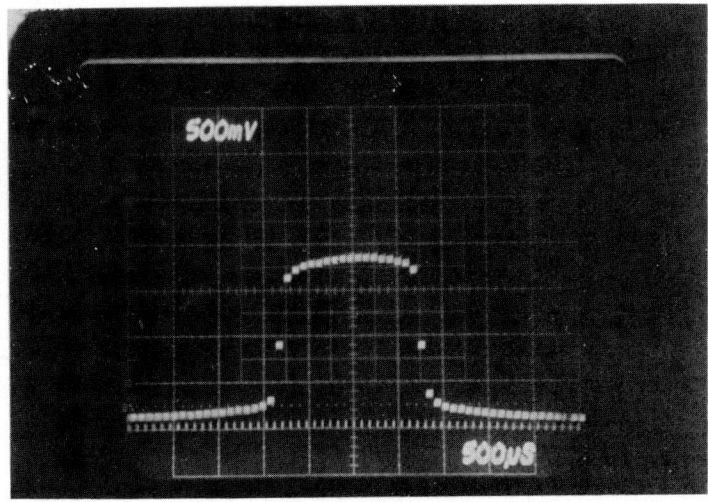

FIG. 8. Response of a typical portion of the 256 element IRCCD line sensor as illuminated by a 230°C source (using 3·4 to 4·2 μm band pass filter and a 50% neutral density filter).

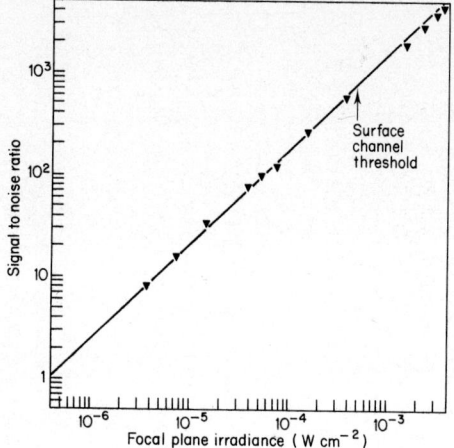

FIG. 9. Transfer characteristics of the 256 element IRCCD line sensor.

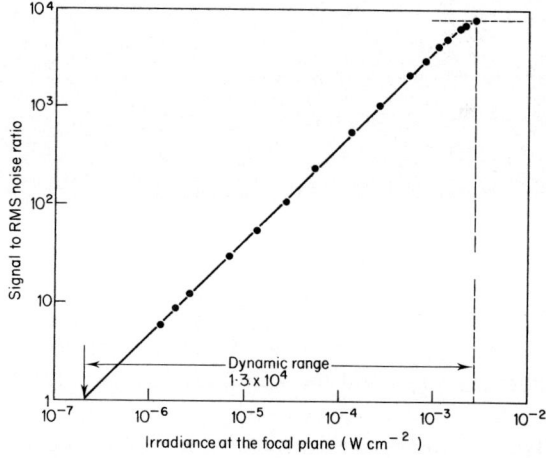

FIG. 10. Transfer characteristics of the 25 × 50 element IRCCD area sensor.

sensor is shown in Fig. 10. In this case NEH = $2 \cdot 1 \times 10^{-7}$ W cm^{-2}, NEP = 4×10^{-12} W, and the dynamic range is 13 000.

Since the line sensor has been operated in the continuous skimming mode, its dynamic range is limited by the capacity of the 50 μm wide CCD output register. The area sensor is operated in the voltage reset mode in which case the dynamic range of the sensor (the saturation level)

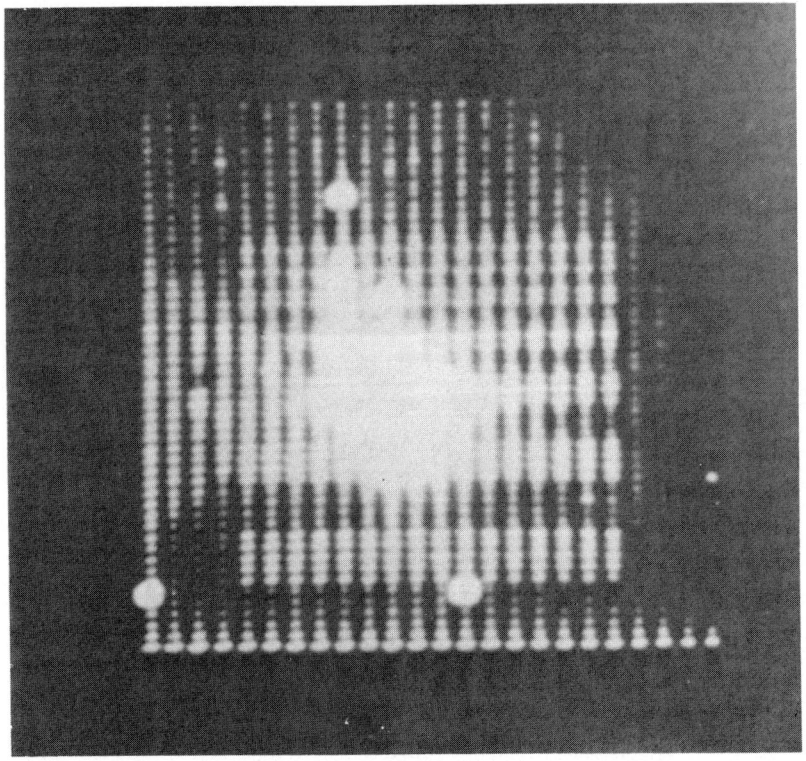

Fig. 11. A test grid 10°C above ambient temperature and a match flame imaged by the 25 × 50 element IRCCD area sensor.

can become limited by the effective capacitance of the detectors (estimated to be about 0·4 pF) and the amplitude of the transfer gate voltage pulse.

The large dynamic range of the area sensor is illustrated in Fig. 11. This figure shows a calibrated test grid 10°C above ambient detected by the area sensor with a match flame in the scene. While the match saturates the sensor where it is imaged, no blooming around the match is observed. Further, there is no image lag when the match is removed; thus sensor recovery occurs in less than one "stare" time (i.e., 30 msec).

Thermal Response

The thermal transfer response of the 256 element IRCCD line sensor is shown in Fig. 12 for 30 msec integration time. The experimental points in

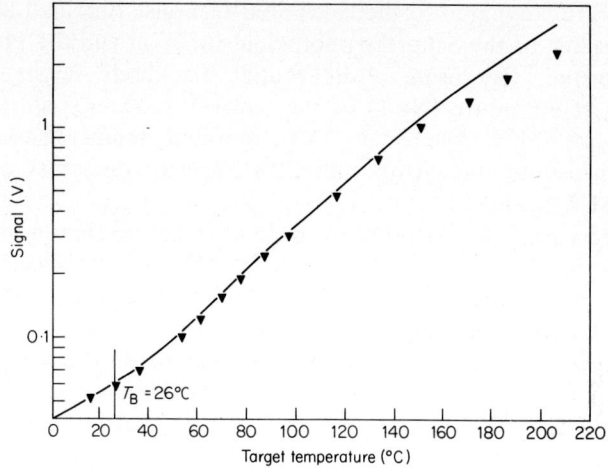

FIG. 12. Thermal response of the 256 element IRCCD line sensor.

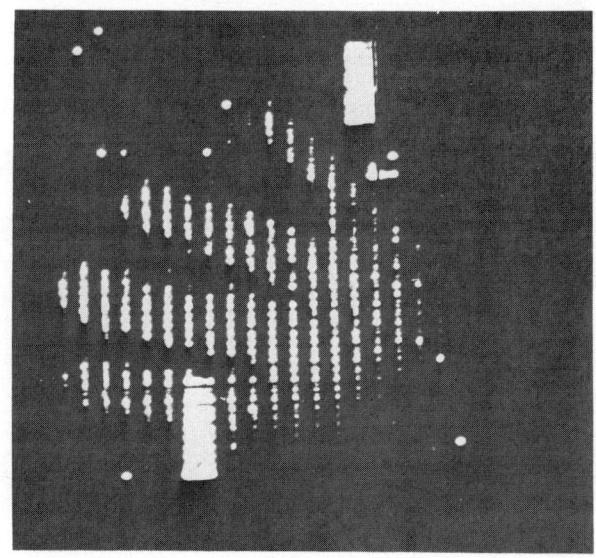

FIG. 13. 25 × 50 element IRCCD camera image of human hand.

this figure are compared to the calculated response obtained by integrating the product of the Schottky photoyield function and the Planck black body spectrum.[2] By using a differential blackbody target, the noise equivalent temperature (NET) of this array for 30 msec stare time was measured as 0·8°C against a 24°C ambient temperature. A more recently measured line array with 20×200 μm^2 detectors exhibited a reduced NET of 0·4°C.

Figure 13 shows the area sensor image of radiation from a human hand (~32°C against 26°C ambient temperature). The present measurements are limited by excess noise and low diode breakdown voltages. Correcting the above problems should improve sensitivity by at least a factor of two. Further, recent processing advances have led to thin diodes that are a factor of two more sensitive at all wavelengths. Thus we believe that both present devices could be improved so as to have NET in the range of 0·1 to 0·2°C.

Table 1 summarizes and compares parameters measured for both line and area sensors.

TABLE I

Parameters for 256 element linear and 25×50 element area PtSi IRCCD: 3·4 to 4·2 μm, 30 msec stare time

Parameter	Value Line	Value Area	Comment
NET (°C)	0·8	0·5	0·4°C for line arrays with larger area diodes; ultimately 0·1°C.
NEH (W cm^{-2})	$4·5 \times 10^{-7}$	$2·1 \times 10^{-7}$	
NEP (W)	8×10^{-12}	4×10^{-12}	
Linear dynamic range	5000	13 000	
Noise level (mV p–p)	~2	2	1·5 mV random; 1·5 mV coherent pickup.
Photoresponse uniformity (%)	0·5	2	Reticulation-limited for line array.
Quantum efficiency coefficient (% per eV)	~5	~5	
Operating temperature (K)	80	80	40–103 K
Transfer inefficiency (per transfer)	5×10^{-5}	5×10^{-5}	77 K, 10% bias charge.

Conclusions

The advantages of the Schottky IRCCDs are

(i) Monolithic silicon construction with standard IC processing.

(ii) High uniformity limited only by the geometric definition of the detectors.

(iii) High detector densities possible.

(iv) Modest cooling requirements (in the range of 80 to 90K).

(v) No optical crosstalk.

(vi) Staring mode of operating.

(vii) No lag.

(viii) Very large dynamic range.

(ix) Antiblooming capacity.

We have developed two types of IRCCDs with Schottky barrier platinum silicide detectors, a 256 element line sensor and a 25×50 element area sensor and have demonstrated thermal imaging with a temperature discrimination of $0.5°C$ above ambient for the 25×50 element area sensor and $0.4°C$ for the 256 element line sensor.

With new designs and processing improvements, Schottky IRCCDs with higher density and capable of temperature discrimination in the range of 0.1 to $0.2°C$ above ambient are anticipated. Such IRCCD sensors, with resolution approaching that of commercial television, are expected to have applications in IR surveillance, reliability studies, and medical diagnostics.

References

1. Shepherd, F. D. and Yang, A. C., *In* "Proc. International Electron Devices Meeting, Technical Digest", p. 310 (1973).
2. Shepherd, F. D., Yang, A. C., Roosild, S. A., Bloom, J. H., Capone, B. R., Ludington, C. E. and Taylor, R. W., *In* "Adv. E.E.P." Vol. 40B, p. 981 (1976).
3. Shepherd, F. D., Taylor, R. W., Roosild, S. A., Skolnik, L., Cochrun, B. and Kohn, E., *In* "Proc. of IRIS Thermal Imaging Meeting", El Toro, California (1977).
4. Kohn, E. S., Roosild, S. A., Shepherd, F. D., and Yang, A. C., *In* "Proc. International Conference on Application of CCDs" (1975).
5. Taylor, R. W., Shepherd, F. D., Roosild, S. A., Yang, A. C. and Kohn, E. S., IRIS Detector Speciality Group Meeting, AF Academy, Colorado (1977).
6. Skolnik, L. H., Taylor, R. W., Capone, B. R., Shepherd, F. D., Roosild, S. A. and Kosonocky, W. F., 26th National IRIS, USAFA, Colorado Springs, (1978).
7. Capone, B. R., Skolnik, L. H., Taylor, R. W., Shepherd, F. D., Roosild, S. A., Ewing, W., Kosonocky, W. F. and Kohn, E. S., 22nd International Technical Symposium of the Society of Photo-Optical Instrumentation Engineers, San Diego (1978).

8. Kosonocky, W. F., Sauer, D. J., and Shallcross, F. V., RADC-TR-77-304, September (1977).
9. Kohn, E. S., Kosonocky, W. F. and Shallcross, F. V., RADC-TR-77-303, September (1977).
10. Kosonocky, W. F., Kohn, E. S., and Shallcross, F. V., RADC-TR-78, July (1978).
11. Archer, R. J. and Yep, T. O., *J. Appl. Phys.* **31,** 303 (1970).
12. Dalal, V. L., *J. Appl. Phys.* **32,** 2280 (1971).
13. Vickers, V. E., *Appl. Opt.* **10,** 2190 (1971).
14. Chu Wei-Kan, Low, S. S., Mayer, J. EW. and Nicolet, M., AFCRL-TR-75-0092, January (1975).
15. Chen, C. L. and Boyd, J. T., *IEEE Trans. Electron Devices* **ED–25,** 67 (1978).
16. White, J. M. and Chamberlain, S. G., *IEEE Trans. Electron Devices* **ED–25,** 125 (1978).
17. Hall, J. A., *Appl. Opt.* **10,** 838 (1971).

Author Index

Numbers in parentheses are reference numbers and are included to assist in locating references where the authors' names are not mentioned in the text. Numbers in italics refer to the page on which the reference is listed.

A

Ables, H. D., 146(15), *420*
Adams, M. C., 270(9), *273*
Aihara, S., 39(1), *50*
Airey, R. W., 155(27), *158*, 159(1), *166*, 171(2), 492(16), *494*
Alcaino, G., 459(12), *462*
Amelio, G. F., 425(11), *428*
Ando, K. J., 472(13), 473(13), *480*
Ando, T., 34(7), *37*
Angel, J. R. P., 342(6), *346*, 408(13), *413*, 476(16), 478(17), *480*
Antcliffe, G. A., 449(6), *452*, 469(10), *480*
Archer, R. J., 496(11), *512*
Arndt, U. W., 146(1), *158*, 209(1), (2), (3), 213(3), 214(1), (2), 215(3), *216*
Aslam, M., 155(27), *158*, 159(1), *166*, 171(2), *181*
Audier, M., 369(4), 372(4), *378*
Awcock, M. L., 231(2), *236*
Axon, D. J., 325(9), *327*

B

Bailey, S., 20(14), *22*, 97(14), *99*
Balick, B., 459(10), *462*
Baliunas, S. L., 415(6), *420*
Barton, J. B., 472(12), 473(14), 474(12), *480*
Bateman, J. E., 193(2), 194(2), *200*
Bates, C. W. Jr., 149(6), 150(6), *158*, 201(1), *207*
Barat, C., 370(5), 372(5), *378*
Baum, W. A., 90(1), *98*, 458(9), *461* 492(17), *494*

Beddoes, D. R., 483(5), (7), 489(7), *494*
Beaver, E. A., 101(3), 4, 101(5), *108*, 296(4), *303*, 397(4), *413*, 427(16), (17), (18), *428*, *429*, 463(2), 469(3), 478(17), *480*
Beeley, J. R., British Patent 35396/78, 255(7), *260*
Bergh, S. van der., 459(11), *462*
Beurle, R. L., 119(2), 126(2), *132*, 448(5), *452*
Bhide, G. K., 19(12), *22*
Biberman, L., 63(5), *73*
Bijaoui, A., 380(6), *388*
Bingham, R. G., 325(8), (9), *327*
Bishop, H. E., 159(4), *166*
Blazit, A., 379(3), *388*, 485(9), *494*
Bloom, J. H., 495(2), 496(2), 510(2), *511*
Blouke, M. M., 426(13), *428*
Boerio, A. H., 189(1), *200*
Boksenberg, A., 90(9), *99*, 355(1), (2), (3), (4), 365(7), *367*, 379(1), (2), *388*, 397(2), *413*
Bonneau, D., 379(3), *388*
Bortolot, V. J., *420*
Boulesteix, J., 380(6), *388*
Boutot, J. P., 369(3), (4), 372(3), (4), *378*
Bowen, I. S., 90(2), *98*
Bowen, P. J., 90(9), *99*
Boyd, J. T., *512*
Bradley, D. J., 21(15), *22*, 265(1), (2), (3), 266(4), 267(6), 268(3), 270(9), (10), *273*
Braver, W., 177(3), *181*
Breare, J. M., 437(12), *440*
Breitzmann, J. F., 426(13), *428*
Bril, A., 136(4), *141*

Broadfoot, A. L., 373(9), *378*
Brown, R. A., 415(1), 419(19), *419, 420*
Bruining, H., 127(7), 130(7), *132*, 134(2), *141*
Buchholz, V., 431(6), 436(6), *440*, 453(2), *461*
Burgess, D. E., 17(11), *22*, 134(2), 355(3), *367*, 397(3), *413*
Burkhead, M. S., 329(4), 333(4), 334(4), *338*
Butler, D. J., 159(2), *166*

C

Campbell, A. W., 437(12), *440*
Campbell, B., 431(2), 436(2), *440*, 457(8), *461*
Capone, B. R., 495(2), (6), (7), 496(2), (6), (7), 510(2), *511*
Carlo, J. T., 426(13), *428*
Carroll, J. P., 153(22), *158*
Carruthers, G. R., 90(3), *98*, 284(1), (2), 288(4), 291(6), 292(9), *293*
Carvennec, F. L., 31(2), *37*
Ceckowski, H. D., 92(12), *99*, 101(1), *108*, 339(1), *346*
Cenalmor, V., 380(4), *388*
Chaffee, F. H. Jr., 187(2), *188*, 415(1), (2), (13), (14), 416(16), (17), 417(16), *419, 420*
Chalmeton, V., 281(1), *282*
Chamberlain, S. G., *512*
Chan, W. W., 449(6), *452*, 469(10), 473(14), *480*
Charles, D. R., 31(2), *37*
Chen, C. L., *512*
Cheng, I., 427(22), *429*
Chodil, G. J., 426(14), *428*
Choisser, J. P., 101(4), *108*, 397(5), *412*, 427(17), (19), (20), *428*, 470(11), *480*
Chou, L. W., 128(8), *132*
Choudry, A., 19(13), *22*
Chu, Wei-Kan, 496(14), *512*
Cioffi, P. P., 91(11), *99*
Cochrun, B., 495(3), 496(3), *511*
Coleman, C. I., 90(8), (9), (10), *99*, 103(6), *108*, 158(28), *158*, 359(5), 362(6), 365(7), *367*
Coleman, G. D., 476(16), *480*
Coleman, L., 427(22), *428*
Collins, D. R., 449(6), *452*, 469(10), 473(14), *480*
Combes, M., 445(2), 446(2), 447(2), *452*
Condal, A., 431(6), 436(6), *440*, 453(2), *461*
Conder, P. C., 51(3), *61*, 150(12), *158*
Cope, A. D., 153(22), *158*
Craine, E. R., 340(2), 344(2), (7), (10), *346*
Cromwell, R. H., 341(3), (4), (5), 342(6), *346*, 398(11), *413*, 476(16), *480*
Currie, D. G., 469(8), *480*, 489(15), *494*
Curtis, N. A., 109(2), 112(2), *117*, 315(1), (4), *327*

D

Dahn, C. C., 408(15), *413*
Dainty, J. C., 481(1), 483(5), (7), 484(8), 489(7), (8), *493, 494*
Dalal, V. L., 496(12), *512*
Deasley, P. J., 153(23), *158*
Deharveng, J. M., 380(7), (8), *388*
Delmotte, J. C., 369(4), 372(4), *378*
Delamere, W. A., 20(14), *22*, 90(5), 97(14), *98, 99*, 101(2), 102(2), *108*, 469(3), *480*
Dennison, E. W., 90(2), *98*
de Vaucouleurs, A., 329(2), 333 (10), 336(2), *338*
de Vaucouleurs, G., 329(2), (6), 333 (10), 336(2), *338*, 454(4), 461(13), (14), *461, 462*
Dionne, N. J., 20(14), *22*, 97(14), *99*
Dickson, J., 431(9), *440*
DiStefano, T. H., 447(4), *452*
Dolizy, P., 51(2), *61*
Dravins, D., 431(8), 436(8), *440*
Dreux, M., 426(15), *428*
Driard, B., 147(3), 148(4), 148(5), *158*, 227(1), *236*
Duchenois, V., 369(3), 370(7), 372(3), (7), *378*

Duerr, R., 344(8), *346*
Dunham, T. Jr., 415(13)
Dupree, A. K., 415(6), *420*
Du Toit, A. G., 136(3), *141*
Duck, R. H., 425(11), *428*, 485(13), *494*
Dyvig, R. R., 341(3), (4), *346*, 398(11), *413*

E

Edgecumbe, J., 147(2), *158*
Enck, R. S. Jr., 442(1), *452*
Eschard, G., 253(1), *263*, 369(3), 372(3), *378*
Evans, G. B., 150(10), *158*
Evrard, R., 133(1), 134(1), *141*
Ewing, W., 495(7), 496(7), *511*

F

Faber, S. M., 459(10), *462*
Fahlman, G. G., 431(6), 436(6), *440*, 453(2), *461*
Faivre, J. C., 193(3), *200*
Fanet, H., 193(3), *200*
Fauconnier, T., 426(5), *428*
Faulkner, K. R., 153(23), *158*
Felenbok, P., 445(2), 446(2), 447(2), *452*
Fish, R. A., 329(9), *338*
Folkes, J. R., 152(17), *158*
Fort, B., 365(7), *367*, 426(15), *428*, 445(2), 446(2), 447(2), *452*
Fouassier, M., 369(2), 370(2), 372(2), *377*
Francken, J. C., 127(7), 130(7), *132*, 134(2), *141*
Frank, K., *61*
Freeman, K. C., 454(3), *461*

G

Gallagher, J. S., 459(10), *462*
Ganson, A., 493(17), *494*

Garfield, B. R. C., 159(2), *166*
Gardier, S., 369(1), *377*
Garn, L., 32(5), *37*
Garwin, E. L., 147(2), *158*
Gavin, A., 193(3), *200*
Geary, J. C., 397(8), *413*, 431(1), 432(1), 436(1), *440*
Geppert, D. V., 153(18), *158*
Gersho, S., 469(9), *480*
Gilbert, G. R., 476(16), *480*
Gillespie, L. F., 258(8), *263*
Gilmore, D. J., 146(1), *158*, 209(1), (3), 213(3), 214(1), 215(3), *216*
Glaspey, J., 431(6), 436(6), *440*, 453(2), *461*
Goetz, G. W., 13(5), *22*, 189(1), 193(4), 199(4), *200*
Gomer, R., 138(6), *141*
Goodson, J. H., 159(2), *166*
Graf, J., 253(1), (2), *263*
Grandi, S. A., 476(16), *480*
Green, E. Jr., 217(1), *225*
Green, M., 189(1), *200*
Greenberg, G. A., 122(3), *132*
Gresham, M., 408(15), *413*
Griboval, P., 305(1), 307(1), *314*
Grover, C. G., 253(4), *263*
Guyot, L. F., 148(4), (5), *158*

H

Hachenberg, O., 177(3), *181*
Hagelbarger, D. W., 91(11), *99*
Hagino, M., 154(25), *158*
Hall, J. A., 505(17), *512*
Hall, J. E., 426(13), *428*
Hallam, K. L., 110(3), *117*, 445(3), *452*
Harao, N., 201(2), *207*
Harms, R. J., 101(5), *108*, 397(4), *413*, 427(18), *429*, 463(2), *480*
Harper, B., 17(11), *22*
Harris, R. B. A., British Patent 14700/65, 254(5), *263*
Hartley, K. F., 111(4), 116(5), *117*, 126(5), *132*, 315(3), *327*
Harvey, J., 397(7), *413*, 431(4), 436(4), *440*

Hatanaka, Y., 31(4), 31(6), 34(7), *37*
Hayes, D., 398(12), 403(12), *413*
Haynes, K. A. F., British Patent 45295/77, 254(6), *263*
Hazelwood, J., 376(10), *378*
Hege, K., 408(13), (14), (15), 412(16), *413*
Hearnshaw, J. G., 415(4), 415(5), *420*
Hedge, A. R., 437(12), *440*
Heinrich, K. F. J., 159(5), *166*
Helbrough, K., British Patent 35396/78, 255(7), *263*
Herbig, G. H. Z., *420*
Heritage, J. P., 269(7), *273*
Herrmannsfeldt, W. B., 106(7), *108*
Herstel, W., 150(7), *158*
Hewitt, A. V., 416(15), *420*
Hicks, G. T., 291(6), *293*
Hier, R. G., 463(1), *480*
Hiltner, W. A., 389(1), *395*, 397(3), *412*, 416(18), *420*
Holliday, J. E., 159(3), *166*
Hopkins, G. P., 152(17), *158*
Hopkinson, G. R., 437(12), *440*
Holeman, B. R., 17(9), *22*, 23(1), *30*, 31(1), *37*, 51(3), *61*, 150(12), 151(14), *158*
Horlick, G., 397(10), *413*, 435(10), *440*
Hornbeck, L. J., 449(6), *452*, 469(10), *480*
Hoshiko, H. H., 426(14), *428*
Hounsfield, G. N., 150(8), *158*, (1), *251*
Howorth, J. R., 152(16), (17), 153(24), *158*
Hubble, E., 329(1), (7), 334(7), *338*
Humrich, A., 437(12), *440*
Huston, A. E., British Patent 35396/78, 255(7), *263*

I

Ibrahim, A. A., 426(12), *428*
Izatt, J. R., 63(1), *72*, 75(4), *88*

J

Jack, M. D., 485(13), *494*
Jackson, D. A., 151(13), *158*

Jain, R. K., 269(7), *273*
Jamar, J., 369(1), *377*
Jensen, E. B., 476(16), *480*
Johnson, C. B., 110(3), *117*, 445(3), *452*
Johnson, J. J., 426(14), *428*
Jones, K. W., 21(15), *22*
Jones, R. E., 193(2), 194(2), *200*
Jorden, A. R., 116(5), *117*

K

Kalibjian, R., 266(5), *273*
Kalinowski, J. K., 329(4), 333(4), 334(4), *338*
Kamminga, W., 94(13), *99*
Kan, H., 154(25), *158*
Kaneda, E., 15(8), 16(8), *22*
Kato, T., 31(4), *37*
Katsuna, H., 154(25), *158*
Kelton, P., 397(9), *413*, 431(5), 436(5), 437(5), *440*
Kervitsky, J., 291(6), *293*
Khogali, A., 493(17), *494*
King, D., 116(5), *117*
King, I., 336(15), *338*
King, W. L., 253(4), *263*
Kinoshita, K., 15(8), 16(8), *22*
Kiuchi, Y., 39(1), *50*
Klasens, H. A., 136(4), *141*
Koechlin, L., 379(3), *388*, 485(9), *494*
Koester, D., 408(14), *413*
Kohn, E. S., 151(15), 153(20), (21) *158*, 495(3), (4), (5), 495(7), (9), (10), 496(3), (4), (5), (7), (9), (10), 501(4), (9), 503(9), *511, 512*
Kopriva, D., 412(16), *413*
Korff, D., 482(4), *494*
Kosonocky, W. F., 495(6), (7), (8), (9), (10), 496(6), (7), (8), (9), (10), 501(9), 503(9), *511, 512*
Kron, G. E., 416(15), *420*
Kuike, N., 423(9), *428*
Kurashige, M., 75(5), *88*

L

van Laar, J., 150(9), *158*
Labeyrie, A., 379(3), *388*, 482(3), *493*

Lalak, J., 23(3), *30*
Lallemand, A., 295(1), (2), (3), *303*
Lampton, M., 427(21), *429*
Lamy, Ph., 380(5), *388*
Lane-Wright, D., 431(6), 436(6), *440*, 453(2), *461*
Lapp, H. S., 92(12), *99*, 101(1), *108*
Larson, R. B., 337(16), (17), *338*
Lecomte, P., 376(10), *378*
Legoux, R., 51(2), *61*
Lester, J. B., 415(6), *420*
Liddy, B., 266(4), *273*
Lieber, A. J., 253(3), *263*
Liebert, J., 408(13), (14), (15), *413*
Linden, S. R., 130(9), 131(9), *132*
Livingstone, W. C., 397(7), *413*, 431(3), (4), 436(3), (4), *440*, 485(11), *494*
Lo, C. C., 376(10), *378*
Lodge, J. A., 12(2), *22*
Loh, E. D., 485(11), *494*
Long, D. C., 90(7), *98, 99*
Low, S. S., 496(14), *512*
Lowrance, J. L., 90(7), *99, 98*, 469(4), (5), *480*
Lubszynski, H. G., 12(2), (3), *22*
Ludington, C. E., 495(2), 496(2), 510(2), *511*
Luedicke, E., 153(22), *158*
Lutz, B. L., 416(17), *420*

M

Macau, J. P., 369(1), *377*
Macleod, N. A., 202(3), *207*
Makamaya, M., 15(8), 16(8), *22*
Manchester, R. N., 116(5), *117*
Mastinelli, R. U., 151(15), *158*
Mayer, J. E. W., 496(14), *512*
McAlister, H. A., 483(6),
McCollough, W. V., 63(4), 72(9), *73*
McConaughy, R. M., 426(14), *428*
McGee, J. D., 155(27), *158*, 159(1), *166*, 167(1), 171(2), *181*, 493(17), *494*
McIllwain, C. E., 101(3), (4), *108*, 427(16)(17), *428, 429*
McMullan, D., 58(6), *61*, 109(2), 112(2), *117*, 315(1), (2), (3), (4), (5), (6), 316(6), 318(5), 322(6), 325(9), 327, 431(9), *446*
Meltzer, B., 64(6), *73*
Milch, J. R., 213(4), *216*
Miller, A., 63(1), 72, 75(4), *88*
Miller, H. R., 344(9), *346*
Miller, J., 397(1), *413*
Miller, J. S., 187(1), *188*
Miller, R. H., 329(5), 334(5), *338*
Minami, H., 201(2), *207*
Mochnacki, S. W., 431(6), 436(6), *440*, 453(1), (2), 454(1), (6), *461*
Moore, C. H., 488(14), *494*
Moore, R., 408(15), 412(16), *413*
Morgan, B. L., 483(5), (7), 489(7), 492(16), *494*
Morgan, W. W., 329(3), *338*
Moss, H., 75(2), *88*
Muller, K. F., 63(3), *72*
Murdin, P. G., 116(5), *117*, 325(8), 327

N

Nakamura, T., 154(25), *158*
N'Guyen-Trong 380(5), *388*
Nicolet, M., 496(14), *512*
Nishida, R., 31(4), (6), 34(7), *37*
Niwa, N., 15(8), 16(8), *22*
Nudelman, S., 63(5), *73*

O

Oemler, A., 334(14), *338*
Okamoto, S., 31(4), (6), 34(7), *37*
Oneto, J. L., 485(9), *494*
Opal, C. B., 291(6), *293*
Ormerod, J., 431(9), *440*
Owen, R. B., 231(2), *236*

P

Pallister, W. S., 325(9), *327*
Palmer, I. C., 152(17), 153(24), *158*
Pellet, A., 380(8), *388*
Penny, A. J., 116(5), *117*
Perez-Mendez, V., 376(10), *378*
Pern, J. M., 380(5), *388*

Peterson, B. A., 116(5), *117*
Peterson, B. M., 296(4), *303*
Peterson, R. C., 415(3), (7), (8), (9), (10), (11), (12), 416(3), *420*
Petford, A. D., 90(9), *99*
Piaget, C., 150(11), *158*
Picat, J. P., 126(6), *132*, 286(3), *293* 445(2), 446(2), 447(2), *452*
Pierce, J. R., 64(8), *73*
Polaert, R., 150(11), *158*, 253(1), *263*, 369(3), 370(5), (6), 372(3), (5), (6), *378*
Powell, J. R., 109(2), 112(2), *117*, 315(1), (3), (4), (5), (6), 316(6), 318(5), 322(6), *327*
Prendergast, K. H., 329(5), 334(5), *338*

R

Rangarajan, L. M., 19(12), *22*
Redman, R. O., 329(8), *338*
Renda, G., 469(4), *480*
Renard, L., 295(1), (2), (3), *303*
Reynolds, G. T., 213(4), *216*
Reynolds, J. H., 334(12), (13), *338*
Rhines, W. C., 449(6), *452*, 469(10), 473(14), *480*
Richard, J. C., 150(11)
Ricodeau, M., 147(3), *158*
Ring, J., 492(16), *494*
Robert, J. P., 193(3), *200*
Roberts, C. G., 473(14), *480*
Robbins, C. D., 442(1), *452*
Robinson, L. B., 187(1), *188*, 397(1), *413*
Roddie, A. G., 266(4), *273*
Rodgers 13(6), *22*
Rodway, D. C., 151(14), *158*
Romanishan, W., 408(15), *413*
Roosild, S. A., 495(2), (3), (4), (5), (6), (7), 496(2), (3), (4), (5), (6), (7), 501(4), 510(2), *511*
Rosier, J. C., 369(2), 370(2), (5), (6), (7), 372(2), (5), (6), (7), *377*, *378*
Rougeot, G., 147(3), *158*
Rouger, M., 193(3), *200*
Rozière, G., 148(5), *158*, 227(1), *236*
Ruddock, I. S., 265(2), 270(10), *273*
Ryan, J. P., 267(6), 269(8), *273*

S

Sackinger, J. P., 442(1), *452*
Sadowski, M., 155(26), *158*
Sandage, A., 303(5), *303*, 310(2), *314*, 333(11), *338*
Sandel, B. R., 373(9), *378*
Santilli, V. J., 14(7), 15(7), *22*
Sato, K., 58(5), *61*
Saudinos, J., 193(3), *200*
Sauer, D. J., 495(8), 496(8), *512*
Sauerman, G. O., 63(3), *72*
Scaddan, R, J., 483(5), (7), 489(7), *494*
Scarrott, S. M., 325(8), (9), *327*
Schade, O. H., 75(1), *88*
Schagen, P., 127(7), 130(7), *132*, 134(2), *141*
Schectman, S., 397(3), *413*
Scheer, J. J., 150(9), *158*
Schmidt, G. D., 296(4), *303*, 342(6), *346*, 478(17), *480*
Schmidt, G. W., 101(5), *108*, 397(4), *413*, 427(18), *429*, 463(1), *480*
Schmidt, M., 90(2), *98*
Schrana-de Pauw, A. D. M., 202(4), *207*
Schroeder, D. J., 290(5), *293*, 416(16), 417(16), *420*
Selke, L. A., 63(2), *72*
Serkowski, K., 344(7), *346*
Servan, B., 295(1), (2), (3), *303*
Shalicross, F. V., 495(8), (9), (10), 496(8), (9), (10), 501(9), 503(9), *512*
Sheetman, S. A., 389(1), *395*, 416(18), *420*
Shemansky, D. E., 373(9), *348*
Shepherd, F. D., 495(1), (2), (3), (4), (5), (6), (7), 496(1), (2), (3), (4), (5), (6), (7), 501(4), 510(2), *511*
Sheppard, C. J. R., 152(16), *158*
Shimizo, K., 39(1), (2), *50*
Shirley, E. G., 329(8), *338*
Shulman, S., *420*
Sibbett, W., 21(15), *22*, 265(1), (2), 266(4), 270(9), *273*
Simkin, S. M., 454(5), *461*
Simon, R. E., 153(19), *158*
Singer, B., 23(3), *30*
Singh, S., 19(12), *22*

Skingsley, J. D., 51(3), *61*, 150(12), *158*
Skolnik, L. M., 495(3), (6), (7), 496(3), (6), (7), *511*
Slaughter, C., 397(7), *413*, 431(4), 436(4), *440*
Sleat, W. E., 266(4), *273*
Smith, G. H., 341(5), *346*
Smithson, R. C., 431(7), 436(7), *440*
Sneden, C., 415(12), *420*
Snell, P. A., 130(9), 131(9), *132*
Snow, E. H., 397(5), *413*, 427(20), *428*
Soares, R. A., 436(11), *440*
Sobieski, S., 469(6), 470(6), 473(14), *480*
Spangenberg, K. R., 64(7), *73*
Spicer, W. E., 447(4), *452*
Sternglass, E. J., 159(3), *166*
Stevels, A. L. N., 202(4), *207*
Steward, J. C., 151(14), *158*
Stockman, H. S., 408(13), *413*
Strittmatter, P. A., 408(15), *413*, 476(16), *480*
Sturrock, P. A., 122(4), *132*
Sukegawa, T., 154(25), *158*
Surridge, R. K., 153(24), *158*
Suzuki, Y., 15(8), 16(8), *22*
Swank, R. K., 219(2), *225*

T

Tapia, S., 344(8), (10), *346*
Tarenghi, M., 344(10), *346*
Taylor, R. W., 495(2), (3), (5), (6), (7), 496(2), (3), (5), (6), (7), 510(2), *511*
Taylor, S., 12(3), *22*
Teranishi, A., 31(3), *37*
Thaddeus, P., *420*
Tinsley, B. M., 337(17), *338*
Title, A. M., 415(3), 416(3), *420*
Towler, G. O., 58(6), *61*
Townslee, A. C., 349(2), *353*
Toyonaga, R., 58(5), *61*
Trawny, E. W. L., 152(16), *158*
Tripp, G., 427(22), *429*
Trumbo, D., 397(7), *413*, 431(4), 436(4), *440*
Turnbull, A. A., 150(10), *158*

Tull, R. G., 397(5), (9), *413*, 427(20), *429*, 431(5), 436(5), 437(5), *440*

V

Van der Polder, L. J., 75(3), *88*
Van Huyssteen, C. F., 51(4), *61*, 138(5), *141*
Van Roosmalen, J. H. T., 12(1), *22*
van Schooneveld, C., 223(3), *225*
van Zuylen, P., 90(6), 94(6), *99*
Verat, M., 148(4), (5), *158*, 227(1), *236*
Vibrans, G. E., 372(8), *378*
Vickers, V. E., 496(13), *512*
Vogt, S. S., 397(9), *413*, 431(5), 436(5), 437(5), *440*

W

Walker, G. A. H., 431(6), 436(6), *440*, 453(2), 457(8), *461*
Walker, M. F., 109(1), *117*
Walker, J. W., 449(6), *452*, 469(10), 476(15), *480*
Wallace, P., 116(5), *117*
Wallace, R., 376(10), *378*
Wampler, E. J., 187(1), *188*, 347(1), *353*, 397(1), *413*,
Wardley, J., 12(3), *22*
Warren-Smith, R. F., 325(8), *327*
Waters, M. W., 193(2), 194(2), *200*
Watton, R., 17(10, 11), *22*, 23(1), 24(4), *30*
Webinger, *327*
Webley, R. S., 12(2), *22*
Weckler, G., 397(6), *413*
Wehinger, P. A., (7), *327*
Wellgate, G. B., 431(9), *440*
Wells, D. C., 478(18), *480*
Westphal, J. A., 457(7), *461*
White, C., 325(9), *327*
White, J. M., *512*
Wilkinson, D. T., 485(11), *494*
Williams, B. F., 151(15), 153(19), *158*
Williams, J. T., 469(7), 470(7), *480*
Williams, R., 412(16), *413*
Wilson, R. J. F., 159(2), *166*

Wooley, R. P., 90(5), *98*, 101(2), 102(2), *108*
Woolf, N. J., 408(13), 412(16), *413*
Worden, S. P., 482(2), *493*
Wreathall, W. M., 17(9), *22*, 23(1), *30*, 31(1), *37*, 119(2), 126(2), *132*, 448(5), *452*
Wysoczanski, W., 101(4), *108*, 427(17), *428*

Y

Yang, A. C., 495(1), (2), (4), (5), 496(1), (2), (4), (5), 501(4), 510(2), *511*
Yamaka, E., 31(3), *37*
Yee, E. M., 151(13), *158*
Yoshida, O., 39(1), (2), *39*, *50*

Z

Zacharov, B., 119(1), 120(1), 126(1), *132*
Zucchino, P., 90(7), *99*, 469(4), (5), *480*

Subject Index

A

Aberrations, see also Electron optics,
  astigmatic in electron beam focus, 78
  geometric, 112, 172, 180
  spherochromatic, 119, 124–126
  in X-ray TV system, 213
Ablebond 190-8, conductive silicone rubber, 186
Alnico V magnets for image tube focusing, 91
Aluminium,
  anodized for X-ray phosphor screen, 203
  anti-reflection coating, 58, 159, 160
  backing, on phosphor screen, 137
  ion etching mask, 24
Aluminium oxide, mosaic as substrate for X-ray phosphor, 202
Anode,
  spherical, in electron optical system, 123
  virtual, in catadioptric image tube, 133
Antihalation layers on phosphor screens, 155, 159, 160
Antimony trioxide vidicon storage target, 349
Antireflection coating,
  Aluminium, 58
  on phosphor screen, 137
Arsenic triselenide,
  layer, porous in chalnicon target, 13, 47
  layer, resistive, 34
Arsenic trisulphide, in chalnicon target, 39
Astronomical observations, see Astronomical objects observed
Astronomical objects observed,
  BD+33 2642 star, 410
  BL Lacertae objects, 343
  Crab pulsar, 116
  Cygnus A radiogalaxy DA 240, 478

Astronomical objects observed—(cont.)
  DQ Her Nova, 410
  Eta Carinae nebula, 323
  M16, 344
  M33, 381, 385
  M51, 385
  M82, 342
  MWC 349, 412
  NGC 157, 478
  NGC 604, 384
  NGC 1068 Scyfert galaxy, 453
  NGC 1275, 478
  NGC 2081, 325
  NGC 2168 (M35), 353
  NGC 2217, 325
  NGC 3379, 329
  NGC 3384, 329
  NGC 4736, 388
  NGC 5102 S0 galaxy, 458
  NGC 6643, 375
  NGC 6853, 374
  NGC 6888, 375
  N3C 390, 3, 302
  ζ0ph, 417
  Orion nebula, 456
  PKS 0830+115, 303
  PKS 1004+141, 303
  Shakbazian 78, 297
  Shakbazian 82, 297
  Vega, 439
  Vela pulsar, 355
  Virgo A radiogalaxy, 296
  Zwicky 1ZW 181, 302
Autocorrelation function, in speckle interferometry, 487

B

Background, see also Dark current
  light induced, 175
Backprojection in X-ray tomography, 244

Bakeout temperature, maximum electron permeable membrane, 292
Beam,
  landing characteristics, 77
  spreading, vidicon, 79, 82
  -target, interaction curve, 62
Binary star, analysis by speckle interferometry, 481
Black coating,
  bismuth, infrared absorber, 31
  aluminium antihalation, 160
  silicon antihalation, 160
Black level stabilization in X-ray TV system, 213
Buffer store,
  use in photon counting system, 390
  use in speckle interferometry, 487, 489.
Buried channel CCD, 495

C

Cadmium Selenide,
  target photoconductive, 39, 43
  target with arsenic triselenide, 46
Caesium Iodide,
  phosphor screen, 203
  X-ray photocathode, 189
Caesium Telluride opaque photocathode for UV, 291
Calcite,
  crystal for artificial double star, 489
  plate in astronomical polarimetry, 344
Calcium Fluoride windows, 287
CAMAC computer interface, 437, 467, 488
Caméra électronique, see electronic camera, electronographic camera.
Camera electronographic, see Electronographic camera.
Camera tube, see SEC vidicon, SIT vidicon, Television camera tube, vidicon.
Capacitance, image localizer target diode, 230
Cascade intensifier, see image intensifier.

Cassegrain,
  Schmidt camera, Steward Observatory, 399
  Spectrograph, Haute Provence observatory, 392
Cataphoresis, for phosphor screen deposition, 171
Cathode, see also Photocathode.
Cathode, cold emitting
  GaAsP, 154
  GaP/GaAlP, 154
Cathode potential stabilization (CPS), 31
Cathode ray tube, fibre optic for convolution processing, 249
Cathode, spherical, 133
Centroiding, in photon counting detector, 359, 373, 380, 390, 416
Ceramic rods, in image tube structure, 171
Chalnicon, TV camera tube, 39, 42
Chamber,
  degassing for electronographic emulsion, 308
  processing for photocathodes, 53, 307
Channel electron multiplier, see Microchannel plate.
Channel plate, see Microchannel plate.
Charge
  amplifier, on self scanned photodiode array, 432
  saturation in microchannel plate, 369
  skimming in IRCCD, 498–500
  storage tube in convolution processing, 245
  spreading in CCD, 470
Charge coupled device (CCD),
  buried channel, 422
  Fairchild 100×100, 481
  Fairchild 244×180, 394
  for speckle interferometry, 482
  intensified, 108
  optically coupled to intensifier, 476
  surface channel, 421
  Texas 160×100 back illuminated, 463
  Texas 400×400, 455
Charge injection device (CID), 422
Chromium oxide coating to reduce secondary emission, 316

SUBJECT INDEX

Circle of confusion in electron optics, 126
Circuits, driving, for magnetic deflection, 112
Coating, chromium oxide, to reduce secondary emission, 316
black, 31, 160
Cockcroft-Walton voltage multiplier, 168
Cold electron emitter GaP–GaAlP junction, 154
Cold weld, copper, 170
Compact galaxy, 297
Computation of electron trajectories, 103
Computer,
  Amdahl, 460
  CDC 6600, 106
  IBM 370, 106
  Interdata 70, 488
  NOVA, 467
  NOVA 2, 417
  NOVA 800, 400
  PDP 11, 391
  PDP 11/03, 437
  PDP 11–40, 209
Computer language, see Programming language.
Computer model of oblique magnetic focusing, 446
Computer simulation of intensified CCD, 472
Control system, for electronographic camera, 320
Convolution, in optical image processing, 239
Convolution function in image processing, 246
Convolution system, using image intensifier, 237–250
Cooling,
  of astronomical detector, 380
  of CCD, 426, 502
Copper, OFHC, in image tube structure, 169
Corona electrical, in image tubes, 183–188
Correlator, hard wired for speckle interferometry, 488
Crossover,
  electron optical 79, 133

Crossover—(cont.)
  potential, 58, 75
Current density distribution in vidicon electron beam, 76–79

D

Dark current, see also Background, Noise.
Dark current,
  Chalnicon camera tube, 40, 43
  platinum silicide detector, 502
  solid state arrys, 372, 426
Dead layer, in electron bombarded CCD, 444
Deflecting field,
  effect of nonuniformities, 105
  used with magnetic focusing, 102
Deflection coils,
  for electronographic camera, 109–111
  for image intensifier, 401
Degassing,
  during tube evacuation, 33
  of electronographic emulsion, 289
Demagnification, in intensified CCD tube, 439
Detective quantum efficiency (DQE), 3, 406
  of X-ray TV system, 214
Dielectric materials for high voltage insulation, 183
Digicon image tube, 427
  demountable, 469
  magnetically deflected, 101
Digitization of images, 382
Diode array, see Solid state imaging detectors.
Distortion, see also Aberrations, Electron optics.
Distortion, geometrical,
  due to image deflection, 112
  in miniature image tube, 172, 180
  in X-ray TV system, 213
Dye laser, mode locked, 265
Dynamic range,
  of CCD, 487
  of intensified CCD, 442

E

Electron,
 equations of motion, 103
 trajectories in image tube, 134
 trajectories in magnetically focused electron gun, 85
Electron emission, see also Photo cathode.
Electron emission,
 Auger, 190
 cold, 154
 Compton, 190
Electron focusing, see Electron optics, Magnetic focusing.
Electron gun, radial initial velocity, 81
Electronic camera, see also Electronographic camera.
Electronic camera, 4, 295–303
Electrodes, titanium for electronographic image tube, 314
Electronographic camera,
 Kron, 416
 Lallemand, 295
 large format, 283
 RGO 40 mm, 315–318
 RGO 85 mm, 318
 Spectracon, 330
 with stepped deflection, 113
 ultraviolet, 285
 University of Texas 5 cm, 305
 University of Texas 20 cm, 313
Electronographic image tube, see also Electronography, Spectracon.
Electronography, see Electronographic camera, emulsion electronographic, spectracon.
Electron optics, see also Electron focusing, Magnetic focusing.
Electron optics,
 of catadioptric intensifier, 133–141
 of concentric spherical system, 119–132
 of intense field magnetic focusing, 286
 of magnetically deflected digicon, 101–108
 of magnetically focused electron beam, 75–88
 of miniature image intensifier, 167

Electron optics—(cont.)
 of stepped deflection system, 110
 tri-electrode electrostatic, 130
Electron pulse height distribution, see Pulse height distribution.
Electron secondary emission, see Secondary emission.
Electron sensitive emulsions, see Emulsion electronographic.
Electrostatic focusing,
 concentric spherical, 127
Emulsion electronographic,
 Ilford G5, 116 323, 330
 Ilford L4, 296
 Kodak electron image, 340
 Kodak Industrex A, 295
 Kodak NTB2, 340
 Kodak NTB3, 340
 outgassing, 308
Emulsion photographic,
 Kodak IIa–D, 341
Energy gap, of CdSe, 39
Energy linearity of gamma scintillation camera, 234
Envelope, image tube,
 fused silica, 316
 pyrex glass, 318
ERF function, 69
Etching,
 by ion beam, 23
 masks for, 24
Exposure control automatic, for image tube camera, 255

F

Fibre optic,
 coupling at high voltage, 187
 face plate, 165, 169, 339
 output window, 370
Field mesh in vidicon, 81
Field penetration, in miniature image tube, 172
Film,
 plastic thin, for IR camera, 24
 tin oxide conducting, 138, 186
Filter,
 compensating for gamma camera, 234
 spatial, 239

SUBJECT INDEX

Flare optical, reduction, 58
Floppy disc, for astronomical data storage, 417
Fluorescent screen, *see* Phosphor screen.
Flux compensated permanent magnet focusing assembly, 94
Focal reducer, Haute Provence, 380
Focusing,
 electromagnetic, 75, 126, 167
 spherical concentric, 119–132
 proximity, in X-ray intensifier, 195
Fountain intensifier tube, 133–141
Frame TV digital store for X-ray crystallography, 209
Framing camera, *see* High speed photography, streak camera.
Frequency, chopping for TGS vidicon, 27
Fringe patterns, interference in thinned CCD's, 455, 467

G

Gain,
 automatic control in image tube, 168
 of electron bombarded CCD, 442
 enhancement in miniature image tube, 176
 over direct photography of image tube camera, 257
 of photon counting system, 380
Galaxy,
 light, 332
 luminosity profiles, 297, 302, 332
 compact, 297
Gallium Arsenide photocathode, 150
Gamma, light transfer characteristic of SEC vidicon, 59
Gamma ray camera, 148, 227–236
Gating, of intensified CCD, 442
Germanium lens, for the infrared, 34
Getters, barium in image tube, 493
Glass,
 pyrex vacuum envelope, 318
 soda-lime for image tube, 169
 pyroceram, 169

Gold wire seal, for image tube, 318
Guiding telescope by TV system, 347

H

Hera, machinable magnetic material, 94
High speed photography using MCP image tube, 156
Hubble's law, 334
Humidity, and voltage breakdown, 184

I

Image intensifier, *see also* Image tube, microchannel plate image tube.
Image intensifier,
 for convolution processing, 237–249
 diode electrostatic VLI-116, 209
 double proximity focus, 369
 four stage EMI, 90, 483
 four stage Varo 8605, 398
 gated inverter, Nitec R-6340, 253
 large photocathode ITT F-4094, 339
 microchannel plate, 153
 microchannel plate TH9304, 380
 microchannel plate X-ray TH9303, 380
 miniature magnetic focus, 167–181
 proximity focus ITT-4109, 399
 six stage electrostatic, 389
 stray light in, 159–166
 three stage Varo 8585, 476
 two stage magnetic RCA C3306, 399
 with 146 mm photocathode, 346
 with scintillator for X-rays, 217–225
Image localizer tube with silicon target, 227–236
Image tube,
 Digicon, 427
 Photochron I streak tube, 265
 Photochron II streak tube, 273
 scintillation camera THX1427, 228–235
 wafer type, 369
 with catadoptric electron optics, 133

Imaging, real time in X-ray topography, 223
Indium seal,
 for camera tube, 53, 56
 for image tube, 102, 318
Infra red,
 camera tube, 23, 31, 42
 schottky barrier CCD imager, 495–512
 sky survey, 344
Integration of image, using photodiode array, 401
Intensified solid state detectors, 101, 108, 426, 441–452
Interactive Picture Processing System (IPPS), 478
Ion events,
 in image tube, 402
 in intensified CCD, 478
 in photon counting system, 359
Ion pump, appendage, on electronographic camera, 312, 316
Isophotes, 296, 332

K

Kapton plastic film for electronographic camera, 293, 307
Kepler's law applied to electron optics, 134
Kovar alloy, disturbance of magnetic field by, 93
Kron, electronographic camera, 416

L

Lallemand electronographic camera, 275, 293–301
Lag, in TV camera tubes, 11, 13, 39
Laser,
 Argon ion, 267
 CW Dye, mode locked, 266
 He–Ne, 33
Leakage current, thermal in solid state arrays, 433
Lenard Window, *see also* Membrane

Lenard Window—(*cont.*)
 electron permeable, Mica window
Lenard window image tube, 315
Lens coupling,
 Nikon $f/1\cdot2$, 390
 Repro-Nikkor $f/1\cdot0$, 400
Light induced background,
 in miniature image tube, 180
 in proximity focused image tube, 143, 159–167
Light scatter, in image tube, 175
Light transfer characteristic of TV camera tube, 12, 59
Limiting resolution, of image tube, 219
Linearity, of CCD imager, 487

M

Magazine, roll film for electronographic camera, 313
Magnesium fluoride,
 secondary emitter, 177
 window, for MCP image tube, 376
 windows, on TV camera tube, 16
 with silver, resistive target, 51, 56
Magnesium oxide, secondary emitter, 58, 177
Magnetic focusing,
 field non-uniformities, 106
 permanent magnet, 89–99
 vidicon camera tube, 75
Magnetic focus solenoid,
 power dissipation, 91
Magnetic material,
 Alnico V, 91
 Co-netic, 91
 Hera, machinable, 94
 Mu-metal, 93
 Samarium-cobalt, 92
 Soft, 92
Magnetic scanning concentrated flux, 82
Magnet, radially polarized, 93
Magnification factor, of catadioptric image tube, 138
Matching factor spectral, multalkali photocathode to X-ray phosphor, 220, 222

SUBJECT INDEX

Membrane, electron permeable, 290
Mesh, metal,
  evaporation mask, 24
  field, in TV camera, 81
  in image intensifier, 167
  suppressor, elimination of, 51
Metal coating transparent, for photocathode substrate, 102
Metallization, with nichrome, 170
Mica windows, for electronographic image tube, 5, 315
Microchannel plate,
  as gating electrode, 253
  chevron configuration, 492
  in TV camera tube, 15
  Mullard J27, 191
  with curved channels, 369
  X-ray detector, 380
Microchannel plate image tube,
  aging effects, 373
  gated, 157, 253–263
  TV camera, 15
  wafer type, 369
  X-ray, 380
Microdensitometer,
  Joyce-Loebl, 331
  P.D.S., 416
Modular construction of image tube, 169
Modulation Transfer Function (MTF),
  as imaging device parameter, 11
  of CsI X-ray screen, 204
  of pyroelectric vidicon, 23, 36
  of silicon target camera tube, 66
Multichannel analyser, optical PAR 1205D, 268
Mu metal shield, 93
Mylar, electron permeable membrane, 292

N

Negative electron affinity, 18
Nichrome, metallized layer, 170
Nitrocellulose laquer, for phosphor screen, 164
Noise, *see also* Background, Dark current.

Noise,
  characteristics of solid state imaging device, 454, 485
  fixed pattern in solid state imaging device, 403
  in photon counting system, 374, 381
  in X-ray topography system, 223
  light induced, 175
  white, integrated out in TV system, 209
Noise Equivalent Power (NEP) of Schottky barrier detector, 495
Noise Equivalent Temperature (NET) of Schottky barrier detectors, 495, 510

O

Oblique magnetic focus, image tube, 445
Optical feedback in image tube, 175, 370
Optical flat quartz, for image sampling, 402
Optical Transfer Function, *see* Modulation Transfer Function.
Optical,
  feedback in image tube, 175, 370
  flat, quartz for image sampling, 402
  multichannel analyser, 268
  processing using image tube, 237
Orientation of CsI phosphor layer, 204

P

Paint, conductive, silver, 186
Paraxial trajectories in spherical electromagnetic focusing, 121
Pedastel current, in pyroelectric vidicon, 27, 33
Permanent magnet focusing array, 19, 89–99, 399
Phase Transfer Function (PTF), 66
Phosphor, *see also* Scintillator
Phosphor screen
  $CaWO_4$ for X-rays, 222

Phosphor screen—(cont.)
CsI·Tl evaporated, for X-rays, 149, 203, 222
deposited by cataphoresis, 171
energy conversion efficiency, 5
halo effect in, 137
high efficiency uncoated, 136
high resolution for X-rays, 201–207
P·11, 371
P·11 performance study, 155
P·20, 371
P·20 coupling to CCD, 489
P·43 Cd_2O_2S, 220, 222
time decay in photon counting, 390, 491
unevenness effects on X-ray TV system, 214
wet settling technique, 138, 155
Yttrium silicate fast decay, 144
ZnS·Ag for X-rays, 145
ZnCdS, 220, 222
Photocathode,
CsI for UV
electron emission energy, 107
semi transparent, 284, 287, 447
CsI, for X-rays, 146, 189
CsSb(O) in electronographic camera, 307
CsTe opaque, 291
electron emission energy, 107
curved substrate, 291
GaAs semi transparent, 151
KCsSb(O) in electronographic camera, 312
multialkali in miniature image tube, 173
negative electron affinity, 152
non-flat with magnetic focus, 286
palladium semitransparent, 469
S·20
activation process, 53
extended red, 138
in MCP intensifier, 370
on sapphire window, 399
optical transmission, 163
sensitivity, 56
Spherical, 123
substrate plate silica, 316
transfer procedure, 51, 493
X-ray sensitive, 189–199

Photochron I streak tube, 265
Photochron II streak tube, 273
Photoconductor, see Target, Television camera tube, Vidicon.
Photodiode array, self scanned
for astronomy, 431–440
intensified, 416
Photodiode, vacuum, fast
TIXL 56, 267
Photoelectron energy distribution from CsI photocathode, 447
Photoemission, see also Photocathode.
Photoemission, from mesh, 175
Photographic emulsion, see Emulsion photographic.
Photography, high speed
using MCP image tube, 156
Photometry U,B,V, 296, 329, 334
Photon counting system,
electron optical design for, 101
space borne, 355–378
using CCD, 355–378, 379–388, 394, 489
using photodiode array, 389, 416
with MCP image tube, 369, 373
Photoresist technique, 25
Photoyield curve for Schottky barrier detectors, 496
Pins platinum, metal-to-glass seal, 169
Plate holder, for astronomical image tube, 339
Platinum silicide, Schottky barrier detectors, 496
Pockels cell, in astronomical polarimetry, 478
Point spread function,
in electron optics, 19, 69, 179
of convolution TV system, 243
Poisson statistics in image tube system, 406
Polarimetry in astronomy, 325, 342, 478
Polyvinylidene fluoride (PVF_2) film, for pyroelectric vidicon, 31
Potassium chloride low density layer, 58, 191
Power supply, for stepped deflection, 111
Preamplifier, on chip, for CCD, 425
Primary electrons, backscattered, 175

Profiled reluctance permanent magnet focusing, 91–92
Programming language,
CATY 2, 437
FORTH, 403, 467, 488
FORTRAN, 69, 391
Projectile photography, camera for, 258
Proximity focus,
electron optics, 19, 129
in SEC vidicon, 51
in wafer image tube, 371
in X-ray intensifier, 195
Pulsar,
Crab nebula, 116
optical, 99
Pulse generator, for gated image tube, 254
Pulse height distribution,
of demountable digicon tube, 469
of microchannel plate, 370
of X-ray detector, 191
Pyroceram cement, in image tube structure, 169
Pyroelectric vidicon, *see also*, Target, Television camera
Pyroelectric vidicon,
PVF_2, 31–37
TGS, 23–30

Q

Quantum efficiency,
as detector parameter, 1, 3
of commercial photocathodes, 405
of photon counting system, 374
of S·20 photocathode, 443
of Schottky barrier detector, 496
of X-ray photocathode, 190
Quantum yield,
of CdSe target, 39
of phosphor screens, 221, 222
Quartz faceplate, on UV camera tube, 42
Quasar,
PKS 0830+115
PKS 1004+ 141, 303

R

Radiogalaxy,
Cygnus A, 301

Radiogalaxy—(*cont.*)
Virgo A, 296
Readout,
dark, subtraction of, 453
double sampling with diode array, 432
noise, in diode array, 434
non-destructive in diode array, 431–433
variable rate in CCD, 463
Rear illuminated CCD, 455
Reluctor, permanent focusing, 92–93, 101
Reprocessing, electronographic image tube, 322
Reset mode, voltage of CCD operation, 501
Resistors, metal oxide glaze, for image tube, 316
Resolution, *see also* Modulation Transfer Function.
Resolution,
energy, of gamma ray camera, 234
limiting, of image tube, 219
of intensified diode array, 405
of miniature image tube, 168, 171
of photon counting system, 380
of pyroelectric vidicon, 27
of KCl dynode, 193
of X-ray intensifier, 193
of X-ray TV system, 219
Response,
flat field of CCD, 458
linearity of X-ray TV system, 215
Responsivity, *see also* Quantum efficiency.
Responsivity of diode array, 433
Reticulation, in pyroelectric target, 23
Ripple in magnetic focus current, 114
Ritchey-Chrétien spectrograph 120 cm at Steward Observatory, 399
Rubber, silicone, conductive, 186

S

Samarium Cobalt alloy in permanent magnet focusing, 92
Saturated mode operation of microchannel plate, 376, 492

SUBJECT INDEX

Scan converter, for astronomical TV system, 347
Schottky barrier, infrared detector, 12, 496
Schmidt camera at Steward Observatory, 399
Scintillations, *see* Background, Noise, Pulse height distribution.
Scintillator,
 coupled to image tube, 227
 for X-ray topography, 217
Screen fluorescent, *see* Fluorescent screen.
S-distortion in miniature image tube, 172
Seal,
 demountable, 318
 expansion matched, 170
 Housekeeper, 170
 indium, 53, 56, 102, 318
 metal to glass platinum pin, 169
Sealing technique, hydraulic ram, 169
SEC vidicon camera tube, with Mg F_2-Ag target, 14, 51
Secondary electrons,
 from mesh, 175
 from phosphor screen, 165
Self sharpening effect in electron beam scanning, 75
Sensitivity, *see* Quantum efficiency Responsivity.
Shielding magnetic, 90
Signal induced background in image tube system, 404
Signal processing,
 for diode array, 437
 on chip for CCD, 425
Signal to noise ratio, *see also* Background, Dark current, Noise, Pulse height distribution.
Silicone rubber,
 conductive Ablebond 190-8, 186
 potting compound RTV-511, 186
 Sealant 732 RTV, 185
Silica substrate for photocathode, 287, 316
Silicon,
 camera tube target, 63
 image localizer target, 23
 transmission secondary emission

Silicon—(*cont.*)
 dynode, 152
Silver,
 copper alloy braze, 169
 low secondary emission layer, 58
Single electron response,
 of intensified CCD, 449
SIT vidicon camera tube, 15
 RCA 4532, 63
 RCA 4826, 63
 TH 9651, 218
 TH 9655, 380
Sky brightness, 332
SLAC electron trajectory computer programme, 106
Sleeve, silicone rubber insulating, 318
Soda-lime glass in image tube structure, 169
Solid state photodiode array,
 cooling, 401
 for astronomy, 15, 453-462
 for spectroscopy, 397-413
 Reticon
  CP1001, 400
  dual 1024 element, 390
  50×50, 453
Space borne, photon counting system, 362
Space shuttle missions, 283
Space Telescope Faint Object camera, 90
Space telescope instruments, 101
Spatial Frequency Response *see also* Modulation Transfer Function, Resolution.
Spatial frequency response of solid state detector array, 424
Speckle interferometry, 481-489
Spectracon electronographic image tube, 330
Spectrograph,
 Boler and Chivens, 398
 Cassegrain, 410, 415
 Fabry-Perot, 116
 Objective prism, 303
 Pellet-Deherveng, 374
 Ritchey-Chretien, 399
 Thomson scattering, 441
Spectrophotometer photoelectric, 1
Spectroscopy, using photon counting

Spectroscopy—(cont.)
  system, 361
Spherical electrodes concentric, 119
Streak camera,
  developments, 157
  Photochron I, 265
  Photochron II, 273
  picosecond, 265–271
  temporal resolution, 268
Substrate,
  with mosaic pattern for X-ray phosphor, 202
Surveillance camera, 254
Sync generator integrated, 352
Synchroscan camera, 265

T

Target, *see also* Television camera tube.
Target,
  capacitance,
    effect of, 18
    in chalnicon, 45
    in SEC vidicon, 58, 59
    in SIT vidicon, 64
  chalnicon camera tube
    CdSe, 39
    As_2Se_3, 47
  electron bombarded silicon, 63
  image localizer, 228
  low lag in Chalnicon, 46
  metallic in vidicon, 76
  photoconductive, leakage in, 75
  pyroelectric
    PVF_2, 31
    reticulated, 17, 23
    TGS, 17, 23
  storage SbO_3, 349
Teflon film, electron permeable, 292
Telescope,
  Anglo-Australian 3.9 m, 325, 355
  Dominion Astrophysical Observatory 1·8 m, 453
  E.S.O. 3·6 m, 355
  Hale 5 m, 355
  Haute Provence 1·93 m, 277, 295, 374, 380, 392
  Hawaii C.F.H. 3·6 m, 277, 295
  Helwan 74 in, 330
  La Silla, Chile 1·5 m, 315, 319

Telescope—(cont.)
  McDonald Observatory 0·76 m, 277, 308
  Mount Hopkins
    1·5 m, 415, 417
    multimirror, 348
  RGO
    30 in, 437
    36 in, 116
  Spacelab Wide Angle (SWAT) 0.8 m, 283
  Starlab 1 m, 283
  Steward Observatory
    0·5 m, 340
    2·3 m, 340, 352, 398, 478
  television guidance system, 347
Television camera tube,
  Chalnicon, 39
  image isocon, 213
  image orthicon EEV P887, 209
  Plumbicon, 12, 371
  Pyroelectric vidicon, 23, 31
  SEC vidicon, 14, 51–61
  SIT vidicon, 15, 63
    RCA 4532, 63
    RCA 4826, 63
    TH 9651, 218
    TH 9655, 380
  with X-ray scintillator, 217–225
Television finder for astronomical telescope, 347
Temperature, minimum resolvable (MRT) of pyroelectric vidicon, 27
Temporal resolution, of X-ray intensifier, 194, 198
Test mask, photo etched, 196
Test pattern, USAF 1951, 287
Thermal response of infrared CCD, 508
Time-bandwidth product of image intensifier, 239
Time delay integration (TDI) in solid state photodetectors, 426
Titanium, in image tube structure, 169
Tomography,
  transverse analogue, 248
  X-ray, 150, 237
Transfer characteristic of infrared CCD, 505

Transit time fluctuation in X-ray intensifier, 194
Transmission secondary emission, 6
Transverse Analogue Tomography, 248
Tri-electrode focusing system, 130
Triglycine sulphate (TGS),
  deuterated, 23
  pyroelectric target, 17

U

Ultraviolet,
  camera tube solar blind, 15
  chalnicon camera tube, 39, 42
  electronographic camera, large format, 284
Uniformity of response of infrared CCD, 505

V

Vacuum envelope,
  fused silica, 316
  pyrex glass, 318
Vidicon, see also SEC vidicon, SIT vidicon, Target, Television camera tube.
Vidicon camera tube,
  chalnicon, 39
  infrared sensitive, 23, 31, 42
  pyroelectric, 23, 31
  SEC with MgF_2-Ag target, 51
  SIT, 15, 63, 218, 380
  Storage, Teltron 9300 ST/fo, 347–353
  X-ray sensitive, 217
Voltage
  multiplier, Cockcroft-Walton, 168
  reset mode for CCD, 501

W

Wafer image tube, 370
Wavelength,
  converter fluorescent, 423, 452
  response of CCD, 458
Weld,
  argon arc, 170
  cold, copper, 169
Windowless intensified CCD, 441, 445
Window,
  CaF_2 as photocathode substrate, 287
  fibre optic, 165, 169, 339, 370
  mica for electrography, 5, 315
  plastic for electronography, 290, 307
Work function, in silicon target camera tube, 64

X

X-ray
  crystallography, 145
  cross sectional imaging, 241
  diffraction studies of Chalnicon target, 43
  generator, 224
  intensifier magnetic focus, 199
  medical radiology, 147
  phosphor screen high resolution, 201
  phosphor screen ZnCdS 159, 163
  photocathode, 191
  topography, 217
  Transverse Analogue Tomography, 237

Y

Yttrium silicate, fast decay phosphor, 144